I0033555

Dragonflies and Damselflies (Odonata) of Texas

Volume 3
2008

Edited by John C. Abbott

Abbott, J.C. 2008. Dragonflies and Damselflies (Odonata) of Texas. Odonata Survey of Texas. Vol. 3. Austin, Texas.

Publication date: February 2008
ISBN 978-0-6151-9494-3
Copyright © J.C. Abbott, 2008

Cover Photo: Jade-striped Sylph (*Macrothemis inequiunguis*) from Lockhart Municipal Park in Lockhard, Texas; photo by John C. Abbott.
Cover design by John C. Abbott.

All rights reserved. No part of this publication may be reproduced or used in any form by any means - graphic, electronic or mechanical, including photocopying, recording, taping or information storage and retrieval systems without written permission. Critics or reviewers may quote brief passages in connection with a review or critical article in any media.

The Odonata Survey of Texas (OST), centered at The University of Texas at Austin (UT), includes a group of people with a shared interest in the study of the distribution, biology, behavior, and enjoyment of the dragonflies and damselflies occurring in Texas. The purpose of the OST is to act as an official organization whose job it will be to encourage, solicit, and maintain the Texas database for dragonfly and damselfly distributional information. The goal is to maintain a high level of scientific integrity while involving as many individuals as possible in an effort to thoroughly survey the state's Odonata fauna. Membership is free and encouraged. Information about OST and up-to-date announcements can be found at the official website for the survey, http://www.odonatacentral.org.

My hope is that the OST will provide direction for members interested in the Texas Odonata fauna. The annual series of volumes act as a resource guide, containing, in a concise, summarized version, everything we know about the distribution and seasonality of Texas Odonata. Each year new records will be published, and you are encouraged to submit articles describing your discoveries and adventures in Texas from the previous year.

I believe the formation of this survey could not be timelier, with public interest in dragonflies and damselflies at an all time high. A new group of enthusiasts are providing the motivation for new field guides, books, and surveys all over North America. My hope is that you, as a volunteer participant in the OST will be able to not only help the OST in whatever capacity you can offer, but be able to make use of all the information that is available for this fascinating and wonderful group of insects. You will find a summary of OST on the last page of this volume. I encourage you to copy and distribute it to friends, fellow odonate enthusiasts, and others who might be interested in joining the survey.

This is the third volume of the Dragonflies and Damselflies (Odonata) of Texas to be published and serves as an update of records including all those reported in 2007. As with earlier volumes, this book is meant to serve as a guide to the distributions and seasonality of all 224 species occurring in the state. The interest in dragonflies and damselflies in North America, and Texas specifically, continues to grow as does our knowledge of the fauna. Judging by the increasing number of records submitted since 2005, The Odonata Survey of Texas (OST) appears to be a success. I hope this volume continues to increase interest and excitement for the Odonata fauna in Texas.

Contact information –
Odonata Survey of Texas
c/o John C. Abbott, Ph.D.
Section of Integrative Biology
1 University Station #L7000
The University of Texas at Austin
Austin, Texas 78712 USA
Office Phone: (512) 471-5467
Lab Phone: (512) 232-1896
Fax: (512) 475-6286
jcabbott@mail.utexas.edu

Acknowledgements

This book would not have been possible without the help of a number of individuals. I would like to thank the contributing authors for taking the time to write up their observations and allowing me to include them. I would like to thank the invaluable assistance of Damon Broglie in producing the distribution maps and seasonal graphs. Sara Pratt provided substantial input and editorial assistance throughout the process of creating this volume. Greg Lasley graciously read through an early draft and helped with editing.

I would like to thank the continually growing community of dragonfly watchers whose enthusiasm for dragonflies ultimately motivated me to establish the Odonata Survey of Texas and create this series of resource guides. It is truly exciting to see so many individuals developing an enthusiasm for a group of insects that are so spectacular.

TABLE OF CONTENTS

MAP of
TEXAS
— Showing County Boundaries —

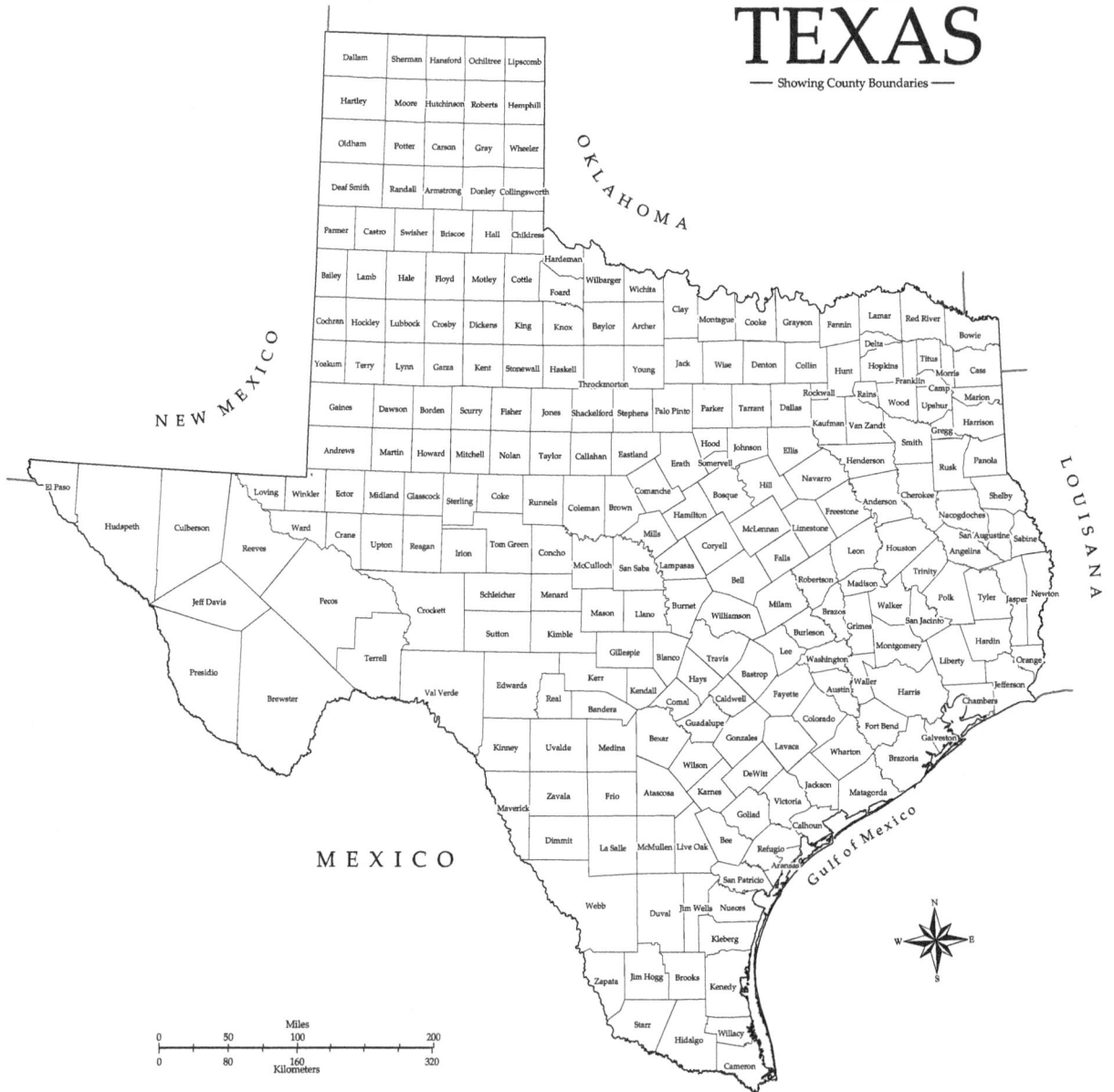

OKLAHOMA

NEW MEXICO

LOUISIANA

MEXICO

Gulf of Mexico

Dallam | Sherman | Hansford | Ochiltree | Lipscomb
Hartley | Moore | Hutchinson | Roberts | Hemphill
Oldham | Potter | Carson | Gray | Wheeler
Deaf Smith | Randall | Armstrong | Donley | Collingsworth
Parmer | Castro | Swisher | Briscoe | Hall | Childress
Bailey | Lamb | Hale | Floyd | Motley | Cottle | Hardeman | Wilbarger | Wichita
Foard
Cochran | Hockley | Lubbock | Crosby | Dickens | King | Knox | Baylor | Archer | Clay | Montague | Cooke | Grayson | Fannin | Lamar | Red River
Bowie
Yoakum | Terry | Lynn | Garza | Kent | Stonewall | Haskell | Young | Jack | Wise | Denton | Collin | Hunt | Hopkins | Delta | Titus | Cass
Throckmorton | Rockwall | Rains | Franklin | Morris
Camp | Marion
Gaines | Dawson | Borden | Scurry | Fisher | Jones | Shackelford | Stephens | Palo Pinto | Parker | Tarrant | Dallas | Wood | Upshur | Harrison
Kaufman | Van Zandt | Smith | Gregg
Andrews | Martin | Howard | Mitchell | Nolan | Taylor | Callahan | Eastland | Hood | Johnson | Ellis | Rusk | Panola
Erath | Somervell | Henderson
Comanche | Hill | Navarro | Anderson | Cherokee | Shelby
El Paso | Loving | Winkler | Ector | Midland | Glasscock | Sterling | Coke | Runnels | Coleman | Brown | Hamilton | Bosque | Freestone | Nacogdoches
McLennan | Limestone | Leon | Houston | San Augustine | Sabine
Hudspeth | Culberson | Ward | Crane | Upton | Reagan | Irion | Tom Green | Concho | Mills | Coryell | Falls | Trinity | Angelina
McCulloch | San Saba | Lampasas | Robertson | Madison | Polk | Tyler | Jasper | Newton
Reeves | Burnet | Milam | Brazos | Walker | San Jacinto
Jeff Davis | Pecos | Crockett | Schleicher | Menard | Mason | Llano | Williamson | Burleson | Grimes | Montgomery | Hardin
Sutton | Kimble | Lee | Washington | Liberty | Orange
Presidio | Terrell | Gillespie | Blanco | Travis | Bastrop | Austin | Waller | Harris | Jefferson
Brewster | Val Verde | Edwards | Kerr | Hays | Caldwell | Fayette | Colorado | Fort Bend | Chambers
Kendall | Comal | Guadalupe | Gonzales | Lavaca | Wharton | Galveston
Bandera | Bexar | Wilson | DeWitt | Jackson | Brazoria
Kinney | Uvalde | Medina | Matagorda
Zavala | Frio | Atascosa | Karnes | Victoria | Calhoun
Goliad
Maverick | Dimmit | La Salle | McMullen | Live Oak | Bee | Refugio | Aransas
San Patricio
Webb | Duval | Jim Wells | Nueces | Kleberg
Zapata | Jim Hogg | Brooks | Kenedy
Starr | Hidalgo | Willacy | Cameron

Miles
0 | 50 | 100 | 200
0 | 80 | 160 | 320
Kilometers

N
W   E
S

Odonata of the Lower Rio Grande Valley
2007 Summary

Joshua S. Rose

In contrast with 2006, in which several species were observed in the Lower Rio Grande Valley on New Year's Day, 2007 started a bit slower. The Caribbean Yellowface (*Neoerythromma cultellatum*) was seen on January 8, 2007 and marked the beginning of a cool, cloudy, and wet January. Damselflies were generally absent until the last week of February and dragonfly observations were very sparse during that time. Red Saddlebags (*Tramea onusta*) began to appear in appreciable numbers in the last week of February, as did several common damselfly species.

The year's first sighting of an unusual species was a vagrant from the north, rather than the south; perhaps appropriate given the cool weather. On March 2 an *Epitheca* (*Tetragoneuria*) baskettail was observed patrolling over an irrigation canal at Bentsen-Rio Grande Valley State Park (Hidalgo Co.). While *E.* (*Epicordulia*) *princeps* is a regularly occurring species in the LRGV, no other species of *Epitheca* has been documented further south than Bee County, roughly 150 miles from the northern rim of the Valley. Unfortunately, the dragon was in a location physically inaccessible for collecting, photos were not sufficient to determine which species of *Epitheca* it was, and no further individuals were observed. Photos were consistent with Dot-winged Baskettail (*E. petechialis*), the species which had been previously collected in Bee County, but other species could not ruled out.

On March 10, an amberwing was observed at the Valley Nature Center in Weslaco (Hidalgo), as well as a large number of Desert Firetails (*Telebasis salva*). The amberwing, or another like it, was photographed on March 27 and confirmed to be Slough (*Perithemis domitia*), a species common at this location during the summer months but not previously observed in Texas at such an early date. The large number of *Telebasis*, and the fact that some had acquired full adult coloration and commenced mating, hints that this species was likely on the wing earlier than detected.

A number of different species were observed for the first time of the year between March 16 and 30 at Bentsen-RGV State Park. Perhaps most noteworthy was Caribbean Yellowface, observed on the 16th and photographed on the 23rd. Neotropical Bluet (*Enallagma novahispaniae*) was also seen on the 16th. A Thornbush Dasher (*Micrathyria hagenii*) on the 24th was much earlier than expected. A pair of male Amelia's Threadtails (*Neoneura amelia*) at Bentsen on the 30th exhibited full adult coloration and were behaving territorially, perhaps indicating that the species may have emerged some time

before this date. The first Spot-tailed Dasher (*Micrathyria aequalis*) and Red-tailed Pennant (*Brachymesia furcata*) of the year were observed on the same day.

On April 26, just beyond the Valley's boundaries at Pollywog Pond in Corpus Christi, Ann Johnson and Jim Bangma photographed a Turquoise-tipped Darner (*Rhionaeschna psilus*) for a first Nueces County record. Observations of this rarely reported species came in from a few LRGV locations during the year, including Santa Ana National Wildlife Refuge (Hidalgo) (Martin Reid, June 15) and Bentsen-RGV SP (June 21). One sighting even included oviposition (Dave Czaplak, Santa Ana National Wildlife Refuge, May 13).

May 9, Terry Fuller observed a Fragile Forktail (*Ischnura posita*) in his yard in San Benito (Cameron County). Another turned up July 21. This species is rarely observed in the LRGV, but was just the tip of the iceberg in terms of species in Terry's yard…

Dave Czaplak visited the LRGV from Maryland May 12 & 13 and observed a number of noteworthy species, all in Hidalgo County. He noted Tawny Pennant (*Brachymesia herbida*) and Carmine Skimmer (*Orthemis discolor*) at Santa Ana NWR, and observed the female darner ovipositing. At Anzalduas County Park (Hidalgo) he spotted Coral-fronted Threadtail (*N. aaroni*) as well as Amelia's, just above the dam.

When "Dragonfly Days" was called off for 2007 it cast a bit of gloom over late May odonate-watching in the LRGV, but a group from the Cleveland Museum of Natural History persisted in their plans to visit despite the festival's cancellation. They spotted 27 odonate species on May 19 between Bentsen-RGV SP and the adjacent NABA International Butterfly Park, the most remarkable probably being a Pin-tailed Pondhawk (*Erythemis plebeja*) at Bentsen. On May 21, this group visited the McAllen Nature Center (Hidalgo Co.) and tallied Gray-waisted Skimmer (*Cannaphila insularis*) and Golden-winged Dancer (*Argia rhoadsi*), two species recorded regularly at this site but rarely if ever from elsewhere in the LRGV. Other uncommon sightings included Chalky Spreadwing (*Lestes sigma*), Striped Saddlebags (*Tramea calverti*), and another Pin-tailed Pondhawk. Striped Saddlebags is much more abundant and widespread in some years but was in very short supply in the LRGV this year.

The summer was a good one for the Ringed Forceptail (*Phyllocycla breviphylla*), a species documented in the US for the first time only a few years ago. One photographed on June 8 at Bentsen-RGV, and possibly eaten by a Crablike Spiny Orb Weaver (*Gasteracantha cancriformis*) that same morning, was followed by several other observations, including a pair photographed in the wheel at Bentsen-RGV on July 10, and both male and female individuals separately observed at Anzalduas County Park August 24. Another female was found at Bentsen-RGV September 7.

Martin Reid photographed a Three-striped Dasher (*M. didyma*) at Santa Ana NWR on June 15. He also got photos of a Bar-sided Darner (*Gynacantha mexicana*). Then on June 24 he discovered a small population of Three-striped Dashers, a species very rarely detected in Texas, at the Sabal Palm Audubon Preserve (Cameron Co.), where at least six males were on territory, and one female was mating and ovipositing. Terry Fuller's yard harbored a later individual October 6.

Terry Fuller left his yard and photographed a Cream-tipped Swamp-damsel (*Leptobasis melinogaster*) at the World Birding Center at Harlingen's Arroyo Colorado, a.k.a. Hugh Ramsey Nature Preserve, on July 8. This was not only a first record for Cameron County, but only the fourth US location where this species has been documented, joining the McAllen Nature Center, Santa Ana NWR, and King Ranch.

The biggest story of the year in LRGV odonatology, however, was Terry's backyard. From August 11 through 19 it hosted a Pale-green Darner (*Triacanthagyna septima*), the first record for Cameron County, third for Texas, and fourth for the US. He ultimately noted up to three individuals; one of them, or a fourth, appeared on September 13. He also had Blue-faced (*Coryphaeschna adnexa*) and Bar-sided Darners in residence around the same time. In anyone else's yard, this would have been far and away the high point of the year, if not their life; but Terry was just getting warmed up! An Evening Skimmer (*Tholymis citrina*) appeared on Sepember 30; a different individual of the same species showed up October 9, and another November 18.

Then, on October 6, Terry found not only the aforementioned Three-striped Dasher but also a Claret Pondhawk (*Erythemis mithroides*)! Terry related that the pondhawk had eluded him for two days before he finally managed to photograph it. The exact classification of this taxon remains unresolved, as Dennis Paulson has suggested that LRGV and northern Mexico populations are most likely a cryptic and as-yet unnamed species separate from true *E. mithroides*. And, just to take the situation from incredible to unbelievable, the very next day Terry photographed what he thought was the female, only to have it turn out to be a female Black Pondhawk (*Erythemis attala*)! The Claret Pondhawk is only the second known US record, following one at Santa Ana NWR in 2004; the Black has been found breeding on the King Ranch in Kleberg County but before Terry's find the only other US record was a lone specimen from

Alabama. And Terry's yard may be the first place in the country with five different documented species of Pondhawk …

Gil Quintanilla photographed another Black Pondhawk on November 9 near La Lomita Mission, just northwest of Anzalduas County Park.

Several species of dragonfly which prefer to breed in clear rivers or streams with rocky or gravelly beds are uncommon vagrants to the LRGV, presumably from populations in the Hill Country. A Bronzed River Cruiser (*Macromia annulata*) was seen May 2 at Santa Ana NWR by Bangma & Johnson. Tom Pendleton photographed another hanging in the Bentsen-RGV SP butterfly garden on September 5. On October 26, Lawrence Duhon reported a Pale-faced Clubskimmer (*Brechmorhoga mendax*) at the NABA International Butterfly Park near Mission (Hidalgo), and Josh Rose noted one at Bentsen-RGV SP the same day. Martin Reid noted two Straw-colored Sylphs (*Macrothemis inacuta*) at Falcon SP on October 27; Bob Behrstock described two males at Santa Ana NWR on November 18, one patrolling territorially over a road, the other feeding on a swarm of tiny dipterans.

The autumn was very productive for Tawny Pennant. Multiple males were on territory at Bentsen-RGV September 7, and one was re-found the following day. After a lull, another individual was found at Falcon State Park (Starr) by David and Jan Dauphin on October 21, and several more were seen there over the week or two that followed. A particularly late one was photographed by Rick Nirschl on December 26 at the NABA park.

The Texas Butterfly Festival featured "Dragonfly Treks" October 20 & 21, both of which visited Anzalduas CP and Bentsen-RGV SP. Among the sightings were a pair of Marl Pennants (*Macrodiplax balteata*), unusual for this area; a Filigree Skimmer (*Pseudoleon superbus*), less unusual but always thrilling; the year's first Variegated Meadowhawks (*Sympetrum corruptum*); and late-flying male Carmine Skimmer (*Orthemis discolor*), Caribbean Yellowface, and Amelia's Threadtail. All were at Anzalduas CP except for the *Orthemis* which were at Bentsen-RGV SP.

Another Filigree Skimmer turned up at NABA on November 12, but the latest reports came from Bentsen-RGV December 11-15.

A Bar-sided Darner was found and photographed at Bentsen-RGV SP on November 30—somewhat late for an observation, though this species most likely overwinters in the LRGV as an adult.

November 29 two visitors from California, David Edwards and John Hall photographed a Golden-winged Dancer at Bentsen-RGV. The same damsel or another of its species was located on December 2. In recent years the only reports of this extremely localized species had come from the McAllen Nature Center. A number of other sightings of local interest took place at Bentsen that day, including Kiowa Dancer (*Argia immunda*), not typically found in the park; Little Blue Dragonlet (*Erythrodiplax minuscula*), uncommon anywhere in the LRGV; and a late-flying Amelia's Threadtail.

David and John also visited Santa Ana NWR in late November and photographed Evening Skimmer and Blue-faced Darner.

The year concluded with warm, mild weather, so odonates were seen at least through December 30, when a walk at Bentsen-RGV SP encountered 9 species including a Rainpool Spreadwing (*Lestes forficula*).

Joshua S. Rose, Ph.D.
Natural Resource Specialist
World Birding Center
Bentsen-Rio Grande Valley State Park
http://www.worldbirdingcenter.org/sites/mission
joshua.rose@tpwd.state.tx.us

WESLACO

Rio Grande Valley
Onionfest
April 5th, 2008

www.weslaco.com · 956-968-2102

Entertainment

Food

Variety of Music

Kids Activities

Fun with Friends

Crafts and Shopping

WESLACO

9TH ANNUAL
Dragonfly
D A Y S

Brought to you by
Estero Llano Grande State Park
and
The Valley Nature Center

Friday
May 15 –
Sunday
May 18,
2008

Speakers
·
Field Trips
·
Seminars
·
Kids Events
·
Family Day
May 10th
·
Banquet

www.valleynaturecenter.org

Damselflies and Dragonflies at Wright Patman Lake

Mike Dillon

Wright Patman Lake is southwest of Texarkana in extreme northeast Texas. The dam on the Sulphur River straddles the Bowie-Cass county line just west of US Highway 59. Habitat includes mixed pine and hardwood forests, grasslands, sandy beaches and mudflats. Access is easy from the dam, Corps of Engineer properties, and state parks surrounding the lake as far west as Texas Highway 8, north of Douglassville. Atlanta State Park is on the south shore of the lake on Farm to Market 96.

As a birder, I have observed over 200 species of birds around the lake, including such rare birds as Red-necked Grebe, Little Gull, and Black-headed Gull. I have found many great birding areas to be just as productive for odonates. Although the list is growing, over thirty species of dragonflies and ten species of damselflies have been found around the lake.

My new-found interest in odonates (in addition to birds) has led me to search many spots around the lake area. Photographs have helped with identification and documentation of species new to the two-county area (Bowie and Cass). John Abbott has been quite helpful (and patient) with identifications, especially with damselflies. Although Wright Patman Lake has not been as productive (yet) as Millwood Lake in Arkansas and Red Slough in Oklahoma, many of the species found in those two nearby locations are expected around the lake area.

In Bowie County, the most productive area has been around the observation platform and at the boat ramp further along the road behind the dam. At lower lake levels, sandy beaches are easy walking and several species of damselflies can be seen on the beach itself as well as in the bushes and trees around the beach. Bushes around the platform seem to offer shelter and foraging perches for dragonflies. North of the platform is a grassy field where several species of dragonflies are often numerous.

In Cass County, two places have been productive. The two parks directly south behind the dam give access to the beach and the grassy overflow spillway area. However, most species in Cass County have been found west of the dam in Atlanta State Park at Knight's Bluff and the Wilkins Boat Ramp. Trees and grassy areas at Knight's Bluff harbor several species of dragonflies and damselflies, but more can be found around the Wilkins Boat Ramp. Clubtails, pennants, skimmers and Prince Baskettails (*Epitheca princeps*) are common there, along with several species of damselflies. These are found by walking along the shore on both sides of the boat ramp among the bushes and higher grass.

Below is a list of the species I have observed so far at Wright Patman Lake.

Damselflies:
Ebony Jewelwing (*Calopteryx maculata*)
Powdered Dancer (*Argia moesta*)
Familiar Bluet (*Enallagma civile*)
Orange Bluet (*Enallagma signatum*)
Vesper Bluet (*Enallagma vesperum*)
Citrine Forktail (*Ischnura hastata*)
Fragile Forktail (*Ischnura posita*)
Rambur's Forktail (*Ischnura ramburii*)

Dragonflies:
Common Green Darner (*Anax junius*)
Jade Clubtail (*Arigomphus submedianus*)
Flag-tailed Spinyleg (*Dromogomphus spoliatus*)
Pronghorn Clubtail (*Gomphus graslinellus*)
Prince Baskettail (*Epitheca princeps*)
Four-spotted Pennant (*Brachymesia gravida*)
Halloween Pennant (*Celithemis eponina*)
Banded Pennant (*Celithemis fasciata*)
Eastern Pondhawk (*Erythemis simplicicollis*)
Spangled Skimmer (*Libellula cyanea*)
Yellow-sided Skimmer (*Libellula flavida*)
Slaty Skimmer (*Libellula incesta*)
Widow Skimmer (*Libellula luctuosa*)
Great Blue Skimmer (*Libellula vibrans*)
Roseate Skimmer (*Orthemis ferruginea*)
Blue Dasher (*Pachydiplax longipennis*)
Wandering Glider (*Pantala flavescens*)
Spot-winged Glider (*Pantala hymenaea*)
Eastern Amberwing (*Perithemis tenera*)
Common Whitetail (*Plathemis lydia*)
Variegated Meadowhawk (*Sympetrum corruptum*)
Black Saddlebags (*Tramea lacerata*)
Red Saddlebags (*Tramea onusta*)

Although the list is rather limited now, I expect to add more species with further study of the area. A project for 2008 will be to start early in the year and document as many species as possible establishing a baseline of flight season data for the species in northeast Texas.

Mike Dillon
Marshall, Texas
mdillon444@charter.net

The Odonata of Kerr County and the Guadalupe River System of Texas

Tony Gallucci

Introduction

In 1971, I began a concerted effort to compile a complete Flora and Fauna of a single Texas county – Trinity County on the northwestern corner of the Big Thicket of east Texas (Gallucci 2000d,e, 2005e,f). This resulted in publications on the plants (Gallucci 1972) and birds (Gallucci 1993). This effort has since been expanded to two Texas Hill Country Counties – Kerr (begun in 1986; Gallucci 1987, 1989, 2000a,b,c,f, 2005a,b,c,d, 2006b, 2007a) and Real (begun in 1988; Gallucci 2000f, 2005b, 2006c, 2007b).

Here I update the first report on the Odonate Fauna of Kerr County (Gallucci 2006a), whose county seat, Kerrville, is located about 95 km northwest of San Antonio, and about 160 km west-southwest of Austin. Although cursory efforts were made to photograph and collect invertebrates, until 1998 most of my efforts in that regard were limited to the class Mollusca in addition to large-scale cataloguing of the vertebrates and plants. In 1998, with the vertebrates largely completed, work with the class Insecta began in earnest, and focused largely on the Lepidoptera and Odonata. A number of other orders have followed since, but Odonata has remained a group of particular interest to me.

Regional Overview

It's easy enough to describe a lot of Texas regions as crossroads – the Lower Rio Grande Valley, the Trans-Pecos, the Big Thicket – but no region, perhaps anywhere in the U.S., can match the Hill Country for its range of habitats... from deeply wooded riparian stretches to open desert, prairie, and wooded hills, and with fauna and flora ranging from ice-age relict Bigtooth Maples, *Acer grandidentatum,* to rare butterflies such as Rawson's Metalmark, *Calephelis rawsoni,* and Lacey's Scrub-Hairstreak, *Strymon alea laceyi,* along with mixes of southern and northern, eastern and western bird pairs.

Yet the area is also known worldwide for its high level of endemism, which includes a dozen species of unique salamanders, most in the highly local subterranean genera *Eurycea* and *Typhlomolge,* at least that many snails, many plants including the critically endangered Tobusch's Fishhook Cactus, *Ancistrocactus tobuschii,* and Texas Snowbells, *Styrax texanus,* four turtles including the extremely rare Cagle's Map Turtle, *Graptemys caglei,* known only from the Guadalupe River System, and two endangered and highly sought after birds, the Golden-cheeked Warbler, *Dendroica chrysoparia,* and the Black-capped Vireo, *Vireo atricapillus.*

This is due in large part to the extreme changes in topography resulting from a five million year old uplift, the Balcones Fault Zone, which created a ~700-meter plateau, essentially giving birth to a biological island surrounded by strongly differing faunal and floral complexes.

The far western reaches of the plateau skirt the northeastern edge of the Chihuahuan Desert and feature indicator species such as *Agave lechuguilla* and *Larrea tridentata.* Sharply delineated canyonlands in the central portion, created by millennia of erosion, give rise to a multitude of springfed rivers that course through the eastern third, falling off the plateau at an escarpment that runs roughly from San Antonio to Austin and northward, and west to north of Uvalde.

It is this eroded canyon/springs component spreading into a wide agricultural valley that makes Kerr County special in terms of providing habitat for odonates. Springs in the county feed five river systems as the central divide terminates here. The Frio, Sabinal, Medina, Pedernales and Guadalupe Rivers all owe at least a part of their flow to the hills of the western part of the county. Of these, virtually all the upper flow of the Guadalupe River, the quintessential Hill Country River, originates in the county via four major tributaries – the North Fork, South Fork, Johnson Creek and Turtle Creek.

That the area is also a well-known artists' and recreational mecca, and is thus creating a growing residential sprawl, has led to manipulation of the river and its forks, with the consequent creation of a number of microhabitats not otherwise naturally available. Coupled with the wide range of niches available in a river system that ranges from bubbling springs and falls to a 60-meter wide river, the opportunities for establishment of a number of odonate species is evident. Included in this retinue are spring pools, riffle zones, boulder-field rapids, sedge marshes, wooded swamps, wide flat two-inch silt-free runs with tiny, intricate channels, pools approaching 15-meter depths with significantly thick muddy bottoms, beaver and roadway dammed pondings, regulatory dam-produced lakes, densely Baldcypress (*Taxodium distichum*) shaded to completely uncovered stretches of river, all in Kerr and adjacent Hill Country counties, and long sandy or muddy stretches, and wide estuarine bay entries below the escarpment. In addition, there are numerous stock impoundments ranging from small, concrete troughs to large, earthen-dammed lakes of several hectares in size.

The Fauna

A total of 61 species of Odonates were known from Kerr County prior to 1998 when this effort to document them here began. This total was derived from literature records, from data as posted on the websites of Forrest Mitchell and James Lasswell at Tarleton State University, and John Abbott's Odonata Central at the University of Texas-Austin (Abbott 2000). Since then another 16 species have been documented (Abbott 2005, 2007, 2008 and personal records). The official total for the known Odonate fauna of Kerr County now stands at 77. Three additional species, Filigree Skimmer (*Pseudoleon superbus*), Carmine Skimmer (*Orthemis discolor),* and Red Rock Skimmer *(Paltothemis lineatipes),* have been observed in the county but not yet adequately documented, and would bring the county total to 80, enough to rank it among the top ten counties in Texas, most of which are either metropolitan/university in nature, or are areas of intense research such as San Jacinto County in the Big Thicket and Hidalgo County in the Lower Rio Grande Valley.

Since the initial report (Gallucci 2006a), four species have been documented in Kerr County: Comet Darner, *Anax longipes,* Bronzed River Cruiser, *Macromia annulata* (found and photographed by Martin Reid), Thornbush Dasher, *Micrathyria hagenii* (previously listed as hypothetical based on sight records), and Marl Pennant, *Macrodiplax balteata* One species, Black Setwing, *Dythemis nigrescens,* previously listed with poor documentation was re-found and documented anew. An additional species, Carmine Skimmer, *Orthemis discolor,* has been added to the hypothetical list in the last year based on observations not supported by adequate documentation.

There is the potential for more as an additional 21 species have been found in neighboring counties whose habitat regimes are much the same, if not generally poorer, than Kerr County's. That gives us an area total of 101 species.

In the Guadalupe Watershed, comprised of the Guadalupe, Medina, San Antonio, Comal, Blanco and San Marcos Rivers, the fauna is enhanced by the addition of eastern woodland, southern brushland, coastal lowland, and estuarine species. The 42 additional species found in the overall watershed bring the total for this system to 122 species.

When describing the odofauna of Kerr County it would be easy to focus on the damselflies, as they are not only extremely diverse, but notably abundant.

Of chief interest in Kerr county is the large contingent of species in the genus *Argia.* Ten are known for the county. These include the relatively restricted and poorly known Comanche Dancer, *Argia barretti* (with one known population), and the recently described (Garrison 1994) Leonora's Dancer, *Argia leonorae* (two known populations in the county). Also present are four known populations of Springwater Dancer, *Argia plana* and five known

populations of Aztec Dancer, *Argia nahuana.*

Populations of three other species are located within a few miles of the county line. These include Lavender Dancer, *Argia hinei* (Bandera and Real Counties), Coppery Dancer, *Argia cuprea* (Bandera, Edwards, Kendall and Real Counties), and Sooty Dancer, *Argia lugens* (Kimble County).

The genus *Enallagma* is also well-represented in the county with seven species, including Stream Bluet, *Enallagma exsulans,* Arroyo Bluet, *Enallagma praevarum,* Neotropical Bluet, *Enallagma novaehispaniae,* and Vesper Bluet, *Enallagma vesperum.*

Perhaps most notable among the Kerr County and Hill Country Zygoptera is the presence of two of the three U.S. Threadtails: Orange-striped Threadtail, *Protoneura cara,* and Coral-fronted Threadtail, *Neoneura aaroni.* It was recently thought that these two species were limited roughly to the Hill Country and central Texas in the U.S., but recent discoveries by Bob Behrstock, Dennis Paulson, Martin Reid, and others have extended the known range of Coral-fronted Threadtail to south Texas. Nevertheless, they remain representative of the uniqueness of the odonate fauna of the Hill Country.

In the Anisoptera, individuals are far less common than in some areas of the state, yet the diversity is high. Typical Kerr County species would include Pale-faced Clubskimmer, *Brechmorhoga mendax,* Five-striped Leaftail, *Phyllogomphoides albrighti,* Dot-winged Baskettail, *Epitheca petechialis,* and Comanche Skimmer, *Libellula comanche.*

The area has also produced records of such south Texas specialties as Pin-tailed Pondhawk, *Erythemis plebeja,* Black Setwing, *Dythemis nigrescens,* Great Pondhawk, *Erythemis vesiculosa,* Tawny Pennant, *Brachymesia herbida,* and Striped Saddlebags, *Tramea calverti.*

Species typical of eastern low montane areas such as the Ozarks of Arkansas and Oklahoma can be found in the Hill Country. Examples include Pronghorn Clubtail, *Gomphus graslinellus,* and Orange Shadowdragon, *Neurocordulia xanthosoma.* Western species include the aforementioned Filigree and Red Rock Skimmers, and an eastern contingent is represented by many species, but an excellent single example is the Yellow-sided Skimmer, *Libellula flavida.*

Beyond Kerr County the Guadalupe River System is home to significant populations of such rarities as Blue-faced Ringtail (*Erpetogomphus eutainia*), whose entire U.S. range is limited to here plus one spot on the Rio Grande; Carmine Skimmer (*Orthemis discolor*) and Slough Amberwing (*Perithemis domitia*), both considered extreme rarities in this country until explorations of the last five years located a number of new populations; and to coastal populations of the more common but saltwater-restricted Seaside Dragonlet (*Erythrodiplax berenice*). However among the dragonflies none are more special, nor representative of the Guadalupe drainage than two sylphs of the genus *Macrothemis*: the Ivory-striped Sylph (*Macrothemis imitans leucozona*), and the Jade-striped Sylph (*Macrothemis inequiunguis*), both of which are known in the U.S. only from the Texas Hill Country and just off the escarpment in southern Hill Country-fed rivers. In this regard they are much like the threadtails. The delicateness and retiring nature of those four special insects are indicative of what makes central Texas incomparable as a wildlife refugium.

Research and Other Pursuits

The phenomenal wildlife paradise that is Kerr County has led to a focus on it as a research location. W.C. Young and C.W. Bayer (1977) are at least partly responsible for the large odofaunal list that predates my work in the county via their work on Guadalupe River System larval cohorts.

In addition there has been a significant amount of recent research in the county. John Abbott and Travis Tidwell (UT-Austin) have worked here on the unknown pre-adult stages of Comanche Dancer. The wing-darkening progression of Smoky Rubyspot, *Hetaerina titia,* has been examined by Abbott and Alexandra Shepherd (UT-Austin). And the three have been involved in examining Kerr populations of Leonora's Dancer as well.

Chris Anderson of UCLA has spent four field seasons working on the display dynamics of Smoky Rubyspot and American Rubyspot, *Hetaerina americana,* near Hunt, Ingram and Kerrville. Ryan Caesar, Ohio State University, is using Kerr County specimens of *Argia* in a DNA-based systematic look at the genus, and Mark McPeek of Dartmouth will be looking at copulatory attachment mechanisms and acceptance in species of the genus *Enallagma*.

Interest in the odonates of the county also extends to other pursuits. In the past two years a growing number of observers have come merely to look at the local consortium in a pursuit akin to birding, whetting the appetites of those who see ecotourism as a boon to conservation efforts. Professional photographers Greg Lasley and Erland Nielsen, and others with a serious photographic bent, Abbott for instance, have spent time photographing the Kerr County fauna, notably Comanche and Leonora's Dancers. And I produced a documentary film, *Ode to a River,* focused on research and discovery of Kerr County and Guadalupe River System odonates, which premiered at Dragonfly Days 2005 in Weslaco.

Thanks to the foresight of Bob Behrstock and Ted Eubanks, odonates have been featured in the Great Texas Wildlife Trail Map series, published by Texas Parks and Wildlife, including a number of sites in Kerr County on the Heart of the Hills and Little Deutschland loops on the Heart of Texas West Trail Map (Behrstock 2004).

Another result of burgeoning interest in the group has been my founding of a list-serv for Odistas, folks interested in all aspects of odonate biology and distribution, with a focus on the insects of Texas and immediately surrounding states. *TexOdes* can be found at http://groups.yahoo.com/group/TexOdes and is open to anyone with a sincere interest. Those interested in my continuing research and filmwork with Kerr County and Guadalupe River odonates are referred to my online journal *milkriverblog* at http://milkriver.blogspot.com, where amongst the many field entries and photos, and myriad unrelated posts, can be found an extensive list of insect and odonate-related weblinks.

Acknowledgments

Thanks to Bob Behrstock for instilling an initial interest and much help with references and identifications; to Greg Lasley and Kelly Bryan, my field partners for thirty years, for virtually everything, but most importantly enthusiasm and friendship; to Mark Lockwood, Dave and Jan Dauphin, Keith Hackland, and Barbara Ribble for their hospitality and friendship; and to Dr. John Abbott for encouraging all this amateur mucking about, for the incredible database that is Odonata Central, and for sending research this way; to Martin Hagne and his crew at the Valley Nature Center for the wonderful festival that is Dragonfly Days; to a number of folks with whom I have shared odonate field time and correspondence, including Tom Langscheid, Mike Quinn, Martin Hagne, Brush Freeman, Giff Beaton, Sheryl Chacon, Jim Bangma, Gayle and Jeanelle Strickland, Mitch Heindel, Martin Reid, Sid Dunkle, Dennis Paulson, Nick Donnelly, Josh Rose, Chris Anderson, Forrest Mitchell, James Lasswell, Sibyl Deacon, David Poteet, Derek Muschalek, Kreg Ellzey, Mike Overton, Ken Cave, Jason Penney, Sage Kawecki, Marcy Dorman, Susan Sander, and certainly others that I have egregiously failed to mention; and to all the many other correspondents, especially via the *TexOdes* listserv and at the Valley Nature Center's Dragonfly Days.

The Odonate Fauna of Kerr County, Texas

Nomenclature after Abbott, *OdonataCentral* web database
* Documented as new by this project; bracketed records are hypothetical

Calopterygidae
American Rubyspot, *Hetaerina americana*
Smoky Rubyspot, *Hetaerina titia*

Lestidae
*Great Spreadwing, *Archilestes grandis*
Plateau Spreadwing, *Lestes alacer*
Chalky Spreadwing, *Lestes sigma*

Protoneuridae
Coral-fronted Threadtail, *Neoneura aaroni*
Orange-striped Threadtail, *Protoneura cara*

Coenagrionidae
Blue-fronted Dancer, *Argia apicalis*
Comanche Dancer, *Argia barretti*
Violet Dancer, *Argia fumipennis violacea*
Kiowa Dancer, *Argia immunda*
*Leonora's Dancer, *Argia leonorae*
Powdered Dancer, *Argia moesta*
Aztec Dancer, *Argia nahuana*
*Springwater Dancer, *Argia plana*
Blue-ringed Dancer, *Argia sedula*
Dusky Dancer, *Argia translata*
Double-striped Bluet, *Enallagma basidens*
Familiar Bluet, *Enallagma civile*
Stream Bluet, *Enallagma exsulans*
*Neotropical Bluet, *Enallagma novaehispaniae*
*Arroyo Bluet, *Enallagma praevarum*
Orange Bluet, *Enallagma signatum*
Vesper Bluet, *Enallagma vesperum*
Citrine Forktail, *Ischnura hastata*
Fragile Forktail, *Ischnura posita*
Rambur's Forktail, *Ishcnura ramburii*
Desert Firetail, *Telebasis salva*

Aeshnidae
Common Green Darner, *Anax junius*
*Comet Darner, *Anax longipes*
Springtime Darner, *Basiaeschna janata*

Gomphidae
Black-shouldered Spinyleg, *Dromogomphus spinosus*
Flag-tailed Spinyleg, *Dromogomphus spoliatus*
Eastern Ringtail, *Erpetogomphus designatus*
Cobra Clubtail, *Gomphus (Gomphurus) vastus*
Sulphur-tipped Clubtail, *Gomphus (Gomphus) militaris*
*Pronghorn Clubtail, *Gomphus (Gomphus) graslinellus*
Dragonunter, *Hagenius brevistylus*
Five-striped Leaftail, *Phyllogomphoides albrighti*
Four-striped Leaftail, *Phyllogomphoides stigmatus*

Macromiidae
Bronzed River Cruiser, *Macromia annulata*

Corduliidae
Prince Baskettail, *Epitheca (Epicordulia) princeps*
*Dot-winged Baskettail, *Epitheca (Tetragoneuria) petechialis*
Orange Shadowdragon, *Neurocordulia xanthosoma*

Libellulidae
Red-tailed Pennant, *Brachymesia furcata*
Four-spotted Pennant, *Brachymesia gravida*
Tawny Pennant, *Brachymesia herbida*
Pale-faced Clubskimmer, *Brechmorhoga mendax*
Halloween Pennant, *Celithemis eponina*
*Banded Pennant, *Celithemis fasciata*
Checkered Setwing, *Dythemis fugax*
*Black Setwing, *Dythemis nigrescens*
Swift Setwing, *Dythemis velox*
Pin-tailed Pondhawk, *Erythemis plebeja*
Common (Eastern) Pondhawk, *Erythemis simplicicollis simplicicollis*

Great Pondhawk, *Erythemis vesiculosa*
Little Blue Dragonlet, *Erythrodiplax minuscula*
Band-winged Dragonlet, *Erythrodiplax umbrata*
Comanche Skimmer, *Libellula comanche*
*Neon Skimmer, *Libellula croceipennis*
Yellow-sided Skimmer, *Libellula flavida*
Widow Skimmer, *Libellula luctuosa*
Twelve-spotted Skimmer, *Libellula pulchella*
*Flame Skimmer, *Libellula saturata*
*Marl Pennant, *Macrodiplax balteata*
*Thornbush Dasher, *Micrathyria hagenii*
[*Carmine Skimmer, *Orthemis discolor*]
Roseate Skimmer, *Orthemis ferruginea*
Blue Dasher, *Pachydiplax longipennis*
[*Red Rock Skimmer, *Paltothemis lineatipes*]
Wandering Glider, *Pantala flavescens*
Spot-winged Glider, *Pantala hymenaea*
Eastern Amberwing, *Perithemis tenera*
Common Whitetail, *Plathemis lydia*
[*Filigree Skimmer, *Pseudoleon superbus*]
*Variegated Meadowhawk, *Sympetrum corruptum*
Autumn Meadowhawk, *Sympetrum vicinum*
Striped Saddlebags, *Tramea calverti*
Black Saddlebags, *Tramea lacerata*
Red Saddlebags, *Tramea onusta*

Species found in surrounding counties, but not yet documented in Kerr
Bandera, Edwards, Gillespie, Kendall, Kimble, Real
* Documented as new by this project

Coenagrionidae
*Coppery Dancer, *Argia cuprea* (Bandera, Edwards, Kendall, Real)
*Lavender Dancer, *Argia hinei* (Bandera, Real)
Sooty Dancer, *Argia lugens* (Kimble)

Lestidae
Rainpool Spreadwing, *Lestes forficula* (Kendall)

Aeshnidae
Amazon Darner, *Anax amazili* (Edwards)
Cyrano Darner, *Nasiaeschna pentacantha* (Real)

Gomphidae
Plains Clubtail, *Gomphus (Gomphurus) externus* (Gillespie, Edwards)
Common Sanddragon, *Progomphus obscurus* (Bandera, Gillespie, Kendall)
Russet-tipped Clubtail, *Stylurus plagiatus* (Kimble)

Macromiidae
Stream Cruiser, *Didymops transversa* (Kimble)
Gilded River Cruiser, *Macromia pacifica* (Kimble)

Corduliidae
Stripe-winged Baskettail, *Epitheca (Tetragoneuria) costalis* (Bandera, Kimble)

Libellulidae

Flame-tailed Pondhawk, *Erythemis peruviana* (Kimble)
Red-faced Dragonlet, *Erythrodiplax fusca* (Edwards)
*Ivory-striped Sylph, *Macrothemis imitans leucozona* (Bandera, Real)
*Jade-striped Sylph, *Macrothemis inequiunguis* (Bandera, Real)
Carmine Skimmer, *Orthemis discolor* (Bandera)
*Red Rock Skimmer, *Paltothemis lineatipes* (Bandera, Real)
*Slough Amberwing, *Perithemis domitia* (Real)
*Filigree Skimmer, *Pseudoleon superbus* (Bandera, Kendall)
Carolina Saddlebags, *Tramea carolina* (Bandera)

Species of the Guadalupe/San Antonio River System, not yet found in Kerr County

[Guadalupe River – Kendall, Comal, Guadalupe, Gonzales, DeWitt, Victoria, Real and Calhoun Counties; San Antonio River – Bexar, Wilson, Karnes, Goliad, Refugio Counties; Medina River – Bandera, Medina, Atascosa and Bexar Counties; Comal River – Comal and Guadalupe Counties; San Marcos River – Hays, Caldwell and Guadalupe Counties; Blanco River – Blanco, Hays, Caldwell and Guadalupe Counties]
* Documented as new by this project

Calopterygidae

Canyon Rubyspot, *Hetaerina vulnerata* (Bexar)

Lestidae

Southern Spreadwing, *Lestes australis* (Blanco, Gonzales, Hays Victoria, Wilson)
*Rainpool Spreadwing, *Lestes forficula* (Atascosa, Bexar, DeWitt, Hays, Kendall)

Coenagrionidae

*Coppery Dancer, *Argia cuprea* (Bandera, Kendall, Real)
*Lavender Dancer, *Argia hinei* (Bandera, Blanco, Real)

Aeshnidae

Amazon Darner, *Anax amazili* (Hays)
Giant Darner, *Anax walsinghami* (Blanco)
Regal Darner, *Coryphaeschna ingens* (Caldwell)
Swamp Darner, *Epiaeschna heros* (Gonzales)
Cyrano Darner, *Nasiaeschna pentacantha* (Comal, Gonzales, Guadalupe, Medina, Real, Wilson)
Blue-eyed Darner, *Rhionaeschna multicolor* (Bexar)
Turquoise-tipped Darner, *Rhionaeschna psilus* (Bexar, Comal)

Gomphidae

Broad-striped Forceptail, *Aphylla angustifolia* (Atascosa, Bexar, Caldwell, Calhoun, Gonzales, Hays)
Narrow-striped Forceptail, *Aphylla protracta* (Bexar, Hays, Wilson)
Stillwater Clubtail, *Arigomphus lentulus* (Gonzales, Victoria)
Jade Clubtail, *Arigomphus submedianus* (Bexar, Caldwell, Gonzales, Victoria)
Blue-faced Ringtail, *Erpetogomphus eutainia* (Caldwell, Gonzales)
Plains Clubtail, *Gomphus (Gomphurus) externus* (Bexar, Caldwell, Comal, Gonzales, Guadalupe, Wilson)
Common Sanddragon, *Progomphus obscurus* (Atascosa, Bandera, Bexar, Blanco, Caldwell, Goliad, Guadalupe, Kendall, Victoria, Wilson)
Russet-tipped Clubtail, *Stylurus plagiatus* (Bexar, Caldwell, Calhoun, DeWitt, Goliad, Gonzales, Guadalupe, Hays, Refugio, Victoria, Wilson)

Macromiidae

Georgia River Cruiser, *Macromia illinoensis georgina* (Comal, Gonzales, Hays)
Stream Cruiser, *Didymops transversa* (Bexar, Blanco, Gonzales, Guadalupe, Hays)
Gilded River Cruiser, *Macromia pacifica* (Comal, Gonzales)

Corduliidae

Stripe-winged Baskettail, *Epitheca (Tetragoneuria) costalis* (Bandera, Bexar, Blanco, Gonzales)
Mantled Baskettail, *Epitheca (Tetragoneuria) semiaquea* (Gonzales)

Libellulidae

Gray-waisted Skimmer, *Cannaphila insularis funerea* (Bexar, Caldwell, Gonzales)
Calico Pennant, *Celithemis elisa* (Bexar, Gonzales, Medina)
Plateau Dragonlet, *Erythrodiplax basifusca* (Blanco, Medina)
Seaside Dragonlet, *Erythrodiplax berenice* (Calhoun, Medina, Refugio)
Black-winged Dragonlet, *Erythrodiplax funerea* (Bexar)
Red-faced Dragonlet, *Erythrodiplax fusca* (Blanco, Medina)
Spangled Skimmer, *Libellula cyanea* (Bexar)
Slaty Skimmer, *Libellula incesta* (Gonzales, Hays, Medina)
Great Blue Skimmer, *Libellula vibrans* (Atascosa, Gonzales)
*Ivory-striped Sylph, *Macrothemis imitans leucozona* (Bandera, Bexar, Caldwell, Guadalupe, Hays, Real)
*Jade-striped Sylph, *Macrothemis inequiunguis* (Bandera, Caldwell, Comal, Real)
Hyacinth Glider, *Miathyria marcella* (Atascosa, Bexar, Calhoun, Gonzales, Hays, Refugio)
*Carmine Skimmer, *Orthemius discolor* (Bandera, Bexar, Caldwell, Comal, Gonzales, Medina, Wilson)
*Red Rock Skimmer, *Paltothemis lineatipes* (Bandera, Real)
*Slough Amberwing, *Perithemis domitia* (Bexar, Real)
*Filigree Skimmer, *Pseudoleon superbus* (Bandera, Kendall)
Carolina Saddlebags, *Tramea carolina* (Bandera)

References Cited

Abbott, J.C. 2000. OdonataCentral: An online resource for the distribution and identification of Odonata. Texas Natural Science Center, The University of Texas at Austin. Available at http://www.odonatacentral.org. (Accessed: October 2000).

Abbott, J.C. 2005. OdonataCentral: An online resource for the distribution and identification of Odonata. Texas Natural Science Center, The University of Texas at Austin. Available at http://www.odonatacentral.org. (Accessed: November-December 2005).

Abbott, J.C. 2007. OdonataCentral: An online resource for the distribution and identification of Odonata. Texas Natural Science Center, The University of Texas at Austin. Available at http://www.odonatacentral.org. (Accessed: May 2007).

Abbott, J.C. 2008. OdonataCentral: An online resource for the distribution and identification of Odonata. Texas Natural Science Center, The University of Texas at Austin. Available at http://www.odonatacentral.org. (Accessed: February 2008).

Behrstock, R.A. 2004. *Heart of Texas Wildlife Trail West*. Great Texas Wildlife Trail Map series, Texas Parks and Wildlife Department.

Gallucci, T. 1972. *The Flora of Trinity County*. National Science Foundation Student Grant report. Department of Wildlife and Fisheries Sciences, Texas A&M University, 110 pp.

Gallucci, T. 1987. *The Birds of The Texas Lions Camp*. Texas Lions Camp, Kerrville, Texas, and subsequent editions.

Gallucci, T. 1989. *The Flora and Fauna of the Riverside Nature Center Project*. Report for the Riverside Nature Center Association, Kerrville, Texas.

Gallucci, T. 1993. *The Birds of Lake Livingston*. Texas Parks and Wildlife Department. (covering Lake Livingston State Park; Trinity, Walker, San Jacinto and Polk Counties).

Gallucci, T. 2000a.*The Vertebrate Fauna of Kerr County, Texas* website. http://www.fortunecity.com/greenfield/egret/290/

Gallucci, T. 2000b.*The Invertebrate Fauna of Kerr County, Texas* website. http://www.fortunecity.com/victorian/margaret/203/

Gallucci, T. 2000c.*The Flora of Kerr County, Texas* website. http://www.fortunecity.com/victorian/sculptured/36/

Gallucci, T. 2000d.*The Fauna of Trinity County, Texas* website. http://www.fortunecity.com/victorian/kapoor/96/

Gallucci, T. 2000e.*The Flora of Trinity County, Texas* website. http://www.fortunecity.com/victorian/charcoal/84/

Gallucci, T. 2000f. *milkriverblog* journal website. http://milkriver.blogspot.com

Gallucci, T. 2005a. *Ode To A River*. Documentary by Milk River Film.

Gallucci, T. 2005b.*The Flora and Fauna of the Texas Hill Country* website. http://hillcountryfauna.blogspot.com/

Gallucci, T. 2005c. *The Flora of Kerr County, Texas* website. http://kerrflora.blogspot.com/

Gallucci, T. 2005d. *The Fauna of Kerr County, Texas* website. http://kerrfauna.blogspot.com/

Gallucci, T, 2005e, *The Flora and Fauna of Trinity County, Texas* website. http://trinityfauna.blogspot.com/

Gallucci, T. 2005f. *The Flora and Fauna of The Texas Big Thicket* website. http://easttexasfauna.blogspot.com/

Gallucci, T. 2006a. *The Odonata of Kerr County and the Guadalupe River System of Texas. in* The Dragonflies and Damselflies (Odonata) of Texas, Volume 1, John C. Abbott, ed., pp. 18-22.

Gallucci, T 2006b. *Checklist of the Fauna and Flora of the Hill Country Youth Ranch*. HCYR Publications. May 2006 and subsequent revisions.

Gallucci, T 2006c. *Checklist of the Fauna and Flora of the Big Springs Ranch for Children*. HCYR Publications. August 2006 and subsequent revisions.

Gallucci, T 2007a. *Checklist of the Fauna and Flora of the Hill Country Youth Ranch*. HCYR Publications. April 2007 and subsequent revisions.

Gallucci, T 2007b. *Checklist of the Fauna and Flora of the Big Springs Ranch for Children*. HCYR Publications. April 2007 and subsequent revisions.

Garrison, R.W. 1994. *A Synopsis of the Genus Argia of the United States with Keys and Descriptions of New Species. Argia sabino, A. leonorae, and A. pima (Odonata: Coenagrionidae)*. Transactions of the American Entomological Society 120:287-368.

Young, W.C. and C.W. Bayer. 1977. *The Dragonfly Nymphs (Odonata: Anisoptera) of The Guadalupe River Basin, Texas*. The Texas Journal of Science 31:85-98.

Tony Gallucci
Gulf Coast Laboratory for Wildlife Research *and* Milk River Film
P.O. Box 6, Camp Verde, Texas 78010-5006
http://milkriver.blogspot.com
milkrivermusic@hotmail.com

The Odonata of Real County and the Frio-Nueces River System of Texas

Tony Gallucci

Introduction

Beginning in 1971 I began a concerted effort to compile a complete Flora and Fauna of a single Texas county – Trinity County, on the northwestern corner of the Big Thicket of east Texas (Gallucci 2000d,e, 2005e,f). This resulted in publications on the plants (Gallucci 1972) and birds (Gallucci 1993). This effort has since been expanded to two Texas Hill Country Counties – Kerr (begun in 1986; Gallucci 1987, 1989, 2000a,b,c,f, 2005a,b,c,d, 2006b, 2007a) and Real (begun in 1988; Gallucci 2000f, 2005b, 2006c, 2007b,c,d, Gallucci and Freeman 2007). In addition, Greg Lasley (2006) and I began a project to accumulate new county records of Odonates for under-reported counties in Texas in 2004, over the years we have amassed a few hundred new records and some of those are represented here in the watershed lists following the Real County checklist.

Here I report for the first time on the Odonate Fauna of Real County, whose county seat, Leakey, is located about 110 km northwest of San Antonio on US Highway 83, and about 200 km west-southwest of Austin. Efforts have been made to document the fauna and flora of the county since 1987, but these were largely focused on the molluscs, amphibians, reptiles, birds and plants. In 1998 work with the class Hexapoda began in earnest in both Real and Kerr Counties, and focused largely on the Lepidoptera and Odonata. A number of other orders have followed since.

Regional Overview

It's easy enough to describe a lot of Texas regions as crossroads – the Lower Rio Grande Valley, the Trans-Pecos, the Big Thicket – but no region, perhaps anywhere in the U.S., can match the Hill Country for its range of habitats, from deeply wooded riparian stretches to open desert, and prairie to wooded hills, and with fauna and flora ranging from ice-age relict Bigtooth Maples, *Acer grandidentatum,* to rare butterflies such as Rawson's Metalmark, *Calephelis rawsoni,* and Lacey's Scrub-Hairstreak, *Strymon alea laceyi,* and mixes of southern and northern, eastern and western bird pairs.
And yet the area is also known worldwide for its high level of endemism, which includes a dozen species of unique salamanders, most in the highly local subterranean genera *Eurycea* and *Typhlomolge,* many of them as yet undescribed to science. There are at least as many snails, many rare plants including the critically endangered Tobusch's Fishhook Cactus, *Ancistrocactus tobuschii,* and Texas Snowbells, *Styrax texanus,* and such isolated globally rare species as Texas Mock-Orange, *Philadelphus texensis.* Four turtles including the Cagle's Map Turtle, *Graptemys caglei,* known only from the

Guadalupe River System, and two endangered and highly sought after birds, the Golden-cheeked Warbler, *Dendroica chrysoparia,* and the Black-capped Vireo, *Vireo atricapillus* are also present.

This is due in large part to the extreme changes in topography resulting from a five million year old uplift, the Balcones Fault, which created the ~700-meter Edwards Plateau, essentially giving birth to a biological island surrounded by strongly differing faunal and floral complexes which creep into the Edwards faunal zones at its several boundaries.

The far western reaches of the plateau skirt the northeastern edge of the Chihuahuan Desert and feature indicator species such as *Agave lechuguilla* and *Larrea tridentata.* Sharply delineated canyonlands in the central portion, created by millennia of erosion, give rise to a multitude of springfed rivers that course through the eastern third, falling off the plateau at an escarpment that runs roughly from San Antonio to Austin and northward, and west to north of Uvalde.

It is this eroded canyon/springs component cascading off the escarpment into the broad South Texas Brushlands that helps make Real County special in terms of providing habitat for odonates. Drainages in the county feed four river systems: the Frio, Sabinal, Nueces and Guadalupe Rivers all owe at least a part of their flow to the deeply cut hills and divides of the county. Of these, virtually all the flow of the Frio River originates in the county via several large headwaters springs, most of them protected on large acreages owned by two non-profit organizations – H.E. Butt Foundation Camp (HEBFC) and the Big Springs Ranch for Children (BSRC; of the Hill Country Youth Ranch). On this latter site is the spring made famous as the label logo of Pearl Beer, "From the Country of 1100 Springs."

The area is a popular recreational mecca, famous as a spring break destination for high school and college students and for families looking for summer cooling-off spots. This has led to significant manipulation of the river, notably downstream of the county (especially in Uvalde County, with its hotspots of Garner State Park, Rio Frio, and Concan).

The wide range of niches available in a river system that features bubbling springs, shallow white water rapids and deep limestone-based pools, provides opportunities for the establishment of a number of odonate species. Included in this retinue are spring pools, riffle zones, boulder-field rapids, sedge marshes, swampy woodlands, wide, flat, two-inch deep, silt-free runs with intricate channeling, pools approaching 15-meter depths, and roadway dammed ponds. In addition there are numerous stock impoundments ranging from small concrete troughs to small earthen-dammed lakes.

The Fauna

A total of 56 species of Odonates were known from Real County outside this project. This total was derived from literature records, and the data as posted on the websites of Forrest Mitchell and James Lasswell (1998a, 1998b, 2000a, 2000b) at Tarleton State University, and John Abbott's Odonata Central at the University of Texas (Abbott 2000, 2005, 2007, 2008). Since then another seven species have been documented (Abbott 2007, 2008, and personal records). The total known Odonate fauna for Real County now stands at 63. Three additional species, Fragile Forktail, *Ischnura posita*, Cobra Clubtail, *Gomphus vastus*, and Band-winged Dragonlet, *Erythrodiplax umbrata*, have been observed in the county but not yet adequately documented, and would bring the county total to 66.

There is the potential for more as an additional 36 species have been found in neighboring counties whose habitat regimes are much the same. That gives us an area total of 102 species.

In the Frio Watershed, comprised largely of the Frio, Nueces, West Nueces and Sabinal Rivers, the fauna is enhanced by the addition of southern brushland, coastal lowland and estuarine species. The 44 additional species found in the overall watershed bring the total for this system to 110 species (plus one additional undocumented species).

Of interest in Real county is the large contingent of species in the genus *Argia*; ten are known for the county. These include the relatively restricted and poorly known Comanche Dancer, *Argia barretti*, the Hill Country specialty Coppery Dancer, *Argia cuprea*, and an eastern station for Lavender Dancer, *Argia hinei*. Also present are large populations of Springwater Dancer, *Argia plana* and Aztec Dancer, *Argia nahuana* in springs areas.

Populations of another species are located in adjacent counties – the recently described (Garrison 1994) Leonora's Dancer, *Argia leonorae* (Kerr & Bandera Counties).

Perhaps most notable among the Real County and Hill Country Zygoptera is the presence of two of the three U.S. Threadtails: Orange-striped Threadtail, *Protoneura cara*, and Coral-fronted Threadtail, *Neoneura aaroni*. It was recently thought that these two species were limited roughly to the Hill Country and central Texas in the U.S., but recent discoveries by Bob Behrstock, Dennis Paulson, Martin Reid and others involved in the North American Dot Map Project (Donnelly) have extended their known ranges to south Texas. Nevertheless they

remain representative of the uniqueness of the odonate fauna of the Hill Country. A third species, Amelia's Threadtail, *Neoneura amelia*, has recently been found just off the plateau in the Frio River system in Frio and Jim Wells Counties.

In the Anisoptera, individuals are less common than in some areas of the state, yet the diversity is high. Typical Real County species would include Pale-faced Clubskimmer, *Brechmorhoga mendax*, Five-striped Leaftail, *Phyllogomphoides albrighti*, Prince Baskettail, *Epitheca princeps*, and Comanche Skimmer, *Libellula comanche*.

But among the dragonflies none are more special, nor representative of the area, than two sylphs of the genus *Macrothemis*: the Ivory-striped Sylph (*Macrothemis imitans leucozona*), and the Jade-striped Sylph (*Macrothemis inequiunguis*), both of which are known in the U.S. only from the Texas Hill Country and just off the escarpment in Hill Country-fed rivers. In this regard they are similar to threadtails. Like the threadtails, a third U.S. species, Straw-colored Sylph, *Macrothemis inacuta*, has been found just off the plateau in the Nueces watershed of Uvalde County, the adjacent Rio Grande watershed in Kinney County, and in San Patricio County where the Nueces empties into the Gulf of Mexico.

These six delicate and retiring insects are indicative of what makes central Texas incomparable as a wildlife refugium.

Research and Other Pursuits

In addition to my work, there has been significant recent study in the county. Chris Anderson of UCLA looked at the display dynamics of American Rubyspot, *Hetaerina americana*, at HEBFC; and Ryan Caesar, Ohio State University, is using Real County specimens of *Argia* in a DNA-based systematic look at the genus.

Interest in the odonates of the county also extends to other pursuits. In the past few years growing numbers of observers have come merely to look at the local odofauna in a pursuit akin to birding, whetting the appetites of those who see ecotourism as a boon to conservation efforts. Professional photographer Greg Lasley, and others with a serious photographic bent, Jason Penney, Ken Cave, Mike Overton, and Bob Thomas for instance, have spent time photographing the Real County fauna, notably the sylph pair.

Another result of my burgeoning interest in the group has been the founding of a list-serv for Odistas, folks interested in all aspects of odonate biology and distribution, with a focus on the insects of Texas and immediately surrounding states. *TexOdes* can be found at http://groups.yahoo.com/group/TexOdes and is open to anyone with a sincere interest. Those interested in my continuing research and filmwork (Gallucci 2005, 2007a) with Real County and Frio-Nueces River watershed odonates are referred to my online journal *milkriverblog* at http://milkriver.blogspot.com where, amongst the many field entries and photos and unrelated posts, there can be found an extensive list of insect and odonate-related weblinks.

Acknowledgments
Thanks to Bob Behrstock for instilling an initial interest and much help with references and identifications; to Greg Lasley and Kelly Bryan, my field partners for thirty-plus years, for virtually everything, but most importantly enthusiasm and friendship; to Mark Lockwood, Dave and Jan Dauphin, and John & Barbara Ribble for their hospitality and friendship on oding and other trips; and to Dr. John Abbott for encouraging all this amateur mucking about, for the incredible database that is his OdonataCentral, and for sending research this way; to Martin Hagne and his crew at the Valley Nature Center, now run by the World Birding Center, for the wonderful festival that is Dragonfly Days, and to Keith Hackland for his hospitality there; to a number of folks with whom I have shared odonate field time and correspondence, including Cheryl Johnson, Charles Bryant, Tom Langscheid, Mike Quinn, Martin Hagne, Brush Freeman, Giff Beaton, Sheryl Chacon, Jim Bangma, Gayle and Jeanelle Strickland, Mitch and Kathy Heindel, Martin Reid, Sid Dunkle, Dennis Paulson, Nick Donnelly, Josh Rose, Chris Anderson, Forrest Mitchell, James Lasswell, Sibyl Deacon, David Poteet, Derek Muschalek, Kreg Ellzey, Mike Overton, Ken Cave, Jason Penney, Sage Kawecki, Marcy Dorman, Susan Sander, Bob Thomas, Brandon Best, Tom Collins, Mike Gray, Bob Barber, Bob Rasa, Scott Young, members of several years' worth of Nature Quest Field Trips, and certainly others that I have egregiously failed to mention; and to all the many other correspondents, especially via the *TexOdes* listserv and at the Valley Nature Center's Dragonfly Days.

The Odonate Fauna of Real County, Texas (63 species documented, plus 3 hypothetical in brackets)
* Recent project additions

Calopterygidae
American Rubyspot, *Hetaerina americana*

Protoneuridae
Coral-fronted Threadtail, *Neoneura aaroni*
Orange-striped Threadtail, *Protoneura cara*

Lestidae
*Great Spreadwing, *Archilestes grandis*

Coenagrionidae
Citrine Forktail, *Ischnura hastata*
[*Fragile Forktail, *Ischnura posita*]
Rambur's Forktail, *Ischnura ramburii*
Desert Firetail, *Telebasis salva*
Arroyo Bluet, *Enallagma praevarum*
Double-striped Bluet, *Enallagma basidens*
*Orange Bluet, *Enallagma signatum*
Familiar Bluet, *Enallagma civile*

Stream Bluet, *Enallagma exsulans*
Neotropical Bluet, *Enallagma novaehispaniae*
Comanche Dancer, *Argia barretti*
Lavender Dancer, *Argia hinei*
Kiowa Dancer, *Argia immunda*
Coppery Dancer, *Argia cuprea*
Aztec Dancer, *Argia nahuana*
Springwater Dancer, *Argia plana*
Violet Dancer, *Argia fumipennis violacea*
Blue-ringed Dancer, *Argia sedula*
Powdered Dancer, *Argia moesta*
Dusky Dancer, *Argia translata*

Aeshnidae
Common Green Darner, *Anax junius*
Springtime Darner, *Basiaeschna janata*
Cyrano Darner, *Nasiaeschna pentacantha*

Gomphidae
Black-shouldered Spinyleg, *Dromogomphus spinosus*
Flag-tailed Spinyleg, *Dromogomphus spoliatus*
Eastern Ringtail, *Erpetogomphus designatus*
Pronghorn Clubtail, *Gomphus graslinellus*
Sulphur-tipped Clubtail, *Gomphus militaris*
[*Cobra Clubtail, *Gomphus vastus*]
Dragonhunter, *Hagenius brevistylus*
Five-striped Leaftail, *Phyllogomphoides albrighti*
Four-striped Leaftail, *Phyllogomphoides stigmatus*

Macromiidae
Bronzed River Cruiser, *Macromia annulata*

Corduliidae
*Dot-winged Baskettail, *Epitheca petechialis*
Prince Baskettail, *Epitheca princeps*

Libellulidae
Pale-faced Clubskimmer, *Brechmorhoga mendax*
Halloween Pennant, *Celithemis eponina*
Banded Pennant, *Celithemis fasciata*
Checkered Setwing, *Dythemis fugax*
Black Setwing, *Dythemis nigrescens*
Swift Setwing, *Dythemis velox*
Eastern Pondhawk, *Erythemis simplicicollis simplicicollis*
[*Band-winged Dragonlet, *Erythrodiplax umbrata*]
Comanche Skimmer, *Libellula comanche*
Neon Skimmer, *Libellula croceipennis*
Widow Skimmer, *Libellula luctuosa*
Twelve-spotted Skimmer, *Libellula pulchella*
Flame Skimmer, *Libellula saturata*
*Jade-striped Sylph, *Macrothemis inequiunguis*
*Ivory-striped Sylph, *Macrothemis imitans leucozona*
Roseate Skimmer, *Orthemis ferruginea*
Blue Dasher, *Pachydiplax longipennis*
*Red Rock Skimmer, *Paltothemis lineatipes*
Wandering Glider, *Pantala flavescens*
Spot-winged Glider, *Pantala hymenaea*
Eastern Amberwing, *Perithemis tenera*
*Slough Amberwing, *Perithemis domitia*
Common Whitetail, *Plathemis lydia*

Variegated Meadowhawk, *Sympetrum corruptum*
Autumn Meadowhawk, *Sympetrum vicinum*
Black Saddlebags, *Tramea lacerata*
Red Saddlebags, *Tramea onusta*

Additional Odonate Fauna from Counties adjacent to Real
including Edwards, Kinney, Uvalde, Bandera, Kerr

Calopterygidae
Smoky Rubyspot, *Hetaerina titia* (Bandera, Kerr, Kinney, Uvalde)

Lestidae
Plateau Spreadwing, *Lestes alacer* (Bandera, Kerr, Kinney, Uvalde)
Chalky Spreadwing, *Lestes sigma* (Bandera, Kerr, Kinney)

Coenagrionidae
Blue-fronted Dancer, *Argia apicalis* (Kerr, Kinney, Uvalde)
Leonora's Dancer, *Argia leonorae* (Bandera, Kerr, Kinney, Uvalde)
Golden-winged Dancer, *Argia rhoadsi* (Kinney)
Vesper Bluet, *Enallagma vesperum* (Kerr)
Fragile Forktail, *Ischnura posita posita* (Kerr, Kinney)

Aeshnidae
Amazon Darner, *Anax amazili* (Edwards, Uvalde)
Comet Darner, *Anax longipes* (Kerr, Uvalde)

Gomphidae
Broad-striped Forceptail, *Aphylla angustifolia* (Kinney, Uvalde)
Narrow-striped Forceptail, *Aphylla protracta* (Uvalde)
Jade Clubtail, *Arigomphus submedianus* (Uvalde)
Cobra Clubtail, *Gomphus (Gomphurus) vastus* (Kerr)
Plains Clubtail, *Gomphus (Gomphurus) externus* (Edwards)
Common Sanddragon, *Progomphus obscurus* (Bandera)
Russet-tipped Clubtail, *Stylurus plagiatus* (Kinney)

Macromiidae
Stream Cruiser, *Didymops transversa* (Uvalde)
Georgia River Cruiser, *Macromia illinoensis georgina* (Kerr)

Corduliidae
Stripe-winged Baskettail, *Epitheca (Tetragoneuria) costalis* (Bandera, Uvalde)
Orange Shadowdragon, *Neurcordulia xanthosoma* (Kerr)

Libellulidae
Red-tailed Pennant, *Brachymesia furcata* (Kerr, Kinney, Uvalde)
Four-spotted Pennant, *Brachymesia gravida* (Edwards, Kerr, Uvalde)
Tawny Pennant, *Brachymesia herbida* (Kerr)
Gray-waisted Skimmer, *Cannaphila insularis funerea* (Kinney)
Pin-tailed Pondhawk, *Erythemis plebeja* (Kerr, Uvalde)
Great Pondhawk, *Erythemis vesiculosa* (Kerr, Kinney, Uvalde)

Red-faced Dragonlet, *Erythrodiplax fusca* (Edwards)
Little Blue Dragonlet, *Erythrodiplax minuscula* (Kerr, Uvalde)
Band-winged Dragonlet, *Erythrodiplax umbrata* (Bandera, Edwards, Kerr, Uvalde)
Yellow-sided Skimmer, *Libellula flavida* (Kerr)
Marl Pennant, *Macrodiplax balteata* (Edwards, Kerr)
Straw-colored Sylph, *Macrothemis inacuta* (Kinney, Uvalde)
Thornbush Dasher, *Micrathyria hagenii* (Kerr, Uvalde)
Carmine Skimmer, *Orthemis discolor* (Bandera, [Kerr], Uvalde)
Filigree Skimmer, *Pseudoleon superbus* (Bandera, Edwards,[Kerr], Kinney, Uvalde)
Striped Saddlebags, *Tramea calverti* (Kerr, Kinney, Uvalde)
Carolina Saddlebags, *Tramea carolina* (Bandera)
Antillean Saddlebags, *Tramea insularis* (Kinney)

Additional Odonate Fauna from the Frio-Nueces Watershed System
including Zavala, LaSalle, Nueces, Live Oak, McMullen, Frio, Uvalde, Jim Wells, San Patricio

Calopterygidae
Smoky Rubyspot, *Hetaerina titia* (Frio, Jim Wells, LaSalle, Live Oak, McMullen, Nueces, San Patricio, Uvalde, Zavala)

Lestidae
Plateau Spreadwing, *Lestes alacer* (San Patricio, Uvalde)
Southern Spreadwing, *Lestes australis* (Jim Wells, Nueces, San Patricio)
Rainpool Spreadwing, *Lestes forficula* (Nueces, San Patricio)
Chalky Spreadwing, *Lestes sigma* (LaSalle, Nueces, San Patricio)

Protoneuridae
Amelia's Threadtail, *Neoneura amelia* (Frio, Jim Wells)

Coenagrionidae
Blue-fronted Dancer, *Argia apicalis* (Frio, Jim Wells, LaSalle, Live Oak, McMullen, Nueces, San Patricio, Uvalde)
Leonora's Dancer, *Argia leonorae* (Uvalde)
Big Bluet, *Enallagma durum* (Live Oak, San Patricio)

Aeshnidae
Amazon Darner, *Anax amazili* (Uvalde)
Comet Darner, *Anax longipes* (Uvalde)
Regal Darner, *Coryphaeschna ingens* (San Patricio)
Swamp Darner, *Epiaeschna heros* (San Patricio)
Turquoise-tipped Darner, *Rhionaeschna psilus* (Frio, Nueces)

Gomphidae
Broad-striped Forceptail, *Aphylla angustifolia* (Jim Wells, LaSalle, Uvalde)
Narrow-striped Forceptail, *Aphylla protracta* (San Patricio, Uvalde)

Plains Clubtail, *Gomphus (Gomphurus) externus* (Jim Wells, LaSalle, Live Oak, McMullen, San Patricio)

Cobra Clubtail, *Gomphus vastus* (Jim Wells)

Jade Clubtail, *Arigomphus submedianus* (McMullen, Uvalde)

Common Sanddragon, *Progomphus obscurus* (Jim Wells, McMullen, San Patricio)

Russet-tipped Clubtail, *Stylurus plagiatus* (Jim Wells, San Patricio)

Macromiidae

Stream Cruiser, *Didymops transversa* (Frio, Uvalde)

Georgia River Cruiser, *Macromia illinoensis georgina* (Frio, Live Oak, San Patricio)

Gilded River Cruiser, *Macromia pacifica* (San Patricio)

Royal River Cruiser, *Macromia taeniolata* (Jim Wells)

Corduliidae

Stripe-winged Baskettail, *Epitheca (Tetragoneuria) costalis* (Uvalde)

Libellulidae

Red-tailed Pennant, *Brachymesia furcata* (Frio, LaSalle, McMullen, Nueces, San Patricio, Uvalde, Zavala)

Four-spotted Pennant, *Brachymesia gravida* (Jim Wells, Live Oak, Nueces, San Patricio, Uvalde, Zavala)

Pin-tailed Pondhawk, *Erythemis plebeja* (Frio, Nueces, Uvalde)

Great Pondhawk, *Erythemis vesiculosa* (Jim Wells, McMullen, Nueces, San Patricio, Uvalde)

Plateau Dragonlet, *Erythrodiplax basifusca* (Frio)

Seaside Dragonlet, *Erythrodiplax berenice berenice* (LaSalle, Nueces, San Patricio)

Red-faced Dragonlet, *Erythrodiplax fusca* (Frio)

Little Blue Dragonlet, *Erythrodiplax minuscula* (Jim Wells, McMullen, San Patricio, Uvalde)

Band-winged Dragonlet, *Erythrodiplax umbrata* (Jim Wells, Live Oak, McMullen, Nueces, San Patricio, Uvalde)

Golden-winged Skimmer, *Libellula auripennis* (Jim Wells, San Patricio)

Needham's Skimmer, *Libellula needhami* (Jim Wells, Nueces, San Patricio)

Marl Pennant, *Macrodiplax balteata* (Frio, Nueces, San Patricio)

Straw-colored Sylph, *Macrothemis inacuta* (San Patricio, Uvalde)

Hyacinth Glider, *Miathyria marcella* (Jim Wells, Nueces, San Patricio)

Three-striped Dasher, *Micrathyria didyma* (Nueces)

Thornbush Dasher, *Micrathyria hagenii* (Frio, LaSalle, Nueces, Uvalde, Zavala)

Carmine Skimmer, *Orthemis discolor* (Frio, Uvalde)

Filigree Skimmer, *Pseudoleon superbus* (Uvalde, Zavala)

Striped Saddlebags, *Tramea calverti* (Nueces, San Patricio, Uvalde)

Antillean Saddlebags, *Tramea insularis* (Frio, Zavala)

Additional Odonate Fauna from Counties shared with

the Guadalupe River System
including Kerr, Bandera, Medina, Bexar, Edwards, Wilson, Karnes, Bee, Atascosa

Calopterygidae

Smoky Rubyspot, *Hetaerina titia* (Kerr, Bandera, Bexar, Karnes, Medina, Wilson)

Canyon Rubyspot, *Hetaerina vulnerata* (Bexar)

Lestidae

Plateau Spreadwing, *Lestes alacer* (Bandera, Bexar, Kerr)

Southern Spreadwing, *Lestes australis* (Wilson)

Rainpool Spreadwing, *Lestes forficula* (Atascosa, Bexar)

Chalky Spreadwing, *Lestes sigma* (Bandera, Bexar, Kerr)

Coenagrionidae

Blue-fronted Dancer, *Argia apicalis* (Atascosa, Bee, Bexar, Kerr, Karnes, Medina, Wilson)

Leonora's Dancer, *Argia leonorae* (Bandera, Kerr, Medina)

Vesper Bluet, *Enallagma vesperum* (Kerr)

Fragile Forktail, *Ischnura posita posita* (Karnes, Kerr)

Aeshnidae

Amazon Darner, *Anax amazili* (Bexar, Edwards)

Comet Darner, *Anax longipes* (Kerr)

Swamp Darner, *Epiaeschna heros* (Bee)

Blue-eyed Darner, *Rhionaeschna multicolor* (Bexar)

Turquoise-tipped Darner, *Rhionaeschna psilus* (Bexar)

Gomphidae

Broad-striped Forceptail, *Aphylla angustifolia* (Atascosa, Bexar)

Narrow-striped Forceptail, *Aphylla protracta* (Bexar, Wilson)

Jade Clubtail, *Arigomphus submedianus* (Bexar)

Cobra Clubtail, *Gomphus (Gomphurus) vastus* (Bexar, Kerr)

Plains Clubtail, *Gomphus (Gomphurus) externus* (Bexar, Edwards, Wilson)

Common Sanddragon, *Progomphus obscurus* (Atascosa, Bandera, Bexar, Wilson)

Russet-tipped Clubtail, *Stylurus plagiatus* (Bexar, Wilson)

Macromiidae

Stream Cruiser, *Didymops transversa* (Bexar)

Georgia River Cruiser, *Macromia illinoensis georgina* (Kerr, Karnes, Wilson)

Corduliidae

Stripe-winged Baskettail, *Epitheca (Tetragoneuria) costalis* (Bandera, Bexar)

Orange Shadowdragon, *Neurocordulia xanthosoma* (Bexar, Kerr)

Libellulidae

Red-tailed Pennant, *Brachymesia furcata* (Atascosa, Bee, Bexar, Kerr)

Four-spotted Pennant, *Brachymesia gravida* (Atascosa, Bee, Bexar, Edwards, Kerr)

Tawny Pennant, *Brachymesia herbida* (Kerr)

Gray-waisted Skimmer, *Cannaphila insularis funerea* (Bexar)

Calico Pennant, *Celithemis elisa* (Bexar, Medina)

Pin-tailed Pondhawk, *Erythemis plebeja* (Bexar, Kerr)

Great Pondhawk, *Erythemis vesiculosa* (Bexar, Kerr, Wilson)

Plateau Dragonlet, *Erythrodiplax basifusca* (Medina)

Seaside Dragonlet, *Erythrodiplax berenice* (Medina)

Black-winged Dragonlet, *Erythrodiplax funerea* (Bexar)

Red-faced Dragonlet, *Erythrodiplax fusca* (Edwards, Medina)

Little Blue Dragonlet, *Erythrodiplax minuscula* (Bexar, Kerr)

Band-winged Dragonlet, *Erythrodiplax umbrata* (Bandera, Bexar, Edwards, Kerr)

Spangled Skimmer, *Libellula cyanea* (Bexar)

Yellow-sided Skimmer, *Libellula flavida* (Kerr)

Slaty Skimmer, *Libellula incesta* (Medina)

Great Blue Skimmer, *Libellula vibrans* (Atascosa)

Marl Pennant, *Macrodiplax balteata* (Atascosa, Bee, Bexar, Edwards, Kerr)

Hyacinth Glider, *Miathyria marcella* (Atascosa, Bee, Bexar)

Thornbush Dasher, *Micrathyria hagenii* (Atascosa, Bee, Bexar, Kerr, Karnes)

Carmine Skimmer, *Orthemis discolor* (Bandera, Bexar, Bee, [Kerr], Medina, Wilson)

Filigree Skimmer, *Pseudoleon superbus* (Bandera, Edwards, [Kerr])

Striped Saddlebags, *Tramea calverti* (Bexar, Kerr)

Carolina Saddlebags, *Tramea carolina* (Bandera)

Additional Odonate Fauna from Counties shared with the Rio Grande River System
including Edwards, Kinney, Maverick, Dimmit, Webb, Duval

Calopterygidae
Smoky Rubyspot, *Hetaerina titia* (Kinney, Maverick, Webb)

Lestidae
Chalky Spreadwing, *Lestes sigma* (Dimmit, Kinney)

Plateau Spreadwing, *Lestes alacer* (Duval, Kinney)

Coenagrionidae
Blue-fronted Dancer, *Argia apicalis* (Dimmit, Kinney, Maverick, Webb)

Leonora's Dancer, *Argia leonorae* (Kinney)

Golden-winged Dancer, *Argia rhoadsi* (Kinney)

Fragile Forktail, *Ischnura posita posita* (Dimmit, Kinney, Maverick)

Aeshnidae
Amazon Darner, *Anax amazili* (Edwards)

Gomphidae
Broad-striped Forceptail, *Aphylla angustifolia* (Kinney, Webb)

Jade Clubtail, *Arigomphus submedianus* (Dimmit)

White-belted Ringtail, *Erpetogomphus compositus* (Maverick)

Blue-faced Ringtail, *Erpetogomphus eutainia* (Webb)

Plains Clubtail, *Gomphus (Gomphurus) externus* (Edwards)

Russet-tipped Clubtail, *Stylurus plagiatus* (Dimmit, Kinney)

Corduliidae
Orange Shadowdragon, *Neurocordulia xanthosoma* (Dimmit)

Libellulidae
Red-tailed Pennant, *Brachymesia furcata* (Dimmit, Duval, Kinney)

Four-spotted Pennant, *Brachymesia gravida* (Dimmit, Edwards)

Gray-waisted Skimmer, *Cannaphila insularis funerea* (Kinney)

Great Pondhawk, *Erythemis vesiculosa* (Kinney)

Red-faced Dragonlet, *Erythrodiplax fusca* (Edwards)

Little Blue Dragonlet, *Erythrodiplax minuscula* (Dimmit, Webb)

Band-winged Dragonlet, *Erythrodiplax umbrata* (Dimmit, Edwards, Webb)

Marl Pennant, *Macrodiplax balteata* (Edwards, Maverick)

Straw-colored Sylph, *Macrothemis inacuta* (Kinney, Webb)

Thornbush Dasher, *Micrathyria hagenii* (Dimmit, Duval, Webb)

Carmine Skimmer, *Orthemis discolor* (Duval, Webb)

Filigree Skimmer, *Pseudoleon superbus* (Edwards, Kinney, Maverick, Webb)

Striped Saddlebags, *Tramea calverti* (Kinney)

Antillean Saddlebags, *Tramea insularis* (Kinney)

Important county records for watershed species of special interest:
Coral-fronted Threadtail, *Neoneura aaroni* (Bandera, Bexar, Blanco, Caldwell, Comal, Goliad, Gonzales, Guadalupe, Hays, Hidalgo, Kerr, Medina, Nueces, Real, San Patricio, Val Verde, Victoria, Zapata, Zavala)

Amelia's Threadtail, *Neoneura amelia* (Cameron, Frio, Hidalgo, Jim Wells, Zapata)

Orange-striped Threadtail, *Protoneura cara* (Comal, Edwards, Harris, Hays, Hidalgo, Kendall, Kerr, Kimble, Medina, Menard, Real, Terrell, Uvalde, Val Verde)

Ivory-striped Sylph, *Macrothemis imitans leucozona* (Bandera, Bexar, Caldwell, Frio, Hays, Kinney, Real, Uvalde, Val Verde)

Straw-colored Sylph, *Macrothemis inacuta* (Hidalgo, Kinney, San Patricio, Starr, Travis, Uvalde, Webb, Zapata)

Jade-striped Sylph, *Macrothemis inequiunguis* (Bandera, Caldwell, Comal, Real, Travis)

Literature Cited
Abbott, J.C. 2000. OdonataCentral: An online resource for the distribution and identification of Odonata. Texas Natural Science Center, The University of Texas

at Austin. Available at http://www.odonatacentral.org. (Accessed: October 2000).

Abbott, J.C. 2005. OdonataCentral: An online resource for the distribution and identification of Odonata. Texas Natural Science Center, The University of Texas at Austin. Available at http://www.odonatacentral.org. (Accessed: November-December 2005).

Abbott, J.C. 2007. OdonataCentral: An online resource for the distribution and identification of Odonata. Texas Natural Science Center, The University of Texas at Austin. Available at http://www.odonatacentral.org. (Accessed: May 2007).

Abbott, J.C. 2008. OdonataCentral: An online resource for the distribution and identification of Odonata. Texas Natural Science Center, The University of Texas at Austin. Available at http://www.odonatacentral.org. (Accessed: February 2008).

Gallucci, T. 1972. *The Flora of Trinity County*. National Science Foundation Student Grant report. Department of Wildlife and Fisheries Sciences, Texas A&M University, 110 pp.

Gallucci, T. 1987. *The Birds of The Texas Lions Camp*. Texas Lions Camp, Kerrville, Texas, and subsequent editions.

Gallucci, T. 1989. *The Flora and Fauna of the Riverside Nature Center Project*. Report for the Riverside Nature Center Association, Kerrville, Texas.

Gallucci, T. 1993. *The Birds of Lake Livingston*. Texas Parks and Wildlife Department. (covering Lake Livingston State Park; Trinity, Walker, San Jacinto and Polk Counties).

Gallucci, T. 2000a. *The Vertebrate Fauna of Kerr County, Texas* website. http://www.fortunecity.com/greenfield/egret/290/

Gallucci, T. 2000b. *The Invertebrate Fauna of Kerr County, Texas* website. http://www.fortunecity.com/victorian/margaret/203/

Gallucci, T. 2000c. *The Flora of Kerr County, Texas* website. http://www.fortunecity.com/victorian/sculptured/36/

Gallucci, T. 2000d. *The Fauna of Trinity County, Texas* website. http://www.fortunecity.com/victorian/kapoor/96/

Gallucci, T. 2000e. *The Flora of Trinity County, Texas* website. http://www.fortunecity.com/victorian/charcoal/84/

Gallucci, T. 2000f. *milkriverblog* journal website. http://milkriver.blogspot.com

Gallucci, T. 2005a. *Ode To A River*. Documentary by Milk River Film.
Gallucci, T. 2005b. *The Flora and Fauna of the Texas Hill Country* website. http://hillcountryfauna.blogspot.com/

Gallucci, T. 2005c. *The Flora of Kerr County, Texas* website. http://kerrflora.blogspot.com/

Gallucci, T. 2005d. *The Fauna of Kerr County, Texas* website. http://kerrfauna.blogspot.com/

Gallucci, T, 2005e, *The Flora and Fauna of Trinity County, Texas* website. http://trinityfauna.blogspot.com/

Gallucci, T. 2005f. *The Flora and Fauna of The Texas Big Thicket* website. http://easttexasfauna.blogspot.com/

Gallucci, T. 2006a. *The Odonata of Kerr County and the Guadalupe River System of Texas. in* The Dragonflies and Damselflies (Odonata) of Texas, Volume 1, John C. Abbott, ed., pp. 18-22.

Gallucci, T 2006b. *Checklist of the Fauna and Flora of the Hill Country Youth Ranch*. HCYR Publications. May 2006 and subsequent revisions.

Gallucci, T 2006c. *Checklist of the Fauna and Flora of the Big Springs Ranch for Children*. HCYR Publications. August 2006 and subsequent revisions.

Gallucci, T 2007a. *Checklist of the Fauna and Flora of the Hill Country Youth Ranch*. HCYR Publications. April 2007 and subsequent revisions.

Gallucci, T 2007b. *Checklist of the Fauna and Flora of the Big Springs Ranch for Children*. HCYR Publications. April 2007 and subsequent revisions.

Gallucci, T. 2007c. Red Wasp (*Polistes carolina*) Predation on Pale-faced Clubskimmer (*Brechmorhoga mendax*). Argia 19(2): 20-21.

Gallucci, T. 2007d. Observations on the Behavior of *Macrothemis inequiunguis* (Jade-striped Sylph). Argia 19(3): 14-17.

Gallucci, T., and B. Freeman. 2007. Notes on Avian Predators of Odonata. Argia 18(2): 21-23.

Garrison, Rosser W. 1994. *A Synopsis of the Genus Argia of the United States with Keys and Descriptions of New Species*. Argia sabino, A. leonorae, *and* A. pima (*Odonata: Coenagrionidae*). Transactions of the American Entomological Society 120:287-368.

Lasley, G.W. 2006. *Hornsby Bend - It's not just for birds*. In Dragonflies and Damselflies (Odonata) of Texas. Austin, TX, J. C. Abbott, ed., pp. 13-15.

Mitchell, F. and J. Lasswell. 1998a. Digital Dragonflies website. Tarleton State University, Texas A&M University System, Texas Agricultural Experiment Station. http://www.dragonflies.org/

Mitchell, F. and J. Lasswell. 1998b. Damselflies of Texas website. Tarleton State University, Texas A&M University System, Texas Agricultural Experiment Station. http://stephenville.tamu.edu/~fmitchel/damselfly/index.html

Mitchell, F. and J. Lasswell. 2000a. Digital Dragonflies website. Tarleton State University, Texas A&M University System, Texas Agricultural Experiment Station. http://www.dragonflies.org/

Mitchell, F. and J. Lasswell. 2000b. Damselflies of Texas website. Tarleton State University, Texas A&M University System, Texas Agricultural Experiment Station. http://stephenville.tamu.edu/~fmitchel/damselfly/index.html

Tony Gallucci
Gulf Coast Laboratory for Wildlife Research *and* Milk River Film
P.O. Box 6, Camp Verde, Texas 78010-5006
http://milkriver.blogspot.com
hurricanetg@hotmail.com

Statistical Summary of Odonata in Texas

224 Species Reported for Texas

Suborder	# spp.
Anisoptera	150
Zygoptera	74

Family	# spp.
Aeshnidae	19
Calopterygidae	5
Coenagrionidae	57
Cordulegastridae	2
Corduliidae	15
Gomphidae	32
Lestidae	9
Libellulidae	76
Macromiidae	5
Petaluridae	1
Protoneurida	3

Family	Genus	# spp.
Aeshnidae	Aeshna	2
	Anax	5
	Basiaeschna	1
	Boyeria	1
	Coryphaeschna	2
	Epiaeschna	1
	Gomphaeschna	1
	Gynacantha	1
	Nasiaeschna	1
	Rhionaeschna	3
	Triacanthagyna	1
Calopterygidae	Calopteryx	2
	Hetaerina	3
Coenagrionidae	Acanthagrion	1
	Argia	21
	Enallagma	18
	Hesperagrion	1
	Ischnura	10
	Leptobasis	1
	Nehalennia	2
	Neoerythromma	1
	Telebasis	2
Cordulegastridae	Cordulegaster	2
Corduliidae	Epitheca	6
Corduliidae	Helocordulia	1
	Neurocordulia	3
	Somatochlora	5
Gomphidae	Aphylla	3
	Arigomphus	3
	Dromogomphus	2
	Erpetogomphus	5
	Gomphus	10
	Hagenius	1
	Phyllocycla	1
	Phyllogomphoides	2
	Progomphus	2
	Stylurus	3
Lestidae	Archilestes	1
	Lestes	8
Libellulidae	Brachymesia	3
	Brechmorhoga	1
	Cannaphila	1
	Celithemis	6
	Dythemis	4
	Erythemis	7
	Erythrodiplax	6
	Ladona	1
	Libellula	14
	Macrodiplax	1
	Macrothemis	3
	Miathyria	1
	Micrathyria	3
	Orthemis	2
	Pachydiplax	1
	Paltothemis	1
	Pantala	2
	Perithemis	2
	Plathemis	2
	Pseudoleon	1
	Sympetrum	6
	Tauriphila	1
	Tholymis	1
	Tramea	6
Macromiidae	Didymops	1
	Macromia	4
Petaluridae	Tachopteryx	1
	Neoneura	2
Protoneuridae	Protoneura	1

Abundance & Distribution of Texas Odonata

John C. Abbott

I plotted the abundance and distribution for all species occurring in Texas based on specimens in the University of Texas Insect collection (Fig. 1). I then divided each axis into quadrants and applied appropriate descriptors to these quadrants. This graph represents all the species and records in the UT collection and database as of 2005. Each species is plotted based on the number of adult specimens known for the species (abundance, y-axis) and the number of counties from which the species is known (distribution, x-axis).

Most species appear to be restricted or limited in their distribution. Those species that are more widely distributed however, are generally more abundant in the collection. In some cases, a species is very restricted in its distribution, but may be rated as common in its abundance. Though this may seem unexpected at first, it is

actually quite logical. These are species that are restricted to a specific, possibly rare habitat, but are numerous at that habitat.

A system like this has the advantage of being unbiased by an individual's impressions. It is however, biased by the ease of capture of a species. Other factors such as accessibility of habitat, time of day, season, flight speed and agility can all affect the number of specimens captured and thus the way a species falls out in the plot. Additional bias may involve the individual preference of collectors for certain habitats or species.

I propose that as the OST gathers data we begin to use terms like these to describe the general abundance and distribution of the Texas fauna.

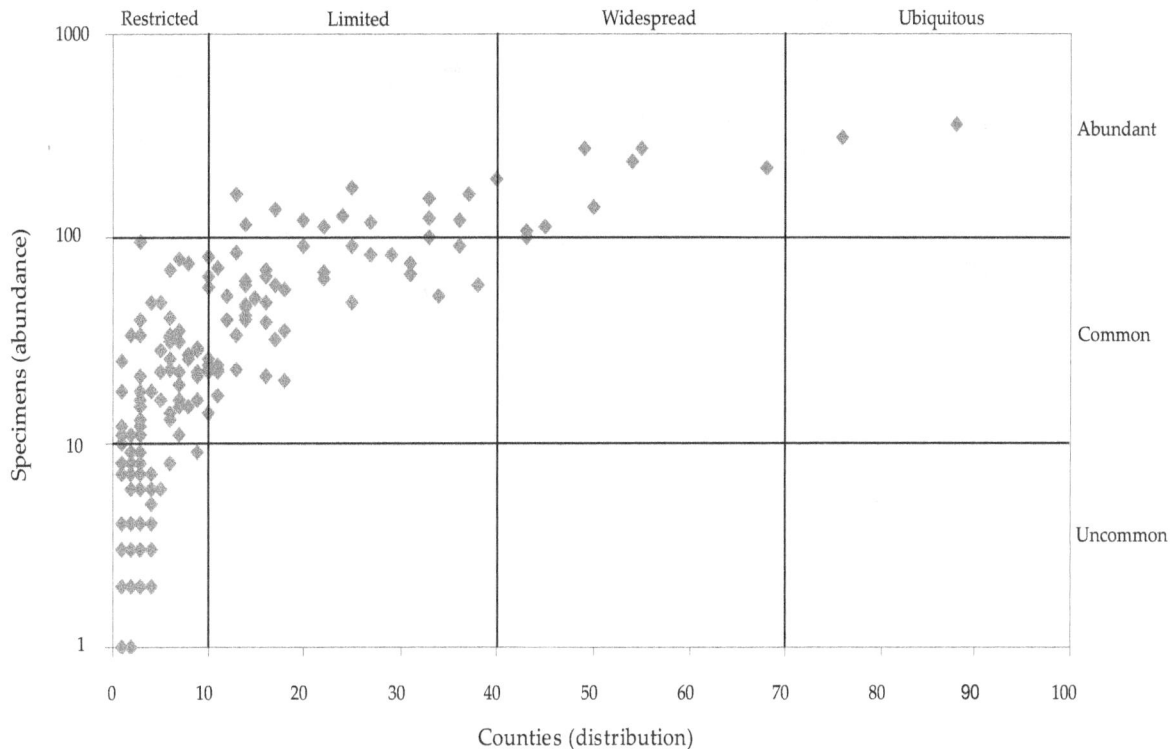

Fig. 1 - Abundance and distribution plot for Texas Odonata based on specimens in the University of Texas Insect Collection.

Diversity of Texas Odonata by County

The map below is shaded to show the number odonate species known for individual counties. The darker the county, the more species known. We broke the 100 mark in 2007 with 102 species now known from Travis County. San Jancinto lead the pack for years with 95 species, but is now boasts a respectable 97 species. Nine counties now have 75 or more species. In addition to the two mentioned above, they are: Hidalgo, Bexar, Val Verde, Harris, Uvalde, Collin, and Kerr. There are only 10 counties in Texas that have not had a single odonate reported from them. They are: Andrews, Armstrong, Castro, Cochran, Ector, Lamb, Nolan, Parmer, Sterling, and Terry. Hopefully we can say that odonates have been recorded in every county by the end of 2008.

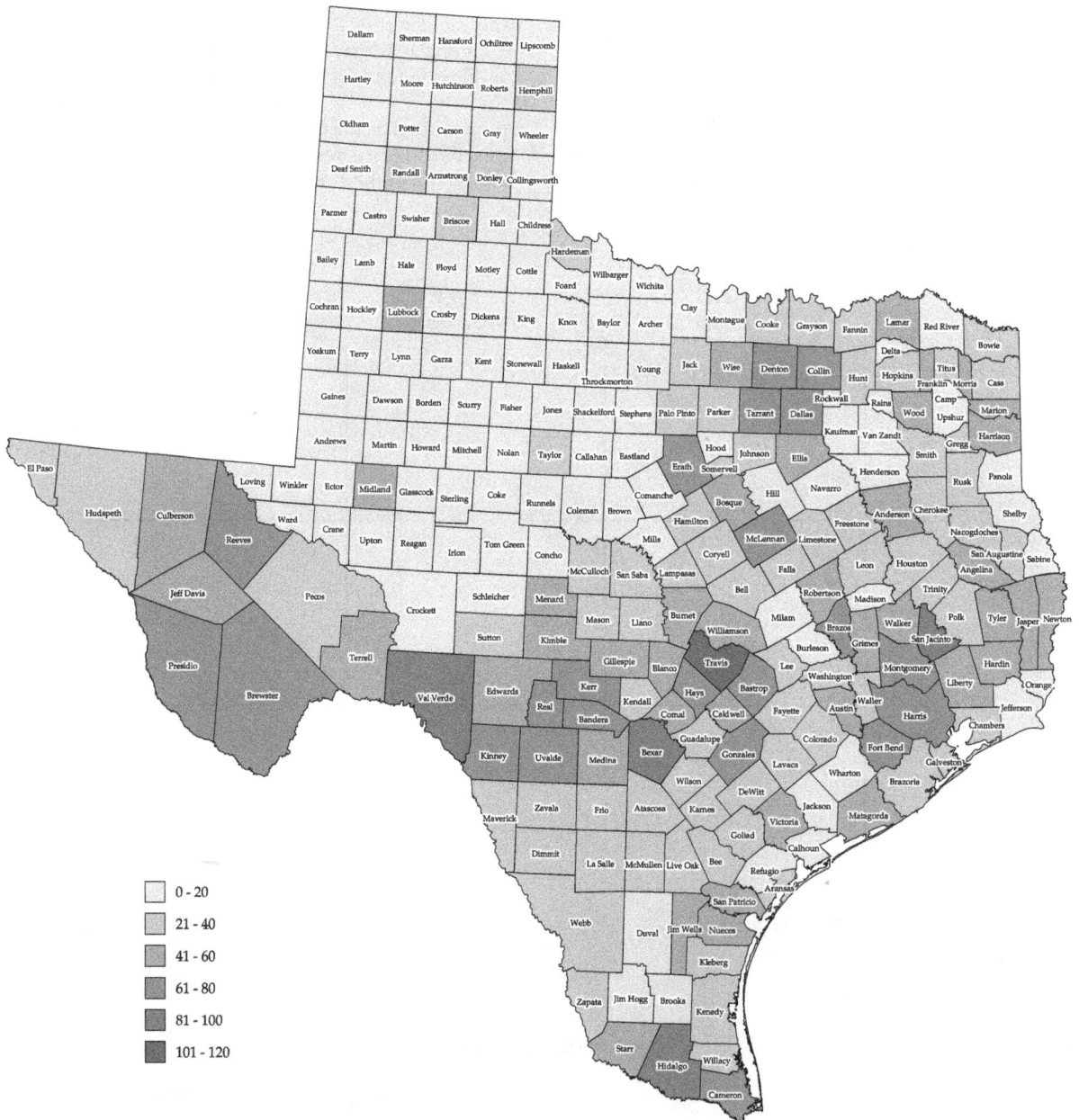

Dragonflies & Damselflies of Texas

————————————————— John C. Abbott —————————————————

A single species was added to the state list in 2007. I photographed *Argia tezpi* (Tezpi Dancer) in May of 2005 at ZH Canyon (Presidio County) thinking it was the very similar *Argia translata* (Dusky Dancer). Martin Reid actually noticed it on my website labeld as *A. translata* and called my attention to it. Individuals out in west Texas should keep their eyes out for this species. The Texas odonate fauna now stands at 224 species.

DAMSELFLIES - ZYGOPTERA (74)

Broad-winged Damsels - Calopterygidae (5)
Sparkling Jewelwing [*Calopteryx dimidiata* Burm.]
Ebony Jewelwing [*C. maculata* (Beauvois)]

American Rubyspot [*Hetaerina americana* (Fab.)]
Smoky Rubyspot [*H. titia* (Drury)]
Canyon Rubyspot [*H. vulnerata* Hagen *in* Selys]

Spreadwings - Lestidae (9)
Great Spreadwing [*Archilestes grandis* (Rambur)]

Plateau Spreadwing [*Lestes alacer* Hagen]
Southern Spreadwing [*L. australis* Walker]
Rainpool Spreadwing [*L. forficula* Rambur]
Elegant Spreadwing [*L. inaequalis* Walsh]
Slender Spreadwing [*L. rectangularis* Say]
Chalky Spreadwing [*L. sigma* Calvert]
Lyre-tipped Spreadwing [*L. unguiculatus* Hagen]
Swamp Spreadwing [*L. vigilax* Hagen *in* Selys]

Threadtails - Protoneuridae (3)
Coral-fronted Threadtail [*Neoneura aaroni* Calvert]
Amelia's Threadtail [*N. amelia* Calvert]

Orange-Striped Threadtail [*Protoneura cara* Calvert]

Pond Damsels - Coenagrionidae (57)
Mexican Wedgetail [*Acanthagrion quadratum* Selys]

Paiute Dancer [*Argia alberta* Kennedy]
Blue-fronted Dancer [*A. apicalis* (Say)]
Comanche Dancer [*A. barretti* Calvert]
Seepage Dancer [*A. bipunctulata* (Hagen)]
Coppery Dancer [*A. cuprea* (Hagen)]
Violet Dancer [*A. fumipennis violacea* (Hagen)]
Lavender Dancer [*A. hinei* Kennedy]
Kiowa Dancer [*A. immunda* (Hagen)]
Leonora's Dancer [*A. leonorae* Garrison]
Sooty Dancer [*A. lugens* (Hagen)]
Powdered Dancer [*A. moesta* (Hagen)]
Apache Dancer [*A. munda* Calvert]
Aztec Dancer [*A. nahuana* Calvert]
Fiery-eyed Dancer [*A. oenea* Hagen *in* Selys]

Amethyst Dancer [*A. pallens* Calvert]
Springwater Dancer [*A. plana* Calvert]
Golden-winged Dancer [*A. rhoadsi* Calvert]
Blue-ringed Dancer [*A. sedula* (Hagen)]
Tezpi Dancer [*A. tezpi* Calvert]
Blue-tipped Dancer [*A. tibialis* (Rambur)]
Dusky Dancer [*A. translata* Hagen]

Rainbow Bluet [*Enallagma antennatum* (Say)]
Azure Bluet [*E. aspersum* (Hagen)]
Double-Striped Bluet [*E. basidens* Calvert]
Tule Bluet [*E. carunculatum* Morse]
Familiar Bluet [*E. civile* (Hagen)]
Alkali Bluet [*E. clausum* Morse]
Attenuated Bluet [*E. daeckii* (Calvert)]
Turquoise Bluet [*E. divagans* Selys]
Atlantic Bluet [*E. doubledayi* (Selys)]
Burgundy Bluet [*E. dubium* Root]
Big Bluet [*E. durum* (Hagen)]
Stream Bluet [*E. exsulans* (Hagen)]
Skimming Bluet [*E. geminatum* Kellicott]
Neotropical Bluet [*E. novaehispaniae* Calvert]
Arroyo Bluet [*E. praevarum* (Hagen)]
Orange Bluet [*E. signatum* (Hagen)]
Slender Bluet [*E. traviatum westfalli* Donnelly]
Vesper Bluet [*E. vesperum* Calvert]

Painted Damsel [*Hesperagrion heterodoxum* (Selys)]

Desert Forktail [*Ischnura barberi* Currie]
Plains Forktail [*I. damula* Calvert]
Mexican Forktail [*I. demorsa* (Hagen)]
Black-fronted Forktail [*I. denticollis* (Burmeister)]
Citrine Forktail [*I. hastata* (Say)]
Lilypad Forktail [*I. kellicotti* Williamson]
Fragile Forktail [*I. posita posita* (Hagen)]
Furtive Forktail [*I. prognata* (Hagen)]
Rambur's Forktail [*I. ramburii* (Selys)]
Eastern Forktail [*I. verticalis* (Say)]

Cream-tipped Swampdamsel [*Leptobasis melinogaster* Gonzalez-Soriano]

Southern Sprite [*Nehalennia integricollis* Calvert]
Everglades Sprite [*N. pallidula* Calvert]

Caribbean Yellowface [*Neoerythromma cultellatum* (Selys)]

Duckweed Firetail [*Telebasis byersi* Westfall]
Desert Firetail [*T. salva* (Hagen)]

DRAGONFLIES - ANISOPTERA (150)
Petaltails - Petaluridae (1)
Gray Petaltail [*Tachopteryx thoreyi* (Hagen *in* Selys)]

Darners - Aeshnidae (19)
Persephone's Darner [*Aeshna persephone* Donnelly]
Shadow Darner [*A. umbrosa* Walker]

Amazon Darner [*Anax amazili* (Burmeister)]
Blue-Spotted Comet Darner [*A. concolor* Brauer]
Common Green Darner [*A. junius* (Drury)]
Comet Darner [*A. longipes* (Hagen)]
Giant Darner [*A. walsinghami* McLachlan]

Springtime Darner [*Basiaeschna janata* (Say)]

Fawn Darner [*Boyeria vinosa* (Say)]

Blue-faced Darner [*Coryphaeshna adnexa* (Hagen)]
Regal Darner [*C. ingens* (Rambur)]

Swamp Darner [*Epiaeschna heros* (Fabricius)]

Harlequin Darner [*Gomphaeschna furcillata* (Say)]

Bar-sided Darner [*Gynacantha mexicana* Selys]

Cyrano Darner [*Nasiaeschna pentacantha* (Rambur)]

Arroyo Darner [*Rhionaeschna dugesi* Calvert]
Blue-eyed Darner [*R. multicolor* Hagen]
Turquoise-tipped Darner [*R. psilus* (Calvert)]

Pale-green Darner [*Triacanthagyna septima* (Selys in Sagra)]

Clubtails - Gomphidae (32)
Broad-striped Forceptail [*Aphylla angustifolia* Garrison]
Narrow-striped Forecptail [*A. protracta* (Hagen)]
Two-striped Forceptail [*A. williamsoni* (Gloyd)]

Stillwater Clubtail [*Arigomphus lentulus* (Needham)]
Bayou Clubtail [*A. maxwelli* (Ferguson)]
Jade Clubtail [*A. submedianus* (Hagen)]

Black-shouldered Spinyleg [*Dromogomphus spinosus* Selys]
Flag-tailed Spinyleg [*D. spoliatus* (Hagen *in* Selys)]

White-belted Ringtail [*Erpetogomphus compositus* Hagen *in* Selys]
Eastern Ringtail [*E. designatus* Hagen *in* Selys]
Blue-faced Ringtail [*E. eutainia* Calvert]
Dashed Ringtail [*E. heterodon* Garrison]
Serpent Ringtail [*E. lampropeltis* Wmsn. & Wmsn.]
Plains Clubtail [*Gomphus (Gomphurus) externus* Hagen *in* Selys]
Tamaulipan Clubtail [*G. (G.) gonzalezi* Dunkle]
Cocoa Clubtail [*G. (G.) hybridus* Williamson]

Gulf Coast Clubtail [*G. (G.) modestus* Needham]
Cobra Clubtail [*G. (G.) vastus* Walsh]

Banner Clubtail [*G. (s.str.) apomyius* Donnelly]
Pronghorn Clubtail [*G. (G.) graslinellus* (Walsh)]
Ashy Clubtail [*G. (G.) lividus* (Selys)]
Sulphur-Tipped Clubtail [*G. (G.) militaris* Hag. *in* Selys]
Oklahoma Clubtail [*G. (G.) oklahomensis* Pritchard]

Dragonhunter [*Hagenius brevistylus* Selys]

Dark-tailed Forceptail [*Phyllocycla breviphylla* Belle]
Five-striped Leaftail [*Phyllogomphoides albrighti* (Needham)]
Four-striped Leaftail [*P. stigmatus* (Say)]

Gray Sanddragon [*Progomphus borealis* McL. *in* Selys]
Common Sanddragon [*P. obscurus* (Rambur)]

Brimstone Clubtail [*Stylurus intricatus* (Hagen *in* Selys)]
Laura's Clubtail [*S. laurae* Williamson]
Russet-tipped Clubtail [*S. plagiatus* (Selys)]

Spiketails - Cordulegastridae (2)
Twin-spotted Spiketail [*Cordulegaster maculata* Selys]
Arrowhead Spiketail [*C. obliqua obliqua* (Say)]

Cruisers - Corduliidae: Macromiinae (5)
Stream Cruiser [*Didymops transversa* (Say)]

Bronzed River Cruiser [*Macromia annulata* Hagen]
Georgia River Cruiser [*M. illinoiensis georgina* (Selys)]
Gilded River Cruiser [*M. pacifica* Hagen]
Royal River Cruiser [*M. taeniolata* Rambur]

Emeralds - Corduliidae: Corduliinae (15)
Prince Baskettail [*Epitheca (Epicordullia) princeps* (Hagen)]
Stripe-winged Baskettail [*E. (Tetragoneuria) costalis* (Selys)]
Common Baskettail [*E. (T.) cynosura* (Say)]
Dot-winged Baskettail [*E. (T.) petechialis* (Mutt.)]
Mantled Baskettail [*E. (T.) semiaquea* (Burmeister)]
Robust Baskettail [*E. (T.) spinosa* (Hagen *in* Selys)]
Selys' Sundragon [*Helocordulia selysii* (Hagen *in* Selys)]

Alabama Shadowdragon [*Neurocordulia alabamensis* Hodges *in* Need. & West.]
Smoky Shadowdragon [*N. molesta* (Walsh)]
Orange Shadowdragon [*N. xanthosoma* (Wmsn.)]

Fine-lined Emerald [*Somatochlora filosa* (Hagen)]
Coppery Emerald [*S. georgiana* Walker]
Mocha Emerald [*S. linearis* (Hagen)]
Texas Emerald [*S. margarita* Donnelly]
Clamp-tipped Emerald [*S. tenebrosa* (Say)]

Skimmers - Libellulidae (76)

Red-tailed Pennant [*Brachymesia furcata* (Hagen)]
Four-spotted Pennant [*B. gravida* (Calvert)]
Tawny Pennant [*B. herbida* (Gundlach)]

Pale-faced Clubskimmer [*Brechmorhoga mendax* (Hagen)]

Gray-waisted Skimmer [*Cannaphila insularis funerea* (Carpenter)]
Amanda's Pennant [*Celithemis amanda* (Hagen)]
Calico Pennant [*C. elisa* (Hagen)]
Halloween Pennenat [*C. eponina* (Drury)]
Banded Pennant [*C. fasciata* Kirby]
Faded Pennant [*C. ornata* (Rambur)]
Double-ringed Pennant [*C. verna* Pritchard]
Checkered Setwing [*Dythemis fugax* Hagen]
Mayan Setwing [*D. maya* Calvert]
Black Setwing [*D. nigrescens* Calvert]
Swift Setwing [*D. velox* Hagen]

Black Pondhawk [*Erythemis attala* (Selys *in* Sagra)]
Western Pondhawk [*E. collocata* (Hagen)]
Claret Pondhawk [*E. mithroides* Brauer]
Flame-tailed Pondhawk [*E. peruviana* (Rambur)]
Pin-tailed Pondhawk [*E. plebja* (Burmeister)]
Eastern Pondhawk [*E. simplicicollis* (Say)]
Great Pondhawk [*E. vesiculosa* (Fabricius)]

Plateau Dragonlet [*Erythrodiplax basifusca*]
Seaside Dragonlet [*E. berenice berenice* (Drury)]
Black-winged Dragonlet [*E. funerea* (Hagen)]
Red-faced Dragonlet [*E. fusca* (Rambur)]
Little Blue Dragonlet [*E. minuscula* (Rambur)]
Band-winged Dragonlet [*E. umbrata* (Linnaeus)]

Blue Corporal [*Ladona deplanata* (Rambur)]

Golden-winged Skimmer [*Libellula auripennis* Burm.]
Bar-winged Skimmer [*L. axilena* Westwood]
Comanche Skimmer [*L. comanche* Calvert]
Bleached Skimmer [*L. composita* (Hagen)]
Neon Skimmer [*L. croceipennis* Selys]
Spangled Skimmer [*L. cyanea* Fabricius]
Yellow-sided Skimmer [*L. flavida* Rambur]
Slaty Skimmer [*L. incesta* Hagen]
Widow Skimmer [*L. luctuosa* Burmeister]
Needham's Skimmer [*L. needhami* Westfall]
Twelve-spotted Skimmer [*L. pulchella* Drury]
Flame Skimmer [*L. saturata* Uhler]
Painted Skimmer [*L. semifasciata* Burmeister]
Great Blue Skimmer [*L. vibrans* Fabricius]

Marl Pennant [*Macrodiplax balteata* (Hagen)]

Ivory-striped Sylph [*Macrothemis imitans leucozona* Ris]
Straw-colored Sylph [*M. inacuta* Calvert]
Jade-striped Sylph [*M. inequiunguis* Calvert]

Hyacinth Glider [*Miathyria marcella* (Selys *in* Sagra)]

Spot-tailed Dasher [*Micrathyria aequalis* (Hagen)]
Three-striped Dasher [*M. didyma* (Selys)
Thornbush Dasher [*M. hagenii* Kirby]

Carmine Skimmer [*Orthemis discolor* (Burmeister)]
Roseate Skimmer [*O. ferruginea* (Fabricius)]

Blue Dasher [*Pachydiplax longipennis* (Burmeister)]
Red Rock Skimmer [*Paltothemis lineatipes* Karsch]

Wandering Glider [*Pantala flavescens* (Fabricius)]
Spot-winged Glider [*P. hymenaea* (Say)]

Slough Amberwing [*Perithemis domitia* (Drury)]
Eastern Amberwing [*P. tenera* (Say)]

Common Whitetail [*Plathemis lydia* (Drury)]
Desert Whitetail [*P. subornata* Hagen]

Filigree Skimmer [*Pseudoleon superbus* (Hagen)]

Blue-faced Meadowhawk [*Sympetrum ambiguum* Rambur)]
Variegated Meadowhawk [*S. corruptum* (Hagen)]
Cardinal Meadowhawk [*S. illotum* (Hagen)]
Cherry-faced Meadowhawk [*S. internum* Mont.]
Band-winged Meadowhawk [*S. semicinctum* (Say)]
Autumn Meadowhawk [*S. vicinum* (Hagen)]
Aztec Glider [*Tauriphila azteca* Calvert]

Evening Skimmer [*Tholymis citrina* Hagen]

Vermilion Saddlebags [*Tramea abdominalis* (Ramb.)]
Striped Saddlebags [*T. calverti* Muttkowski]
Carolina Saddlebags [*T. carolina* (Linnaeus)]
Antillean Saddlebags [*T. insularis* Hagen]
Black Saddlebags [*T. lacerata* Hagen]
Red Saddlebags [*T. onusta* Hagen]

Seasonality of Odonata in Texas

John C. Abbott

The flight season of most species of Odonata in this region extends from the spring through the summer months, and occasionally persists into the fall. The onset and duration of emergence, however, are both variable. Many species of temperate origin (e.g. Attenuated Bluet, Springtime Darner, Harlequin Darner, Banner Clubtail, Twin-spotted Spiketail, Stream Cruiser, Blue Corporal) have early (March - May) and explosive emergences in this area and then soon disappear. In the southern portions of the region, the year-round temperatures averaging 23° C (73° F) and the subtropical climate allow several species to fly year round. Fifteen species (including Familiar Bluet, Fragile Forktail, Common Green Darner, Eastern Pondhawk) have been encountered as adults in every month. Other species (Fine-lined Emerald and Blue-faced Meadowhawk) are seen flying only later in the year. Because of the latitudinal gradient in temperature seen in the region, emergence occurs one to several weeks later in the more northern areas than in the subtropical southern areas.

Damselflies generally emerge as soon as temperatures permit in the spring and continue to emerge throughout much of the summer. This results in a heterogeneous age structure, often allowing more than one generation per year. Many of the smaller pond damsels (e.g. forktails) have multiple (2 or 3) generations per year. The larger pond damsels (e.g. the spreadwings,) and the broad-winged damsels generally require a full year for development.

Many dragonflies differ from damselflies in having an obligate larval diapause (a required period of arrested development) followed by a synchronous spring or early-summer emergence. This pattern results in a homogenous age structure and a sudden disappearance later in the summer or fall. Most species require at least a year to develop, and some (e.g. spiketails) require longer (several years).

Development time is generally longer for those species restricted to running-water situations than it is for those found in ponds or lakes. A general seasonal progression, with the peak months in June and July, as seen for the entire south-central United States, was observed by Bick (1957) in Louisiana. Species present early in the year (January-February) were also generally present later in the year (November-December).

The following pages contain the seasonal distribution for all 224 species found in Texas ordered by first appearance in the year. Please report new early and late dates to John Abbott (jcabbott@mail.utexas.edu) or OdonataCentral.org.

Literature Cited

Bick, G.H. 1957. The Odonata of Louisiana. Tulane Studies in Zoology 5:71-135.

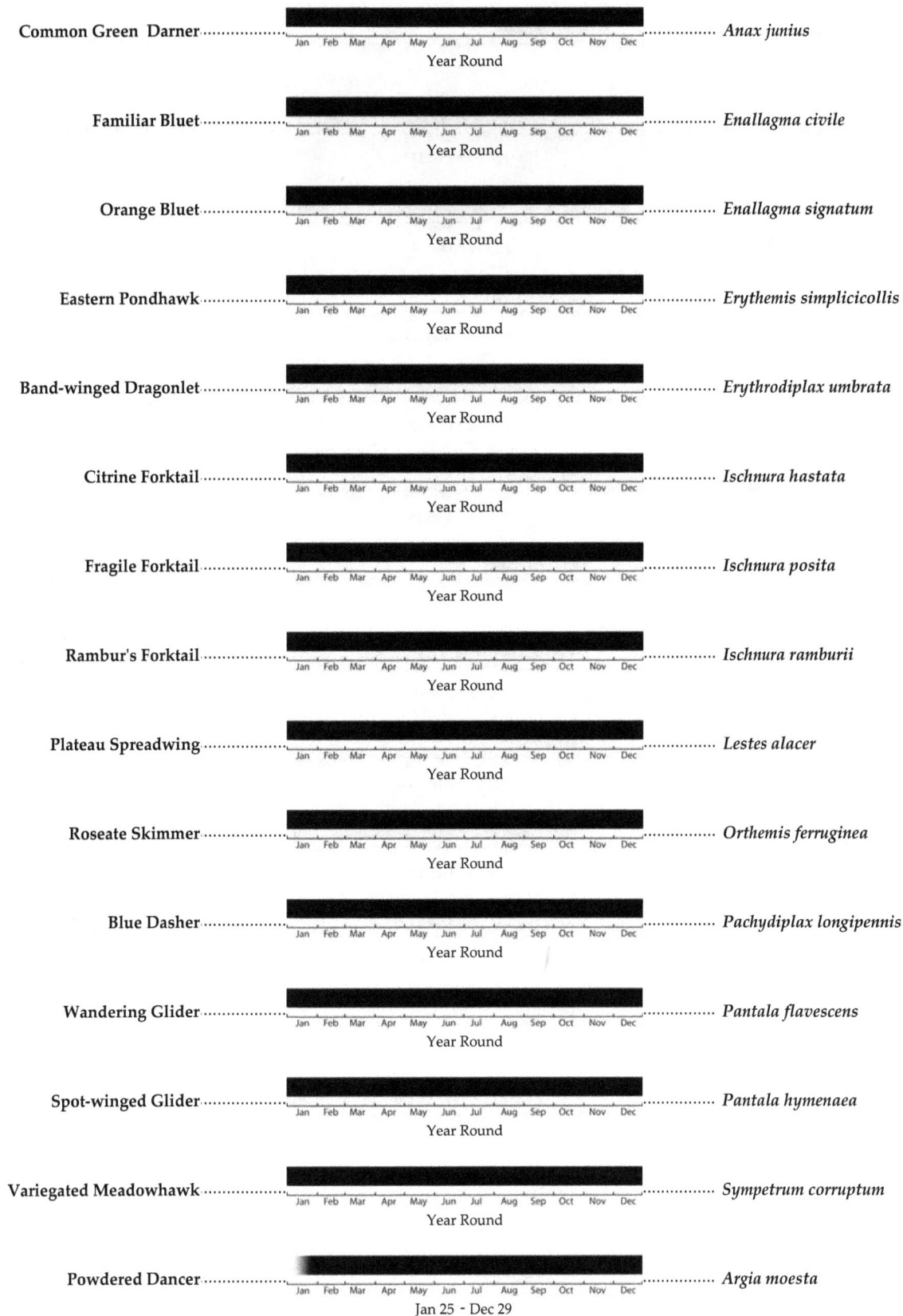

Common Green Darner .... | Jan Feb Mar Apr May Jun Jul Aug Sep Oct Nov Dec |............................... *Anax junius*
Year Round

Familiar Bluet .... | Jan Feb Mar Apr May Jun Jul Aug Sep Oct Nov Dec |............................... *Enallagma civile*
Year Round

Orange Bluet .... | Jan Feb Mar Apr May Jun Jul Aug Sep Oct Nov Dec |............................... *Enallagma signatum*
Year Round

Eastern Pondhawk .... | Jan Feb Mar Apr May Jun Jul Aug Sep Oct Nov Dec |............................... *Erythemis simplicicollis*
Year Round

Band-winged Dragonlet .... | Jan Feb Mar Apr May Jun Jul Aug Sep Oct Nov Dec |............................... *Erythrodiplax umbrata*
Year Round

Citrine Forktail .... | Jan Feb Mar Apr May Jun Jul Aug Sep Oct Nov Dec |............................... *Ischnura hastata*
Year Round

Fragile Forktail .... | Jan Feb Mar Apr May Jun Jul Aug Sep Oct Nov Dec |............................... *Ischnura posita*
Year Round

Rambur's Forktail .... | Jan Feb Mar Apr May Jun Jul Aug Sep Oct Nov Dec |............................... *Ischnura ramburii*
Year Round

Plateau Spreadwing .... | Jan Feb Mar Apr May Jun Jul Aug Sep Oct Nov Dec |............................... *Lestes alacer*
Year Round

Roseate Skimmer .... | Jan Feb Mar Apr May Jun Jul Aug Sep Oct Nov Dec |............................... *Orthemis ferruginea*
Year Round

Blue Dasher .... | Jan Feb Mar Apr May Jun Jul Aug Sep Oct Nov Dec |............................... *Pachydiplax longipennis*
Year Round

Wandering Glider .... | Jan Feb Mar Apr May Jun Jul Aug Sep Oct Nov Dec |............................... *Pantala flavescens*
Year Round

Spot-winged Glider .... | Jan Feb Mar Apr May Jun Jul Aug Sep Oct Nov Dec |............................... *Pantala hymenaea*
Year Round

Variegated Meadowhawk .... | Jan Feb Mar Apr May Jun Jul Aug Sep Oct Nov Dec |............................... *Sympetrum corruptum*
Year Round

Powdered Dancer .... | Jan Feb Mar Apr May Jun Jul Aug Sep Oct Nov Dec |............................... *Argia moesta*
Jan 25 - Dec 29

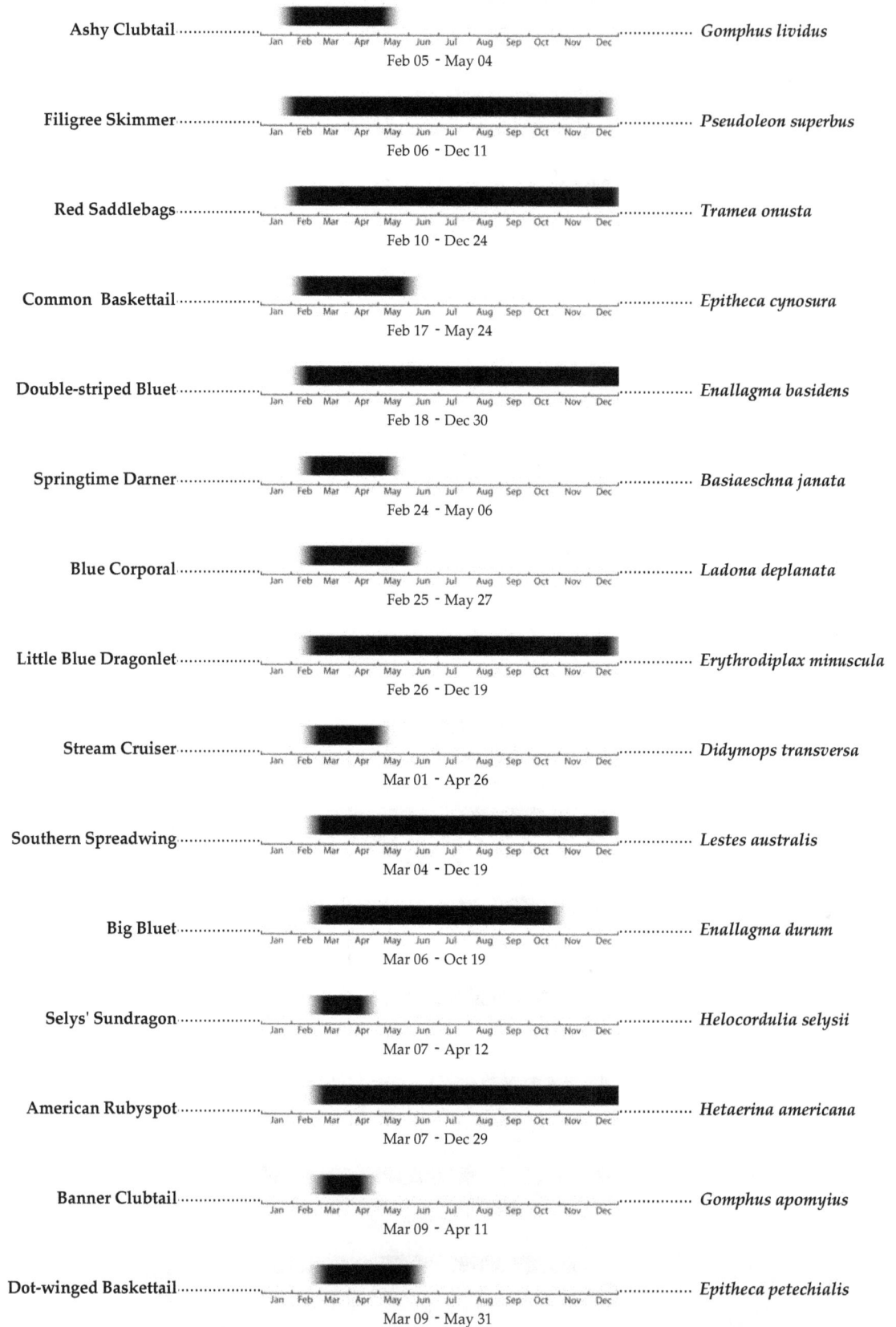

Ashy Clubtail ·················

Feb 05 - May 04

Gomphus lividus

Filigree Skimmer ···············

Feb 06 - Dec 11

Pseudoleon superbus

Red Saddlebags ···············

Feb 10 - Dec 24

Tramea onusta

Common Baskettail ···············

Feb 17 - May 24

Epitheca cynosura

Double-striped Bluet ···············

Feb 18 - Dec 30

Enallagma basidens

Springtime Darner ···············

Feb 24 - May 06

Basiaeschna janata

Blue Corporal ···············

Feb 25 - May 27

Ladona deplanata

Little Blue Dragonlet ···············

Feb 26 - Dec 19

Erythrodiplax minuscula

Stream Cruiser ···············

Mar 01 - Apr 26

Didymops transversa

Southern Spreadwing ···············

Mar 04 - Dec 19

Lestes australis

Big Bluet ···············

Mar 06 - Oct 19

Enallagma durum

Selys' Sundragon ···············

Mar 07 - Apr 12

Helocordulia selysii

American Rubyspot ···············

Mar 07 - Dec 29

Hetaerina americana

Banner Clubtail ···············

Mar 09 - Apr 11

Gomphus apomyius

Dot-winged Baskettail ···············

Mar 09 - May 31

Epitheca petechialis

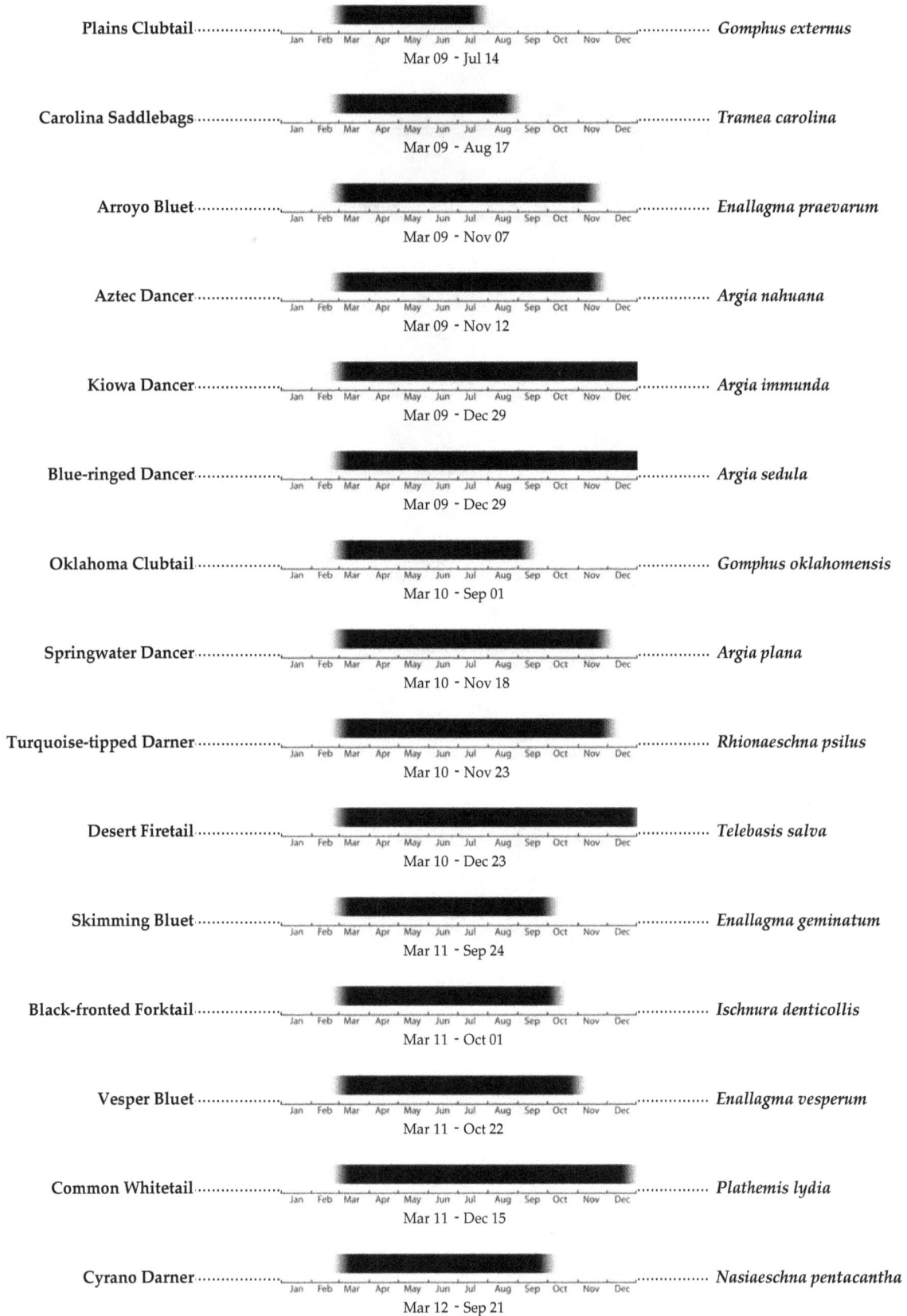

Plains Clubtail .... Jan Feb Mar Apr May Jun Jul Aug Sep Oct Nov Dec .... *Gomphus externus*
Mar 09 - Jul 14

Carolina Saddlebags .... Jan Feb Mar Apr May Jun Jul Aug Sep Oct Nov Dec .... *Tramea carolina*
Mar 09 - Aug 17

Arroyo Bluet .... Jan Feb Mar Apr May Jun Jul Aug Sep Oct Nov Dec .... *Enallagma praevarum*
Mar 09 - Nov 07

Aztec Dancer .... Jan Feb Mar Apr May Jun Jul Aug Sep Oct Nov Dec .... *Argia nahuana*
Mar 09 - Nov 12

Kiowa Dancer .... Jan Feb Mar Apr May Jun Jul Aug Sep Oct Nov Dec .... *Argia immunda*
Mar 09 - Dec 29

Blue-ringed Dancer .... Jan Feb Mar Apr May Jun Jul Aug Sep Oct Nov Dec .... *Argia sedula*
Mar 09 - Dec 29

Oklahoma Clubtail .... Jan Feb Mar Apr May Jun Jul Aug Sep Oct Nov Dec .... *Gomphus oklahomensis*
Mar 10 - Sep 01

Springwater Dancer .... Jan Feb Mar Apr May Jun Jul Aug Sep Oct Nov Dec .... *Argia plana*
Mar 10 - Nov 18

Turquoise-tipped Darner .... Jan Feb Mar Apr May Jun Jul Aug Sep Oct Nov Dec .... *Rhionaeschna psilus*
Mar 10 - Nov 23

Desert Firetail .... Jan Feb Mar Apr May Jun Jul Aug Sep Oct Nov Dec .... *Telebasis salva*
Mar 10 - Dec 23

Skimming Bluet .... Jan Feb Mar Apr May Jun Jul Aug Sep Oct Nov Dec .... *Enallagma geminatum*
Mar 11 - Sep 24

Black-fronted Forktail .... Jan Feb Mar Apr May Jun Jul Aug Sep Oct Nov Dec .... *Ischnura denticollis*
Mar 11 - Oct 01

Vesper Bluet .... Jan Feb Mar Apr May Jun Jul Aug Sep Oct Nov Dec .... *Enallagma vesperum*
Mar 11 - Oct 22

Common Whitetail .... Jan Feb Mar Apr May Jun Jul Aug Sep Oct Nov Dec .... *Plathemis lydia*
Mar 11 - Dec 15

Cyrano Darner .... Jan Feb Mar Apr May Jun Jul Aug Sep Oct Nov Dec .... *Nasiaeschna pentacantha*
Mar 12 - Sep 21

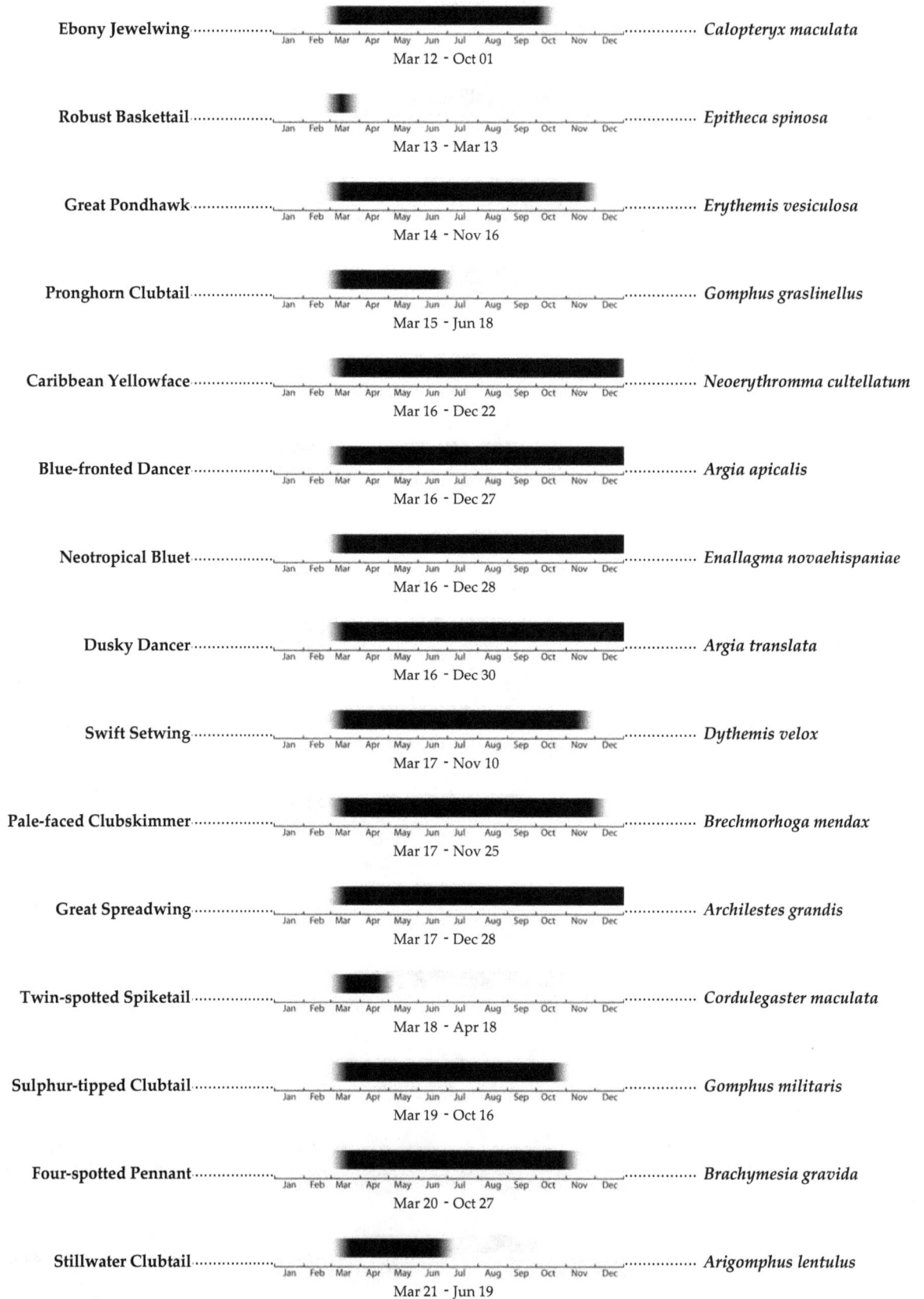

Ebony Jewelwing ⋯⋯⋯⋯ | Jan Feb Mar Apr May Jun Jul Aug Sep Oct Nov Dec | ⋯⋯⋯⋯ *Calopteryx maculata*
Mar 12 - Oct 01

Robust Baskettail ⋯⋯⋯⋯ | Jan Feb Mar Apr May Jun Jul Aug Sep Oct Nov Dec | ⋯⋯⋯⋯ *Epitheca spinosa*
Mar 13 - Mar 13

Great Pondhawk ⋯⋯⋯⋯ | Jan Feb Mar Apr May Jun Jul Aug Sep Oct Nov Dec | ⋯⋯⋯⋯ *Erythemis vesiculosa*
Mar 14 - Nov 16

Pronghorn Clubtail ⋯⋯⋯⋯ | Jan Feb Mar Apr May Jun Jul Aug Sep Oct Nov Dec | ⋯⋯⋯⋯ *Gomphus graslinellus*
Mar 15 - Jun 18

Caribbean Yellowface ⋯⋯⋯⋯ | Jan Feb Mar Apr May Jun Jul Aug Sep Oct Nov Dec | ⋯⋯⋯⋯ *Neoerythromma cultellatum*
Mar 16 - Dec 22

Blue-fronted Dancer ⋯⋯⋯⋯ | Jan Feb Mar Apr May Jun Jul Aug Sep Oct Nov Dec | ⋯⋯⋯⋯ *Argia apicalis*
Mar 16 - Dec 27

Neotropical Bluet ⋯⋯⋯⋯ | Jan Feb Mar Apr May Jun Jul Aug Sep Oct Nov Dec | ⋯⋯⋯⋯ *Enallagma novaehispaniae*
Mar 16 - Dec 28

Dusky Dancer ⋯⋯⋯⋯ | Jan Feb Mar Apr May Jun Jul Aug Sep Oct Nov Dec | ⋯⋯⋯⋯ *Argia translata*
Mar 16 - Dec 30

Swift Setwing ⋯⋯⋯⋯ | Jan Feb Mar Apr May Jun Jul Aug Sep Oct Nov Dec | ⋯⋯⋯⋯ *Dythemis velox*
Mar 17 - Nov 10

Pale-faced Clubskimmer ⋯⋯⋯⋯ | Jan Feb Mar Apr May Jun Jul Aug Sep Oct Nov Dec | ⋯⋯⋯⋯ *Brechmorhoga mendax*
Mar 17 - Nov 25

Great Spreadwing ⋯⋯⋯⋯ | Jan Feb Mar Apr May Jun Jul Aug Sep Oct Nov Dec | ⋯⋯⋯⋯ *Archilestes grandis*
Mar 17 - Dec 28

Twin-spotted Spiketail ⋯⋯⋯⋯ | Jan Feb Mar Apr May Jun Jul Aug Sep Oct Nov Dec | ⋯⋯⋯⋯ *Cordulegaster maculata*
Mar 18 - Apr 18

Sulphur-tipped Clubtail ⋯⋯⋯⋯ | Jan Feb Mar Apr May Jun Jul Aug Sep Oct Nov Dec | ⋯⋯⋯⋯ *Gomphus militaris*
Mar 19 - Oct 16

Four-spotted Pennant ⋯⋯⋯⋯ | Jan Feb Mar Apr May Jun Jul Aug Sep Oct Nov Dec | ⋯⋯⋯⋯ *Brachymesia gravida*
Mar 20 - Oct 27

Stillwater Clubtail ⋯⋯⋯⋯ | Jan Feb Mar Apr May Jun Jul Aug Sep Oct Nov Dec | ⋯⋯⋯⋯ *Arigomphus lentulus*
Mar 21 - Jun 19

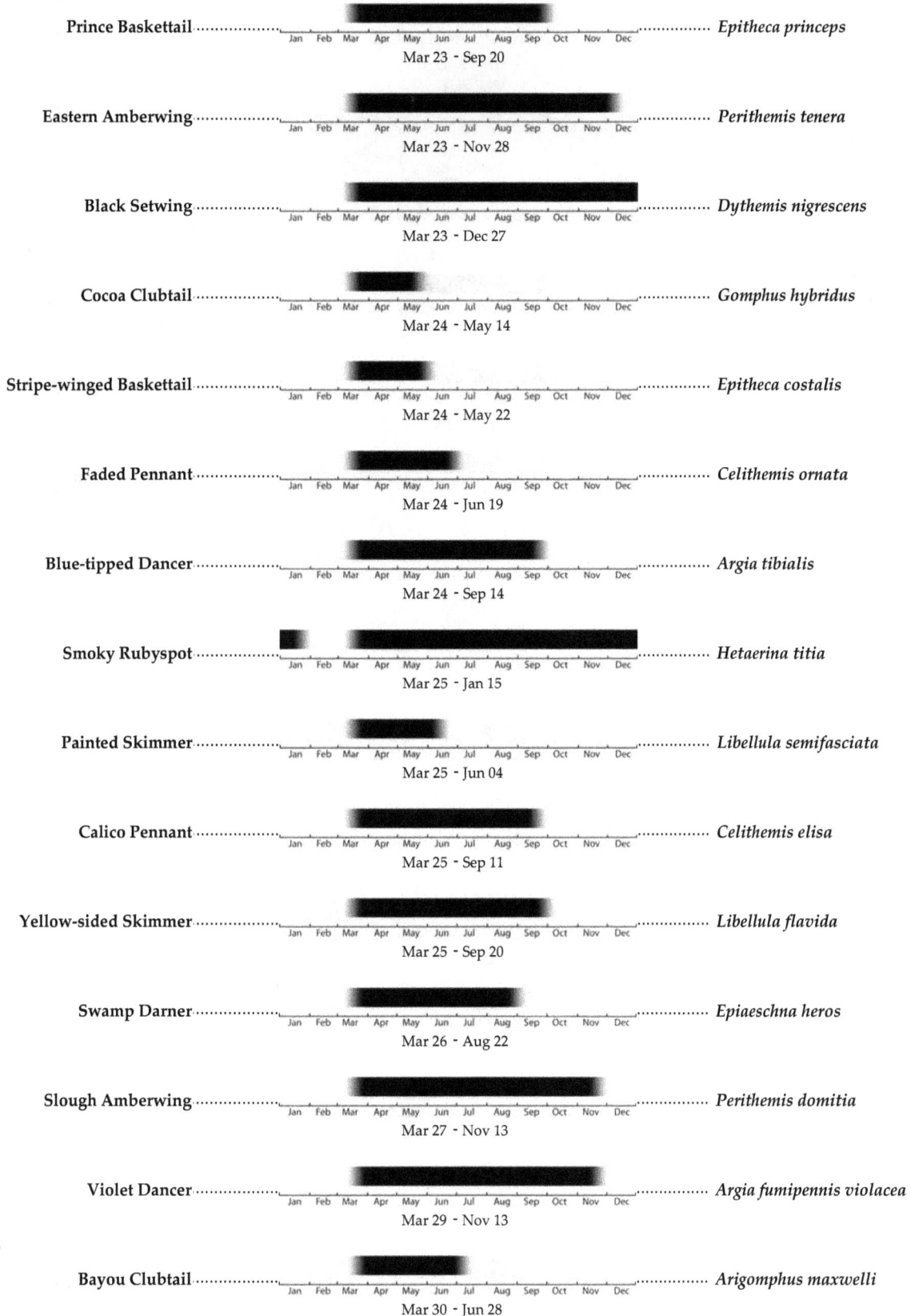

Prince Baskettail ·· *Epitheca princeps*
Jan Feb Mar Apr May Jun Jul Aug Sep Oct Nov Dec
Mar 23 - Sep 20

Eastern Amberwing ··· *Perithemis tenera*
Jan Feb Mar Apr May Jun Jul Aug Sep Oct Nov Dec
Mar 23 - Nov 28

Black Setwing ··· *Dythemis nigrescens*
Jan Feb Mar Apr May Jun Jul Aug Sep Oct Nov Dec
Mar 23 - Dec 27

Cocoa Clubtail ·· *Gomphus hybridus*
Jan Feb Mar Apr May Jun Jul Aug Sep Oct Nov Dec
Mar 24 - May 14

Stripe-winged Baskettail ··· *Epitheca costalis*
Jan Feb Mar Apr May Jun Jul Aug Sep Oct Nov Dec
Mar 24 - May 22

Faded Pennant ··· *Celithemis ornata*
Jan Feb Mar Apr May Jun Jul Aug Sep Oct Nov Dec
Mar 24 - Jun 19

Blue-tipped Dancer ··· *Argia tibialis*
Jan Feb Mar Apr May Jun Jul Aug Sep Oct Nov Dec
Mar 24 - Sep 14

Smoky Rubyspot ··· *Hetaerina titia*
Jan Feb Mar Apr May Jun Jul Aug Sep Oct Nov Dec
Mar 25 - Jan 15

Painted Skimmer ··· *Libellula semifasciata*
Jan Feb Mar Apr May Jun Jul Aug Sep Oct Nov Dec
Mar 25 - Jun 04

Calico Pennant ·· *Celithemis elisa*
Jan Feb Mar Apr May Jun Jul Aug Sep Oct Nov Dec
Mar 25 - Sep 11

Yellow-sided Skimmer ·· *Libellula flavida*
Jan Feb Mar Apr May Jun Jul Aug Sep Oct Nov Dec
Mar 25 - Sep 20

Swamp Darner ··· *Epiaeschna heros*
Jan Feb Mar Apr May Jun Jul Aug Sep Oct Nov Dec
Mar 26 - Aug 22

Slough Amberwing ··· *Perithemis domitia*
Jan Feb Mar Apr May Jun Jul Aug Sep Oct Nov Dec
Mar 27 - Nov 13

Violet Dancer ··· *Argia fumipennis violacea*
Jan Feb Mar Apr May Jun Jul Aug Sep Oct Nov Dec
Mar 29 - Nov 13

Bayou Clubtail ·· *Arigomphus maxwelli*
Jan Feb Mar Apr May Jun Jul Aug Sep Oct Nov Dec
Mar 30 - Jun 28

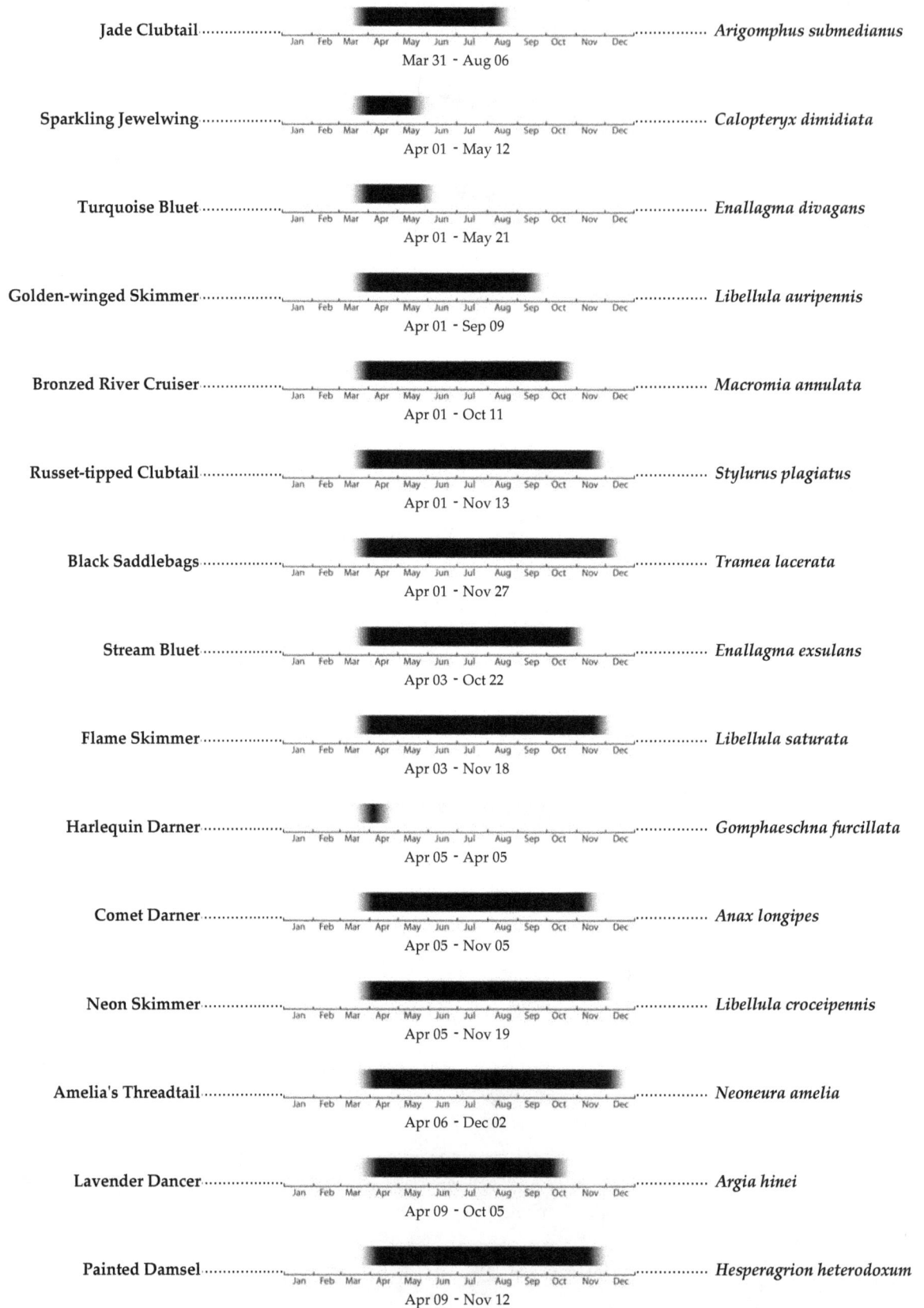

Jade Clubtail .... Jan Feb Mar Apr May Jun Jul Aug Sep Oct Nov Dec .... *Arigomphus submedianus*
Mar 31 - Aug 06

Sparkling Jewelwing .... Jan Feb Mar Apr May Jun Jul Aug Sep Oct Nov Dec .... *Calopteryx dimidiata*
Apr 01 - May 12

Turquoise Bluet .... Jan Feb Mar Apr May Jun Jul Aug Sep Oct Nov Dec .... *Enallagma divagans*
Apr 01 - May 21

Golden-winged Skimmer .... Jan Feb Mar Apr May Jun Jul Aug Sep Oct Nov Dec .... *Libellula auripennis*
Apr 01 - Sep 09

Bronzed River Cruiser .... Jan Feb Mar Apr May Jun Jul Aug Sep Oct Nov Dec .... *Macromia annulata*
Apr 01 - Oct 11

Russet-tipped Clubtail .... Jan Feb Mar Apr May Jun Jul Aug Sep Oct Nov Dec .... *Stylurus plagiatus*
Apr 01 - Nov 13

Black Saddlebags .... Jan Feb Mar Apr May Jun Jul Aug Sep Oct Nov Dec .... *Tramea lacerata*
Apr 01 - Nov 27

Stream Bluet .... Jan Feb Mar Apr May Jun Jul Aug Sep Oct Nov Dec .... *Enallagma exsulans*
Apr 03 - Oct 22

Flame Skimmer .... Jan Feb Mar Apr May Jun Jul Aug Sep Oct Nov Dec .... *Libellula saturata*
Apr 03 - Nov 18

Harlequin Darner .... Jan Feb Mar Apr May Jun Jul Aug Sep Oct Nov Dec .... *Gomphaeschna furcillata*
Apr 05 - Apr 05

Comet Darner .... Jan Feb Mar Apr May Jun Jul Aug Sep Oct Nov Dec .... *Anax longipes*
Apr 05 - Nov 05

Neon Skimmer .... Jan Feb Mar Apr May Jun Jul Aug Sep Oct Nov Dec .... *Libellula croceipennis*
Apr 05 - Nov 19

Amelia's Threadtail .... Jan Feb Mar Apr May Jun Jul Aug Sep Oct Nov Dec .... *Neoneura amelia*
Apr 06 - Dec 02

Lavender Dancer .... Jan Feb Mar Apr May Jun Jul Aug Sep Oct Nov Dec .... *Argia hinei*
Apr 09 - Oct 05

Painted Damsel .... Jan Feb Mar Apr May Jun Jul Aug Sep Oct Nov Dec .... *Hesperagrion heterodoxum*
Apr 09 - Nov 12

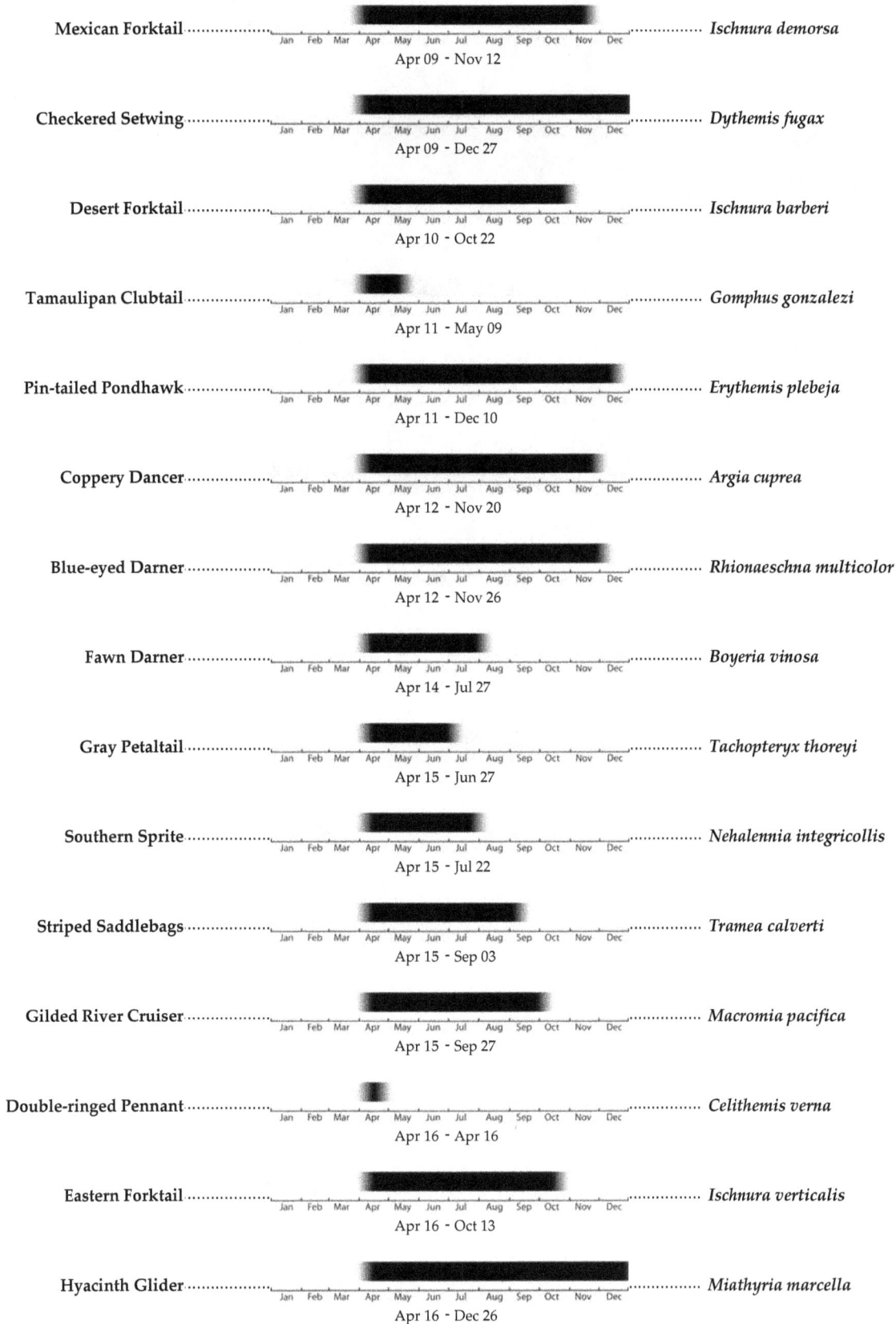

Mexican Forktail .... Jan Feb Mar Apr May Jun Jul Aug Sep Oct Nov Dec .... *Ischnura demorsa*
Apr 09 - Nov 12

Checkered Setwing .... Jan Feb Mar Apr May Jun Jul Aug Sep Oct Nov Dec .... *Dythemis fugax*
Apr 09 - Dec 27

Desert Forktail .... Jan Feb Mar Apr May Jun Jul Aug Sep Oct Nov Dec .... *Ischnura barberi*
Apr 10 - Oct 22

Tamaulipan Clubtail .... Jan Feb Mar Apr May Jun Jul Aug Sep Oct Nov Dec .... *Gomphus gonzalezi*
Apr 11 - May 09

Pin-tailed Pondhawk .... Jan Feb Mar Apr May Jun Jul Aug Sep Oct Nov Dec .... *Erythemis plebeja*
Apr 11 - Dec 10

Coppery Dancer .... Jan Feb Mar Apr May Jun Jul Aug Sep Oct Nov Dec .... *Argia cuprea*
Apr 12 - Nov 20

Blue-eyed Darner .... Jan Feb Mar Apr May Jun Jul Aug Sep Oct Nov Dec .... *Rhionaeschna multicolor*
Apr 12 - Nov 26

Fawn Darner .... Jan Feb Mar Apr May Jun Jul Aug Sep Oct Nov Dec .... *Boyeria vinosa*
Apr 14 - Jul 27

Gray Petaltail .... Jan Feb Mar Apr May Jun Jul Aug Sep Oct Nov Dec .... *Tachopteryx thoreyi*
Apr 15 - Jun 27

Southern Sprite .... Jan Feb Mar Apr May Jun Jul Aug Sep Oct Nov Dec .... *Nehalennia integricollis*
Apr 15 - Jul 22

Striped Saddlebags .... Jan Feb Mar Apr May Jun Jul Aug Sep Oct Nov Dec .... *Tramea calverti*
Apr 15 - Sep 03

Gilded River Cruiser .... Jan Feb Mar Apr May Jun Jul Aug Sep Oct Nov Dec .... *Macromia pacifica*
Apr 15 - Sep 27

Double-ringed Pennant .... Jan Feb Mar Apr May Jun Jul Aug Sep Oct Nov Dec .... *Celithemis verna*
Apr 16 - Apr 16

Eastern Forktail .... Jan Feb Mar Apr May Jun Jul Aug Sep Oct Nov Dec .... *Ischnura verticalis*
Apr 16 - Oct 13

Hyacinth Glider .... Jan Feb Mar Apr May Jun Jul Aug Sep Oct Nov Dec .... *Miathyria marcella*
Apr 16 - Dec 26

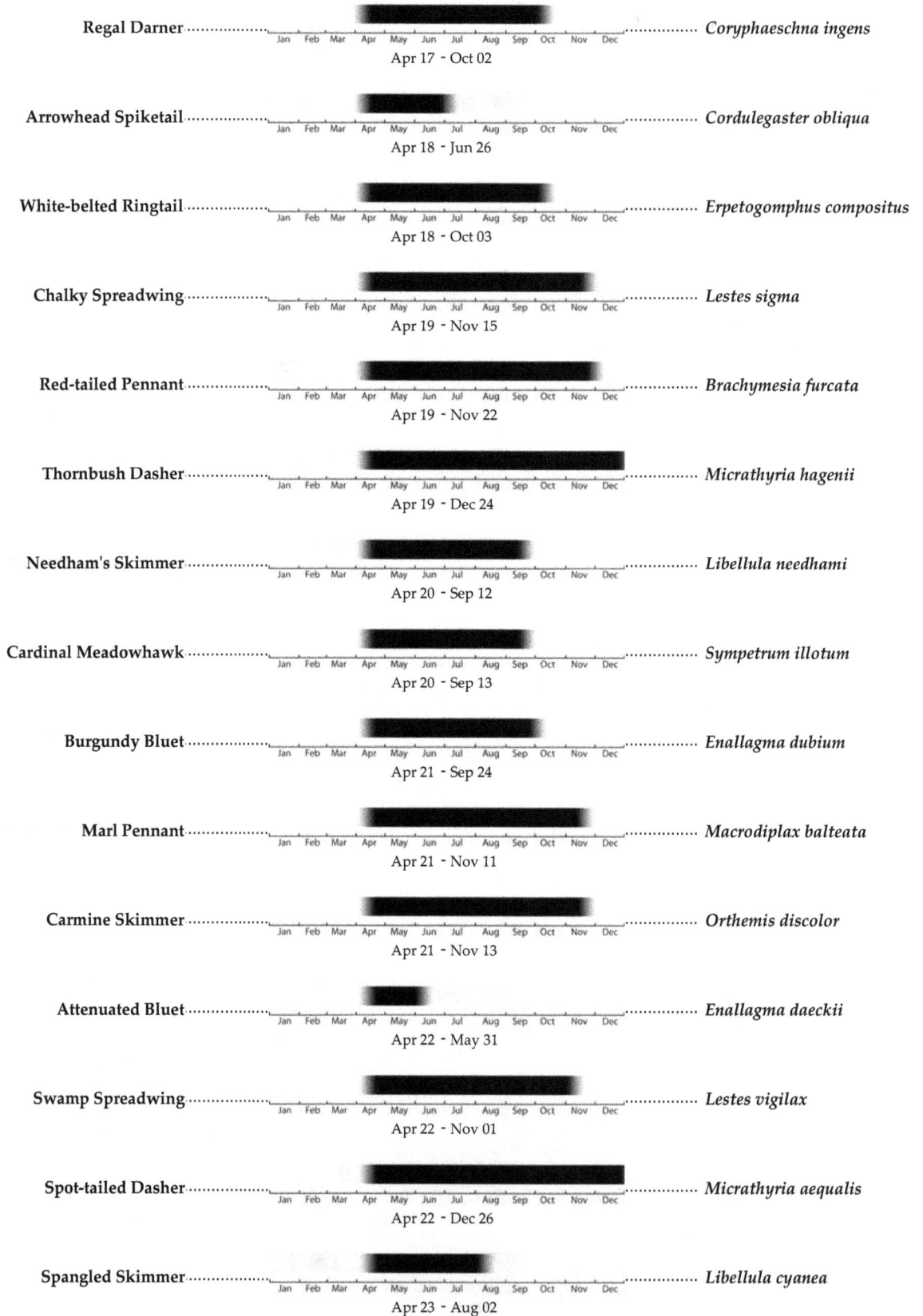

Regal Darner .... Jan Feb Mar Apr May Jun Jul Aug Sep Oct Nov Dec .... *Coryphaeschna ingens*
Apr 17 – Oct 02

Arrowhead Spiketail .... Jan Feb Mar Apr May Jun Jul Aug Sep Oct Nov Dec .... *Cordulegaster obliqua*
Apr 18 – Jun 26

White-belted Ringtail .... Jan Feb Mar Apr May Jun Jul Aug Sep Oct Nov Dec .... *Erpetogomphus compositus*
Apr 18 – Oct 03

Chalky Spreadwing .... Jan Feb Mar Apr May Jun Jul Aug Sep Oct Nov Dec .... *Lestes sigma*
Apr 19 – Nov 15

Red-tailed Pennant .... Jan Feb Mar Apr May Jun Jul Aug Sep Oct Nov Dec .... *Brachymesia furcata*
Apr 19 – Nov 22

Thornbush Dasher .... Jan Feb Mar Apr May Jun Jul Aug Sep Oct Nov Dec .... *Micrathyria hagenii*
Apr 19 – Dec 24

Needham's Skimmer .... Jan Feb Mar Apr May Jun Jul Aug Sep Oct Nov Dec .... *Libellula needhami*
Apr 20 – Sep 12

Cardinal Meadowhawk .... Jan Feb Mar Apr May Jun Jul Aug Sep Oct Nov Dec .... *Sympetrum illotum*
Apr 20 – Sep 13

Burgundy Bluet .... Jan Feb Mar Apr May Jun Jul Aug Sep Oct Nov Dec .... *Enallagma dubium*
Apr 21 – Sep 24

Marl Pennant .... Jan Feb Mar Apr May Jun Jul Aug Sep Oct Nov Dec .... *Macrodiplax balteata*
Apr 21 – Nov 11

Carmine Skimmer .... Jan Feb Mar Apr May Jun Jul Aug Sep Oct Nov Dec .... *Orthemis discolor*
Apr 21 – Nov 13

Attenuated Bluet .... Jan Feb Mar Apr May Jun Jul Aug Sep Oct Nov Dec .... *Enallagma daeckii*
Apr 22 – May 31

Swamp Spreadwing .... Jan Feb Mar Apr May Jun Jul Aug Sep Oct Nov Dec .... *Lestes vigilax*
Apr 22 – Nov 01

Spot-tailed Dasher .... Jan Feb Mar Apr May Jun Jul Aug Sep Oct Nov Dec .... *Micrathyria aequalis*
Apr 22 – Dec 26

Spangled Skimmer .... Jan Feb Mar Apr May Jun Jul Aug Sep Oct Nov Dec .... *Libellula cyanea*
Apr 23 – Aug 02

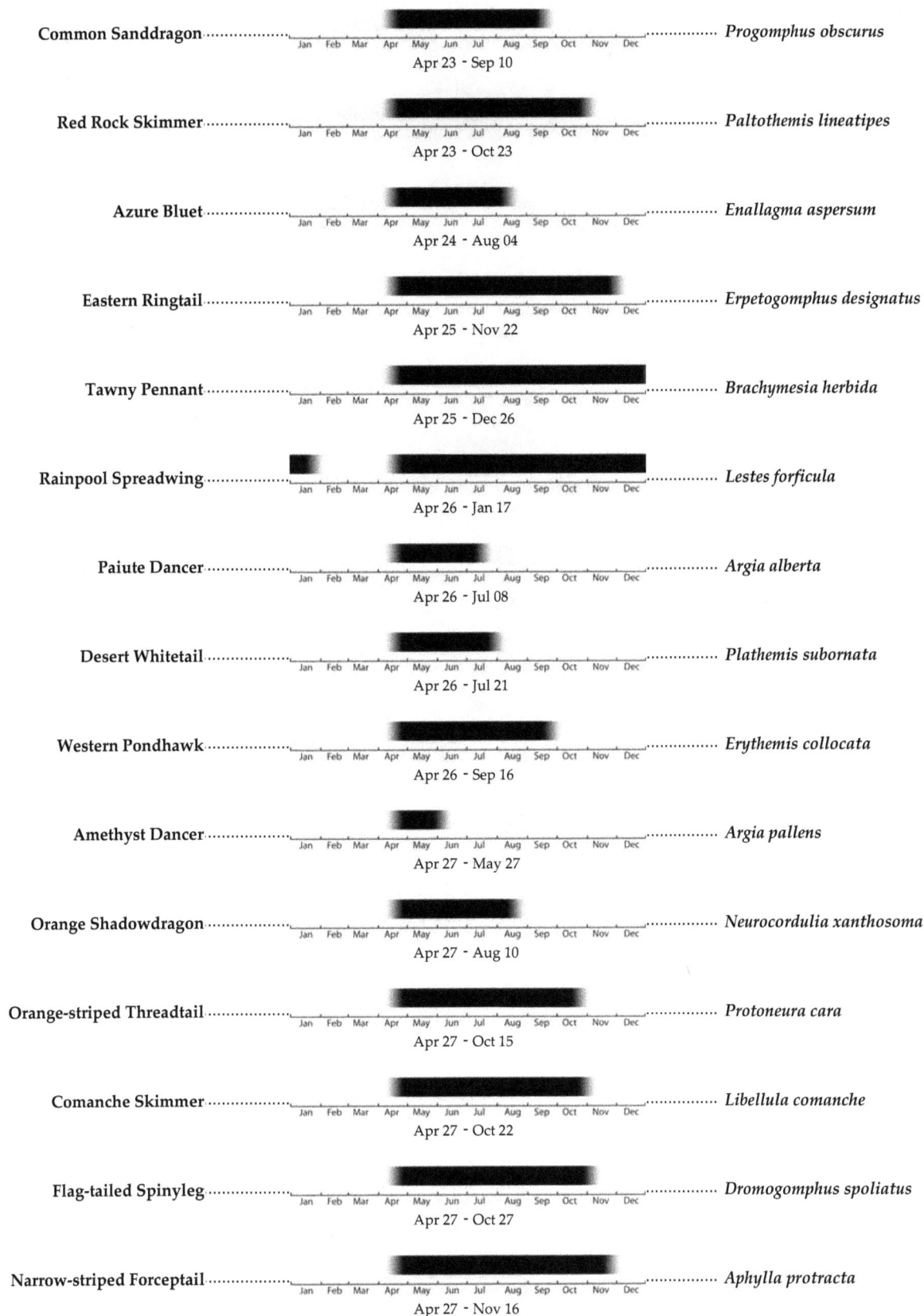

Common Sanddragon ·············· Jan Feb Mar Apr May Jun Jul Aug Sep Oct Nov Dec ··············· *Progomphus obscurus*
Apr 23 – Sep 10

Red Rock Skimmer ·············· Jan Feb Mar Apr May Jun Jul Aug Sep Oct Nov Dec ··············· *Paltothemis lineatipes*
Apr 23 – Oct 23

Azure Bluet ·············· Jan Feb Mar Apr May Jun Jul Aug Sep Oct Nov Dec ··············· *Enallagma aspersum*
Apr 24 – Aug 04

Eastern Ringtail ·············· Jan Feb Mar Apr May Jun Jul Aug Sep Oct Nov Dec ··············· *Erpetogomphus designatus*
Apr 25 – Nov 22

Tawny Pennant ·············· Jan Feb Mar Apr May Jun Jul Aug Sep Oct Nov Dec ··············· *Brachymesia herbida*
Apr 25 – Dec 26

Rainpool Spreadwing ·············· Jan Feb Mar Apr May Jun Jul Aug Sep Oct Nov Dec ··············· *Lestes forficula*
Apr 26 – Jan 17

Paiute Dancer ·············· Jan Feb Mar Apr May Jun Jul Aug Sep Oct Nov Dec ··············· *Argia alberta*
Apr 26 – Jul 08

Desert Whitetail ·············· Jan Feb Mar Apr May Jun Jul Aug Sep Oct Nov Dec ··············· *Plathemis subornata*
Apr 26 – Jul 21

Western Pondhawk ·············· Jan Feb Mar Apr May Jun Jul Aug Sep Oct Nov Dec ··············· *Erythemis collocata*
Apr 26 – Sep 16

Amethyst Dancer ·············· Jan Feb Mar Apr May Jun Jul Aug Sep Oct Nov Dec ··············· *Argia pallens*
Apr 27 – May 27

Orange Shadowdragon ·············· Jan Feb Mar Apr May Jun Jul Aug Sep Oct Nov Dec ··············· *Neurocordulia xanthosoma*
Apr 27 – Aug 10

Orange-striped Threadtail ·············· Jan Feb Mar Apr May Jun Jul Aug Sep Oct Nov Dec ··············· *Protoneura cara*
Apr 27 – Oct 15

Comanche Skimmer ·············· Jan Feb Mar Apr May Jun Jul Aug Sep Oct Nov Dec ··············· *Libellula comanche*
Apr 27 – Oct 22

Flag-tailed Spinyleg ·············· Jan Feb Mar Apr May Jun Jul Aug Sep Oct Nov Dec ··············· *Dromogomphus spoliatus*
Apr 27 – Oct 27

Narrow-striped Forceptail ·············· Jan Feb Mar Apr May Jun Jul Aug Sep Oct Nov Dec ··············· *Aphylla protracta*
Apr 27 – Nov 16

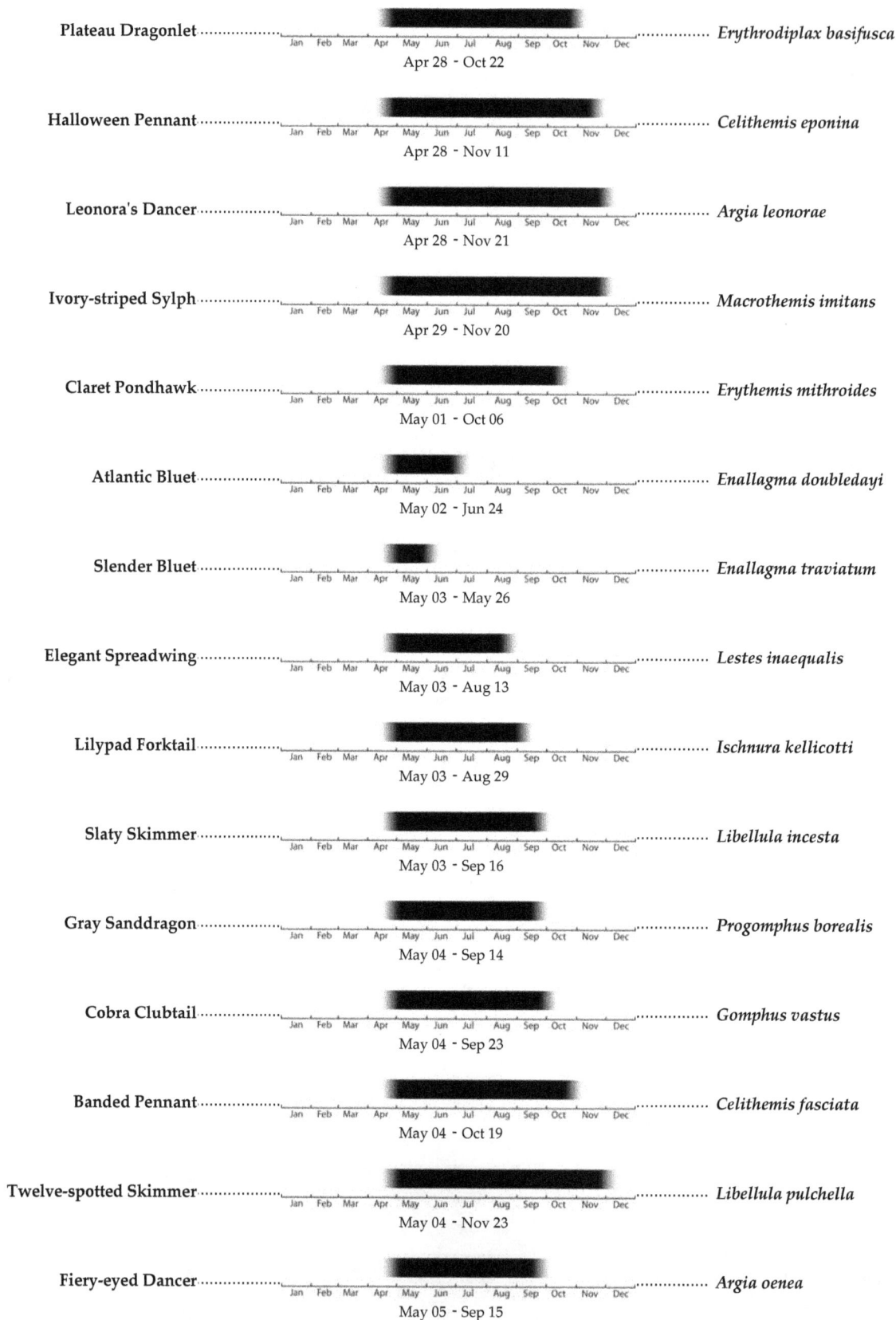

Plateau Dragonlet ·················· Jan Feb Mar Apr May Jun Jul Aug Sep Oct Nov Dec ·············· *Erythrodiplax basifusca*
Apr 28 - Oct 22

Halloween Pennant ·················· Jan Feb Mar Apr May Jun Jul Aug Sep Oct Nov Dec ·············· *Celithemis eponina*
Apr 28 - Nov 11

Leonora's Dancer ·················· Jan Feb Mar Apr May Jun Jul Aug Sep Oct Nov Dec ·············· *Argia leonorae*
Apr 28 - Nov 21

Ivory-striped Sylph ·················· Jan Feb Mar Apr May Jun Jul Aug Sep Oct Nov Dec ·············· *Macrothemis imitans*
Apr 29 - Nov 20

Claret Pondhawk ·················· Jan Feb Mar Apr May Jun Jul Aug Sep Oct Nov Dec ·············· *Erythemis mithroides*
May 01 - Oct 06

Atlantic Bluet ·················· Jan Feb Mar Apr May Jun Jul Aug Sep Oct Nov Dec ·············· *Enallagma doubledayi*
May 02 - Jun 24

Slender Bluet ·················· Jan Feb Mar Apr May Jun Jul Aug Sep Oct Nov Dec ·············· *Enallagma traviatum*
May 03 - May 26

Elegant Spreadwing ·················· Jan Feb Mar Apr May Jun Jul Aug Sep Oct Nov Dec ·············· *Lestes inaequalis*
May 03 - Aug 13

Lilypad Forktail ·················· Jan Feb Mar Apr May Jun Jul Aug Sep Oct Nov Dec ·············· *Ischnura kellicotti*
May 03 - Aug 29

Slaty Skimmer ·················· Jan Feb Mar Apr May Jun Jul Aug Sep Oct Nov Dec ·············· *Libellula incesta*
May 03 - Sep 16

Gray Sanddragon ·················· Jan Feb Mar Apr May Jun Jul Aug Sep Oct Nov Dec ·············· *Progomphus borealis*
May 04 - Sep 14

Cobra Clubtail ·················· Jan Feb Mar Apr May Jun Jul Aug Sep Oct Nov Dec ·············· *Gomphus vastus*
May 04 - Sep 23

Banded Pennant ·················· Jan Feb Mar Apr May Jun Jul Aug Sep Oct Nov Dec ·············· *Celithemis fasciata*
May 04 - Oct 19

Twelve-spotted Skimmer ·················· Jan Feb Mar Apr May Jun Jul Aug Sep Oct Nov Dec ·············· *Libellula pulchella*
May 04 - Nov 23

Fiery-eyed Dancer ·················· Jan Feb Mar Apr May Jun Jul Aug Sep Oct Nov Dec ·············· *Argia oenea*
May 05 - Sep 15

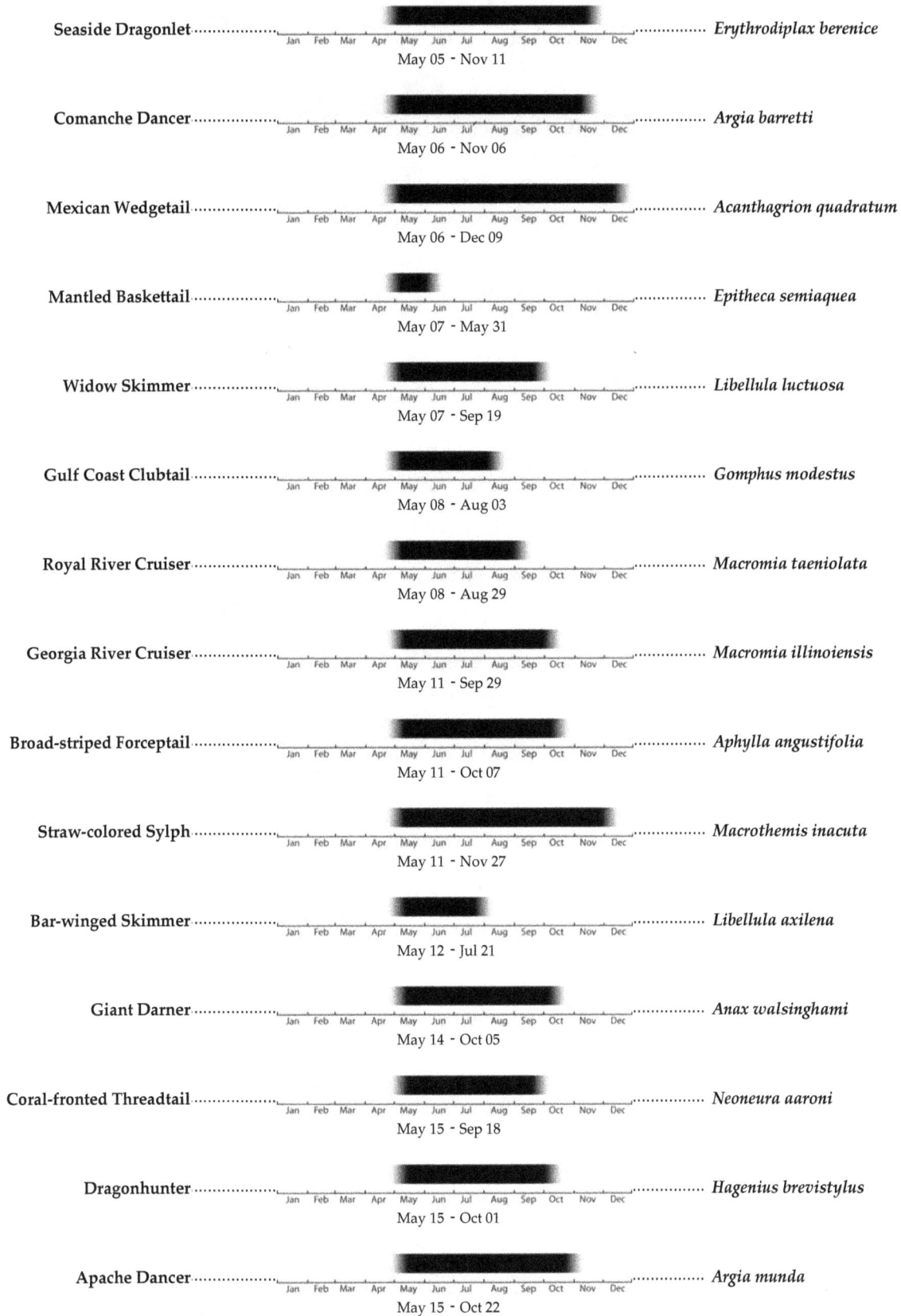

Seaside Dragonlet ················· Jan Feb Mar Apr May Jun Jul Aug Sep Oct Nov Dec ··········· *Erythrodiplax berenice*
May 05 - Nov 11

Comanche Dancer ················· Jan Feb Mar Apr May Jun Jul Aug Sep Oct Nov Dec ··········· *Argia barretti*
May 06 - Nov 06

Mexican Wedgetail ················· Jan Feb Mar Apr May Jun Jul Aug Sep Oct Nov Dec ··········· *Acanthagrion quadratum*
May 06 - Dec 09

Mantled Baskettail ················· Jan Feb Mar Apr May Jun Jul Aug Sep Oct Nov Dec ··········· *Epitheca semiaquea*
May 07 - May 31

Widow Skimmer ················· Jan Feb Mar Apr May Jun Jul Aug Sep Oct Nov Dec ··········· *Libellula luctuosa*
May 07 - Sep 19

Gulf Coast Clubtail ················· Jan Feb Mar Apr May Jun Jul Aug Sep Oct Nov Dec ··········· *Gomphus modestus*
May 08 - Aug 03

Royal River Cruiser ················· Jan Feb Mar Apr May Jun Jul Aug Sep Oct Nov Dec ··········· *Macromia taeniolata*
May 08 - Aug 29

Georgia River Cruiser ················· Jan Feb Mar Apr May Jun Jul Aug Sep Oct Nov Dec ··········· *Macromia illinoiensis*
May 11 - Sep 29

Broad-striped Forceptail ················· Jan Feb Mar Apr May Jun Jul Aug Sep Oct Nov Dec ··········· *Aphylla angustifolia*
May 11 - Oct 07

Straw-colored Sylph ················· Jan Feb Mar Apr May Jun Jul Aug Sep Oct Nov Dec ··········· *Macrothemis inacuta*
May 11 - Nov 27

Bar-winged Skimmer ················· Jan Feb Mar Apr May Jun Jul Aug Sep Oct Nov Dec ··········· *Libellula axilena*
May 12 - Jul 21

Giant Darner ················· Jan Feb Mar Apr May Jun Jul Aug Sep Oct Nov Dec ··········· *Anax walsinghami*
May 14 - Oct 05

Coral-fronted Threadtail ················· Jan Feb Mar Apr May Jun Jul Aug Sep Oct Nov Dec ··········· *Neoneura aaroni*
May 15 - Sep 18

Dragonhunter ················· Jan Feb Mar Apr May Jun Jul Aug Sep Oct Nov Dec ··········· *Hagenius brevistylus*
May 15 - Oct 01

Apache Dancer ················· Jan Feb Mar Apr May Jun Jul Aug Sep Oct Nov Dec ··········· *Argia munda*
May 15 - Oct 22

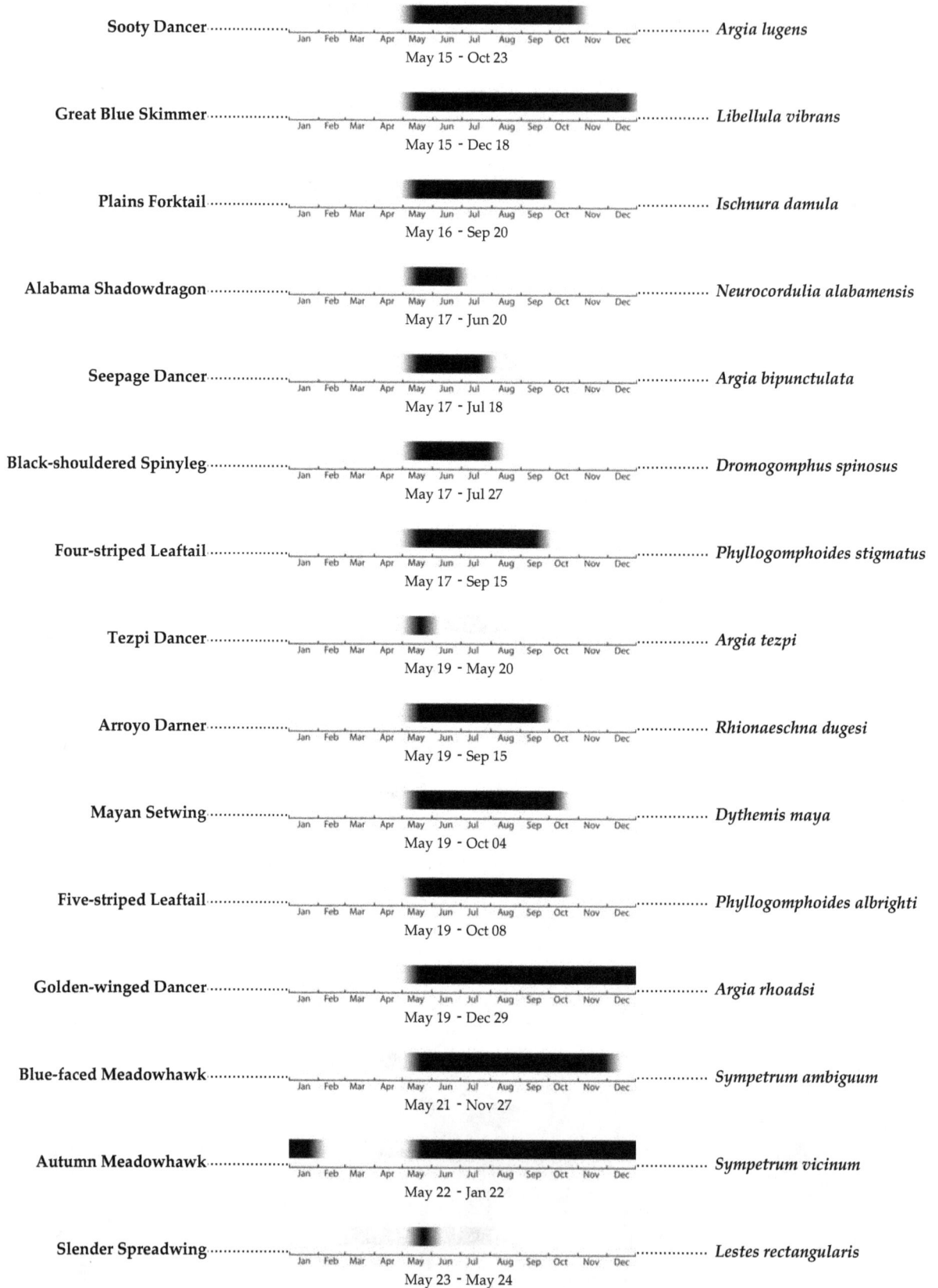

Sooty Dancer .... Jan Feb Mar Apr May Jun Jul Aug Sep Oct Nov Dec .... *Argia lugens*
May 15 - Oct 23

Great Blue Skimmer .... Jan Feb Mar Apr May Jun Jul Aug Sep Oct Nov Dec .... *Libellula vibrans*
May 15 - Dec 18

Plains Forktail .... Jan Feb Mar Apr May Jun Jul Aug Sep Oct Nov Dec .... *Ischnura damula*
May 16 - Sep 20

Alabama Shadowdragon .... Jan Feb Mar Apr May Jun Jul Aug Sep Oct Nov Dec .... *Neurocordulia alabamensis*
May 17 - Jun 20

Seepage Dancer .... Jan Feb Mar Apr May Jun Jul Aug Sep Oct Nov Dec .... *Argia bipunctulata*
May 17 - Jul 18

Black-shouldered Spinyleg .... Jan Feb Mar Apr May Jun Jul Aug Sep Oct Nov Dec .... *Dromogomphus spinosus*
May 17 - Jul 27

Four-striped Leaftail .... Jan Feb Mar Apr May Jun Jul Aug Sep Oct Nov Dec .... *Phyllogomphoides stigmatus*
May 17 - Sep 15

Tezpi Dancer .... Jan Feb Mar Apr May Jun Jul Aug Sep Oct Nov Dec .... *Argia tezpi*
May 19 - May 20

Arroyo Darner .... Jan Feb Mar Apr May Jun Jul Aug Sep Oct Nov Dec .... *Rhionaeschna dugesi*
May 19 - Sep 15

Mayan Setwing .... Jan Feb Mar Apr May Jun Jul Aug Sep Oct Nov Dec .... *Dythemis maya*
May 19 - Oct 04

Five-striped Leaftail .... Jan Feb Mar Apr May Jun Jul Aug Sep Oct Nov Dec .... *Phyllogomphoides albrighti*
May 19 - Oct 08

Golden-winged Dancer .... Jan Feb Mar Apr May Jun Jul Aug Sep Oct Nov Dec .... *Argia rhoadsi*
May 19 - Dec 29

Blue-faced Meadowhawk .... Jan Feb Mar Apr May Jun Jul Aug Sep Oct Nov Dec .... *Sympetrum ambiguum*
May 21 - Nov 27

Autumn Meadowhawk .... Jan Feb Mar Apr May Jun Jul Aug Sep Oct Nov Dec .... *Sympetrum vicinum*
May 22 - Jan 22

Slender Spreadwing .... Jan Feb Mar Apr May Jun Jul Aug Sep Oct Nov Dec .... *Lestes rectangularis*
May 23 - May 24

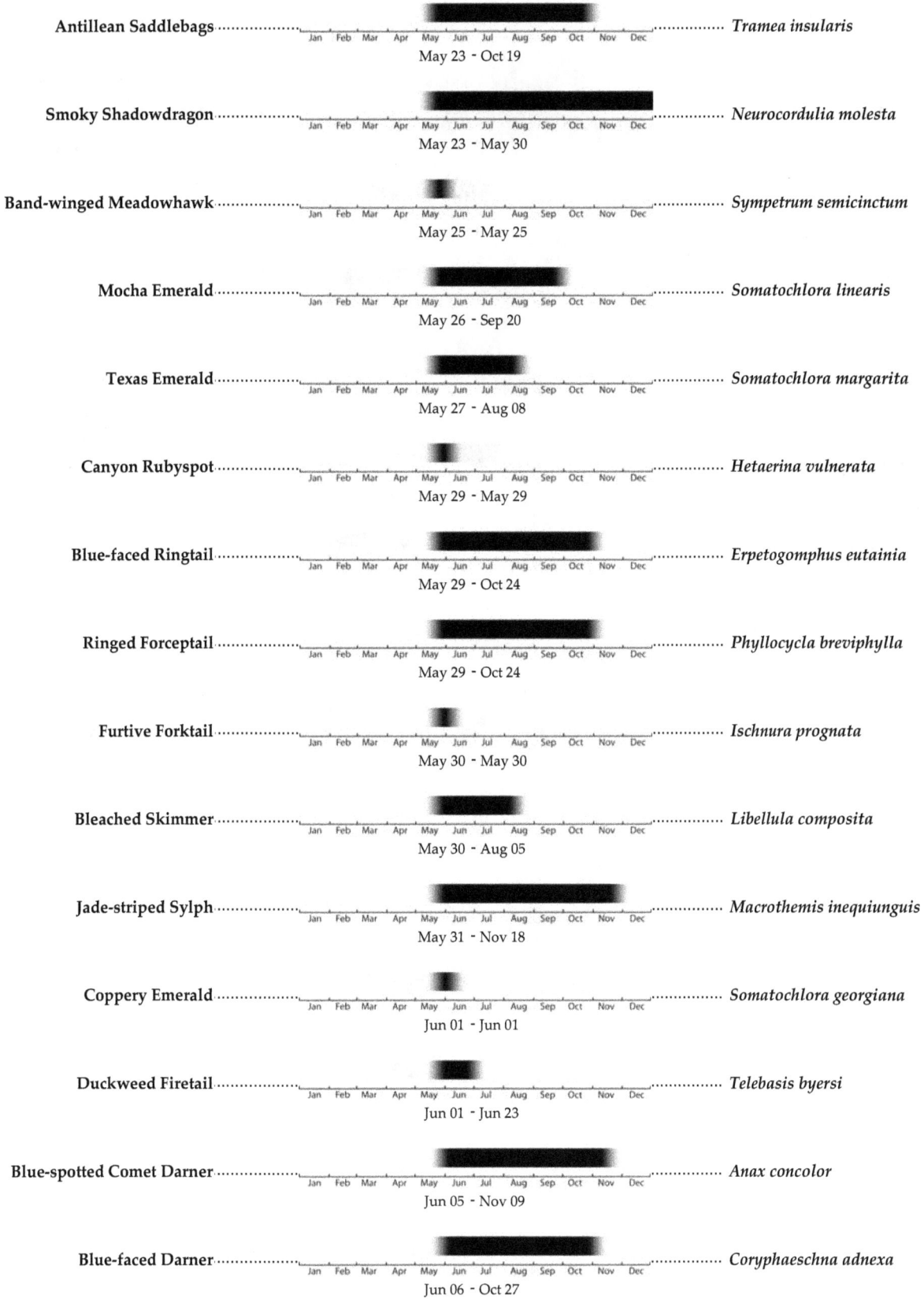

Antillean Saddlebags ················· Jan Feb Mar Apr May Jun Jul Aug Sep Oct Nov Dec ············· *Tramea insularis*
May 23 - Oct 19

Smoky Shadowdragon ················· Jan Feb Mar Apr May Jun Jul Aug Sep Oct Nov Dec ············· *Neurocordulia molesta*
May 23 - May 30

Band-winged Meadowhawk ················· Jan Feb Mar Apr May Jun Jul Aug Sep Oct Nov Dec ············· *Sympetrum semicinctum*
May 25 - May 25

Mocha Emerald ················· Jan Feb Mar Apr May Jun Jul Aug Sep Oct Nov Dec ············· *Somatochlora linearis*
May 26 - Sep 20

Texas Emerald ················· Jan Feb Mar Apr May Jun Jul Aug Sep Oct Nov Dec ············· *Somatochlora margarita*
May 27 - Aug 08

Canyon Rubyspot ················· Jan Feb Mar Apr May Jun Jul Aug Sep Oct Nov Dec ············· *Hetaerina vulnerata*
May 29 - May 29

Blue-faced Ringtail ················· Jan Feb Mar Apr May Jun Jul Aug Sep Oct Nov Dec ············· *Erpetogomphus eutainia*
May 29 - Oct 24

Ringed Forceptail ················· Jan Feb Mar Apr May Jun Jul Aug Sep Oct Nov Dec ············· *Phyllocycla breviphylla*
May 29 - Oct 24

Furtive Forktail ················· Jan Feb Mar Apr May Jun Jul Aug Sep Oct Nov Dec ············· *Ischnura prognata*
May 30 - May 30

Bleached Skimmer ················· Jan Feb Mar Apr May Jun Jul Aug Sep Oct Nov Dec ············· *Libellula composita*
May 30 - Aug 05

Jade-striped Sylph ················· Jan Feb Mar Apr May Jun Jul Aug Sep Oct Nov Dec ············· *Macrothemis inequiunguis*
May 31 - Nov 18

Coppery Emerald ················· Jan Feb Mar Apr May Jun Jul Aug Sep Oct Nov Dec ············· *Somatochlora georgiana*
Jun 01 - Jun 01

Duckweed Firetail ················· Jan Feb Mar Apr May Jun Jul Aug Sep Oct Nov Dec ············· *Telebasis byersi*
Jun 01 - Jun 23

Blue-spotted Comet Darner ················· Jan Feb Mar Apr May Jun Jul Aug Sep Oct Nov Dec ············· *Anax concolor*
Jun 05 - Nov 09

Blue-faced Darner ················· Jan Feb Mar Apr May Jun Jul Aug Sep Oct Nov Dec ············· *Coryphaeschna adnexa*
Jun 06 - Oct 27

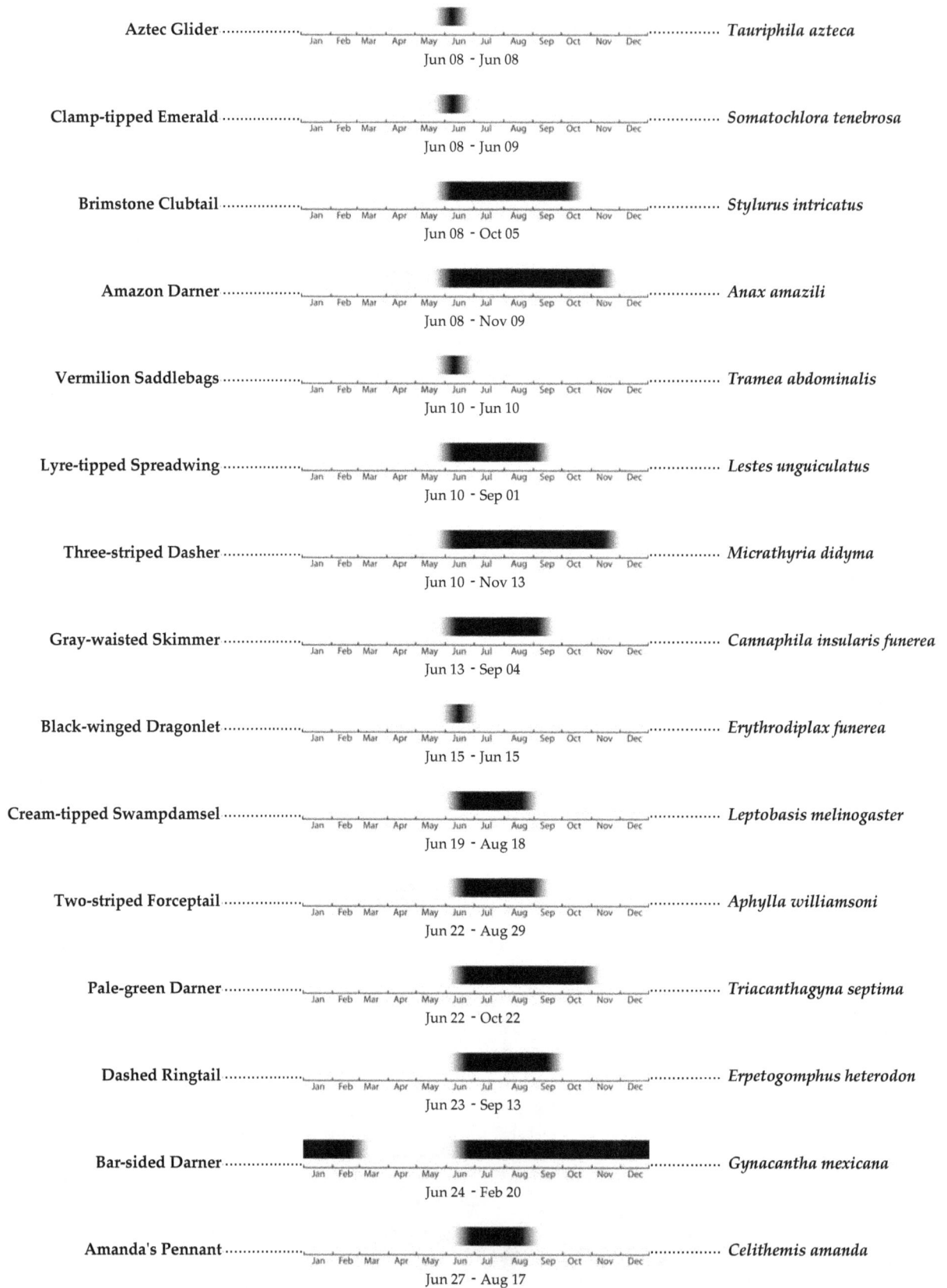

Aztec Glider .... Jan Feb Mar Apr May Jun Jul Aug Sep Oct Nov Dec .... *Tauriphila azteca*
Jun 08 - Jun 08

Clamp-tipped Emerald .... Jan Feb Mar Apr May Jun Jul Aug Sep Oct Nov Dec .... *Somatochlora tenebrosa*
Jun 08 - Jun 09

Brimstone Clubtail .... Jan Feb Mar Apr May Jun Jul Aug Sep Oct Nov Dec .... *Stylurus intricatus*
Jun 08 - Oct 05

Amazon Darner .... Jan Feb Mar Apr May Jun Jul Aug Sep Oct Nov Dec .... *Anax amazili*
Jun 08 - Nov 09

Vermilion Saddlebags .... Jan Feb Mar Apr May Jun Jul Aug Sep Oct Nov Dec .... *Tramea abdominalis*
Jun 10 - Jun 10

Lyre-tipped Spreadwing .... Jan Feb Mar Apr May Jun Jul Aug Sep Oct Nov Dec .... *Lestes unguiculatus*
Jun 10 - Sep 01

Three-striped Dasher .... Jan Feb Mar Apr May Jun Jul Aug Sep Oct Nov Dec .... *Micrathyria didyma*
Jun 10 - Nov 13

Gray-waisted Skimmer .... Jan Feb Mar Apr May Jun Jul Aug Sep Oct Nov Dec .... *Cannaphila insularis funerea*
Jun 13 - Sep 04

Black-winged Dragonlet .... Jan Feb Mar Apr May Jun Jul Aug Sep Oct Nov Dec .... *Erythrodiplax funerea*
Jun 15 - Jun 15

Cream-tipped Swampdamsel .... Jan Feb Mar Apr May Jun Jul Aug Sep Oct Nov Dec .... *Leptobasis melinogaster*
Jun 19 - Aug 18

Two-striped Forceptail .... Jan Feb Mar Apr May Jun Jul Aug Sep Oct Nov Dec .... *Aphylla williamsoni*
Jun 22 - Aug 29

Pale-green Darner .... Jan Feb Mar Apr May Jun Jul Aug Sep Oct Nov Dec .... *Triacanthagyna septima*
Jun 22 - Oct 22

Dashed Ringtail .... Jan Feb Mar Apr May Jun Jul Aug Sep Oct Nov Dec .... *Erpetogomphus heterodon*
Jun 23 - Sep 13

Bar-sided Darner .... Jan Feb Mar Apr May Jun Jul Aug Sep Oct Nov Dec .... *Gynacantha mexicana*
Jun 24 - Feb 20

Amanda's Pennant .... Jan Feb Mar Apr May Jun Jul Aug Sep Oct Nov Dec .... *Celithemis amanda*
Jun 27 - Aug 17

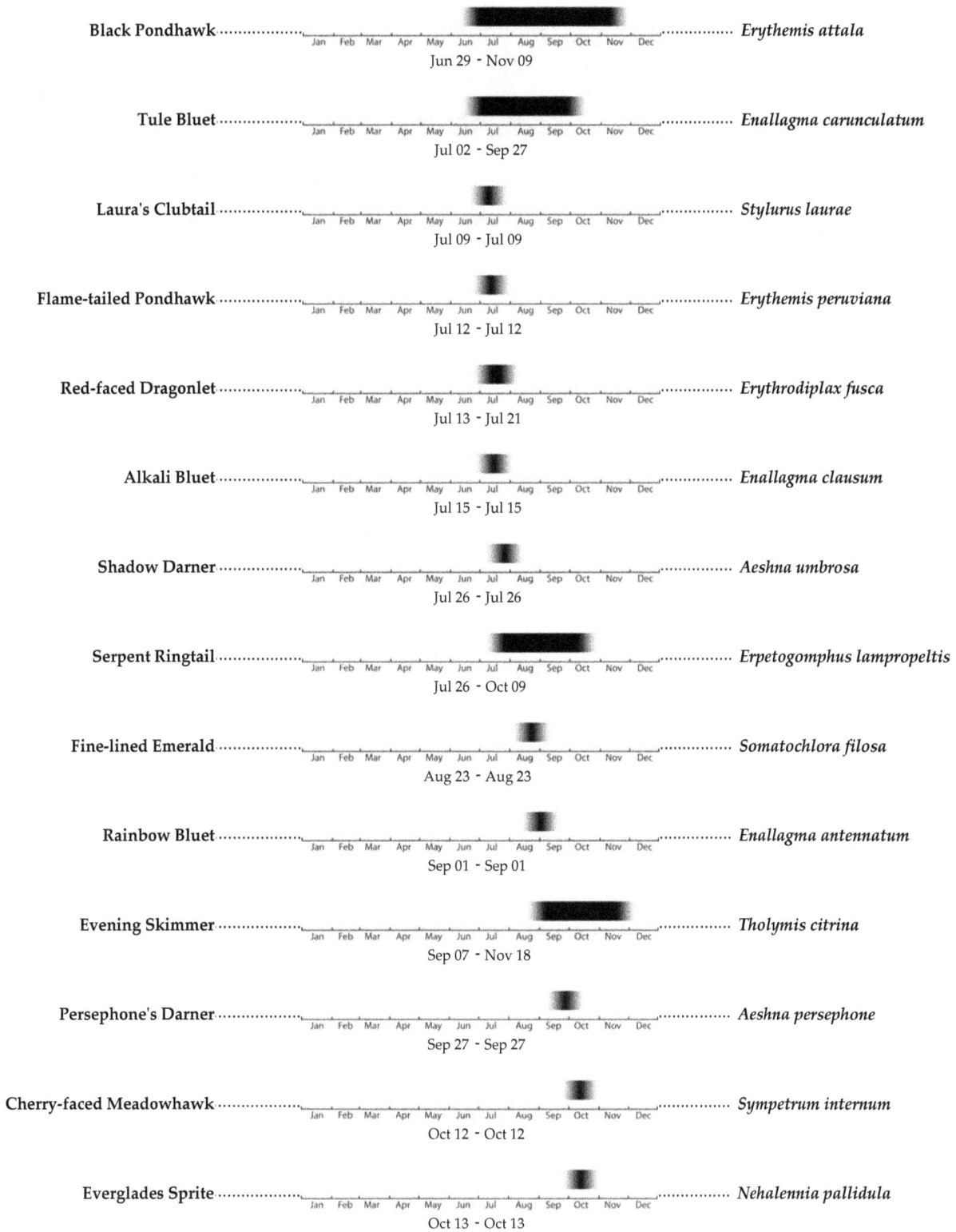

Black Pondhawk ················· Jan Feb Mar Apr May Jun Jul Aug Sep Oct Nov Dec ················· *Erythemis attala*
Jun 29 - Nov 09

Tule Bluet ················· Jan Feb Mar Apr May Jun Jul Aug Sep Oct Nov Dec ················· *Enallagma carunculatum*
Jul 02 - Sep 27

Laura's Clubtail ················· Jan Feb Mar Apr May Jun Jul Aug Sep Oct Nov Dec ················· *Stylurus laurae*
Jul 09 - Jul 09

Flame-tailed Pondhawk ················· Jan Feb Mar Apr May Jun Jul Aug Sep Oct Nov Dec ················· *Erythemis peruviana*
Jul 12 - Jul 12

Red-faced Dragonlet ················· Jan Feb Mar Apr May Jun Jul Aug Sep Oct Nov Dec ················· *Erythrodiplax fusca*
Jul 13 - Jul 21

Alkali Bluet ················· Jan Feb Mar Apr May Jun Jul Aug Sep Oct Nov Dec ················· *Enallagma clausum*
Jul 15 - Jul 15

Shadow Darner ················· Jan Feb Mar Apr May Jun Jul Aug Sep Oct Nov Dec ················· *Aeshna umbrosa*
Jul 26 - Jul 26

Serpent Ringtail ················· Jan Feb Mar Apr May Jun Jul Aug Sep Oct Nov Dec ················· *Erpetogomphus lampropeltis*
Jul 26 - Oct 09

Fine-lined Emerald ················· Jan Feb Mar Apr May Jun Jul Aug Sep Oct Nov Dec ················· *Somatochlora filosa*
Aug 23 - Aug 23

Rainbow Bluet ················· Jan Feb Mar Apr May Jun Jul Aug Sep Oct Nov Dec ················· *Enallagma antennatum*
Sep 01 - Sep 01

Evening Skimmer ················· Jan Feb Mar Apr May Jun Jul Aug Sep Oct Nov Dec ················· *Tholymis citrina*
Sep 07 - Nov 18

Persephone's Darner ················· Jan Feb Mar Apr May Jun Jul Aug Sep Oct Nov Dec ················· *Aeshna persephone*
Sep 27 - Sep 27

Cherry-faced Meadowhawk ················· Jan Feb Mar Apr May Jun Jul Aug Sep Oct Nov Dec ················· *Sympetrum internum*
Oct 12 - Oct 12

Everglades Sprite ················· Jan Feb Mar Apr May Jun Jul Aug Sep Oct Nov Dec ················· *Nehalennia pallidula*
Oct 13 - Oct 13

Dragonflies & Damselflies of Texas
Listed by County

Anderson *(45)*
Blue-fronted Dancer (*Argia apicalis*)
Seepage Dancer (*Argia bipunctulata*)
Violet Dancer (*Argia fumipennis violacea*)
Powdered Dancer (*Argia moesta*)
Blue-tipped Dancer (*Argia tibialis*)
Fawn Darner (*Boyeria vinosa*)
Ebony Jewelwing (*Calopteryx maculata*)
Calico Pennant (*Celithemis elisa*)
Halloween Pennant (*Celithemis eponina*)
Banded Pennant (*Celithemis fasciata*)
Black-shouldered Spinyleg (*Dromogomphus
    spinosus*)
Familiar Bluet (*Enallagma civile*)
Atlantic Bluet (*Enallagma doubledayi*)
Stream Bluet (*Enallagma exsulans*)
Orange Bluet (*Enallagma signatum*)
Vesper Bluet (*Enallagma vesperum*)
Swamp Darner (*Epiaeschna heros*)
Stripe-winged Baskettail (*Epitheca costalis*)
Common Baskettail (*Epitheca cynosura*)
Mantled Baskettail (*Epitheca semiaquea*)
Eastern Pondhawk (*Erythemis simplicicollis*)
Little Blue Dragonlet (*Erythrodiplax minuscula*)
Ashy Clubtail (*Gomphus lividus*)
Sulphur-tipped Clubtail (*Gomphus militaris*)
Oklahoma Clubtail (*Gomphus oklahomensis*)
Dragonhunter (*Hagenius brevistylus*)
Citrine Forktail (*Ischnura hastata*)
Fragile Forktail (*Ischnura posita*)
Rambur's Forktail (*Ischnura ramburii*)
Blue Corporal (*Ladona deplanata*)
Golden-winged Skimmer (*Libellula auripennis*)
Spangled Skimmer (*Libellula cyanea*)
Yellow-sided Skimmer (*Libellula flavida*)
Slaty Skimmer (*Libellula incesta*)
Widow Skimmer (*Libellula luctuosa*)
Great Blue Skimmer (*Libellula vibrans*)
Royal River Cruiser (*Macromia taeniolata*)
Roseate Skimmer (*Orthemis ferruginea*)
Blue Dasher (*Pachydiplax longipennis*)
Spot-winged Glider (*Pantala hymenaea*)
Eastern Amberwing (*Perithemis tenera*)
Common Whitetail (*Plathemis lydia*)
Common Sanddragon (*Progomphus obscurus*)
Texas Emerald (*Somatochlora margarita*)
Gray Petaltail (*Tachopteryx thoreyi*)

Angelina *(44)*
Common Green Darner (*Anax junius*)
Blue-fronted Dancer (*Argia apicalis*)
Violet Dancer (*Argia fumipennis violacea*)
Powdered Dancer (*Argia moesta*)
Blue-tipped Dancer (*Argia tibialis*)
Four-spotted Pennant (*Brachymesia gravida*)
Tawny Pennant (*Brachymesia herbida*)
Ebony Jewelwing (*Calopteryx maculata*)
Calico Pennant (*Celithemis elisa*)
Halloween Pennant (*Celithemis eponina*)
Banded Pennant (*Celithemis fasciata*)
Black-shouldered Spinyleg (*Dromogomphus
    spinosus*)
Swift Setwing (*Dythemis velox*)
Familiar Bluet (*Enallagma civile*)
Stream Bluet (*Enallagma exsulans*)
Swamp Darner (*Epiaeschna heros*)
Prince Baskettail (*Epitheca princeps*)
Eastern Pondhawk (*Erythemis simplicicollis*)
Little Blue Dragonlet (*Erythrodiplax minuscula*)

Band-winged Dragonlet (*Erythrodiplax umbrata*)
Oklahoma Clubtail (*Gomphus oklahomensis*)
Smoky Rubyspot (*Hetaerina titia*)
Citrine Forktail (*Ischnura hastata*)
Rambur's Forktail (*Ischnura ramburii*)
Elegant Spreadwing (*Lestes inaequalis*)
Golden-winged Skimmer (*Libellula auripennis*)
Spangled Skimmer (*Libellula cyanea*)
Slaty Skimmer (*Libellula incesta*)
Widow Skimmer (*Libellula luctuosa*)
Great Blue Skimmer (*Libellula vibrans*)
Smoky Shadowdragon (*Neurocordulia molesta*)
Roseate Skimmer (*Orthemis ferruginea*)
Blue Dasher (*Pachydiplax longipennis*)
Wandering Glider (*Pantala flavescens*)
Spot-winged Glider (*Pantala hymenaea*)
Eastern Amberwing (*Perithemis tenera*)
Common Whitetail (*Plathemis lydia*)
Common Sanddragon (*Progomphus obscurus*)
Mocha Emerald (*Somatochlora linearis*)
Variegated Meadowhawk (*Sympetrum
    corruptum*)
Gray Petaltail (*Tachopteryx thoreyi*)
Carolina Saddlebags (*Tramea carolina*)
Black Saddlebags (*Tramea lacerata*)
Red Saddlebags (*Tramea onusta*)

Aransas *(22)*
Common Green Darner (*Anax junius*)
Gray-waisted Skimmer (*Cannaphila insularis
    funerea*)
Familiar Bluet (*Enallagma civile*)
Eastern Pondhawk (*Erythemis simplicicollis*)
Seaside Dragonlet (*Erythrodiplax berenice*)
Little Blue Dragonlet (*Erythrodiplax minuscula*)
Band-winged Dragonlet (*Erythrodiplax umbrata*)
Citrine Forktail (*Ischnura hastata*)
Fragile Forktail (*Ischnura posita*)
Rambur's Forktail (*Ischnura ramburii*)
Rainpool Spreadwing (*Lestes forficula*)
Needham's Skimmer (*Libellula needhami*)
Twelve-spotted Skimmer (*Libellula pulchella*)
Marl Pennant (*Macrodiplax balteata*)
Roseate Skimmer (*Orthemis ferruginea*)
Blue Dasher (*Pachydiplax longipennis*)
Wandering Glider (*Pantala flavescens*)
Spot-winged Glider (*Pantala hymenaea*)
Eastern Amberwing (*Perithemis tenera*)
Variegated Meadowhawk (*Sympetrum
    corruptum*)
Black Saddlebags (*Tramea lacerata*)
Red Saddlebags (*Tramea onusta*)

Archer *(8)*
Blue-fronted Dancer (*Argia apicalis*)
Powdered Dancer (*Argia moesta*)
Familiar Bluet (*Enallagma civile*)
Eastern Pondhawk (*Erythemis simplicicollis*)
Rambur's Forktail (*Ischnura ramburii*)
Widow Skimmer (*Libellula luctuosa*)
Eastern Amberwing (*Perithemis tenera*)
Black Saddlebags (*Tramea lacerata*)

Atascosa *(27)*
Common Green Darner (*Anax junius*)
Broad-striped Forceptail (*Aphylla angustifolia*)
Blue-fronted Dancer (*Argia apicalis*)
Kiowa Dancer (*Argia immunda*)
Powdered Dancer (*Argia moesta*)
Blue-ringed Dancer (*Argia sedula*)
Dusky Dancer (*Argia translata*)

Red-tailed Pennant (*Brachymesia furcata*)
Four-spotted Pennant (*Brachymesia gravida*)
Halloween Pennant (*Celithemis eponina*)
Swift Setwing (*Dythemis velox*)
Familiar Bluet (*Enallagma civile*)
Eastern Pondhawk (*Erythemis simplicicollis*)
American Rubyspot (*Hetaerina americana*)
Rambur's Forktail (*Ischnura ramburii*)
Rainpool Spreadwing (*Lestes forficula*)
Neon Skimmer (*Libellula croceipennis*)
Widow Skimmer (*Libellula luctuosa*)
Great Blue Skimmer (*Libellula vibrans*)
Marl Pennant (*Macrodiplax balteata*)
Hyacinth Glider (*Miathyria marcella*)
Thornbush Dasher (*Micrathyria hagenii*)
Blue Dasher (*Pachydiplax longipennis*)
Wandering Glider (*Pantala flavescens*)
Spot-winged Glider (*Pantala hymenaea*)
Eastern Amberwing (*Perithemis tenera*)
Common Sanddragon (*Progomphus obscurus*)

Austin *(53)*
Common Green Darner (*Anax junius*)
Broad-striped Forceptail (*Aphylla angustifolia*)
Narrow-striped Forceptail (*Aphylla protracta*)
Two-striped Forceptail (*Aphylla williamsoni*)
Blue-fronted Dancer (*Argia apicalis*)
Violet Dancer (*Argia fumipennis violacea*)
Blue-ringed Dancer (*Argia sedula*)
Stillwater Clubtail (*Arigomphus lentulus*)
Red-tailed Pennant (*Brachymesia furcata*)
Ebony Jewelwing (*Calopteryx maculata*)
Calico Pennant (*Celithemis elisa*)
Halloween Pennant (*Celithemis eponina*)
Banded Pennant (*Celithemis fasciata*)
Arrowhead Spiketail (*Cordulegaster obliqua*)
Black-shouldered Spinyleg (*Dromogomphus
    spinosus*)
Flag-tailed Spinyleg (*Dromogomphus spoliatus*)
Checkered Setwing (*Dythemis fugax*)
Black Setwing (*Dythemis nigrescens*)
Swift Setwing (*Dythemis velox*)
Double-striped Bluet (*Enallagma basidens*)
Turquoise Bluet (*Enallagma divagans*)
Orange Bluet (*Enallagma signatum*)
Common Baskettail (*Epitheca cynosura*)
Prince Baskettail (*Epitheca princeps*)
Eastern Pondhawk (*Erythemis simplicicollis*)
Great Pondhawk (*Erythemis vesiculosa*)
Band-winged Dragonlet (*Erythrodiplax umbrata*)
Sulphur-tipped Clubtail (*Gomphus militaris*)
Oklahoma Clubtail (*Gomphus oklahomensis*)
American Rubyspot (*Hetaerina americana*)
Citrine Forktail (*Ischnura hastata*)
Lilypad Forktail (*Ischnura kellicotti*)
Fragile Forktail (*Ischnura posita*)
Blue Corporal (*Ladona deplanata*)
Southern Spreadwing (*Lestes australis*)
Yellow-sided Skimmer (*Libellula flavida*)
Slaty Skimmer (*Libellula incesta*)
Widow Skimmer (*Libellula luctuosa*)
Needham's Skimmer (*Libellula needhami*)
Great Blue Skimmer (*Libellula vibrans*)
Hyacinth Glider (*Miathyria marcella*)
Roseate Skimmer (*Orthemis ferruginea*)
Blue Dasher (*Pachydiplax longipennis*)
Wandering Glider (*Pantala flavescens*)
Spot-winged Glider (*Pantala hymenaea*)

Eastern Amberwing (*Perithemis tenera*)
Four-striped Leaftail (*Phyllogomphoides stigmatus*)
Common Whitetail (*Plathemis lydia*)
Common Sanddragon (*Progomphus obscurus*)
Variegated Meadowhawk (*Sympetrum corruptum*)
Desert Firetail (*Telebasis salva*)
Carolina Saddlebags (*Tramea carolina*)
Red Saddlebags (*Tramea onusta*)

Bailey *(8)*
Double-striped Bluet (*Enallagma basidens*)
Tule Bluet (*Enallagma carunculatum*)
Familiar Bluet (*Enallagma civile*)
Desert Forktail (*Ischnura barberi*)
Plateau Spreadwing (*Lestes alacer*)
Flame Skimmer (*Libellula saturata*)
Variegated Meadowhawk (*Sympetrum corruptum*)
Red Saddlebags (*Tramea onusta*)

Bandera *(63)*
Common Green Darner (*Anax junius*)
Great Spreadwing (*Archilestes grandis*)
Coppery Dancer (*Argia cuprea*)
Violet Dancer (*Argia fumipennis violacea*)
Lavender Dancer (*Argia hinei*)
Kiowa Dancer (*Argia immunda*)
Leonora's Dancer (*Argia leonorae*)
Powdered Dancer (*Argia moesta*)
Aztec Dancer (*Argia nahuana*)
Springwater Dancer (*Argia plana*)
Blue-ringed Dancer (*Argia sedula*)
Dusky Dancer (*Argia translata*)
Springtime Darner (*Basiaeschna janata*)
Pale-faced Clubskimmer (*Brechmorhoga mendax*)
Banded Pennant (*Celithemis fasciata*)
Black-shouldered Spinyleg (*Dromogomphus spinosus*)
Flag-tailed Spinyleg (*Dromogomphus spoliatus*)
Checkered Setwing (*Dythemis fugax*)
Swift Setwing (*Dythemis velox*)
Double-striped Bluet (*Enallagma basidens*)
Familiar Bluet (*Enallagma civile*)
Stream Bluet (*Enallagma exsulans*)
Neotropical Bluet (*Enallagma novaehispaniae*)
Arroyo Bluet (*Enallagma praevarum*)
Orange Bluet (*Enallagma signatum*)
Stripe-winged Baskettail (*Epitheca costalis*)
Dot-winged Baskettail (*Epitheca petechialis*)
Prince Baskettail (*Epitheca princeps*)
Eastern Ringtail (*Erpetogomphus designatus*)
Eastern Pondhawk (*Erythemis simplicicollis*)
Band-winged Dragonlet (*Erythrodiplax umbrata*)
Pronghorn Clubtail (*Gomphus graslinellus*)
Sulphur-tipped Clubtail (*Gomphus militaris*)
Dragonhunter (*Hagenius brevistylus*)
American Rubyspot (*Hetaerina americana*)
Smoky Rubyspot (*Hetaerina titia*)
Citrine Forktail (*Ischnura hastata*)
Rambur's Forktail (*Ischnura ramburii*)
Plateau Spreadwing (*Lestes alacer*)
Chalky Spreadwing (*Lestes sigma*)
Comanche Skimmer (*Libellula comanche*)
Neon Skimmer (*Libellula croceipennis*)
Widow Skimmer (*Libellula luctuosa*)
Flame Skimmer (*Libellula saturata*)
Bronzed River Cruiser (*Macromia annulata*)
Ivory-striped Sylph (*Macrothemis imitans leucozona*)
Jade-striped Sylph (*Macrothemis inequiunguis*)
Coral-fronted Threadtail (*Neoneura aaroni*)
Carmine Skimmer (*Orthemis discolor*)
Roseate Skimmer (*Orthemis ferruginea*)
Red Rock Skimmer (*Paltothemis lineatipes*)
Spot-winged Glider (*Pantala hymenaea*)
Eastern Amberwing (*Perithemis tenera*)

Five-striped Leaftail (*Phyllogomphoides albrighti*)
Four-striped Leaftail (*Phyllogomphoides stigmatus*)
Common Whitetail (*Plathemis lydia*)
Common Sanddragon (*Progomphus obscurus*)
Filigree Skimmer (*Pseudoleon superbus*)
Autumn Meadowhawk (*Sympetrum vicinum*)
Desert Firetail (*Telebasis salva*)
Carolina Saddlebags (*Tramea carolina*)
Black Saddlebags (*Tramea lacerata*)
Red Saddlebags (*Tramea onusta*)

Bastrop *(73)*
Common Green Darner (*Anax junius*)
Blue-fronted Dancer (*Argia apicalis*)
Kiowa Dancer (*Argia immunda*)
Powdered Dancer (*Argia moesta*)
Blue-ringed Dancer (*Argia sedula*)
Blue-tipped Dancer (*Argia tibialis*)
Dusky Dancer (*Argia translata*)
Jade Clubtail (*Arigomphus submedianus*)
Springtime Darner (*Basiaeschna janata*)
Fawn Darner (*Boyeria vinosa*)
Four-spotted Pennant (*Brachymesia gravida*)
Pale-faced Clubskimmer (*Brechmorhoga mendax*)
Ebony Jewelwing (*Calopteryx maculata*)
Calico Pennant (*Celithemis elisa*)
Halloween Pennant (*Celithemis eponina*)
Stream Cruiser (*Didymops transversa*)
Black-shouldered Spinyleg (*Dromogomphus spinosus*)
Flag-tailed Spinyleg (*Dromogomphus spoliatus*)
Checkered Setwing (*Dythemis fugax*)
Swift Setwing (*Dythemis velox*)
Double-striped Bluet (*Enallagma basidens*)
Familiar Bluet (*Enallagma civile*)
Orange Bluet (*Enallagma signatum*)
Vesper Bluet (*Enallagma vesperum*)
Dot-winged Baskettail (*Epitheca petechialis*)
Prince Baskettail (*Epitheca princeps*)
Mantled Baskettail (*Epitheca semiaquea*)
Eastern Ringtail (*Erpetogomphus designatus*)
Eastern Pondhawk (*Erythemis simplicicollis*)
Great Pondhawk (*Erythemis vesiculosa*)
Little Blue Dragonlet (*Erythrodiplax minuscula*)
Band-winged Dragonlet (*Erythrodiplax umbrata*)
Plains Clubtail (*Gomphus externus*)
Pronghorn Clubtail (*Gomphus graslinellus*)
Sulphur-tipped Clubtail (*Gomphus militaris*)
Oklahoma Clubtail (*Gomphus oklahomensis*)
American Rubyspot (*Hetaerina americana*)
Smoky Rubyspot (*Hetaerina titia*)
Citrine Forktail (*Ischnura hastata*)
Fragile Forktail (*Ischnura posita*)
Rambur's Forktail (*Ischnura ramburii*)
Blue Corporal (*Ladona deplanata*)
Plateau Spreadwing (*Lestes alacer*)
Southern Spreadwing (*Lestes australis*)
Rainpool Spreadwing (*Lestes forficula*)
Chalky Spreadwing (*Lestes sigma*)
Neon Skimmer (*Libellula croceipennis*)
Yellow-sided Skimmer (*Libellula flavida*)
Slaty Skimmer (*Libellula incesta*)
Widow Skimmer (*Libellula luctuosa*)
Twelve-spotted Skimmer (*Libellula pulchella*)
Great Blue Skimmer (*Libellula vibrans*)
Bronzed River Cruiser (*Macromia annulata*)
Georgia River Cruiser (*Macromia illinoiensis georgina*)
Hyacinth Glider (*Miathyria marcella*)
Thornbush Dasher (*Micrathyria hagenii*)
Cyrano Darner (*Nasiaeschna pentacantha*)
Carmine Skimmer (*Orthemis discolor*)
Roseate Skimmer (*Orthemis ferruginea*)
Blue Dasher (*Pachydiplax longipennis*)
Wandering Glider (*Pantala flavescens*)

Spot-winged Glider (*Pantala hymenaea*)
Eastern Amberwing (*Perithemis tenera*)
Five-striped Leaftail (*Phyllogomphoides albrighti*)
Common Whitetail (*Plathemis lydia*)
Common Sanddragon (*Progomphus obscurus*)
Mocha Emerald (*Somatochlora linearis*)
Blue-faced Meadowhawk (*Sympetrum ambiguum*)
Variegated Meadowhawk (*Sympetrum corruptum*)
Autumn Meadowhawk (*Sympetrum vicinum*)
Desert Firetail (*Telebasis salva*)
Black Saddlebags (*Tramea lacerata*)
Red Saddlebags (*Tramea onusta*)

Baylor *(3)*
Blue-fronted Dancer (*Argia apicalis*)
Powdered Dancer (*Argia moesta*)
American Rubyspot (*Hetaerina americana*)

Bee *(29)*
Common Green Darner (*Anax junius*)
Blue-fronted Dancer (*Argia apicalis*)
Blue-ringed Dancer (*Argia sedula*)
Red-tailed Pennant (*Brachymesia furcata*)
Four-spotted Pennant (*Brachymesia gravida*)
Halloween Pennant (*Celithemis eponina*)
Checkered Setwing (*Dythemis fugax*)
Black Setwing (*Dythemis nigrescens*)
Swift Setwing (*Dythemis velox*)
Familiar Bluet (*Enallagma civile*)
Orange Bluet (*Enallagma signatum*)
Swamp Darner (*Epiaeschna heros*)
Dot-winged Baskettail (*Epitheca petechialis*)
Eastern Ringtail (*Erpetogomphus designatus*)
Eastern Pondhawk (*Erythemis simplicicollis*)
Sulphur-tipped Clubtail (*Gomphus militaris*)
Citrine Forktail (*Ischnura hastata*)
Widow Skimmer (*Libellula luctuosa*)
Twelve-spotted Skimmer (*Libellula pulchella*)
Flame Skimmer (*Libellula saturata*)
Marl Pennant (*Macrodiplax balteata*)
Hyacinth Glider (*Miathyria marcella*)
Thornbush Dasher (*Micrathyria hagenii*)
Carmine Skimmer (*Orthemis discolor*)
Blue Dasher (*Pachydiplax longipennis*)
Wandering Glider (*Pantala flavescens*)
Desert Firetail (*Telebasis salva*)
Black Saddlebags (*Tramea lacerata*)
Red Saddlebags (*Tramea onusta*)

Bell *(39)*
Common Green Darner (*Anax junius*)
Great Spreadwing (*Archilestes grandis*)
Blue-fronted Dancer (*Argia apicalis*)
Violet Dancer (*Argia fumipennis violacea*)
Kiowa Dancer (*Argia immunda*)
Powdered Dancer (*Argia moesta*)
Aztec Dancer (*Argia nahuana*)
Springwater Dancer (*Argia plana*)
Blue-ringed Dancer (*Argia sedula*)
Dusky Dancer (*Argia translata*)
Jade Clubtail (*Arigomphus submedianus*)
Springtime Darner (*Basiaeschna janata*)
Stream Cruiser (*Didymops transversa*)
Checkered Setwing (*Dythemis fugax*)
Swift Setwing (*Dythemis velox*)
Double-striped Bluet (*Enallagma basidens*)
Familiar Bluet (*Enallagma civile*)
Dot-winged Baskettail (*Epitheca petechialis*)
Prince Baskettail (*Epitheca princeps*)
Eastern Ringtail (*Erpetogomphus designatus*)
Eastern Pondhawk (*Erythemis simplicicollis*)
Little Blue Dragonlet (*Erythrodiplax minuscula*)
Band-winged Dragonlet (*Erythrodiplax umbrata*)
Plains Clubtail (*Gomphus externus*)
Sulphur-tipped Clubtail (*Gomphus militaris*)
American Rubyspot (*Hetaerina americana*)
Smoky Rubyspot (*Hetaerina titia*)

Comanche Skimmer (*Libellula comanche*)
Widow Skimmer (*Libellula luctuosa*)
Roseate Skimmer (*Orthemis ferruginea*)
Blue Dasher (*Pachydiplax longipennis*)
Spot-winged Glider (*Pantala hymenaea*)
Eastern Amberwing (*Perithemis tenera*)
Four-striped Leaftail (*Phyllogomphoides stigmatus*)
Common Whitetail (*Plathemis lydia*)
Common Sanddragon (*Progomphus obscurus*)
Blue-faced Meadowhawk (*Sympetrum ambiguum*)
Desert Firetail (*Telebasis salva*)
Black Saddlebags (*Tramea lacerata*)

Bexar (84)
Common Green Darner (*Anax junius*)
Broad-striped Forceptail (*Aphylla angustifolia*)
Narrow-striped Forceptail (*Aphylla protracta*)
Great Spreadwing (*Archilestes grandis*)
Blue-fronted Dancer (*Argia apicalis*)
Comanche Dancer (*Argia barretti*)
Violet Dancer (*Argia fumipennis violacea*)
Kiowa Dancer (*Argia immunda*)
Powdered Dancer (*Argia moesta*)
Aztec Dancer (*Argia nahuana*)
Springwater Dancer (*Argia plana*)
Blue-ringed Dancer (*Argia sedula*)
Dusky Dancer (*Argia translata*)
Jade Clubtail (*Arigomphus submedianus*)
Springtime Darner (*Basiaeschna janata*)
Red-tailed Pennant (*Brachymesia furcata*)
Four-spotted Pennant (*Brachymesia gravida*)
Pale-faced Clubskimmer (*Brechmorhoga mendax*)
Gray-waisted Skimmer (*Cannaphila insularis funerea*)
Calico Pennant (*Celithemis elisa*)
Halloween Pennant (*Celithemis eponina*)
Banded Pennant (*Celithemis fasciata*)
Stream Cruiser (*Didymops transversa*)
Black-shouldered Spinyleg (*Dromogomphus spinosus*)
Flag-tailed Spinyleg (*Dromogomphus spoliatus*)
Checkered Setwing (*Dythemis fugax*)
Black Setwing (*Dythemis nigrescens*)
Swift Setwing (*Dythemis velox*)
Double-striped Bluet (*Enallagma basidens*)
Familiar Bluet (*Enallagma civile*)
Orange Bluet (*Enallagma signatum*)
Prince Baskettail (*Epitheca princeps*)
Eastern Ringtail (*Erpetogomphus designatus*)
Pin-tailed Pondhawk (*Erythemis plebeja*)
Eastern Pondhawk (*Erythemis simplicicollis*)
Great Pondhawk (*Erythemis vesiculosa*)
Black-winged Dragonlet (*Erythrodiplax funerea*)
Little Blue Dragonlet (*Erythrodiplax minuscula*)
Band-winged Dragonlet (*Erythrodiplax umbrata*)
Plains Clubtail (*Gomphus externus*)
Sulphur-tipped Clubtail (*Gomphus militaris*)
Cobra Clubtail (*Gomphus vastus*)
Dragonhunter (*Hagenius brevistylus*)
American Rubyspot (*Hetaerina americana*)
Smoky Rubyspot (*Hetaerina titia*)
Canyon Rubyspot (*Hetaerina vulnerata*)
Citrine Forktail (*Ischnura hastata*)
Rambur's Forktail (*Ischnura ramburii*)
Plateau Spreadwing (*Lestes alacer*)
Rainpool Spreadwing (*Lestes forficula*)
Chalky Spreadwing (*Lestes sigma*)
Comanche Skimmer (*Libellula comanche*)
Neon Skimmer (*Libellula croceipennis*)
Spangled Skimmer (*Libellula cyanea*)
Widow Skimmer (*Libellula luctuosa*)
Twelve-spotted Skimmer (*Libellula pulchella*)
Flame Skimmer (*Libellula saturata*)
Marl Pennant (*Macrodiplax balteata*)
Ivory-striped Sylph (*Macrothemis imitans*

leucozona)
Hyacinth Glider (*Miathyria marcella*)
Thornbush Dasher (*Micrathyria hagenii*)
Coral-fronted Threadtail (*Neoneura aaroni*)
Orange Shadowdragon (*Neurocordulia xanthosoma*)
Carmine Skimmer (*Orthemis discolor*)
Roseate Skimmer (*Orthemis ferruginea*)
Blue Dasher (*Pachydiplax longipennis*)
Wandering Glider (*Pantala flavescens*)
Spot-winged Glider (*Pantala hymenaea*)
Slough Amberwing (*Perithemis domitia*)
Eastern Amberwing (*Perithemis tenera*)
Five-striped Leaftail (*Phyllogomphoides albrighti*)
Four-striped Leaftail (*Phyllogomphoides stigmatus*)
Common Whitetail (*Plathemis lydia*)
Common Sanddragon (*Progomphus obscurus*)
Filigree Skimmer (*Pseudoleon superbus*)
Blue-eyed Darner (*Rhionaeschna multicolor*)
Turquoise-tipped Darner (*Rhionaeschna psilus*)
Russet-tipped Clubtail (*Stylurus plagiatus*)
Variegated Meadowhawk (*Sympetrum corruptum*)
Autumn Meadowhawk (*Sympetrum vicinum*)
Desert Firetail (*Telebasis salva*)
Striped Saddlebags (*Tramea calverti*)
Black Saddlebags (*Tramea lacerata*)
Red Saddlebags (*Tramea onusta*)

Blanco (58)
Common Green Darner (*Anax junius*)
Giant Darner (*Anax walsinghami*)
Great Spreadwing (*Archilestes grandis*)
Comanche Dancer (*Argia barretti*)
Violet Dancer (*Argia fumipennis violacea*)
Lavender Dancer (*Argia hinei*)
Kiowa Dancer (*Argia immunda*)
Leonora's Dancer (*Argia leonorae*)
Powdered Dancer (*Argia moesta*)
Aztec Dancer (*Argia nahuana*)
Springwater Dancer (*Argia plana*)
Blue-ringed Dancer (*Argia sedula*)
Dusky Dancer (*Argia translata*)
Springtime Darner (*Basiaeschna janata*)
Pale-faced Clubskimmer (*Brechmorhoga mendax*)
Banded Pennant (*Celithemis fasciata*)
Stream Cruiser (*Didymops transversa*)
Flag-tailed Spinyleg (*Dromogomphus spoliatus*)
Checkered Setwing (*Dythemis fugax*)
Swift Setwing (*Dythemis velox*)
Double-striped Bluet (*Enallagma basidens*)
Familiar Bluet (*Enallagma civile*)
Stream Bluet (*Enallagma exsulans*)
Arroyo Bluet (*Enallagma praevarum*)
Stripe-winged Baskettail (*Epitheca costalis*)
Dot-winged Baskettail (*Epitheca petechialis*)
Prince Baskettail (*Epitheca princeps*)
Eastern Ringtail (*Erpetogomphus designatus*)
Eastern Pondhawk (*Erythemis simplicicollis*)
Plateau Dragonlet (*Erythrodiplax basifusca*)
Red-faced Dragonlet (*Erythrodiplax fusca*)
Little Blue Dragonlet (*Erythrodiplax minuscula*)
Band-winged Dragonlet (*Erythrodiplax umbrata*)
Pronghorn Clubtail (*Gomphus graslinellus*)
Sulphur-tipped Clubtail (*Gomphus militaris*)
Dragonhunter (*Hagenius brevistylus*)
American Rubyspot (*Hetaerina americana*)
Smoky Rubyspot (*Hetaerina titia*)
Citrine Forktail (*Ischnura hastata*)
Plateau Spreadwing (*Lestes alacer*)
Comanche Skimmer (*Libellula comanche*)
Neon Skimmer (*Libellula croceipennis*)
Widow Skimmer (*Libellula luctuosa*)
Twelve-spotted Skimmer (*Libellula pulchella*)
Bronzed River Cruiser (*Macromia annulata*)

Thornbush Dasher (*Micrathyria hagenii*)
Coral-fronted Threadtail (*Neoneura aaroni*)
Roseate Skimmer (*Orthemis ferruginea*)
Blue Dasher (*Pachydiplax longipennis*)
Spot-winged Glider (*Pantala hymenaea*)
Five-striped Leaftail (*Phyllogomphoides albrighti*)
Four-striped Leaftail (*Phyllogomphoides stigmatus*)
Common Whitetail (*Plathemis lydia*)
Common Sanddragon (*Progomphus obscurus*)
Variegated Meadowhawk (*Sympetrum corruptum*)
Autumn Meadowhawk (*Sympetrum vicinum*)
Desert Firetail (*Telebasis salva*)
Red Saddlebags (*Tramea onusta*)

Borden (9)
Checkered Setwing (*Dythemis fugax*)
Familiar Bluet (*Enallagma civile*)
Prince Baskettail (*Epitheca princeps*)
Sulphur-tipped Clubtail (*Gomphus militaris*)
Citrine Forktail (*Ischnura hastata*)
Plateau Spreadwing (*Lestes alacer*)
Blue Dasher (*Pachydiplax longipennis*)
Spot-winged Glider (*Pantala hymenaea*)
Eastern Amberwing (*Perithemis tenera*)

Bosque (41)
Common Green Darner (*Anax junius*)
Blue-fronted Dancer (*Argia apicalis*)
Violet Dancer (*Argia fumipennis violacea*)
Kiowa Dancer (*Argia immunda*)
Powdered Dancer (*Argia moesta*)
Aztec Dancer (*Argia nahuana*)
Springwater Dancer (*Argia plana*)
Blue-ringed Dancer (*Argia sedula*)
Dusky Dancer (*Argia translata*)
Springtime Darner (*Basiaeschna janata*)
Pale-faced Clubskimmer (*Brechmorhoga mendax*)
Flag-tailed Spinyleg (*Dromogomphus spoliatus*)
Checkered Setwing (*Dythemis fugax*)
Black Setwing (*Dythemis nigrescens*)
Swift Setwing (*Dythemis velox*)
Double-striped Bluet (*Enallagma basidens*)
Familiar Bluet (*Enallagma civile*)
Stream Bluet (*Enallagma exsulans*)
Dot-winged Baskettail (*Epitheca petechialis*)
Prince Baskettail (*Epitheca princeps*)
Eastern Ringtail (*Erpetogomphus designatus*)
Eastern Pondhawk (*Erythemis simplicicollis*)
Plains Clubtail (*Gomphus externus*)
Sulphur-tipped Clubtail (*Gomphus militaris*)
American Rubyspot (*Hetaerina americana*)
Smoky Rubyspot (*Hetaerina titia*)
Citrine Forktail (*Ischnura hastata*)
Plateau Spreadwing (*Lestes alacer*)
Widow Skimmer (*Libellula luctuosa*)
Twelve-spotted Skimmer (*Libellula pulchella*)
Bronzed River Cruiser (*Macromia annulata*)
Orange Shadowdragon (*Neurocordulia xanthosoma*)
Roseate Skimmer (*Orthemis ferruginea*)
Blue Dasher (*Pachydiplax longipennis*)
Eastern Amberwing (*Perithemis tenera*)
Four-striped Leaftail (*Phyllogomphoides stigmatus*)
Common Whitetail (*Plathemis lydia*)
Common Sanddragon (*Progomphus obscurus*)
Variegated Meadowhawk (*Sympetrum corruptum*)
Autumn Meadowhawk (*Sympetrum vicinum*)
Desert Firetail (*Telebasis salva*)

Bowie (32)
Common Green Darner (*Anax junius*)
Springtime Darner (*Basiaeschna janata*)
Fawn Darner (*Boyeria vinosa*)
Ebony Jewelwing (*Calopteryx maculata*)

Halloween Pennant (*Celithemis eponina*)
Black-shouldered Spinyleg (*Dromogomphus spinosus*)
Flag-tailed Spinyleg (*Dromogomphus spoliatus*)
Double-striped Bluet (*Enallagma basidens*)
Familiar Bluet (*Enallagma civile*)
Orange Bluet (*Enallagma signatum*)
Common Baskettail (*Epitheca cynosura*)
Prince Baskettail (*Epitheca princeps*)
Eastern Pondhawk (*Erythemis simplicicollis*)
Sulphur-tipped Clubtail (*Gomphus militaris*)
Oklahoma Clubtail (*Gomphus oklahomensis*)
Dragonhunter (*Hagenius brevistylus*)
Citrine Forktail (*Ischnura hastata*)
Fragile Forktail (*Ischnura posita*)
Rambur's Forktail (*Ischnura ramburii*)
Blue Corporal (*Ladona deplanata*)
Swamp Spreadwing (*Lestes vigilax*)
Widow Skimmer (*Libellula luctuosa*)
Roseate Skimmer (*Orthemis ferruginea*)
Blue Dasher (*Pachydiplax longipennis*)
Wandering Glider (*Pantala flavescens*)
Spot-winged Glider (*Pantala hymenaea*)
Eastern Amberwing (*Perithemis tenera*)
Common Whitetail (*Plathemis lydia*)
Common Sanddragon (*Progomphus obscurus*)
Variegated Meadowhawk (*Sympetrum corruptum*)
Black Saddlebags (*Tramea lacerata*)
Red Saddlebags (*Tramea onusta*)

Brazoria (30)
Common Green Darner (*Anax junius*)
Blue-fronted Dancer (*Argia apicalis*)
Blue-ringed Dancer (*Argia sedula*)
Blue-tipped Dancer (*Argia tibialis*)
Stillwater Clubtail (*Arigomphus lentulus*)
Jade Clubtail (*Arigomphus submedianus*)
Four-spotted Pennant (*Brachymesia gravida*)
Calico Pennant (*Celithemis elisa*)
Halloween Pennant (*Celithemis eponina*)
Banded Pennant (*Celithemis fasciata*)
Flag-tailed Spinyleg (*Dromogomphus spoliatus*)
Double-striped Bluet (*Enallagma basidens*)
Familiar Bluet (*Enallagma civile*)
Dot-winged Baskettail (*Epitheca petechialis*)
Eastern Pondhawk (*Erythemis simplicicollis*)
Sulphur-tipped Clubtail (*Gomphus militaris*)
Citrine Forktail (*Ischnura hastata*)
Rambur's Forktail (*Ischnura ramburii*)
Widow Skimmer (*Libellula luctuosa*)
Needham's Skimmer (*Libellula needhami*)
Roseate Skimmer (*Orthemis ferruginea*)
Blue Dasher (*Pachydiplax longipennis*)
Wandering Glider (*Pantala flavescens*)
Spot-winged Glider (*Pantala hymenaea*)
Eastern Amberwing (*Perithemis tenera*)
Common Whitetail (*Plathemis lydia*)
Blue-faced Meadowhawk (*Sympetrum ambiguum*)
Variegated Meadowhawk (*Sympetrum corruptum*)
Carolina Saddlebags (*Tramea carolina*)
Black Saddlebags (*Tramea lacerata*)

Brazos (72)
Common Green Darner (*Anax junius*)
Blue-fronted Dancer (*Argia apicalis*)
Kiowa Dancer (*Argia immunda*)
Powdered Dancer (*Argia moesta*)
Springwater Dancer (*Argia plana*)
Blue-ringed Dancer (*Argia sedula*)
Blue-tipped Dancer (*Argia tibialis*)
Dusky Dancer (*Argia translata*)
Fawn Darner (*Boyeria vinosa*)
Ebony Jewelwing (*Calopteryx maculata*)
Calico Pennant (*Celithemis elisa*)
Halloween Pennant (*Celithemis eponina*)
Banded Pennant (*Celithemis fasciata*)

Stream Cruiser (*Didymops transversa*)
Black-shouldered Spinyleg (*Dromogomphus spinosus*)
Flag-tailed Spinyleg (*Dromogomphus spoliatus*)
Checkered Setwing (*Dythemis fugax*)
Swift Setwing (*Dythemis velox*)
Double-striped Bluet (*Enallagma basidens*)
Familiar Bluet (*Enallagma civile*)
Turquoise Bluet (*Enallagma divagans*)
Stream Bluet (*Enallagma exsulans*)
Skimming Bluet (*Enallagma geminatum*)
Orange Bluet (*Enallagma signatum*)
Swamp Darner (*Epiaeschna heros*)
Stripe-winged Baskettail (*Epitheca costalis*)
Common Baskettail (*Epitheca cynosura*)
Dot-winged Baskettail (*Epitheca petechialis*)
Prince Baskettail (*Epitheca princeps*)
Mantled Baskettail (*Epitheca semiaquea*)
Eastern Ringtail (*Erpetogomphus designatus*)
Eastern Pondhawk (*Erythemis simplicicollis*)
Little Blue Dragonlet (*Erythrodiplax minuscula*)
Band-winged Dragonlet (*Erythrodiplax umbrata*)
Plains Clubtail (*Gomphus externus*)
Sulphur-tipped Clubtail (*Gomphus militaris*)
Gulf Coast Clubtail (*Gomphus modestus*)
Oklahoma Clubtail (*Gomphus oklahomensis*)
Cobra Clubtail (*Gomphus vastus*)
Dragonhunter (*Hagenius brevistylus*)
American Rubyspot (*Hetaerina americana*)
Smoky Rubyspot (*Hetaerina titia*)
Citrine Forktail (*Ischnura hastata*)
Rambur's Forktail (*Ischnura ramburii*)
Blue Corporal (*Ladona deplanata*)
Plateau Spreadwing (*Lestes alacer*)
Southern Spreadwing (*Lestes australis*)
Rainpool Spreadwing (*Lestes forficula*)
Comanche Skimmer (*Libellula comanche*)
Slaty Skimmer (*Libellula incesta*)
Widow Skimmer (*Libellula luctuosa*)
Twelve-spotted Skimmer (*Libellula pulchella*)
Great Blue Skimmer (*Libellula vibrans*)
Royal River Cruiser (*Macromia taeniolata*)
Cyrano Darner (*Nasiaeschna pentacantha*)
Smoky Shadowdragon (*Neurocordulia molesta*)
Roseate Skimmer (*Orthemis ferruginea*)
Blue Dasher (*Pachydiplax longipennis*)
Wandering Glider (*Pantala flavescens*)
Spot-winged Glider (*Pantala hymenaea*)
Eastern Amberwing (*Perithemis tenera*)
Four-striped Leaftail (*Phyllogomphoides stigmatus*)
Common Whitetail (*Plathemis lydia*)
Common Sanddragon (*Progomphus obscurus*)
Russet-tipped Clubtail (*Stylurus plagiatus*)
Blue-faced Meadowhawk (*Sympetrum ambiguum*)
Variegated Meadowhawk (*Sympetrum corruptum*)
Autumn Meadowhawk (*Sympetrum vicinum*)
Desert Firetail (*Telebasis salva*)
Carolina Saddlebags (*Tramea carolina*)
Black Saddlebags (*Tramea lacerata*)
Red Saddlebags (*Tramea onusta*)

Brewster (66)
Persephone's Darner (*Aeshna persephone*)
Common Green Darner (*Anax junius*)
Giant Darner (*Anax walsinghami*)
Great Spreadwing (*Archilestes grandis*)
Comanche Dancer (*Argia barretti*)
Violet Dancer (*Argia fumipennis violacea*)
Lavender Dancer (*Argia hinei*)
Kiowa Dancer (*Argia immunda*)
Leonora's Dancer (*Argia leonorae*)
Sooty Dancer (*Argia lugens*)
Powdered Dancer (*Argia moesta*)
Apache Dancer (*Argia munda*)
Aztec Dancer (*Argia nahuana*)

Springwater Dancer (*Argia plana*)
Blue-ringed Dancer (*Argia sedula*)
Pale-faced Clubskimmer (*Brechmorhoga mendax*)
Mayan Setwing (*Dythemis maya*)
Black Setwing (*Dythemis nigrescens*)
Swift Setwing (*Dythemis velox*)
Familiar Bluet (*Enallagma civile*)
Stream Bluet (*Enallagma exsulans*)
Arroyo Bluet (*Enallagma praevarum*)
Orange Bluet (*Enallagma signatum*)
White-belted Ringtail (*Erpetogomphus compositus*)
Eastern Ringtail (*Erpetogomphus designatus*)
Serpent Ringtail (*Erpetogomphus lampropeltis natrix*)
Eastern Pondhawk (*Erythemis simplicicollis*)
Plateau Dragonlet (*Erythrodiplax basifusca*)
Band-winged Dragonlet (*Erythrodiplax umbrata*)
Sulphur-tipped Clubtail (*Gomphus militaris*)
Painted Damsel (*Hesperagrion heterodoxum*)
American Rubyspot (*Hetaerina americana*)
Plains Forktail (*Ischnura damula*)
Mexican Forktail (*Ischnura demorsa*)
Black-fronted Forktail (*Ischnura denticollis*)
Citrine Forktail (*Ischnura hastata*)
Fragile Forktail (*Ischnura posita*)
Rambur's Forktail (*Ischnura ramburii*)
Plateau Spreadwing (*Lestes alacer*)
Comanche Skimmer (*Libellula comanche*)
Neon Skimmer (*Libellula croceipennis*)
Widow Skimmer (*Libellula luctuosa*)
Flame Skimmer (*Libellula saturata*)
Bronzed River Cruiser (*Macromia annulata*)
Thornbush Dasher (*Micrathyria hagenii*)
Roseate Skimmer (*Orthemis ferruginea*)
Blue Dasher (*Pachydiplax longipennis*)
Red Rock Skimmer (*Paltothemis lineatipes*)
Wandering Glider (*Pantala flavescens*)
Slough Amberwing (*Perithemis domitia*)
Eastern Amberwing (*Perithemis tenera*)
Five-striped Leaftail (*Phyllogomphoides albrighti*)
Four-striped Leaftail (*Phyllogomphoides stigmatus*)
Common Whitetail (*Plathemis lydia*)
Desert Whitetail (*Plathemis subornata*)
Gray Sanddragon (*Progomphus borealis*)
Filigree Skimmer (*Pseudoleon superbus*)
Blue-eyed Darner (*Rhionaeschna multicolor*)
Brimstone Clubtail (*Stylurus intricatus*)
Variegated Meadowhawk (*Sympetrum corruptum*)
Cardinal Meadowhawk (*Sympetrum illotum*)
Autumn Meadowhawk (*Sympetrum vicinum*)
Desert Firetail (*Telebasis salva*)
Antillean Saddlebags (*Tramea insularis*)
Black Saddlebags (*Tramea lacerata*)
Red Saddlebags (*Tramea onusta*)

Briscoe (20)
Common Green Darner (*Anax junius*)
Blue-fronted Dancer (*Argia apicalis*)
Sooty Dancer (*Argia lugens*)
Powdered Dancer (*Argia moesta*)
Springwater Dancer (*Argia plana*)
Double-striped Bluet (*Enallagma basidens*)
Familiar Bluet (*Enallagma civile*)
Dot-winged Baskettail (*Epitheca petechialis*)
Eastern Pondhawk (*Erythemis simplicicollis*)
American Rubyspot (*Hetaerina americana*)
Lyre-tipped Spreadwing (*Lestes unguiculatus*)
Widow Skimmer (*Libellula luctuosa*)
Flame Skimmer (*Libellula saturata*)
Blue Dasher (*Pachydiplax longipennis*)
Wandering Glider (*Pantala flavescens*)
Eastern Amberwing (*Perithemis tenera*)
Common Whitetail (*Plathemis lydia*)

Gray Sanddragon (*Progomphus borealis*)
Common Sanddragon (*Progomphus obscurus*)
Black Saddlebags (*Tramea lacerata*)

Brooks *(19)*
Common Green Darner (*Anax junius*)
Leonora's Dancer (*Argia leonorae*)
Red-tailed Pennant (*Brachymesia furcata*)
Four-spotted Pennant (*Brachymesia gravida*)
Familiar Bluet (*Enallagma civile*)
Eastern Pondhawk (*Erythemis simplicicollis*)
Band-winged Dragonlet (*Erythrodiplax umbrata*)
Citrine Forktail (*Ischnura hastata*)
Rambur's Forktail (*Ischnura ramburii*)
Needham's Skimmer (*Libellula needhami*)
Thornbush Dasher (*Micrathyria hagenii*)
Roseate Skimmer (*Orthemis ferruginea*)
Blue Dasher (*Pachydiplax longipennis*)
Wandering Glider (*Pantala flavescens*)
Eastern Amberwing (*Perithemis tenera*)
Variegated Meadowhawk (*Sympetrum corruptum*)
Desert Firetail (*Telebasis salva*)
Black Saddlebags (*Tramea lacerata*)
Red Saddlebags (*Tramea onusta*)

Brown *(17)*
Blue-fronted Dancer (*Argia apicalis*)
Kiowa Dancer (*Argia immunda*)
Powdered Dancer (*Argia moesta*)
Banded Pennant (*Celithemis fasciata*)
Flag-tailed Spinyleg (*Dromogomphus spoliatus*)
Checkered Setwing (*Dythemis fugax*)
Swift Setwing (*Dythemis velox*)
Double-striped Bluet (*Enallagma basidens*)
Eastern Ringtail (*Erpetogomphus designatus*)
Eastern Pondhawk (*Erythemis simplicicollis*)
Sulphur-tipped Clubtail (*Gomphus militaris*)
American Rubyspot (*Hetaerina americana*)
Citrine Forktail (*Ischnura hastata*)
Widow Skimmer (*Libellula luctuosa*)
Blue Dasher (*Pachydiplax longipennis*)
Eastern Amberwing (*Perithemis tenera*)
Four-striped Leaftail (*Phyllogomphoides stigmatus*)

Burleson *(15)*
Common Green Darner (*Anax junius*)
Blue-fronted Dancer (*Argia apicalis*)
Familiar Bluet (*Enallagma civile*)
Big Bluet (*Enallagma durum*)
Eastern Pondhawk (*Erythemis simplicicollis*)
Fragile Forktail (*Ischnura posita*)
Rambur's Forktail (*Ischnura ramburii*)
Twelve-spotted Skimmer (*Libellula pulchella*)
Roseate Skimmer (*Orthemis ferruginea*)
Blue Dasher (*Pachydiplax longipennis*)
Wandering Glider (*Pantala flavescens*)
Spot-winged Glider (*Pantala hymenaea*)
Eastern Amberwing (*Perithemis tenera*)
Common Whitetail (*Plathemis lydia*)
Variegated Meadowhawk (*Sympetrum corruptum*)

Burnet *(56)*
Common Green Darner (*Anax junius*)
Great Spreadwing (*Archilestes grandis*)
Blue-fronted Dancer (*Argia apicalis*)
Violet Dancer (*Argia fumipennis violacea*)
Kiowa Dancer (*Argia immunda*)
Leonora's Dancer (*Argia leonorae*)
Powdered Dancer (*Argia moesta*)
Aztec Dancer (*Argia nahuana*)
Blue-ringed Dancer (*Argia sedula*)
Dusky Dancer (*Argia translata*)
Springtime Darner (*Basiaeschna janata*)
Fawn Darner (*Boyeria vinosa*)
Four-spotted Pennant (*Brachymesia gravida*)
Pale-faced Clubskimmer (*Brechmorhoga mendax*)
Halloween Pennant (*Celithemis eponina*)

Banded Pennant (*Celithemis fasciata*)
Stream Cruiser (*Didymops transversa*)
Flag-tailed Spinyleg (*Dromogomphus spoliatus*)
Checkered Setwing (*Dythemis fugax*)
Black Setwing (*Dythemis nigrescens*)
Swift Setwing (*Dythemis velox*)
Double-striped Bluet (*Enallagma basidens*)
Familiar Bluet (*Enallagma civile*)
Orange Bluet (*Enallagma signatum*)
Dot-winged Baskettail (*Epitheca petechialis*)
Prince Baskettail (*Epitheca princeps*)
White-belted Ringtail (*Erpetogomphus compositus*)
Eastern Ringtail (*Erpetogomphus designatus*)
Eastern Pondhawk (*Erythemis simplicicollis*)
Little Blue Dragonlet (*Erythrodiplax minuscula*)
Band-winged Dragonlet (*Erythrodiplax umbrata*)
American Rubyspot (*Hetaerina americana*)
Citrine Forktail (*Ischnura hastata*)
Rambur's Forktail (*Ischnura ramburii*)
Plateau Spreadwing (*Lestes alacer*)
Comanche Skimmer (*Libellula comanche*)
Neon Skimmer (*Libellula croceipennis*)
Widow Skimmer (*Libellula luctuosa*)
Twelve-spotted Skimmer (*Libellula pulchella*)
Flame Skimmer (*Libellula saturata*)
Marl Pennant (*Macrodiplax balteata*)
Hyacinth Glider (*Miathyria marcella*)
Roseate Skimmer (*Orthemis ferruginea*)
Blue Dasher (*Pachydiplax longipennis*)
Red Rock Skimmer (*Paltothemis lineatipes*)
Wandering Glider (*Pantala flavescens*)
Spot-winged Glider (*Pantala hymenaea*)
Eastern Amberwing (*Perithemis tenera*)
Four-striped Leaftail (*Phyllogomphoides stigmatus*)
Common Whitetail (*Plathemis lydia*)
Common Sanddragon (*Progomphus obscurus*)
Variegated Meadowhawk (*Sympetrum corruptum*)
Autumn Meadowhawk (*Sympetrum vicinum*)
Desert Firetail (*Telebasis salva*)
Black Saddlebags (*Tramea lacerata*)
Red Saddlebags (*Tramea onusta*)

Caldwell *(53)*
Common Green Darner (*Anax junius*)
Broad-striped Forceptail (*Aphylla angustifolia*)
Blue-fronted Dancer (*Argia apicalis*)
Comanche Dancer (*Argia barretti*)
Kiowa Dancer (*Argia immunda*)
Powdered Dancer (*Argia moesta*)
Blue-ringed Dancer (*Argia sedula*)
Dusky Dancer (*Argia translata*)
Jade Clubtail (*Arigomphus submedianus*)
Four-spotted Pennant (*Brachymesia gravida*)
Pale-faced Clubskimmer (*Brechmorhoga mendax*)
Gray-waisted Skimmer (*Cannaphila insularis funerea*)
Halloween Pennant (*Celithemis eponina*)
Regal Darner (*Coryphaeschna ingens*)
Flag-tailed Spinyleg (*Dromogomphus spoliatus*)
Swift Setwing (*Dythemis velox*)
Double-striped Bluet (*Enallagma basidens*)
Familiar Bluet (*Enallagma civile*)
Stream Bluet (*Enallagma exsulans*)
Orange Bluet (*Enallagma signatum*)
Dot-winged Baskettail (*Epitheca petechialis*)
Prince Baskettail (*Epitheca princeps*)
Eastern Ringtail (*Erpetogomphus designatus*)
Blue-faced Ringtail (*Erpetogomphus eutainia*)
Eastern Pondhawk (*Erythemis simplicicollis*)
Little Blue Dragonlet (*Erythrodiplax minuscula*)
Band-winged Dragonlet (*Erythrodiplax umbrata*)
Plains Clubtail (*Gomphus externus*)
Sulphur-tipped Clubtail (*Gomphus militaris*)
Cobra Clubtail (*Gomphus vastus*)

Dragonhunter (*Hagenius brevistylus*)
American Rubyspot (*Hetaerina americana*)
Smoky Rubyspot (*Hetaerina titia*)
Rambur's Forktail (*Ischnura ramburii*)
Plateau Spreadwing (*Lestes alacer*)
Widow Skimmer (*Libellula luctuosa*)
Bronzed River Cruiser (*Macromia annulata*)
Ivory-striped Sylph (*Macrothemis imitans leucozona*)
Jade-striped Sylph (*Macrothemis inequiunguis*)
Thornbush Dasher (*Micrathyria hagenii*)
Coral-fronted Threadtail (*Neoneura aaroni*)
Orange Shadowdragon (*Neurocordulia xanthosoma*)
Carmine Skimmer (*Orthemis discolor*)
Roseate Skimmer (*Orthemis ferruginea*)
Blue Dasher (*Pachydiplax longipennis*)
Eastern Amberwing (*Perithemis tenera*)
Five-striped Leaftail (*Phyllogomphoides albrighti*)
Common Whitetail (*Plathemis lydia*)
Common Sanddragon (*Progomphus obscurus*)
Russet-tipped Clubtail (*Stylurus plagiatus*)
Desert Firetail (*Telebasis salva*)
Black Saddlebags (*Tramea lacerata*)
Red Saddlebags (*Tramea onusta*)

Calhoun *(15)*
Common Green Darner (*Anax junius*)
Broad-striped Forceptail (*Aphylla angustifolia*)
Blue-fronted Dancer (*Argia apicalis*)
Eastern Pondhawk (*Erythemis simplicicollis*)
Seaside Dragonlet (*Erythrodiplax berenice*)
Smoky Rubyspot (*Hetaerina titia*)
Rambur's Forktail (*Ischnura ramburii*)
Hyacinth Glider (*Miathyria marcella*)
Roseate Skimmer (*Orthemis ferruginea*)
Blue Dasher (*Pachydiplax longipennis*)
Wandering Glider (*Pantala flavescens*)
Spot-winged Glider (*Pantala hymenaea*)
Common Whitetail (*Plathemis lydia*)
Russet-tipped Clubtail (*Stylurus plagiatus*)
Black Saddlebags (*Tramea lacerata*)

Callahan *(3)*
Wandering Glider (*Pantala flavescens*)
Variegated Meadowhawk (*Sympetrum corruptum*)
Red Saddlebags (*Tramea onusta*)

Cameron *(63)*
Mexican Wedgetail (*Acanthagrion quadratum*)
Common Green Darner (*Anax junius*)
Narrow-striped Forceptail (*Aphylla protracta*)
Blue-fronted Dancer (*Argia apicalis*)
Kiowa Dancer (*Argia immunda*)
Springwater Dancer (*Argia plana*)
Golden-winged Dancer (*Argia rhoadsi*)
Blue-ringed Dancer (*Argia sedula*)
Dusky Dancer (*Argia translata*)
Red-tailed Pennant (*Brachymesia furcata*)
Four-spotted Pennant (*Brachymesia gravida*)
Gray-waisted Skimmer (*Cannaphila insularis funerea*)
Halloween Pennant (*Celithemis eponina*)
Blue-faced Darner (*Coryphaeschna adnexa*)
Regal Darner (*Coryphaeschna ingens*)
Flag-tailed Spinyleg (*Dromogomphus spoliatus*)
Black Setwing (*Dythemis nigrescens*)
Double-striped Bluet (*Enallagma basidens*)
Familiar Bluet (*Enallagma civile*)
Neotropical Bluet (*Enallagma novaehispaniae*)
Orange Bluet (*Enallagma signatum*)
Prince Baskettail (*Epitheca princeps*)
Black Pondhawk (*Erythemis attala*)
Claret Pondhawk (*Erythemis mithroides*)
Pin-tailed Pondhawk (*Erythemis plebeja*)
Eastern Pondhawk (*Erythemis simplicicollis*)
Great Pondhawk (*Erythemis vesiculosa*)
Seaside Dragonlet (*Erythrodiplax berenice*)

Little Blue Dragonlet (*Erythrodiplax minuscula*)
Band-winged Dragonlet (*Erythrodiplax umbrata*)
Tamaulipan Clubtail (*Gomphus gonzalezi*)
Sulphur-tipped Clubtail (*Gomphus militaris*)
Bar-sided Darner (*Gynacantha mexicana*)
Smoky Rubyspot (*Hetaerina titia*)
Citrine Forktail (*Ischnura hastata*)
Rambur's Forktail (*Ischnura ramburii*)
Cream-tipped Swampdamsel (*Leptobasis melinogaster*)
Plateau Spreadwing (*Lestes alacer*)
Rainpool Spreadwing (*Lestes forficula*)
Chalky Spreadwing (*Lestes sigma*)
Needham's Skimmer (*Libellula needhami*)
Marl Pennant (*Macrodiplax balteata*)
Hyacinth Glider (*Miathyria marcella*)
Spot-tailed Dasher (*Micrathyria aequalis*)
Three-striped Dasher (*Micrathyria didyma*)
Thornbush Dasher (*Micrathyria hagenii*)
Caribbean Yellowface (*Neoerythromma cultellatum*)
Amelia's Threadtail (*Neoneura amelia*)
Roseate Skimmer (*Orthemis ferruginea*)
Blue Dasher (*Pachydiplax longipennis*)
Wandering Glider (*Pantala flavescens*)
Spot-winged Glider (*Pantala hymenaea*)
Slough Amberwing (*Perithemis domitia*)
Eastern Amberwing (*Perithemis tenera*)
Ringed Forceptail (*Phyllocycla breviphylla*)
Filigree Skimmer (*Pseudoleon superbus*)
Turquoise-tipped Darner (*Rhionaeschna psilus*)
Variegated Meadowhawk (*Sympetrum corruptum*)
Desert Firetail (*Telebasis salva*)
Evening Skimmer (*Tholymis citrina*)
Striped Saddlebags (*Tramea calverti*)
Black Saddlebags (*Tramea lacerata*)
Red Saddlebags (*Tramea onusta*)

Camp (15)
Common Green Darner (*Anax junius*)
Blue-tipped Dancer (*Argia tibialis*)
Four-spotted Pennant (*Brachymesia gravida*)
Halloween Pennant (*Celithemis eponina*)
Swamp Darner (*Epiaeschna heros*)
Eastern Pondhawk (*Erythemis simplicicollis*)
Rambur's Forktail (*Ischnura ramburii*)
Slaty Skimmer (*Libellula incesta*)
Widow Skimmer (*Libellula luctuosa*)
Great Blue Skimmer (*Libellula vibrans*)
Blue Dasher (*Pachydiplax longipennis*)
Wandering Glider (*Pantala flavescens*)
Spot-winged Glider (*Pantala hymenaea*)
Eastern Amberwing (*Perithemis tenera*)
Black Saddlebags (*Tramea lacerata*)

Carson (6)
Common Green Darner (*Anax junius*)
Familiar Bluet (*Enallagma civile*)
Black-fronted Forktail (*Ischnura denticollis*)
Twelve-spotted Skimmer (*Libellula pulchella*)
Blue-eyed Darner (*Rhionaeschna multicolor*)
Variegated Meadowhawk (*Sympetrum corruptum*)

Cass (25)
Common Green Darner (*Anax junius*)
Jade Clubtail (*Arigomphus submedianus*)
Four-spotted Pennant (*Brachymesia gravida*)
Ebony Jewelwing (*Calopteryx maculata*)
Calico Pennant (*Celithemis elisa*)
Vesper Bluet (*Enallagma vesperum*)
Prince Baskettail (*Epitheca princeps*)
Eastern Pondhawk (*Erythemis simplicicollis*)
Dragonhunter (*Hagenius brevistylus*)
Citrine Forktail (*Ischnura hastata*)
Fragile Forktail (*Ischnura posita*)
Rambur's Forktail (*Ischnura ramburii*)
Golden-winged Skimmer (*Libellula auripennis*)
Spangled Skimmer (*Libellula cyanea*)

Yellow-sided Skimmer (*Libellula flavida*)
Slaty Skimmer (*Libellula incesta*)
Roseate Skimmer (*Orthemis ferruginea*)
Blue Dasher (*Pachydiplax longipennis*)
Wandering Glider (*Pantala flavescens*)
Spot-winged Glider (*Pantala hymenaea*)
Eastern Amberwing (*Perithemis tenera*)
Common Whitetail (*Plathemis lydia*)
Blue-faced Meadowhawk (*Sympetrum ambiguum*)
Variegated Meadowhawk (*Sympetrum corruptum*)
Black Saddlebags (*Tramea lacerata*)

Chambers (34)
Common Green Darner (*Anax junius*)
Blue-fronted Dancer (*Argia apicalis*)
Blue-ringed Dancer (*Argia sedula*)
Blue-tipped Dancer (*Argia tibialis*)
Bayou Clubtail (*Arigomphus maxwelli*)
Jade Clubtail (*Arigomphus submedianus*)
Four-spotted Pennant (*Brachymesia gravida*)
Halloween Pennant (*Celithemis eponina*)
Familiar Bluet (*Enallagma civile*)
Big Bluet (*Enallagma durum*)
Orange Bluet (*Enallagma signatum*)
Vesper Bluet (*Enallagma vesperum*)
Stripe-winged Baskettail (*Epitheca costalis*)
Common Baskettail (*Epitheca cynosura*)
Prince Baskettail (*Epitheca princeps*)
Eastern Pondhawk (*Erythemis simplicicollis*)
Great Pondhawk (*Erythemis vesiculosa*)
Seaside Dragonlet (*Erythrodiplax berenice*)
Little Blue Dragonlet (*Erythrodiplax minuscula*)
Band-winged Dragonlet (*Erythrodiplax umbrata*)
Oklahoma Clubtail (*Gomphus oklahomensis*)
Citrine Forktail (*Ischnura hastata*)
Rambur's Forktail (*Ischnura ramburii*)
Needham's Skimmer (*Libellula needhami*)
Hyacinth Glider (*Miathyria marcella*)
Roseate Skimmer (*Orthemis ferruginea*)
Blue Dasher (*Pachydiplax longipennis*)
Wandering Glider (*Pantala flavescens*)
Eastern Amberwing (*Perithemis tenera*)
Common Whitetail (*Plathemis lydia*)
Variegated Meadowhawk (*Sympetrum corruptum*)
Carolina Saddlebags (*Tramea carolina*)
Black Saddlebags (*Tramea lacerata*)
Red Saddlebags (*Tramea onusta*)

Cherokee (37)
Common Green Darner (*Anax junius*)
Blue-fronted Dancer (*Argia apicalis*)
Seepage Dancer (*Argia bipunctulata*)
Powdered Dancer (*Argia moesta*)
Blue-ringed Dancer (*Argia sedula*)
Blue-tipped Dancer (*Argia tibialis*)
Ebony Jewelwing (*Calopteryx maculata*)
Halloween Pennant (*Celithemis eponina*)
Stream Cruiser (*Didymops transversa*)
Black-shouldered Spinyleg (*Dromogomphus spinosus*)
Checkered Setwing (*Dythemis fugax*)
Double-striped Bluet (*Enallagma basidens*)
Familiar Bluet (*Enallagma civile*)
Swamp Darner (*Epiaeschna heros*)
Eastern Pondhawk (*Erythemis simplicicollis*)
Little Blue Dragonlet (*Erythrodiplax minuscula*)
Sulphur-tipped Clubtail (*Gomphus militaris*)
Cobra Clubtail (*Gomphus vastus*)
American Rubyspot (*Hetaerina americana*)
Smoky Rubyspot (*Hetaerina titia*)
Citrine Forktail (*Ischnura hastata*)
Rambur's Forktail (*Ischnura ramburii*)
Slaty Skimmer (*Libellula incesta*)
Widow Skimmer (*Libellula luctuosa*)
Twelve-spotted Skimmer (*Libellula pulchella*)
Great Blue Skimmer (*Libellula vibrans*)

Royal River Cruiser (*Macromia taeniolata*)
Roseate Skimmer (*Orthemis ferruginea*)
Blue Dasher (*Pachydiplax longipennis*)
Wandering Glider (*Pantala flavescens*)
Spot-winged Glider (*Pantala hymenaea*)
Eastern Amberwing (*Perithemis tenera*)
Common Whitetail (*Plathemis lydia*)
Mocha Emerald (*Somatochlora linearis*)
Blue-faced Meadowhawk (*Sympetrum ambiguum*)
Black Saddlebags (*Tramea lacerata*)
Red Saddlebags (*Tramea onusta*)

Childress (6)
Powdered Dancer (*Argia moesta*)
Halloween Pennant (*Celithemis eponina*)
Familiar Bluet (*Enallagma civile*)
American Rubyspot (*Hetaerina americana*)
Twelve-spotted Skimmer (*Libellula pulchella*)
Variegated Meadowhawk (*Sympetrum corruptum*)

Clay (15)
Common Green Darner (*Anax junius*)
Blue-fronted Dancer (*Argia apicalis*)
Powdered Dancer (*Argia moesta*)
Four-spotted Pennant (*Brachymesia gravida*)
Halloween Pennant (*Celithemis eponina*)
Familiar Bluet (*Enallagma civile*)
Eastern Pondhawk (*Erythemis simplicicollis*)
Rambur's Forktail (*Ischnura ramburii*)
Widow Skimmer (*Libellula luctuosa*)
Blue Dasher (*Pachydiplax longipennis*)
Eastern Amberwing (*Perithemis tenera*)
Common Whitetail (*Plathemis lydia*)
Variegated Meadowhawk (*Sympetrum corruptum*)
Black Saddlebags (*Tramea lacerata*)
Red Saddlebags (*Tramea onusta*)

Coke (12)
Flag-tailed Spinyleg (*Dromogomphus spoliatus*)
Double-striped Bluet (*Enallagma basidens*)
Familiar Bluet (*Enallagma civile*)
Eastern Pondhawk (*Erythemis simplicicollis*)
Comanche Skimmer (*Libellula comanche*)
Blue Dasher (*Pachydiplax longipennis*)
Spot-winged Glider (*Pantala hymenaea*)
Eastern Amberwing (*Perithemis tenera*)
Four-striped Leaftail (*Phyllogomphoides stigmatus*)
Variegated Meadowhawk (*Sympetrum corruptum*)
Black Saddlebags (*Tramea lacerata*)
Red Saddlebags (*Tramea onusta*)

Coleman (16)
Aztec Dancer (*Argia nahuana*)
Flag-tailed Spinyleg (*Dromogomphus spoliatus*)
Prince Baskettail (*Epitheca princeps*)
Eastern Ringtail (*Erpetogomphus designatus*)
Eastern Pondhawk (*Erythemis simplicicollis*)
Sulphur-tipped Clubtail (*Gomphus militaris*)
American Rubyspot (*Hetaerina americana*)
Comanche Skimmer (*Libellula comanche*)
Slaty Skimmer (*Libellula incesta*)
Widow Skimmer (*Libellula luctuosa*)
Blue Dasher (*Pachydiplax longipennis*)
Spot-winged Glider (*Pantala hymenaea*)
Eastern Amberwing (*Perithemis tenera*)
Common Whitetail (*Plathemis lydia*)
Desert Firetail (*Telebasis salva*)
Black Saddlebags (*Tramea lacerata*)

Collin (76)
Common Green Darner (*Anax junius*)
Comet Darner (*Anax longipes*)
Great Spreadwing (*Archilestes grandis*)
Blue-fronted Dancer (*Argia apicalis*)
Violet Dancer (*Argia fumipennis violacea*)
Kiowa Dancer (*Argia immunda*)
Powdered Dancer (*Argia moesta*)

Aztec Dancer (*Argia nahuana*)
Springwater Dancer (*Argia plana*)
Blue-ringed Dancer (*Argia sedula*)
Blue-tipped Dancer (*Argia tibialis*)
Dusky Dancer (*Argia translata*)
Stillwater Clubtail (*Arigomphus lentulus*)
Jade Clubtail (*Arigomphus submedianus*)
Springtime Darner (*Basiaeschna janata*)
Four-spotted Pennant (*Brachymesia gravida*)
Pale-faced Clubskimmer (*Brechmorhoga mendax*)
Ebony Jewelwing (*Calopteryx maculata*)
Calico Pennant (*Celithemis elisa*)
Halloween Pennant (*Celithemis eponina*)
Banded Pennant (*Celithemis fasciata*)
Stream Cruiser (*Didymops transversa*)
Black-shouldered Spinyleg (*Dromogomphus spinosus*)
Flag-tailed Spinyleg (*Dromogomphus spoliatus*)
Checkered Setwing (*Dythemis fugax*)
Swift Setwing (*Dythemis velox*)
Azure Bluet (*Enallagma aspersum*)
Double-striped Bluet (*Enallagma basidens*)
Familiar Bluet (*Enallagma civile*)
Atlantic Bluet (*Enallagma doubledayi*)
Stream Bluet (*Enallagma exsulans*)
Skimming Bluet (*Enallagma geminatum*)
Orange Bluet (*Enallagma signatum*)
Stripe-winged Basketail (*Epitheca costalis*)
Dot-winged Baskettail (*Epitheca petechialis*)
Prince Baskettail (*Epitheca princeps*)
Eastern Ringtail (*Erpetogomphus designatus*)
Eastern Pondhawk (*Erythemis simplicicollis*)
Great Pondhawk (*Erythemis vesiculosa*)
Little Blue Dragonlet (*Erythrodiplax minuscula*)
Band-winged Dragonlet (*Erythrodiplax umbrata*)
Plains Clubtail (*Gomphus externus*)
Sulphur-tipped Clubtail (*Gomphus militaris*)
American Rubyspot (*Hetaerina americana*)
Smoky Rubyspot (*Hetaerina titia*)
Citrine Forktail (*Ischnura hastata*)
Fragile Forktail (*Ischnura posita*)
Rambur's Forktail (*Ischnura ramburii*)
Blue Corporal (*Ladona deplanata*)
Plateau Spreadwing (*Lestes alacer*)
Comanche Skimmer (*Libellula comanche*)
Neon Skimmer (*Libellula croceipennis*)
Yellow-sided Skimmer (*Libellula flavida*)
Slaty Skimmer (*Libellula incesta*)
Widow Skimmer (*Libellula luctuosa*)
Twelve-spotted Skimmer (*Libellula pulchella*)
Painted Skimmer (*Libellula semifasciata*)
Great Blue Skimmer (*Libellula vibrans*)
Georgia River Cruiser (*Macromia illinoiensis georgina*)
Royal River Cruiser (*Macromia taeniolata*)
Hyacinth Glider (*Miathyria marcella*)
Cyrano Darner (*Nasiaeschna pentacantha*)
Orange Shadowdragon (*Neurocordulia xanthosoma*)
Roseate Skimmer (*Orthemis ferruginea*)
Blue Dasher (*Pachydiplax longipennis*)
Wandering Glider (*Pantala flavescens*)
Spot-winged Glider (*Pantala hymenaea*)
Eastern Amberwing (*Perithemis tenera*)
Four-striped Leaftail (*Phyllogomphoides stigmatus*)
Common Whitetail (*Plathemis lydia*)
Blue-faced Meadowhawk (*Sympetrum ambiguum*)
Variegated Meadowhawk (*Sympetrum corruptum*)
Autumn Meadowhawk (*Sympetrum vicinum*)
Desert Firetail (*Telebasis salva*)
Black Saddlebags (*Tramea lacerata*)
Red Saddlebags (*Tramea onusta*)

Collingsworth *(10)*

Powdered Dancer (*Argia moesta*)
Aztec Dancer (*Argia nahuana*)
Blue-ringed Dancer (*Argia sedula*)
Checkered Setwing (*Dythemis fugax*)
Stream Bluet (*Enallagma exsulans*)
Widow Skimmer (*Libellula luctuosa*)
Flame Skimmer (*Libellula saturata*)
Variegated Meadowhawk (*Sympetrum corruptum*)
Black Saddlebags (*Tramea lacerata*)
Red Saddlebags (*Tramea onusta*)

Colorado *(39)*

Common Green Darner (*Anax junius*)
Blue-fronted Dancer (*Argia apicalis*)
Blue-ringed Dancer (*Argia sedula*)
Stillwater Clubtail (*Arigomphus lentulus*)
Calico Pennant (*Celithemis elisa*)
Halloween Pennant (*Celithemis eponina*)
Banded Pennant (*Celithemis fasciata*)
Flag-tailed Spinyleg (*Dromogomphus spoliatus*)
Checkered Setwing (*Dythemis fugax*)
Black Setwing (*Dythemis nigrescens*)
Swift Setwing (*Dythemis velox*)
Double-striped Bluet (*Enallagma basidens*)
Familiar Bluet (*Enallagma civile*)
Orange Bluet (*Enallagma signatum*)
Swamp Darner (*Epiaeschna heros*)
Dot-winged Baskettail (*Epitheca petechialis*)
Prince Baskettail (*Epitheca princeps*)
Eastern Ringtail (*Erpetogomphus designatus*)
Eastern Pondhawk (*Erythemis simplicicollis*)
Little Blue Dragonlet (*Erythrodiplax minuscula*)
Band-winged Dragonlet (*Erythrodiplax umbrata*)
Plains Clubtail (*Gomphus externus*)
Sulphur-tipped Clubtail (*Gomphus militaris*)
American Rubyspot (*Hetaerina americana*)
Smoky Rubyspot (*Hetaerina titia*)
Citrine Forktail (*Ischnura hastata*)
Rambur's Forktail (*Ischnura ramburii*)
Slaty Skimmer (*Libellula incesta*)
Widow Skimmer (*Libellula luctuosa*)
Royal River Cruiser (*Macromia taeniolata*)
Roseate Skimmer (*Orthemis ferruginea*)
Blue Dasher (*Pachydiplax longipennis*)
Wandering Glider (*Pantala flavescens*)
Spot-winged Glider (*Pantala hymenaea*)
Eastern Amberwing (*Perithemis tenera*)
Four-striped Leaftail (*Phyllogomphoides stigmatus*)
Common Whitetail (*Plathemis lydia*)
Russet-tipped Clubtail (*Stylurus plagiatus*)
Black Saddlebags (*Tramea lacerata*)

Comal *(43)*

Great Spreadwing (*Archilestes grandis*)
Comanche Dancer (*Argia barretti*)
Violet Dancer (*Argia fumipennis violacea*)
Kiowa Dancer (*Argia immunda*)
Powdered Dancer (*Argia moesta*)
Aztec Dancer (*Argia nahuana*)
Springwater Dancer (*Argia plana*)
Blue-ringed Dancer (*Argia sedula*)
Dusky Dancer (*Argia translata*)
Springtime Darner (*Basiaeschna janata*)
Pale-faced Clubskimmer (*Brechmorhoga mendax*)
Black-shouldered Spinyleg (*Dromogomphus spinosus*)
Flag-tailed Spinyleg (*Dromogomphus spoliatus*)
Swift Setwing (*Dythemis velox*)
Double-striped Bluet (*Enallagma basidens*)
Stream Bluet (*Enallagma exsulans*)
Neotropical Bluet (*Enallagma novaehispaniae*)
Prince Baskettail (*Epitheca princeps*)
Eastern Ringtail (*Erpetogomphus designatus*)
Eastern Pondhawk (*Erythemis simplicicollis*)
Plains Clubtail (*Gomphus externus*)
Pronghorn Clubtail (*Gomphus graslinellus*)

Sulphur-tipped Clubtail (*Gomphus militaris*)
Dragonhunter (*Hagenius brevistylus*)
American Rubyspot (*Hetaerina americana*)
Smoky Rubyspot (*Hetaerina titia*)
Plateau Spreadwing (*Lestes alacer*)
Neon Skimmer (*Libellula croceipennis*)
Widow Skimmer (*Libellula luctuosa*)
Georgia River Cruiser (*Macromia illinoiensis georgina*)
Gilded River Cruiser (*Macromia pacifica*)
Jade-striped Sylph (*Macrothemis inequiunguis*)
Thornbush Dasher (*Micrathyria hagenii*)
Cyrano Darner (*Nasiaeschna pentacantha*)
Coral-fronted Threadtail (*Neoneura aaroni*)
Orange Shadowdragon (*Neurocordulia xanthosoma*)
Carmine Skimmer (*Orthemis discolor*)
Roseate Skimmer (*Orthemis ferruginea*)
Common Whitetail (*Plathemis lydia*)
Orange-striped Threadtail (*Protoneura cara*)
Turquoise-tipped Darner (*Rhionaeschna psilus*)
Autumn Meadowhawk (*Sympetrum vicinum*)
Desert Firetail (*Telebasis salva*)

Comanche *(13)*

Common Green Darner (*Anax junius*)
Blue-ringed Dancer (*Argia sedula*)
Calico Pennant (*Celithemis elisa*)
Checkered Setwing (*Dythemis fugax*)
Double-striped Bluet (*Enallagma basidens*)
Familiar Bluet (*Enallagma civile*)
Dot-winged Baskettail (*Epitheca petechialis*)
Eastern Pondhawk (*Erythemis simplicicollis*)
Common Whitetail (*Plathemis lydia*)
Variegated Meadowhawk (*Sympetrum corruptum*)
Desert Firetail (*Telebasis salva*)
Black Saddlebags (*Tramea lacerata*)
Red Saddlebags (*Tramea onusta*)

Concho *(16)*

Great Spreadwing (*Archilestes grandis*)
Kiowa Dancer (*Argia immunda*)
Pale-faced Clubskimmer (*Brechmorhoga mendax*)
Flag-tailed Spinyleg (*Dromogomphus spoliatus*)
Checkered Setwing (*Dythemis fugax*)
Swift Setwing (*Dythemis velox*)
Double-striped Bluet (*Enallagma basidens*)
Familiar Bluet (*Enallagma civile*)
Prince Baskettail (*Epitheca princeps*)
Eastern Ringtail (*Erpetogomphus designatus*)
Eastern Pondhawk (*Erythemis simplicicollis*)
Sulphur-tipped Clubtail (*Gomphus militaris*)
American Rubyspot (*Hetaerina americana*)
Comanche Skimmer (*Libellula comanche*)
Five-striped Leaftail (*Phyllogomphoides albrighti*)
Common Whitetail (*Plathemis lydia*)

Cooke *(24)*

Common Green Darner (*Anax junius*)
Blue-fronted Dancer (*Argia apicalis*)
Violet Dancer (*Argia fumipennis violacea*)
Powdered Dancer (*Argia moesta*)
Aztec Dancer (*Argia nahuana*)
Blue-ringed Dancer (*Argia sedula*)
Dusky Dancer (*Argia translata*)
Checkered Setwing (*Dythemis fugax*)
Double-striped Bluet (*Enallagma basidens*)
Familiar Bluet (*Enallagma civile*)
Stripe-winged Basketail (*Epitheca costalis*)
Plains Clubtail (*Gomphus externus*)
American Rubyspot (*Hetaerina americana*)
Citrine Forktail (*Ischnura hastata*)
Rambur's Forktail (*Ischnura ramburii*)
Eastern Forktail (*Ischnura verticalis*)
Southern Spreadwing (*Lestes australis*)
Slaty Skimmer (*Libellula incesta*)
Roseate Skimmer (*Orthemis ferruginea*)

Blue Dasher (*Pachydiplax longipennis*)
Common Whitetail (*Plathemis lydia*)
Variegated Meadowhawk (*Sympetrum corruptum*)
Autumn Meadowhawk (*Sympetrum vicinum*)
Black Saddlebags (*Tramea lacerata*)

Coryell (25)
Great Spreadwing (*Archilestes grandis*)
Kiowa Dancer (*Argia immunda*)
Aztec Dancer (*Argia nahuana*)
Springwater Dancer (*Argia plana*)
Blue-ringed Dancer (*Argia sedula*)
Dusky Dancer (*Argia translata*)
Four-spotted Pennant (*Brachymesia gravida*)
Pale-faced Clubskimmer (*Brechmorhoga mendax*)
Flag-tailed Spinyleg (*Dromogomphus spoliatus*)
Checkered Setwing (*Dythemis fugax*)
Swift Setwing (*Dythemis velox*)
Azure Bluet (*Enallagma aspersum*)
Dot-winged Baskettail (*Epitheca petechialis*)
Eastern Ringtail (*Erpetogomphus designatus*)
Great Pondhawk (*Erythemis vesiculosa*)
Band-winged Dragonlet (*Erythrodiplax umbrata*)
American Rubyspot (*Hetaerina americana*)
Citrine Forktail (*Ischnura hastata*)
Plateau Spreadwing (*Lestes alacer*)
Neon Skimmer (*Libellula croceipennis*)
Widow Skimmer (*Libellula luctuosa*)
Flame Skimmer (*Libellula saturata*)
Georgia River Cruiser (*Macromia illinoiensis georgina*)
Four-striped Leaftail (*Phyllogomphoides stigmatus*)
Variegated Meadowhawk (*Sympetrum corruptum*)

Cottle (6)
Common Green Darner (*Anax junius*)
Familiar Bluet (*Enallagma civile*)
Dot-winged Baskettail (*Epitheca petechialis*)
Eastern Forktail (*Ischnura verticalis*)
Roseate Skimmer (*Orthemis ferruginea*)
Variegated Meadowhawk (*Sympetrum corruptum*)

Crane (8)
Kiowa Dancer (*Argia immunda*)
Blue-ringed Dancer (*Argia sedula*)
Familiar Bluet (*Enallagma civile*)
Eastern Pondhawk (*Erythemis simplicicollis*)
Desert Forktail (*Ischnura barberi*)
Widow Skimmer (*Libellula luctuosa*)
Flame Skimmer (*Libellula saturata*)
Marl Pennant (*Macrodiplax balteata*)

Crockett (20)
Violet Dancer (*Argia fumipennis violacea*)
Kiowa Dancer (*Argia immunda*)
Leonora's Dancer (*Argia leonorae*)
Powdered Dancer (*Argia moesta*)
Aztec Dancer (*Argia nahuana*)
Dusky Dancer (*Argia translata*)
Pale-faced Clubskimmer (*Brechmorhoga mendax*)
Flag-tailed Spinyleg (*Dromogomphus spoliatus*)
Familiar Bluet (*Enallagma civile*)
Eastern Ringtail (*Erpetogomphus designatus*)
American Rubyspot (*Hetaerina americana*)
Desert Forktail (*Ischnura barberi*)
Mexican Forktail (*Ischnura demorsa*)
Citrine Forktail (*Ischnura hastata*)
Widow Skimmer (*Libellula luctuosa*)
Flame Skimmer (*Libellula saturata*)
Filigree Skimmer (*Pseudoleon superbus*)
Variegated Meadowhawk (*Sympetrum corruptum*)
Desert Firetail (*Telebasis salva*)
Black Saddlebags (*Tramea lacerata*)

Crosby (16)

Great Spreadwing (*Archilestes grandis*)
Kiowa Dancer (*Argia immunda*)
Sooty Dancer (*Argia lugens*)
Powdered Dancer (*Argia moesta*)
Aztec Dancer (*Argia nahuana*)
Springwater Dancer (*Argia plana*)
Familiar Bluet (*Enallagma civile*)
American Rubyspot (*Hetaerina americana*)
Plateau Spreadwing (*Lestes alacer*)
Widow Skimmer (*Libellula luctuosa*)
Blue Dasher (*Pachydiplax longipennis*)
Wandering Glider (*Pantala flavescens*)
Common Whitetail (*Plathemis lydia*)
Variegated Meadowhawk (*Sympetrum corruptum*)
Desert Firetail (*Telebasis salva*)
Black Saddlebags (*Tramea lacerata*)

Culberson (40)
Common Green Darner (*Anax junius*)
Giant Darner (*Anax walsinghami*)
Great Spreadwing (*Archilestes grandis*)
Blue-fronted Dancer (*Argia apicalis*)
Violet Dancer (*Argia fumipennis violacea*)
Lavender Dancer (*Argia hinei*)
Leonora's Dancer (*Argia leonorae*)
Sooty Dancer (*Argia lugens*)
Aztec Dancer (*Argia nahuana*)
Springwater Dancer (*Argia plana*)
Blue-ringed Dancer (*Argia sedula*)
Familiar Bluet (*Enallagma civile*)
Neotropical Bluet (*Enallagma novaehispaniae*)
Arroyo Bluet (*Enallagma praevarum*)
Serpent Ringtail (*Erpetogomphus lampropeltis natrix*)
Eastern Pondhawk (*Erythemis simplicicollis*)
Painted Damsel (*Hesperagrion heterodoxum*)
American Rubyspot (*Hetaerina americana*)
Desert Forktail (*Ischnura barberi*)
Plains Forktail (*Ischnura damula*)
Mexican Forktail (*Ischnura demorsa*)
Black-fronted Forktail (*Ischnura denticollis*)
Plateau Spreadwing (*Lestes alacer*)
Comanche Skimmer (*Libellula comanche*)
Widow Skimmer (*Libellula luctuosa*)
Flame Skimmer (*Libellula saturata*)
Roseate Skimmer (*Orthemis ferruginea*)
Blue Dasher (*Pachydiplax longipennis*)
Red Rock Skimmer (*Paltothemis lineatipes*)
Wandering Glider (*Pantala flavescens*)
Spot-winged Glider (*Pantala hymenaea*)
Common Whitetail (*Plathemis lydia*)
Filigree Skimmer (*Pseudoleon superbus*)
Arroyo Darner (*Rhionaeschna dugesi*)
Blue-eyed Darner (*Rhionaeschna multicolor*)
Variegated Meadowhawk (*Sympetrum corruptum*)
Autumn Meadowhawk (*Sympetrum vicinum*)
Desert Firetail (*Telebasis salva*)
Black Saddlebags (*Tramea lacerata*)
Red Saddlebags (*Tramea onusta*)

Dallam (11)
Violet Dancer (*Argia fumipennis violacea*)
Familiar Bluet (*Enallagma civile*)
Plains Forktail (*Ischnura damula*)
Black-fronted Forktail (*Ischnura denticollis*)
Eastern Forktail (*Ischnura verticalis*)
Southern Spreadwing (*Lestes australis*)
Lyre-tipped Spreadwing (*Lestes unguiculatus*)
Twelve-spotted Skimmer (*Libellula pulchella*)
Desert Whitetail (*Plathemis subornata*)
Blue-eyed Darner (*Rhionaeschna multicolor*)
Variegated Meadowhawk (*Sympetrum corruptum*)

Dallas (65)
Common Green Darner (*Anax junius*)
Great Spreadwing (*Archilestes grandis*)
Blue-fronted Dancer (*Argia apicalis*)

Violet Dancer (*Argia fumipennis violacea*)
Kiowa Dancer (*Argia immunda*)
Powdered Dancer (*Argia moesta*)
Aztec Dancer (*Argia nahuana*)
Springwater Dancer (*Argia plana*)
Blue-ringed Dancer (*Argia sedula*)
Blue-tipped Dancer (*Argia tibialis*)
Dusky Dancer (*Argia translata*)
Four-spotted Pennant (*Brachymesia gravida*)
Pale-faced Clubskimmer (*Brechmorhoga mendax*)
Ebony Jewelwing (*Calopteryx maculata*)
Calico Pennant (*Celithemis elisa*)
Halloween Pennant (*Celithemis eponina*)
Banded Pennant (*Celithemis fasciata*)
Stream Cruiser (*Didymops transversa*)
Flag-tailed Spinyleg (*Dromogomphus spoliatus*)
Checkered Setwing (*Dythemis fugax*)
Black Setwing (*Dythemis nigrescens*)
Swift Setwing (*Dythemis velox*)
Double-striped Bluet (*Enallagma basidens*)
Familiar Bluet (*Enallagma civile*)
Stream Bluet (*Enallagma exsulans*)
Orange Bluet (*Enallagma signatum*)
Swamp Darner (*Epiaeschna heros*)
Dot-winged Baskettail (*Epitheca petechialis*)
Prince Baskettail (*Epitheca princeps*)
Mantled Baskettail (*Epitheca semiaquea*)
White-belted Ringtail (*Erpetogomphus compositus*)
Eastern Ringtail (*Erpetogomphus designatus*)
Eastern Pondhawk (*Erythemis simplicicollis*)
Little Blue Dragonlet (*Erythrodiplax minuscula*)
Band-winged Dragonlet (*Erythrodiplax umbrata*)
Plains Clubtail (*Gomphus externus*)
Ashy Clubtail (*Gomphus lividus*)
Sulphur-tipped Clubtail (*Gomphus militaris*)
American Rubyspot (*Hetaerina americana*)
Smoky Rubyspot (*Hetaerina titia*)
Citrine Forktail (*Ischnura hastata*)
Rambur's Forktail (*Ischnura ramburii*)
Plateau Spreadwing (*Lestes alacer*)
Southern Spreadwing (*Lestes australis*)
Comanche Skimmer (*Libellula comanche*)
Neon Skimmer (*Libellula croceipennis*)
Slaty Skimmer (*Libellula incesta*)
Widow Skimmer (*Libellula luctuosa*)
Twelve-spotted Skimmer (*Libellula pulchella*)
Gilded River Cruiser (*Macromia pacifica*)
Royal River Cruiser (*Macromia taeniolata*)
Thornbush Dasher (*Micrathyria hagenii*)
Cyrano Darner (*Nasiaeschna pentacantha*)
Roseate Skimmer (*Orthemis ferruginea*)
Blue Dasher (*Pachydiplax longipennis*)
Wandering Glider (*Pantala flavescens*)
Spot-winged Glider (*Pantala hymenaea*)
Eastern Amberwing (*Perithemis tenera*)
Common Whitetail (*Plathemis lydia*)
Common Sanddragon (*Progomphus obscurus*)
Blue-faced Meadowhawk (*Sympetrum ambiguum*)
Variegated Meadowhawk (*Sympetrum corruptum*)
Desert Firetail (*Telebasis salva*)
Black Saddlebags (*Tramea lacerata*)
Red Saddlebags (*Tramea onusta*)

Dawson (4)
Familiar Bluet (*Enallagma civile*)
Desert Forktail (*Ischnura barberi*)
Eastern Amberwing (*Perithemis tenera*)
Common Whitetail (*Plathemis lydia*)

Deaf Smith (5)
Plateau Spreadwing (*Lestes alacer*)
Flame Skimmer (*Libellula saturata*)
Blue-eyed Darner (*Rhionaeschna multicolor*)
Variegated Meadowhawk (*Sympetrum corruptum*)

Cherry-faced Meadowhawk (*Sympetrum internum*)

Delta *(16)*
Four-spotted Pennant (*Brachymesia gravida*)
Familiar Bluet (*Enallagma civile*)
Eastern Ringtail (*Erpetogomphus designatus*)
Eastern Pondhawk (*Erythemis simplicicollis*)
Sulphur-tipped Clubtail (*Gomphus militaris*)
Widow Skimmer (*Libellula luctuosa*)
Orange Shadowdragon (*Neurocordulia xanthosoma*)
Roseate Skimmer (*Orthemis ferruginea*)
Blue Dasher (*Pachydiplax longipennis*)
Wandering Glider (*Pantala flavescens*)
Eastern Amberwing (*Perithemis tenera*)
Common Whitetail (*Plathemis lydia*)
Common Sanddragon (*Progomphus obscurus*)
Blue-faced Meadowhawk (*Sympetrum ambiguum*)
Variegated Meadowhawk (*Sympetrum corruptum*)
Black Saddlebags (*Tramea lacerata*)

Denton *(72)*
Common Green Darner (*Anax junius*)
Great Spreadwing (*Archilestes grandis*)
Blue-fronted Dancer (*Argia apicalis*)
Kiowa Dancer (*Argia immunda*)
Powdered Dancer (*Argia moesta*)
Aztec Dancer (*Argia nahuana*)
Blue-ringed Dancer (*Argia sedula*)
Blue-tipped Dancer (*Argia tibialis*)
Dusky Dancer (*Argia translata*)
Jade Clubtail (*Arigomphus submedianus*)
Four-spotted Pennant (*Brachymesia gravida*)
Pale-faced Clubskimmer (*Brechmorhoga mendax*)
Ebony Jewelwing (*Calopteryx maculata*)
Calico Pennant (*Celithemis elisa*)
Halloween Pennant (*Celithemis eponina*)
Banded Pennant (*Celithemis fasciata*)
Stream Cruiser (*Didymops transversa*)
Flag-tailed Spinyleg (*Dromogomphus spoliatus*)
Checkered Setwing (*Dythemis fugax*)
Swift Setwing (*Dythemis velox*)
Azure Bluet (*Enallagma aspersum*)
Double-striped Bluet (*Enallagma basidens*)
Familiar Bluet (*Enallagma civile*)
Stream Bluet (*Enallagma exsulans*)
Orange Bluet (*Enallagma signatum*)
Vesper Bluet (*Enallagma vesperum*)
Swamp Darner (*Epiaeschna heros*)
Stripe-winged Baskettail (*Epitheca costalis*)
Common Baskettail (*Epitheca cynosura*)
Dot-winged Baskettail (*Epitheca petechialis*)
Prince Baskettail (*Epitheca princeps*)
Mantled Baskettail (*Epitheca semiaquea*)
Eastern Ringtail (*Erpetogomphus designatus*)
Eastern Pondhawk (*Erythemis simplicicollis*)
Little Blue Dragonlet (*Erythrodiplax minuscula*)
Band-winged Dragonlet (*Erythrodiplax umbrata*)
Plains Clubtail (*Gomphus externus*)
Pronghorn Clubtail (*Gomphus graslinellus*)
Sulphur-tipped Clubtail (*Gomphus militaris*)
American Rubyspot (*Hetaerina americana*)
Smoky Rubyspot (*Hetaerina titia*)
Citrine Forktail (*Ischnura hastata*)
Fragile Forktail (*Ischnura posita*)
Rambur's Forktail (*Ischnura ramburii*)
Eastern Forktail (*Ischnura verticalis*)
Blue Corporal (*Ladona deplanata*)
Southern Spreadwing (*Lestes australis*)
Comanche Skimmer (*Libellula comanche*)
Neon Skimmer (*Libellula croceipennis*)
Spangled Skimmer (*Libellula cyanea*)
Yellow-sided Skimmer (*Libellula flavida*)
Slaty Skimmer (*Libellula incesta*)
Widow Skimmer (*Libellula luctuosa*)

Twelve-spotted Skimmer (*Libellula pulchella*)
Great Blue Skimmer (*Libellula vibrans*)
Georgia River Cruiser (*Macromia illinoiensis georgina*)
Gilded River Cruiser (*Macromia pacifica*)
Roseate Skimmer (*Orthemis ferruginea*)
Blue Dasher (*Pachydiplax longipennis*)
Wandering Glider (*Pantala flavescens*)
Spot-winged Glider (*Pantala hymenaea*)
Eastern Amberwing (*Perithemis tenera*)
Four-striped Leaftail (*Phyllogomphoides stigmatus*)
Common Whitetail (*Plathemis lydia*)
Common Sanddragon (*Progomphus obscurus*)
Russet-tipped Clubtail (*Stylurus plagiatus*)
Blue-faced Meadowhawk (*Sympetrum ambiguum*)
Variegated Meadowhawk (*Sympetrum corruptum*)
Autumn Meadowhawk (*Sympetrum vicinum*)
Desert Firetail (*Telebasis salva*)
Black Saddlebags (*Tramea lacerata*)
Red Saddlebags (*Tramea onusta*)

DeWitt *(21)*
Common Green Darner (*Anax junius*)
Blue-fronted Dancer (*Argia apicalis*)
Leonora's Dancer (*Argia leonorae*)
Powdered Dancer (*Argia moesta*)
Familiar Bluet (*Enallagma civile*)
Stream Bluet (*Enallagma exsulans*)
Prince Baskettail (*Epitheca princeps*)
Eastern Ringtail (*Erpetogomphus designatus*)
Eastern Pondhawk (*Erythemis simplicicollis*)
American Rubyspot (*Hetaerina americana*)
Smoky Rubyspot (*Hetaerina titia*)
Citrine Forktail (*Ischnura hastata*)
Rainpool Spreadwing (*Lestes forficula*)
Widow Skimmer (*Libellula luctuosa*)
Twelve-spotted Skimmer (*Libellula pulchella*)
Roseate Skimmer (*Orthemis ferruginea*)
Blue Dasher (*Pachydiplax longipennis*)
Eastern Amberwing (*Perithemis tenera*)
Five-striped Leaftail (*Phyllogomphoides albrighti*)
Russet-tipped Clubtail (*Stylurus plagiatus*)
Black Saddlebags (*Tramea lacerata*)

Dickens *(5)*
Great Spreadwing (*Archilestes grandis*)
Familiar Bluet (*Enallagma civile*)
Flame Skimmer (*Libellula saturata*)
Blue Dasher (*Pachydiplax longipennis*)
Eastern Amberwing (*Perithemis tenera*)

Dimmit *(34)*
Common Green Darner (*Anax junius*)
Blue-fronted Dancer (*Argia apicalis*)
Blue-ringed Dancer (*Argia sedula*)
Jade Clubtail (*Arigomphus submedianus*)
Red-tailed Pennant (*Brachymesia furcata*)
Four-spotted Pennant (*Brachymesia gravida*)
Halloween Pennant (*Celithemis eponina*)
Flag-tailed Spinyleg (*Dromogomphus spoliatus*)
Checkered Setwing (*Dythemis fugax*)
Black Setwing (*Dythemis nigrescens*)
Swift Setwing (*Dythemis velox*)
Double-striped Bluet (*Enallagma basidens*)
Familiar Bluet (*Enallagma civile*)
Prince Baskettail (*Epitheca princeps*)
Eastern Pondhawk (*Erythemis simplicicollis*)
Little Blue Dragonlet (*Erythrodiplax minuscula*)
Band-winged Dragonlet (*Erythrodiplax umbrata*)
Sulphur-tipped Clubtail (*Gomphus militaris*)
American Rubyspot (*Hetaerina americana*)
Citrine Forktail (*Ischnura hastata*)
Fragile Forktail (*Ischnura posita*)
Rambur's Forktail (*Ischnura ramburii*)
Chalky Spreadwing (*Lestes sigma*)
Thornbush Dasher (*Micrathyria hagenii*)

Orange Shadowdragon (*Neurocordulia xanthosoma*)
Roseate Skimmer (*Orthemis ferruginea*)
Blue Dasher (*Pachydiplax longipennis*)
Wandering Glider (*Pantala flavescens*)
Eastern Amberwing (*Perithemis tenera*)
Five-striped Leaftail (*Phyllogomphoides albrighti*)
Common Whitetail (*Plathemis lydia*)
Russet-tipped Clubtail (*Stylurus plagiatus*)
Black Saddlebags (*Tramea lacerata*)
Red Saddlebags (*Tramea onusta*)

Donley *(21)*
Common Green Darner (*Anax junius*)
Paiute Dancer (*Argia alberta*)
Blue-fronted Dancer (*Argia apicalis*)
Violet Dancer (*Argia fumipennis violacea*)
Powdered Dancer (*Argia moesta*)
Aztec Dancer (*Argia nahuana*)
Springwater Dancer (*Argia plana*)
Blue-ringed Dancer (*Argia sedula*)
Four-spotted Pennant (*Brachymesia gravida*)
Halloween Pennant (*Celithemis eponina*)
Banded Pennant (*Celithemis fasciata*)
Checkered Setwing (*Dythemis fugax*)
Swift Setwing (*Dythemis velox*)
Double-striped Bluet (*Enallagma basidens*)
Familiar Bluet (*Enallagma civile*)
Eastern Pondhawk (*Erythemis simplicicollis*)
Citrine Forktail (*Ischnura hastata*)
Eastern Forktail (*Ischnura verticalis*)
Comanche Skimmer (*Libellula comanche*)
Blue Dasher (*Pachydiplax longipennis*)
Eastern Amberwing (*Perithemis tenera*)

Duval *(16)*
Common Green Darner (*Anax junius*)
Red-tailed Pennant (*Brachymesia furcata*)
Familiar Bluet (*Enallagma civile*)
Eastern Pondhawk (*Erythemis simplicicollis*)
Citrine Forktail (*Ischnura hastata*)
Rambur's Forktail (*Ischnura ramburii*)
Plateau Spreadwing (*Lestes alacer*)
Neon Skimmer (*Libellula croceipennis*)
Thornbush Dasher (*Micrathyria hagenii*)
Carmine Skimmer (*Orthemis discolor*)
Roseate Skimmer (*Orthemis ferruginea*)
Wandering Glider (*Pantala flavescens*)
Common Whitetail (*Plathemis lydia*)
Desert Firetail (*Telebasis salva*)
Black Saddlebags (*Tramea lacerata*)
Red Saddlebags (*Tramea onusta*)

Eastland *(12)*
Checkered Setwing (*Dythemis fugax*)
Eastern Pondhawk (*Erythemis simplicicollis*)
Sulphur-tipped Clubtail (*Gomphus militaris*)
Comanche Skimmer (*Libellula comanche*)
Spangled Skimmer (*Libellula cyanea*)
Widow Skimmer (*Libellula luctuosa*)
Twelve-spotted Skimmer (*Libellula pulchella*)
Blue Dasher (*Pachydiplax longipennis*)
Eastern Amberwing (*Perithemis tenera*)
Four-striped Leaftail (*Phyllogomphoides stigmatus*)
Common Whitetail (*Plathemis lydia*)
Red Saddlebags (*Tramea onusta*)

Edwards *(43)*
Amazon Darner (*Anax amazili*)
Common Green Darner (*Anax junius*)
Comanche Dancer (*Argia barretti*)
Coppery Dancer (*Argia cuprea*)
Violet Dancer (*Argia fumipennis violacea*)
Kiowa Dancer (*Argia immunda*)
Powdered Dancer (*Argia moesta*)
Blue-ringed Dancer (*Argia sedula*)
Dusky Dancer (*Argia translata*)
Springtime Darner (*Basiaeschna janata*)
Halloween Pennant (*Celithemis eponina*)

Banded Pennant (*Celithemis fasciata*)
Black-shouldered Spinyleg (*Dromogomphus spinosus*)
Flag-tailed Spinyleg (*Dromogomphus spoliatus*)
Checkered Setwing (*Dythemis fugax*)
Black Setwing (*Dythemis nigrescens*)
Swift Setwing (*Dythemis velox*)
Double-striped Bluet (*Enallagma basidens*)
Familiar Bluet (*Enallagma civile*)
Stream Bluet (*Enallagma exsulans*)
Neotropical Bluet (*Enallagma novaehispaniae*)
Arroyo Bluet (*Enallagma praevarum*)
Dot-winged Baskettail (*Epitheca petechialis*)
Eastern Ringtail (*Erpetogomphus designatus*)
Eastern Pondhawk (*Erythemis simplicicollis*)
Red-faced Dragonlet (*Erythrodiplax fusca*)
Band-winged Dragonlet (*Erythrodiplax umbrata*)
Plains Clubtail (*Gomphus externus*)
Dragonhunter (*Hagenius brevistylus*)
American Rubyspot (*Hetaerina americana*)
Citrine Forktail (*Ischnura hastata*)
Rambur's Forktail (*Ischnura ramburii*)
Widow Skimmer (*Libellula luctuosa*)
Marl Pennant (*Macrodiplax balteata*)
Bronzed River Cruiser (*Macromia annulata*)
Roseate Skimmer (*Orthemis ferruginea*)
Wandering Glider (*Pantala flavescens*)
Spot-winged Glider (*Pantala hymenaea*)
Four-striped Leaftail (*Phyllogomphoides stigmatus*)
Orange-striped Threadtail (*Protoneura cara*)
Filigree Skimmer (*Pseudoleon superbus*)
Desert Firetail (*Telebasis salva*)
Red Saddlebags (*Tramea onusta*)

El Paso *(33)*
Common Green Darner (*Anax junius*)
Blue-fronted Dancer (*Argia apicalis*)
Sooty Dancer (*Argia lugens*)
Powdered Dancer (*Argia moesta*)
Aztec Dancer (*Argia nahuana*)
Springwater Dancer (*Argia plana*)
Blue-ringed Dancer (*Argia sedula*)
Four-spotted Pennant (*Brachymesia gravida*)
Double-striped Bluet (*Enallagma basidens*)
Familiar Bluet (*Enallagma civile*)
Arroyo Bluet (*Enallagma praevarum*)
White-belted Ringtail (*Erpetogomphus compositus*)
Western Pondhawk (*Erythemis collocata*)
American Rubyspot (*Hetaerina americana*)
Desert Forktail (*Ischnura barberi*)
Mexican Forktail (*Ischnura demorsa*)
Black-fronted Forktail (*Ischnura denticollis*)
Citrine Forktail (*Ischnura hastata*)
Rambur's Forktail (*Ischnura ramburii*)
Widow Skimmer (*Libellula luctuosa*)
Twelve-spotted Skimmer (*Libellula pulchella*)
Flame Skimmer (*Libellula saturata*)
Roseate Skimmer (*Orthemis ferruginea*)
Blue Dasher (*Pachydiplax longipennis*)
Red Rock Skimmer (*Paltothemis lineatipes*)
Spot-winged Glider (*Pantala hymenaea*)
Eastern Amberwing (*Perithemis tenera*)
Desert Whitetail (*Plathemis subornata*)
Blue-eyed Darner (*Rhionaeschna multicolor*)
Variegated Meadowhawk (*Sympetrum corruptum*)
Desert Firetail (*Telebasis salva*)
Black Saddlebags (*Tramea lacerata*)
Red Saddlebags (*Tramea onusta*)

Ellis *(41)*
Common Green Darner (*Anax junius*)
Blue-fronted Dancer (*Argia apicalis*)
Powdered Dancer (*Argia moesta*)
Blue-ringed Dancer (*Argia sedula*)
Dusky Dancer (*Argia translata*)
Four-spotted Pennant (*Brachymesia gravida*)

Calico Pennant (*Celithemis elisa*)
Halloween Pennant (*Celithemis eponina*)
Banded Pennant (*Celithemis fasciata*)
Checkered Setwing (*Dythemis fugax*)
Swift Setwing (*Dythemis velox*)
Double-striped Bluet (*Enallagma basidens*)
Familiar Bluet (*Enallagma civile*)
Orange Bluet (*Enallagma signatum*)
Prince Baskettail (*Epitheca princeps*)
Eastern Ringtail (*Erpetogomphus designatus*)
Eastern Pondhawk (*Erythemis simplicicollis*)
Great Pondhawk (*Erythemis vesiculosa*)
Little Blue Dragonlet (*Erythrodiplax minuscula*)
Band-winged Dragonlet (*Erythrodiplax umbrata*)
Plains Clubtail (*Gomphus externus*)
Sulphur-tipped Clubtail (*Gomphus militaris*)
Smoky Rubyspot (*Hetaerina titia*)
Citrine Forktail (*Ischnura hastata*)
Fragile Forktail (*Ischnura posita*)
Rambur's Forktail (*Ischnura ramburii*)
Southern Spreadwing (*Lestes australis*)
Neon Skimmer (*Libellula croceipennis*)
Slaty Skimmer (*Libellula incesta*)
Widow Skimmer (*Libellula luctuosa*)
Roseate Skimmer (*Orthemis ferruginea*)
Blue Dasher (*Pachydiplax longipennis*)
Wandering Glider (*Pantala flavescens*)
Spot-winged Glider (*Pantala hymenaea*)
Eastern Amberwing (*Perithemis tenera*)
Common Whitetail (*Plathemis lydia*)
Blue-faced Meadowhawk (*Sympetrum ambiguum*)
Variegated Meadowhawk (*Sympetrum corruptum*)
Desert Firetail (*Telebasis salva*)
Black Saddlebags (*Tramea lacerata*)
Red Saddlebags (*Tramea onusta*)

Erath *(73)*
Common Green Darner (*Anax junius*)
Comet Darner (*Anax longipes*)
Great Spreadwing (*Archilestes grandis*)
Blue-fronted Dancer (*Argia apicalis*)
Violet Dancer (*Argia fumipennis violacea*)
Kiowa Dancer (*Argia immunda*)
Powdered Dancer (*Argia moesta*)
Aztec Dancer (*Argia nahuana*)
Springwater Dancer (*Argia plana*)
Blue-ringed Dancer (*Argia sedula*)
Dusky Dancer (*Argia translata*)
Stillwater Clubtail (*Arigomphus lentulus*)
Jade Clubtail (*Arigomphus submedianus*)
Springtime Darner (*Basiaeschna janata*)
Four-spotted Pennant (*Brachymesia gravida*)
Pale-faced Clubskimmer (*Brechmorhoga mendax*)
Calico Pennant (*Celithemis elisa*)
Halloween Pennant (*Celithemis eponina*)
Banded Pennant (*Celithemis fasciata*)
Stream Cruiser (*Didymops transversa*)
Flag-tailed Spinyleg (*Dromogomphus spoliatus*)
Checkered Setwing (*Dythemis fugax*)
Swift Setwing (*Dythemis velox*)
Azure Bluet (*Enallagma aspersum*)
Double-striped Bluet (*Enallagma basidens*)
Familiar Bluet (*Enallagma civile*)
Stream Bluet (*Enallagma exsulans*)
Skimming Bluet (*Enallagma geminatum*)
Orange Bluet (*Enallagma signatum*)
Swamp Darner (*Epiaeschna heros*)
Dot-winged Baskettail (*Epitheca petechialis*)
Prince Baskettail (*Epitheca princeps*)
Eastern Ringtail (*Erpetogomphus designatus*)
Pin-tailed Pondhawk (*Erythemis plebeja*)
Eastern Pondhawk (*Erythemis simplicicollis*)
Great Pondhawk (*Erythemis vesiculosa*)
Little Blue Dragonlet (*Erythrodiplax minuscula*)
Band-winged Dragonlet (*Erythrodiplax umbrata*)

Plains Clubtail (*Gomphus externus*)
Sulphur-tipped Clubtail (*Gomphus militaris*)
American Rubyspot (*Hetaerina americana*)
Smoky Rubyspot (*Hetaerina titia*)
Citrine Forktail (*Ischnura hastata*)
Rambur's Forktail (*Ischnura ramburii*)
Plateau Spreadwing (*Lestes alacer*)
Comanche Skimmer (*Libellula comanche*)
Neon Skimmer (*Libellula croceipennis*)
Slaty Skimmer (*Libellula incesta*)
Widow Skimmer (*Libellula luctuosa*)
Twelve-spotted Skimmer (*Libellula pulchella*)
Flame Skimmer (*Libellula saturata*)
Great Blue Skimmer (*Libellula vibrans*)
Hyacinth Glider (*Miathyria marcella*)
Thornbush Dasher (*Micrathyria hagenii*)
Cyrano Darner (*Nasiaeschna pentacantha*)
Orange Shadowdragon (*Neurocordulia xanthosoma*)
Roseate Skimmer (*Orthemis ferruginea*)
Blue Dasher (*Pachydiplax longipennis*)
Wandering Glider (*Pantala flavescens*)
Spot-winged Glider (*Pantala hymenaea*)
Eastern Amberwing (*Perithemis tenera*)
Four-striped Leaftail (*Phyllogomphoides stigmatus*)
Common Whitetail (*Plathemis lydia*)
Blue-eyed Darner (*Rhionaeschna multicolor*)
Mocha Emerald (*Somatochlora linearis*)
Blue-faced Meadowhawk (*Sympetrum ambiguum*)
Variegated Meadowhawk (*Sympetrum corruptum*)
Autumn Meadowhawk (*Sympetrum vicinum*)
Desert Firetail (*Telebasis salva*)
Striped Saddlebags (*Tramea calverti*)
Carolina Saddlebags (*Tramea carolina*)
Black Saddlebags (*Tramea lacerata*)
Red Saddlebags (*Tramea onusta*)

Falls *(35)*
Common Green Darner (*Anax junius*)
Blue-fronted Dancer (*Argia apicalis*)
Stillwater Clubtail (*Arigomphus lentulus*)
Jade Clubtail (*Arigomphus submedianus*)
Pale-faced Clubskimmer (*Brechmorhoga mendax*)
Flag-tailed Spinyleg (*Dromogomphus spoliatus*)
Checkered Setwing (*Dythemis fugax*)
Familiar Bluet (*Enallagma civile*)
Dot-winged Baskettail (*Epitheca petechialis*)
Prince Baskettail (*Epitheca princeps*)
Mantled Baskettail (*Epitheca semiaquea*)
Eastern Ringtail (*Erpetogomphus designatus*)
Great Pondhawk (*Erythemis vesiculosa*)
Little Blue Dragonlet (*Erythrodiplax minuscula*)
Band-winged Dragonlet (*Erythrodiplax umbrata*)
Plains Clubtail (*Gomphus externus*)
Sulphur-tipped Clubtail (*Gomphus militaris*)
Cobra Clubtail (*Gomphus vastus*)
Neon Skimmer (*Libellula croceipennis*)
Slaty Skimmer (*Libellula incesta*)
Bronzed River Cruiser (*Macromia annulata*)
Cyrano Darner (*Nasiaeschna pentacantha*)
Roseate Skimmer (*Orthemis ferruginea*)
Wandering Glider (*Pantala flavescens*)
Spot-winged Glider (*Pantala hymenaea*)
Common Whitetail (*Plathemis lydia*)
Common Sanddragon (*Progomphus obscurus*)
Blue-eyed Darner (*Rhionaeschna multicolor*)
Russet-tipped Clubtail (*Stylurus plagiatus*)
Blue-faced Meadowhawk (*Sympetrum ambiguum*)
Variegated Meadowhawk (*Sympetrum corruptum*)
Autumn Meadowhawk (*Sympetrum vicinum*)
Carolina Saddlebags (*Tramea carolina*)
Black Saddlebags (*Tramea lacerata*)

Red Saddlebags (*Tramea onusta*)

Fannin *(30)*

Common Green Darner (*Anax junius*)
Blue-fronted Dancer (*Argia apicalis*)
Powdered Dancer (*Argia moesta*)
Blue-ringed Dancer (*Argia sedula*)
Ebony Jewelwing (*Calopteryx maculata*)
Calico Pennant (*Celithemis elisa*)
Halloween Pennant (*Celithemis eponina*)
Flag-tailed Spinyleg (*Dromogomphus spoliatus*)
Checkered Setwing (*Dythemis fugax*)
Double-striped Bluet (*Enallagma basidens*)
Turquoise Bluet (*Enallagma divagans*)
Stream Bluet (*Enallagma exsulans*)
Orange Bluet (*Enallagma signatum*)
Prince Baskettail (*Epitheca princeps*)
Eastern Pondhawk (*Erythemis simplicicollis*)
Sulphur-tipped Clubtail (*Gomphus militaris*)
Rambur's Forktail (*Ischnura ramburii*)
Spangled Skimmer (*Libellula cyanea*)
Slaty Skimmer (*Libellula incesta*)
Widow Skimmer (*Libellula luctuosa*)
Cyrano Darner (*Nasiaeschna pentacantha*)
Blue Dasher (*Pachydiplax longipennis*)
Wandering Glider (*Pantala flavescens*)
Spot-winged Glider (*Pantala hymenaea*)
Eastern Amberwing (*Perithemis tenera*)
Common Whitetail (*Plathemis lydia*)
Common Sanddragon (*Progomphus obscurus*)
Russet-tipped Clubtail (*Stylurus plagiatus*)
Blue-faced Meadowhawk (*Sympetrum ambiguum*)
Black Saddlebags (*Tramea lacerata*)

Fayette *(26)*

Blue-fronted Dancer (*Argia apicalis*)
Violet Dancer (*Argia fumipennis violacea*)
Kiowa Dancer (*Argia immunda*)
Powdered Dancer (*Argia moesta*)
Blue-ringed Dancer (*Argia sedula*)
Dusky Dancer (*Argia translata*)
Checkered Setwing (*Dythemis fugax*)
Swift Setwing (*Dythemis velox*)
Double-striped Bluet (*Enallagma basidens*)
Swamp Darner (*Epiaeschna heros*)
Eastern Ringtail (*Erpetogomphus designatus*)
Eastern Pondhawk (*Erythemis simplicicollis*)
Sulphur-tipped Clubtail (*Gomphus militaris*)
American Rubyspot (*Hetaerina americana*)
Smoky Rubyspot (*Hetaerina titia*)
Citrine Forktail (*Ischnura hastata*)
Rambur's Forktail (*Ischnura ramburii*)
Slaty Skimmer (*Libellula incesta*)
Widow Skimmer (*Libellula luctuosa*)
Bronzed River Cruiser (*Macromia annulata*)
Carmine Skimmer (*Orthemis discolor*)
Blue Dasher (*Pachydiplax longipennis*)
Four-striped Leaftail (*Phyllogomphoides stigmatus*)
Common Whitetail (*Plathemis lydia*)
Common Sanddragon (*Progomphus obscurus*)
Desert Firetail (*Telebasis salva*)

Fisher *(2)*

Blue-fronted Dancer (*Argia apicalis*)
American Rubyspot (*Hetaerina americana*)

Floyd *(5)*

Common Green Darner (*Anax junius*)
Familiar Bluet (*Enallagma civile*)
Southern Spreadwing (*Lestes australis*)
Eastern Amberwing (*Perithemis tenera*)
Variegated Meadowhawk (*Sympetrum corruptum*)

Foard *(3)*

Familiar Bluet (*Enallagma civile*)
Dot-winged Baskettail (*Epitheca petechialis*)
Variegated Meadowhawk (*Sympetrum corruptum*)

Fort Bend *(60)*

Common Green Darner (*Anax junius*)
Comet Darner (*Anax longipes*)
Blue-fronted Dancer (*Argia apicalis*)
Powdered Dancer (*Argia moesta*)
Blue-ringed Dancer (*Argia sedula*)
Blue-tipped Dancer (*Argia tibialis*)
Stillwater Clubtail (*Arigomphus lentulus*)
Jade Clubtail (*Arigomphus submedianus*)
Four-spotted Pennant (*Brachymesia gravida*)
Calico Pennant (*Celithemis elisa*)
Halloween Pennant (*Celithemis eponina*)
Banded Pennant (*Celithemis fasciata*)
Regal Darner (*Coryphaeschna ingens*)
Flag-tailed Spinyleg (*Dromogomphus spoliatus*)
Checkered Setwing (*Dythemis fugax*)
Black Setwing (*Dythemis nigrescens*)
Familiar Bluet (*Enallagma civile*)
Skimming Bluet (*Enallagma geminatum*)
Orange Bluet (*Enallagma signatum*)
Vesper Bluet (*Enallagma vesperum*)
Swamp Darner (*Epiaeschna heros*)
Common Baskettail (*Epitheca cynosura*)
Dot-winged Baskettail (*Epitheca petechialis*)
Prince Baskettail (*Epitheca princeps*)
Pin-tailed Pondhawk (*Erythemis plebeja*)
Eastern Pondhawk (*Erythemis simplicicollis*)
Great Pondhawk (*Erythemis vesiculosa*)
Little Blue Dragonlet (*Erythrodiplax minuscula*)
Band-winged Dragonlet (*Erythrodiplax umbrata*)
Plains Clubtail (*Gomphus externus*)
Sulphur-tipped Clubtail (*Gomphus militaris*)
Smoky Rubyspot (*Hetaerina titia*)
Citrine Forktail (*Ischnura hastata*)
Rambur's Forktail (*Ischnura ramburii*)
Blue Corporal (*Ladona deplanata*)
Golden-winged Skimmer (*Libellula auripennis*)
Neon Skimmer (*Libellula croceipennis*)
Slaty Skimmer (*Libellula incesta*)
Widow Skimmer (*Libellula luctuosa*)
Needham's Skimmer (*Libellula needhami*)
Twelve-spotted Skimmer (*Libellula pulchella*)
Painted Skimmer (*Libellula semifasciata*)
Great Blue Skimmer (*Libellula vibrans*)
Marl Pennant (*Macrodiplax balteata*)
Hyacinth Glider (*Miathyria marcella*)
Thornbush Dasher (*Micrathyria hagenii*)
Carmine Skimmer (*Orthemis discolor*)
Roseate Skimmer (*Orthemis ferruginea*)
Blue Dasher (*Pachydiplax longipennis*)
Wandering Glider (*Pantala flavescens*)
Spot-winged Glider (*Pantala hymenaea*)
Eastern Amberwing (*Perithemis tenera*)
Common Whitetail (*Plathemis lydia*)
Common Sanddragon (*Progomphus obscurus*)
Blue-eyed Darner (*Rhionaeschna multicolor*)
Russet-tipped Clubtail (*Stylurus plagiatus*)
Variegated Meadowhawk (*Sympetrum corruptum*)
Carolina Saddlebags (*Tramea carolina*)
Black Saddlebags (*Tramea lacerata*)
Red Saddlebags (*Tramea onusta*)

Franklin *(40)*

Common Green Darner (*Anax junius*)
Narrow-striped Forceptail (*Aphylla protracta*)
Two-striped Forceptail (*Aphylla williamsoni*)
Blue-fronted Dancer (*Argia apicalis*)
Violet Dancer (*Argia fumipennis violacea*)
Powdered Dancer (*Argia moesta*)
Blue-tipped Dancer (*Argia tibialis*)
Jade Clubtail (*Arigomphus submedianus*)
Four-spotted Pennant (*Brachymesia gravida*)
Ebony Jewelwing (*Calopteryx maculata*)
Calico Pennant (*Celithemis elisa*)
Halloween Pennant (*Celithemis eponina*)
Banded Pennant (*Celithemis fasciata*)
Double-striped Bluet (*Enallagma basidens*)
Familiar Bluet (*Enallagma civile*)

Turquoise Bluet (*Enallagma divagans*)
Orange Bluet (*Enallagma signatum*)
Slender Bluet {westfalli} (*Enallagma traviatum westfalli*)
Stripe-winged Baskettail (*Epitheca costalis*)
Common Baskettail (*Epitheca cynosura*)
Mantled Baskettail (*Epitheca semiaquea*)
Eastern Pondhawk (*Erythemis simplicicollis*)
Oklahoma Clubtail (*Gomphus oklahomensis*)
Citrine Forktail (*Ischnura hastata*)
Fragile Forktail (*Ischnura posita*)
Elegant Spreadwing (*Lestes inaequalis*)
Slender Spreadwing (*Lestes rectangularis*)
Golden-winged Skimmer (*Libellula auripennis*)
Spangled Skimmer (*Libellula cyanea*)
Slaty Skimmer (*Libellula incesta*)
Widow Skimmer (*Libellula luctuosa*)
Great Blue Skimmer (*Libellula vibrans*)
Blue Dasher (*Pachydiplax longipennis*)
Wandering Glider (*Pantala flavescens*)
Spot-winged Glider (*Pantala hymenaea*)
Eastern Amberwing (*Perithemis tenera*)
Common Whitetail (*Plathemis lydia*)
Variegated Meadowhawk (*Sympetrum corruptum*)
Autumn Meadowhawk (*Sympetrum vicinum*)
Black Saddlebags (*Tramea lacerata*)

Freestone *(26)*

Common Green Darner (*Anax junius*)
Blue-fronted Dancer (*Argia apicalis*)
Powdered Dancer (*Argia moesta*)
Stillwater Clubtail (*Arigomphus lentulus*)
Jade Clubtail (*Arigomphus submedianus*)
Four-spotted Pennant (*Brachymesia gravida*)
Halloween Pennant (*Celithemis eponina*)
Twin-spotted Spiketail (*Cordulegaster maculata*)
Familiar Bluet (*Enallagma civile*)
Orange Bluet (*Enallagma signatum*)
Prince Baskettail (*Epitheca princeps*)
Eastern Pondhawk (*Erythemis simplicicollis*)
Great Pondhawk (*Erythemis vesiculosa*)
Smoky Rubyspot (*Hetaerina titia*)
Rambur's Forktail (*Ischnura ramburii*)
Slaty Skimmer (*Libellula incesta*)
Widow Skimmer (*Libellula luctuosa*)
Great Blue Skimmer (*Libellula vibrans*)
Roseate Skimmer (*Orthemis ferruginea*)
Blue Dasher (*Pachydiplax longipennis*)
Wandering Glider (*Pantala flavescens*)
Spot-winged Glider (*Pantala hymenaea*)
Eastern Amberwing (*Perithemis tenera*)
Common Whitetail (*Plathemis lydia*)
Russet-tipped Clubtail (*Stylurus plagiatus*)
Black Saddlebags (*Tramea lacerata*)

Frio *(37)*

Blue-fronted Dancer (*Argia apicalis*)
Powdered Dancer (*Argia moesta*)
Blue-ringed Dancer (*Argia sedula*)
Dusky Dancer (*Argia translata*)
Red-tailed Pennant (*Brachymesia furcata*)
Pale-faced Clubskimmer (*Brechmorhoga mendax*)
Halloween Pennant (*Celithemis eponina*)
Stream Cruiser (*Didymops transversa*)
Swift Setwing (*Dythemis velox*)
Familiar Bluet (*Enallagma civile*)
Eastern Ringtail (*Erpetogomphus designatus*)
Pin-tailed Pondhawk (*Erythemis plebeja*)
Eastern Pondhawk (*Erythemis simplicicollis*)
Plateau Dragonlet (*Erythrodiplax basifusca*)
Red-faced Dragonlet (*Erythrodiplax fusca*)
Sulphur-tipped Clubtail (*Gomphus militaris*)
American Rubyspot (*Hetaerina americana*)
Smoky Rubyspot (*Hetaerina titia*)
Flame Skimmer (*Libellula saturata*)
Marl Pennant (*Macrodiplax balteata*)
Bronzed River Cruiser (*Macromia annulata*)

Georgia River Cruiser (*Macromia illinoiensis georgina*)
Ivory-striped Sylph (*Macrothemis imitans leucozona*)
Thornbush Dasher (*Micrathyria hagenii*)
Amelia's Threadtail (*Neoneura amelia*)
Carmine Skimmer (*Orthemis discolor*)
Roseate Skimmer (*Orthemis ferruginea*)
Blue Dasher (*Pachydiplax longipennis*)
Spot-winged Glider (*Pantala hymenaea*)
Slough Amberwing (*Perithemis domitia*)
Eastern Amberwing (*Perithemis tenera*)
Five-striped Leaftail (*Phyllogomphoides albrighti*)
Four-striped Leaftail (*Phyllogomphoides stigmatus*)
Common Whitetail (*Plathemis lydia*)
Turquoise-tipped Darner (*Rhionaeschna psilus*)
Desert Firetail (*Telebasis salva*)
Antillean Saddlebags (*Tramea insularis*)

Gaines (1)
Variegated Meadowhawk (*Sympetrum corruptum*)

Galveston (29)
Common Green Darner (*Anax junius*)
Comet Darner (*Anax longipes*)
Broad-striped Forceptail (*Aphylla angustifolia*)
Blue-ringed Dancer (*Argia sedula*)
Red-tailed Pennant (*Brachymesia furcata*)
Four-spotted Pennant (*Brachymesia gravida*)
Regal Darner (*Coryphaeschna ingens*)
Familiar Bluet (*Enallagma civile*)
Orange Bluet (*Enallagma signatum*)
Swamp Darner (*Epiaeschna heros*)
Eastern Pondhawk (*Erythemis simplicicollis*)
Great Pondhawk (*Erythemis vesiculosa*)
Seaside Dragonlet (*Erythrodiplax berenice*)
Little Blue Dragonlet (*Erythrodiplax minuscula*)
Band-winged Dragonlet (*Erythrodiplax umbrata*)
Citrine Forktail (*Ischnura hastata*)
Rambur's Forktail (*Ischnura ramburii*)
Great Blue Skimmer (*Libellula vibrans*)
Marl Pennant (*Macrodiplax balteata*)
Hyacinth Glider (*Miathyria marcella*)
Everglades Sprite (*Nehalennia pallidula*)
Roseate Skimmer (*Orthemis ferruginea*)
Blue Dasher (*Pachydiplax longipennis*)
Wandering Glider (*Pantala flavescens*)
Eastern Amberwing (*Perithemis tenera*)
Variegated Meadowhawk (*Sympetrum corruptum*)
Carolina Saddlebags (*Tramea carolina*)
Black Saddlebags (*Tramea lacerata*)
Red Saddlebags (*Tramea onusta*)

Garza (2)
Blue-fronted Dancer (*Argia apicalis*)
Black-fronted Forktail (*Ischnura denticollis*)

Gillespie (51)
Common Green Darner (*Anax junius*)
Comet Darner (*Anax longipes*)
Great Spreadwing (*Archilestes grandis*)
Violet Dancer (*Argia fumipennis violacea*)
Kiowa Dancer (*Argia immunda*)
Powdered Dancer (*Argia moesta*)
Aztec Dancer (*Argia nahuana*)
Blue-ringed Dancer (*Argia sedula*)
Dusky Dancer (*Argia translata*)
Springtime Darner (*Basiaeschna janata*)
Four-spotted Pennant (*Brachymesia gravida*)
Pale-faced Clubskimmer (*Brechmorhoga mendax*)
Halloween Pennant (*Celithemis eponina*)
Banded Pennant (*Celithemis fasciata*)
Flag-tailed Spinyleg (*Dromogomphus spoliatus*)
Checkered Setwing (*Dythemis fugax*)
Swift Setwing (*Dythemis velox*)
Double-striped Bluet (*Enallagma basidens*)

Familiar Bluet (*Enallagma civile*)
Arroyo Bluet (*Enallagma praevarum*)
Dot-winged Baskettail (*Epitheca petechialis*)
Prince Baskettail (*Epitheca princeps*)
Eastern Ringtail (*Erpetogomphus designatus*)
Eastern Pondhawk (*Erythemis simplicicollis*)
Great Pondhawk (*Erythemis vesiculosa*)
Band-winged Dragonlet (*Erythrodiplax umbrata*)
Plains Clubtail (*Gomphus externus*)
Sulphur-tipped Clubtail (*Gomphus militaris*)
Dragonhunter (*Hagenius brevistylus*)
American Rubyspot (*Hetaerina americana*)
Citrine Forktail (*Ischnura hastata*)
Rambur's Forktail (*Ischnura ramburii*)
Comanche Skimmer (*Libellula comanche*)
Neon Skimmer (*Libellula croceipennis*)
Widow Skimmer (*Libellula luctuosa*)
Twelve-spotted Skimmer (*Libellula pulchella*)
Flame Skimmer (*Libellula saturata*)
Roseate Skimmer (*Orthemis ferruginea*)
Blue Dasher (*Pachydiplax longipennis*)
Wandering Glider (*Pantala flavescens*)
Spot-winged Glider (*Pantala hymenaea*)
Eastern Amberwing (*Perithemis tenera*)
Five-striped Leaftail (*Phyllogomphoides albrighti*)
Four-striped Leaftail (*Phyllogomphoides stigmatus*)
Common Whitetail (*Plathemis lydia*)
Common Sanddragon (*Progomphus obscurus*)
Variegated Meadowhawk (*Sympetrum corruptum*)
Autumn Meadowhawk (*Sympetrum vicinum*)
Desert Firetail (*Telebasis salva*)
Black Saddlebags (*Tramea lacerata*)
Red Saddlebags (*Tramea onusta*)

Glasscock (3)
Familiar Bluet (*Enallagma civile*)
Eastern Pondhawk (*Erythemis simplicicollis*)
Citrine Forktail (*Ischnura hastata*)

Goliad (25)
Common Green Darner (*Anax junius*)
Blue-fronted Dancer (*Argia apicalis*)
Violet Dancer (*Argia fumipennis violacea*)
Kiowa Dancer (*Argia immunda*)
Powdered Dancer (*Argia moesta*)
Blue-ringed Dancer (*Argia sedula*)
Dusky Dancer (*Argia translata*)
Checkered Setwing (*Dythemis fugax*)
Double-striped Bluet (*Enallagma basidens*)
Eastern Ringtail (*Erpetogomphus designatus*)
Eastern Pondhawk (*Erythemis simplicicollis*)
Great Pondhawk (*Erythemis vesiculosa*)
Sulphur-tipped Clubtail (*Gomphus militaris*)
Cobra Clubtail (*Gomphus vastus*)
American Rubyspot (*Hetaerina americana*)
Smoky Rubyspot (*Hetaerina titia*)
Citrine Forktail (*Ischnura hastata*)
Rambur's Forktail (*Ischnura ramburii*)
Coral-fronted Threadtail (*Neoneura aaroni*)
Roseate Skimmer (*Orthemis ferruginea*)
Wandering Glider (*Pantala flavescens*)
Eastern Amberwing (*Perithemis tenera*)
Common Whitetail (*Plathemis lydia*)
Common Sanddragon (*Progomphus obscurus*)
Russet-tipped Clubtail (*Stylurus plagiatus*)

Gonzales (68)
Common Green Darner (*Anax junius*)
Broad-striped Forceptail (*Aphylla angustifolia*)
Blue-fronted Dancer (*Argia apicalis*)
Kiowa Dancer (*Argia immunda*)
Powdered Dancer (*Argia moesta*)
Aztec Dancer (*Argia nahuana*)
Blue-ringed Dancer (*Argia sedula*)
Dusky Dancer (*Argia translata*)
Stillwater Clubtail (*Arigomphus lentulus*)
Jade Clubtail (*Arigomphus submedianus*)

Pale-faced Clubskimmer (*Brechmorhoga mendax*)
Gray-waisted Skimmer (*Cannaphila insularis funerea*)
Calico Pennant (*Celithemis elisa*)
Halloween Pennant (*Celithemis eponina*)
Stream Cruiser (*Didymops transversa*)
Flag-tailed Spinyleg (*Dromogomphus spoliatus*)
Black Setwing (*Dythemis nigrescens*)
Swift Setwing (*Dythemis velox*)
Double-striped Bluet (*Enallagma basidens*)
Familiar Bluet (*Enallagma civile*)
Orange Bluet (*Enallagma signatum*)
Swamp Darner (*Epiaeschna heros*)
Stripe-winged Baskettail (*Epitheca costalis*)
Dot-winged Baskettail (*Epitheca petechialis*)
Prince Baskettail (*Epitheca princeps*)
Mantled Baskettail (*Epitheca semiaquea*)
Eastern Ringtail (*Erpetogomphus designatus*)
Blue-faced Ringtail (*Erpetogomphus eutainia*)
Pin-tailed Pondhawk (*Erythemis plebeja*)
Eastern Pondhawk (*Erythemis simplicicollis*)
Great Pondhawk (*Erythemis vesiculosa*)
Little Blue Dragonlet (*Erythrodiplax minuscula*)
Band-winged Dragonlet (*Erythrodiplax umbrata*)
Plains Clubtail (*Gomphus externus*)
Sulphur-tipped Clubtail (*Gomphus militaris*)
Cobra Clubtail (*Gomphus vastus*)
Dragonhunter (*Hagenius brevistylus*)
American Rubyspot (*Hetaerina americana*)
Smoky Rubyspot (*Hetaerina titia*)
Fragile Forktail (*Ischnura posita*)
Rambur's Forktail (*Ischnura ramburii*)
Plateau Spreadwing (*Lestes alacer*)
Southern Spreadwing (*Lestes australis*)
Chalky Spreadwing (*Lestes sigma*)
Comanche Skimmer (*Libellula comanche*)
Neon Skimmer (*Libellula croceipennis*)
Slaty Skimmer (*Libellula incesta*)
Widow Skimmer (*Libellula luctuosa*)
Great Blue Skimmer (*Libellula vibrans*)
Bronzed River Cruiser (*Macromia annulata*)
Georgia River Cruiser (*Macromia illinoiensis georgina*)
Gilded River Cruiser (*Macromia pacifica*)
Hyacinth Glider (*Miathyria marcella*)
Thornbush Dasher (*Micrathyria hagenii*)
Cyrano Darner (*Nasiaeschna pentacantha*)
Coral-fronted Threadtail (*Neoneura aaroni*)
Carmine Skimmer (*Orthemis discolor*)
Roseate Skimmer (*Orthemis ferruginea*)
Blue Dasher (*Pachydiplax longipennis*)
Wandering Glider (*Pantala flavescens*)
Eastern Amberwing (*Perithemis tenera*)
Five-striped Leaftail (*Phyllogomphoides albrighti*)
Common Whitetail (*Plathemis lydia*)
Russet-tipped Clubtail (*Stylurus plagiatus*)
Variegated Meadowhawk (*Sympetrum corruptum*)
Desert Firetail (*Telebasis salva*)
Black Saddlebags (*Tramea lacerata*)
Red Saddlebags (*Tramea onusta*)

Gray (5)
Common Green Darner (*Anax junius*)
Ebony Jewelwing (*Calopteryx maculata*)
Familiar Bluet (*Enallagma civile*)
Dot-winged Baskettail (*Epitheca petechialis*)
Eastern Forktail (*Ischnura verticalis*)

Grayson (35)
Common Green Darner (*Anax junius*)
Blue-fronted Dancer (*Argia apicalis*)
Kiowa Dancer (*Argia immunda*)
Powdered Dancer (*Argia moesta*)
Blue-ringed Dancer (*Argia sedula*)
Stillwater Clubtail (*Arigomphus lentulus*)
Jade Clubtail (*Arigomphus submedianus*)

Ebony Jewelwing (*Calopteryx maculata*)
Stream Cruiser (*Didymops transversa*)
Azure Bluet (*Enallagma aspersum*)
Double-striped Bluet (*Enallagma basidens*)
Familiar Bluet (*Enallagma civile*)
Turquoise Bluet (*Enallagma divagans*)
Skimming Bluet (*Enallagma geminatum*)
Swamp Darner (*Epiaeschna heros*)
Stripe-winged Baskettail (*Epitheca costalis*)
Prince Baskettail (*Epitheca princeps*)
Eastern Ringtail (*Erpetogomphus designatus*)
Eastern Pondhawk (*Erythemis simplicicollis*)
Great Pondhawk (*Erythemis vesiculosa*)
Band-winged Dragonlet (*Erythrodiplax umbrata*)
Sulphur-tipped Clubtail (*Gomphus militaris*)
Rambur's Forktail (*Ischnura ramburii*)
Eastern Forktail (*Ischnura verticalis*)
Comanche Skimmer (*Libellula comanche*)
Widow Skimmer (*Libellula luctuosa*)
Twelve-spotted Skimmer (*Libellula pulchella*)
Georgia River Cruiser (*Macromia illinoiensis georgina*)
Gilded River Cruiser (*Macromia pacifica*)
Roseate Skimmer (*Orthemis ferruginea*)
Blue Dasher (*Pachydiplax longipennis*)
Common Whitetail (*Plathemis lydia*)
Variegated Meadowhawk (*Sympetrum corruptum*)
Desert Firetail (*Telebasis salva*)
Black Saddlebags (*Tramea lacerata*)

Gregg (21)

Blue-fronted Dancer (*Argia apicalis*)
Seepage Dancer (*Argia bipunctulata*)
Violet Dancer (*Argia fumipennis violacea*)
Powdered Dancer (*Argia moesta*)
Blue-tipped Dancer (*Argia tibialis*)
Sparkling Jewelwing (*Calopteryx dimidiata*)
Ebony Jewelwing (*Calopteryx maculata*)
Double-striped Bluet (*Enallagma basidens*)
Familiar Bluet (*Enallagma civile*)
Turquoise Bluet (*Enallagma divagans*)
Stream Bluet (*Enallagma exsulans*)
Eastern Pondhawk (*Erythemis simplicicollis*)
Dragonhunter (*Hagenius brevistylus*)
American Rubyspot (*Hetaerina americana*)
Citrine Forktail (*Ischnura hastata*)
Rambur's Forktail (*Ischnura ramburii*)
Spangled Skimmer (*Libellula cyanea*)
Slaty Skimmer (*Libellula incesta*)
Widow Skimmer (*Libellula luctuosa*)
Blue Dasher (*Pachydiplax longipennis*)
Eastern Amberwing (*Perithemis tenera*)

Grimes (40)

Blue-fronted Dancer (*Argia apicalis*)
Violet Dancer (*Argia fumipennis violacea*)
Kiowa Dancer (*Argia immunda*)
Powdered Dancer (*Argia moesta*)
Blue-ringed Dancer (*Argia sedula*)
Blue-tipped Dancer (*Argia tibialis*)
Dusky Dancer (*Argia translata*)
Ebony Jewelwing (*Calopteryx maculata*)
Calico Pennant (*Celithemis elisa*)
Halloween Pennant (*Celithemis eponina*)
Black-shouldered Spinyleg (*Dromogomphus spinosus*)
Flag-tailed Spinyleg (*Dromogomphus spoliatus*)
Checkered Setwing (*Dythemis fugax*)
Swift Setwing (*Dythemis velox*)
Double-striped Bluet (*Enallagma basidens*)
Familiar Bluet (*Enallagma civile*)
Stream Bluet (*Enallagma exsulans*)
Orange Bluet (*Enallagma signatum*)
Eastern Ringtail (*Erpetogomphus designatus*)
Eastern Pondhawk (*Erythemis simplicicollis*)
Little Blue Dragonlet (*Erythrodiplax minuscula*)
Sulphur-tipped Clubtail (*Gomphus militaris*)
Cobra Clubtail (*Gomphus vastus*)

American Rubyspot (*Hetaerina americana*)
Smoky Rubyspot (*Hetaerina titia*)
Citrine Forktail (*Ischnura hastata*)
Rambur's Forktail (*Ischnura ramburii*)
Widow Skimmer (*Libellula luctuosa*)
Great Blue Skimmer (*Libellula vibrans*)
Royal River Cruiser (*Macromia taeniolata*)
Blue Dasher (*Pachydiplax longipennis*)
Wandering Glider (*Pantala flavescens*)
Eastern Amberwing (*Perithemis tenera*)
Common Whitetail (*Plathemis lydia*)
Common Sanddragon (*Progomphus obscurus*)
Blue-faced Meadowhawk (*Sympetrum ambiguum*)
Variegated Meadowhawk (*Sympetrum corruptum*)
Desert Firetail (*Telebasis salva*)
Black Saddlebags (*Tramea lacerata*)
Red Saddlebags (*Tramea onusta*)

Guadalupe (40)

Blue-fronted Dancer (*Argia apicalis*)
Comanche Dancer (*Argia barretti*)
Kiowa Dancer (*Argia immunda*)
Powdered Dancer (*Argia moesta*)
Blue-ringed Dancer (*Argia sedula*)
Dusky Dancer (*Argia translata*)
Four-spotted Pennant (*Brachymesia gravida*)
Pale-faced Clubskimmer (*Brechmorhoga mendax*)
Stream Cruiser (*Didymops transversa*)
Flag-tailed Spinyleg (*Dromogomphus spoliatus*)
Checkered Setwing (*Dythemis fugax*)
Swift Setwing (*Dythemis velox*)
Double-striped Bluet (*Enallagma basidens*)
Familiar Bluet (*Enallagma civile*)
Orange Bluet (*Enallagma signatum*)
Prince Baskettail (*Epitheca princeps*)
Eastern Ringtail (*Erpetogomphus designatus*)
Eastern Pondhawk (*Erythemis simplicicollis*)
Plains Clubtail (*Gomphus externus*)
Sulphur-tipped Clubtail (*Gomphus militaris*)
Dragonhunter (*Hagenius brevistylus*)
American Rubyspot (*Hetaerina americana*)
Smoky Rubyspot (*Hetaerina titia*)
Citrine Forktail (*Ischnura hastata*)
Widow Skimmer (*Libellula luctuosa*)
Bronzed River Cruiser (*Macromia annulata*)
Thornbush Dasher (*Micrathyria hagenii*)
Coral-fronted Threadtail (*Neoneura aaroni*)
Orange Shadowdragon (*Neurocordulia xanthosoma*)
Roseate Skimmer (*Orthemis ferruginea*)
Blue Dasher (*Pachydiplax longipennis*)
Spot-winged Glider (*Pantala hymenaea*)
Eastern Amberwing (*Perithemis tenera*)
Five-striped Leaftail (*Phyllogomphoides albrighti*)
Four-striped Leaftail (*Phyllogomphoides stigmatus*)
Common Whitetail (*Plathemis lydia*)
Common Sanddragon (*Progomphus obscurus*)
Russet-tipped Clubtail (*Stylurus plagiatus*)
Black Saddlebags (*Tramea lacerata*)
Red Saddlebags (*Tramea onusta*)

Hale (3)

Eastern Pondhawk (*Erythemis simplicicollis*)
Blue Dasher (*Pachydiplax longipennis*)
Variegated Meadowhawk (*Sympetrum corruptum*)

Hall (6)

Familiar Bluet (*Enallagma civile*)
Desert Forktail (*Ischnura barberi*)
Mexican Forktail (*Ischnura demorsa*)
Fragile Forktail (*Ischnura posita*)
Plateau Spreadwing (*Lestes alacer*)
Variegated Meadowhawk (*Sympetrum corruptum*)

Hamilton (24)

Blue-fronted Dancer (*Argia apicalis*)
Kiowa Dancer (*Argia immunda*)
Powdered Dancer (*Argia moesta*)
Blue-ringed Dancer (*Argia sedula*)
Dusky Dancer (*Argia translata*)
Checkered Setwing (*Dythemis fugax*)
Double-striped Bluet (*Enallagma basidens*)
Familiar Bluet (*Enallagma civile*)
Prince Baskettail (*Epitheca princeps*)
Eastern Ringtail (*Erpetogomphus designatus*)
Eastern Pondhawk (*Erythemis simplicicollis*)
Sulphur-tipped Clubtail (*Gomphus militaris*)
American Rubyspot (*Hetaerina americana*)
Smoky Rubyspot (*Hetaerina titia*)
Citrine Forktail (*Ischnura hastata*)
Rambur's Forktail (*Ischnura ramburii*)
Widow Skimmer (*Libellula luctuosa*)
Orange Shadowdragon (*Neurocordulia xanthosoma*)
Roseate Skimmer (*Orthemis ferruginea*)
Wandering Glider (*Pantala flavescens*)
Spot-winged Glider (*Pantala hymenaea*)
Eastern Amberwing (*Perithemis tenera*)
Four-striped Leaftail (*Phyllogomphoides stigmatus*)
Common Whitetail (*Plathemis lydia*)

Hansford (18)

Common Green Darner (*Anax junius*)
Halloween Pennant (*Celithemis eponina*)
Rainbow Bluet (*Enallagma antennatum*)
Double-striped Bluet (*Enallagma basidens*)
Familiar Bluet (*Enallagma civile*)
Eastern Pondhawk (*Erythemis simplicicollis*)
Black-fronted Forktail (*Ischnura denticollis*)
Eastern Forktail (*Ischnura verticalis*)
Lyre-tipped Spreadwing (*Lestes unguiculatus*)
Widow Skimmer (*Libellula luctuosa*)
Flame Skimmer (*Libellula saturata*)
Roseate Skimmer (*Orthemis ferruginea*)
Blue Dasher (*Pachydiplax longipennis*)
Eastern Amberwing (*Perithemis tenera*)
Common Whitetail (*Plathemis lydia*)
Blue-eyed Darner (*Rhionaeschna multicolor*)
Variegated Meadowhawk (*Sympetrum corruptum*)
Black Saddlebags (*Tramea lacerata*)

Hardeman (25)

Common Green Darner (*Anax junius*)
Violet Dancer (*Argia fumipennis violacea*)
Blue-ringed Dancer (*Argia sedula*)
Four-spotted Pennant (*Brachymesia gravida*)
Halloween Pennant (*Celithemis eponina*)
Banded Pennant (*Celithemis fasciata*)
Checkered Setwing (*Dythemis fugax*)
Swift Setwing (*Dythemis velox*)
Double-striped Bluet (*Enallagma basidens*)
Familiar Bluet (*Enallagma civile*)
Stripe-winged Baskettail (*Epitheca costalis*)
Dot-winged Baskettail (*Epitheca petechialis*)
Prince Baskettail (*Epitheca princeps*)
Eastern Pondhawk (*Erythemis simplicicollis*)
Sulphur-tipped Clubtail (*Gomphus militaris*)
American Rubyspot (*Hetaerina americana*)
Eastern Forktail (*Ischnura verticalis*)
Widow Skimmer (*Libellula luctuosa*)
Twelve-spotted Skimmer (*Libellula pulchella*)
Blue Dasher (*Pachydiplax longipennis*)
Four-striped Leaftail (*Phyllogomphoides stigmatus*)
Common Whitetail (*Plathemis lydia*)
Variegated Meadowhawk (*Sympetrum corruptum*)
Black Saddlebags (*Tramea lacerata*)
Red Saddlebags (*Tramea onusta*)

Hardin (52)

Common Green Darner (*Anax junius*)

Powdered Dancer (*Argia moesta*)
Aztec Dancer (*Argia nahuana*)
Blue-tipped Dancer (*Argia tibialis*)
Bayou Clubtail (*Arigomphus maxwelli*)
Fawn Darner (*Boyeria vinosa*)
Sparkling Jewelwing (*Calopteryx dimidiata*)
Ebony Jewelwing (*Calopteryx maculata*)
Calico Pennant (*Celithemis elisa*)
Halloween Pennant (*Celithemis eponina*)
Regal Darner (*Coryphaeschna ingens*)
Stream Cruiser (*Didymops transversa*)
Black-shouldered Spinyleg (*Dromogomphus spinosus*)
Double-striped Bluet (*Enallagma basidens*)
Familiar Bluet (*Enallagma civile*)
Turquoise Bluet (*Enallagma divagans*)
Stream Bluet (*Enallagma exsulans*)
Swamp Darner (*Epiaeschna heros*)
Stripe-winged Baskettail (*Epitheca costalis*)
Common Baskettail (*Epitheca cynosura*)
Dot-winged Baskettail (*Epitheca petechialis*)
Prince Baskettail (*Epitheca princeps*)
Eastern Pondhawk (*Erythemis simplicicollis*)
Little Blue Dragonlet (*Erythrodiplax minuscula*)
Banner Clubtail (*Gomphus apomyius*)
Cocoa Clubtail (*Gomphus hybridus*)
Oklahoma Clubtail (*Gomphus oklahomensis*)
Dragonhunter (*Hagenius brevistylus*)
Citrine Forktail (*Ischnura hastata*)
Rambur's Forktail (*Ischnura ramburii*)
Blue Corporal (*Ladona deplanata*)
Southern Spreadwing (*Lestes australis*)
Golden-winged Skimmer (*Libellula auripennis*)
Bar-winged Skimmer (*Libellula axilena*)
Slaty Skimmer (*Libellula incesta*)
Widow Skimmer (*Libellula luctuosa*)
Painted Skimmer (*Libellula semifasciata*)
Great Blue Skimmer (*Libellula vibrans*)
Royal River Cruiser (*Macromia taeniolata*)
Hyacinth Glider (*Miathyria marcella*)
Roseate Skimmer (*Orthemis ferruginea*)
Blue Dasher (*Pachydiplax longipennis*)
Wandering Glider (*Pantala flavescens*)
Spot-winged Glider (*Pantala hymenaea*)
Eastern Amberwing (*Perithemis tenera*)
Common Whitetail (*Plathemis lydia*)
Common Sanddragon (*Progomphus obscurus*)
Mocha Emerald (*Somatochlora linearis*)
Texas Emerald (*Somatochlora margarita*)
Blue-faced Meadowhawk (*Sympetrum ambiguum*)
Variegated Meadowhawk (*Sympetrum corruptum*)
Gray Petaltail (*Tachopteryx thoreyi*)

Harris (78)
Common Green Darner (*Anax junius*)
Comet Darner (*Anax longipes*)
Narrow-striped Forceptail (*Aphylla protracta*)
Blue-fronted Dancer (*Argia apicalis*)
Powdered Dancer (*Argia moesta*)
Blue-ringed Dancer (*Argia sedula*)
Blue-tipped Dancer (*Argia tibialis*)
Dusky Dancer (*Argia translata*)
Stillwater Clubtail (*Arigomphus lentulus*)
Jade Clubtail (*Arigomphus submedianus*)
Red-tailed Pennant (*Brachymesia furcata*)
Four-spotted Pennant (*Brachymesia gravida*)
Ebony Jewelwing (*Calopteryx maculata*)
Calico Pennant (*Celithemis elisa*)
Halloween Pennant (*Celithemis eponina*)
Banded Pennant (*Celithemis fasciata*)
Ornate Pennant (*Celithemis ornata*)
Regal Darner (*Coryphaeschna ingens*)
Black-shouldered Spinyleg (*Dromogomphus spinosus*)
Checkered Setwing (*Dythemis fugax*)
Black Setwing (*Dythemis nigrescens*)

Swift Setwing (*Dythemis velox*)
Double-striped Bluet (*Enallagma basidens*)
Familiar Bluet (*Enallagma civile*)
Burgundy Bluet (*Enallagma dubium*)
Skimming Bluet (*Enallagma geminatum*)
Orange Bluet (*Enallagma signatum*)
Vesper Bluet (*Enallagma vesperum*)
Swamp Darner (*Epiaeschna heros*)
Stripe-winged Baskettail (*Epitheca costalis*)
Common Baskettail (*Epitheca cynosura*)
Prince Baskettail (*Epitheca princeps*)
Mantled Baskettail (*Epitheca semiaquea*)
Eastern Pondhawk (*Erythemis simplicicollis*)
Great Pondhawk (*Erythemis vesiculosa*)
Little Blue Dragonlet (*Erythrodiplax minuscula*)
Band-winged Dragonlet (*Erythrodiplax umbrata*)
Sulphur-tipped Clubtail (*Gomphus militaris*)
Oklahoma Clubtail (*Gomphus oklahomensis*)
Dragonhunter (*Hagenius brevistylus*)
American Rubyspot (*Hetaerina americana*)
Smoky Rubyspot (*Hetaerina titia*)
Citrine Forktail (*Ischnura hastata*)
Lilypad Forktail (*Ischnura kellicotti*)
Fragile Forktail (*Ischnura posita*)
Rambur's Forktail (*Ischnura ramburii*)
Blue Corporal (*Ladona deplanata*)
Southern Spreadwing (*Lestes australis*)
Rainpool Spreadwing (*Lestes forficula*)
Swamp Spreadwing (*Lestes vigilax*)
Golden-winged Skimmer (*Libellula auripennis*)
Bar-winged Skimmer (*Libellula axilena*)
Neon Skimmer (*Libellula croceipennis*)
Slaty Skimmer (*Libellula incesta*)
Widow Skimmer (*Libellula luctuosa*)
Needham's Skimmer (*Libellula needhami*)
Twelve-spotted Skimmer (*Libellula pulchella*)
Flame Skimmer (*Libellula saturata*)
Painted Skimmer (*Libellula semifasciata*)
Great Blue Skimmer (*Libellula vibrans*)
Marl Pennant (*Macrodiplax balteata*)
Hyacinth Glider (*Miathyria marcella*)
Thornbush Dasher (*Micrathyria hagenii*)
Cyrano Darner (*Nasiaeschna pentacantha*)
Roseate Skimmer (*Orthemis ferruginea*)
Blue Dasher (*Pachydiplax longipennis*)
Wandering Glider (*Pantala flavescens*)
Spot-winged Glider (*Pantala hymenaea*)
Eastern Amberwing (*Perithemis tenera*)
Common Whitetail (*Plathemis lydia*)
Common Sanddragon (*Progomphus obscurus*)
Orange-striped Threadtail (*Protoneura cara*)
Blue-eyed Darner (*Rhionaeschna multicolor*)
Variegated Meadowhawk (*Sympetrum corruptum*)
Duckweed Firetail (*Telebasis byersi*)
Carolina Saddlebags (*Tramea carolina*)
Black Saddlebags (*Tramea lacerata*)
Red Saddlebags (*Tramea onusta*)

Harrison (54)
Common Green Darner (*Anax junius*)
Two-striped Forceptail (*Aphylla williamsoni*)
Blue-fronted Dancer (*Argia apicalis*)
Violet Dancer (*Argia fumipennis violacea*)
Kiowa Dancer (*Argia immunda*)
Springwater Dancer (*Argia plana*)
Blue-tipped Dancer (*Argia tibialis*)
Stillwater Clubtail (*Arigomphus lentulus*)
Bayou Clubtail (*Arigomphus maxwelli*)
Jade Clubtail (*Arigomphus submedianus*)
Springtime Darner (*Basiaeschna janata*)
Ebony Jewelwing (*Calopteryx maculata*)
Halloween Pennant (*Celithemis eponina*)
Banded Pennant (*Celithemis fasciata*)
Stream Cruiser (*Didymops transversa*)
Black-shouldered Spinyleg (*Dromogomphus spinosus*)
Azure Bluet (*Enallagma aspersum*)

Familiar Bluet (*Enallagma civile*)
Burgundy Bluet (*Enallagma dubium*)
Skimming Bluet (*Enallagma geminatum*)
Orange Bluet (*Enallagma signatum*)
Swamp Darner (*Epiaeschna heros*)
Stripe-winged Baskettail (*Epitheca costalis*)
Common Baskettail (*Epitheca cynosura*)
Dot-winged Baskettail (*Epitheca petechialis*)
Prince Baskettail (*Epitheca princeps*)
Eastern Pondhawk (*Erythemis simplicicollis*)
Band-winged Dragonlet (*Erythrodiplax umbrata*)
Oklahoma Clubtail (*Gomphus oklahomensis*)
Citrine Forktail (*Ischnura hastata*)
Lilypad Forktail (*Ischnura kellicotti*)
Fragile Forktail (*Ischnura posita*)
Rambur's Forktail (*Ischnura ramburii*)
Elegant Spreadwing (*Lestes inaequalis*)
Swamp Spreadwing (*Lestes vigilax*)
Golden-winged Skimmer (*Libellula auripennis*)
Bar-winged Skimmer (*Libellula axilena*)
Spangled Skimmer (*Libellula cyanea*)
Slaty Skimmer (*Libellula incesta*)
Widow Skimmer (*Libellula luctuosa*)
Great Blue Skimmer (*Libellula vibrans*)
Royal River Cruiser (*Macromia taeniolata*)
Hyacinth Glider (*Miathyria marcella*)
Cyrano Darner (*Nasiaeschna pentacantha*)
Orange Shadowdragon (*Neurocordulia xanthosoma*)
Blue Dasher (*Pachydiplax longipennis*)
Wandering Glider (*Pantala flavescens*)
Eastern Amberwing (*Perithemis tenera*)
Five-striped Leaftail (*Phyllogomphoides albrighti*)
Common Whitetail (*Plathemis lydia*)
Blue-faced Meadowhawk (*Sympetrum ambiguum*)
Gray Petaltail (*Tachopteryx thoreyi*)
Black Saddlebags (*Tramea lacerata*)
Red Saddlebags (*Tramea onusta*)

Hartley (12)
Tule Bluet (*Enallagma carunculatum*)
Familiar Bluet (*Enallagma civile*)
Alkali Bluet (*Enallagma clausum*)
Plains Forktail (*Ischnura damula*)
Mexican Forktail (*Ischnura demorsa*)
Black-fronted Forktail (*Ischnura denticollis*)
Eastern Forktail (*Ischnura verticalis*)
Southern Spreadwing (*Lestes australis*)
Roseate Skimmer (*Orthemis ferruginea*)
Blue Dasher (*Pachydiplax longipennis*)
Common Whitetail (*Plathemis lydia*)
Blue-eyed Darner (*Rhionaeschna multicolor*)

Haskell (4)
Common Green Darner (*Anax junius*)
Powdered Dancer (*Argia moesta*)
Familiar Bluet (*Enallagma civile*)
Variegated Meadowhawk (*Sympetrum corruptum*)

Hays (61)
Amazon Darner (*Anax amazili*)
Common Green Darner (*Anax junius*)
Broad-striped Forceptail (*Aphylla angustifolia*)
Narrow-striped Forceptail (*Aphylla protracta*)
Great Spreadwing (*Archilestes grandis*)
Comanche Dancer (*Argia barretti*)
Violet Dancer (*Argia fumipennis violacea*)
Kiowa Dancer (*Argia immunda*)
Powdered Dancer (*Argia moesta*)
Aztec Dancer (*Argia nahuana*)
Springwater Dancer (*Argia plana*)
Blue-ringed Dancer (*Argia sedula*)
Dusky Dancer (*Argia translata*)
Springtime Darner (*Basiaeschna janata*)
Four-spotted Pennant (*Brachymesia gravida*)
Pale-faced Clubskimmer (*Brechmorhoga mendax*)

Halloween Pennant (*Celithemis eponina*)
Flag-tailed Spinyleg (*Dromogomphus spoliatus*)
Checkered Setwing (*Dythemis fugax*)
Black Setwing (*Dythemis nigrescens*)
Swift Setwing (*Dythemis velox*)
Double-striped Bluet (*Enallagma basidens*)
Familiar Bluet (*Enallagma civile*)
Stream Bluet (*Enallagma exsulans*)
Neotropical Bluet (*Enallagma novaehispaniae*)
Orange Bluet (*Enallagma signatum*)
Dot-winged Baskettail (*Epitheca petechialis*)
Prince Baskettail (*Epitheca princeps*)
Eastern Ringtail (*Erpetogomphus designatus*)
Eastern Pondhawk (*Erythemis simplicicollis*)
Great Pondhawk (*Erythemis vesiculosa*)
Little Blue Dragonlet (*Erythrodiplax minuscula*)
Band-winged Dragonlet (*Erythrodiplax umbrata*)
Dragonhunter (*Hagenius brevistylus*)
American Rubyspot (*Hetaerina americana*)
Smoky Rubyspot (*Hetaerina titia*)
Citrine Forktail (*Ischnura hastata*)
Fragile Forktail (*Ischnura posita*)
Rambur's Forktail (*Ischnura ramburii*)
Rainpool Spreadwing (*Lestes forficula*)
Comanche Skimmer (*Libellula comanche*)
Neon Skimmer (*Libellula croceipennis*)
Slaty Skimmer (*Libellula incesta*)
Widow Skimmer (*Libellula luctuosa*)
Bronzed River Cruiser (*Macromia annulata*)
Ivory-striped Sylph (*Macrothemis imitans leucozona*)
Hyacinth Glider (*Miathyria marcella*)
Coral-fronted Threadtail (*Neoneura aaroni*)
Roseate Skimmer (*Orthemis ferruginea*)
Blue Dasher (*Pachydiplax longipennis*)
Eastern Amberwing (*Perithemis tenera*)
Five-striped Leaftail (*Phyllogomphoides albrighti*)
Four-striped Leaftail (*Phyllogomphoides stigmatus*)
Common Whitetail (*Plathemis lydia*)
Orange-striped Threadtail (*Protoneura cara*)
Russet-tipped Clubtail (*Stylurus plagiatus*)
Variegated Meadowhawk (*Sympetrum corruptum*)
Autumn Meadowhawk (*Sympetrum vicinum*)
Desert Firetail (*Telebasis salva*)
Striped Saddlebags (*Tramea calverti*)
Red Saddlebags (*Tramea onusta*)

Hemphill *(30)*
Common Green Darner (*Anax junius*)
Blue-fronted Dancer (*Argia apicalis*)
Blue-ringed Dancer (*Argia sedula*)
Ebony Jewelwing (*Calopteryx maculata*)
Halloween Pennant (*Celithemis eponina*)
Familiar Bluet (*Enallagma civile*)
Dot-winged Baskettail (*Epitheca petechialis*)
Eastern Ringtail (*Erpetogomphus designatus*)
Eastern Pondhawk (*Erythemis simplicicollis*)
Sulphur-tipped Clubtail (*Gomphus militaris*)
American Rubyspot (*Hetaerina americana*)
Rambur's Forktail (*Ischnura ramburii*)
Eastern Forktail (*Ischnura verticalis*)
Slender Spreadwing (*Lestes rectangularis*)
Comanche Skimmer (*Libellula comanche*)
Slaty Skimmer (*Libellula incesta*)
Widow Skimmer (*Libellula luctuosa*)
Twelve-spotted Skimmer (*Libellula pulchella*)
Roseate Skimmer (*Orthemis ferruginea*)
Blue Dasher (*Pachydiplax longipennis*)
Wandering Glider (*Pantala flavescens*)
Eastern Amberwing (*Perithemis tenera*)
Common Whitetail (*Plathemis lydia*)
Desert Whitetail (*Plathemis subornata*)
Common Sanddragon (*Progomphus obscurus*)
Variegated Meadowhawk (*Sympetrum corruptum*)

Band-winged Meadowhawk (*Sympetrum semicinctum*)
Desert Firetail (*Telebasis salva*)
Carolina Saddlebags (*Tramea carolina*)
Black Saddlebags (*Tramea lacerata*)

Henderson *(19)*
Seepage Dancer (*Argia bipunctulata*)
Blue-tipped Dancer (*Argia tibialis*)
Ebony Jewelwing (*Calopteryx maculata*)
Calico Pennant (*Celithemis elisa*)
Halloween Pennant (*Celithemis eponina*)
Eastern Pondhawk (*Erythemis simplicicollis*)
Little Blue Dragonlet (*Erythrodiplax minuscula*)
American Rubyspot (*Hetaerina americana*)
Smoky Rubyspot (*Hetaerina titia*)
Spangled Skimmer (*Libellula cyanea*)
Slaty Skimmer (*Libellula incesta*)
Widow Skimmer (*Libellula luctuosa*)
Needham's Skimmer (*Libellula needhami*)
Roseate Skimmer (*Orthemis ferruginea*)
Blue Dasher (*Pachydiplax longipennis*)
Common Whitetail (*Plathemis lydia*)
Common Sanddragon (*Progomphus obscurus*)
Variegated Meadowhawk (*Sympetrum corruptum*)
Red Saddlebags (*Tramea onusta*)

Hidalgo *(85)*
Amazon Darner (*Anax amazili*)
Blue-spotted Comet Darner (*Anax concolor*)
Common Green Darner (*Anax junius*)
Comet Darner (*Anax longipes*)
Broad-striped Forceptail (*Aphylla angustifolia*)
Narrow-striped Forceptail (*Aphylla protracta*)
Blue-fronted Dancer (*Argia apicalis*)
Kiowa Dancer (*Argia immunda*)
Powdered Dancer (*Argia moesta*)
Golden-winged Dancer (*Argia rhoadsi*)
Blue-ringed Dancer (*Argia sedula*)
Dusky Dancer (*Argia translata*)
Red-tailed Pennant (*Brachymesia furcata*)
Four-spotted Pennant (*Brachymesia gravida*)
Tawny Pennant (*Brachymesia herbida*)
Pale-faced Clubskimmer (*Brechmorhoga mendax*)
Gray-waisted Skimmer (*Cannaphila insularis funerea*)
Halloween Pennant (*Celithemis eponina*)
Blue-faced Darner (*Coryphaeschna adnexa*)
Flag-tailed Spinyleg (*Dromogomphus spoliatus*)
Checkered Setwing (*Dythemis fugax*)
Black Setwing (*Dythemis nigrescens*)
Swift Setwing (*Dythemis velox*)
Double-striped Bluet (*Enallagma basidens*)
Familiar Bluet (*Enallagma civile*)
Big Bluet (*Enallagma durum*)
Neotropical Bluet (*Enallagma novaehispaniae*)
Orange Bluet (*Enallagma signatum*)
Prince Baskettail (*Epitheca princeps*)
Eastern Ringtail (*Erpetogomphus designatus*)
Black Pondhawk (*Erythemis attala*)
Claret Pondhawk (*Erythemis mithroides*)
Pin-tailed Pondhawk (*Erythemis plebeja*)
Eastern Pondhawk (*Erythemis simplicicollis*)
Great Pondhawk (*Erythemis vesiculosa*)
Little Blue Dragonlet (*Erythrodiplax minuscula*)
Band-winged Dragonlet (*Erythrodiplax umbrata*)
Sulphur-tipped Clubtail (*Gomphus militaris*)
Bar-sided Darner (*Gynacantha mexicana*)
American Rubyspot (*Hetaerina americana*)
Smoky Rubyspot (*Hetaerina titia*)
Citrine Forktail (*Ischnura hastata*)
Rambur's Forktail (*Ischnura ramburii*)
Cream-tipped Swampdamsel (*Leptobasis melinogaster*)
Plateau Spreadwing (*Lestes alacer*)
Southern Spreadwing (*Lestes australis*)
Rainpool Spreadwing (*Lestes forficula*)

Chalky Spreadwing (*Lestes sigma*)
Comanche Skimmer (*Libellula comanche*)
Needham's Skimmer (*Libellula needhami*)
Twelve-spotted Skimmer (*Libellula pulchella*)
Flame Skimmer (*Libellula saturata*)
Marl Pennant (*Macrodiplax balteata*)
Bronzed River Cruiser (*Macromia annulata*)
Straw-colored Sylph (*Macrothemis inacuta*)
Hyacinth Glider (*Miathyria marcella*)
Spot-tailed Dasher (*Micrathyria aequalis*)
Three-striped Dasher (*Micrathyria didyma*)
Thornbush Dasher (*Micrathyria hagenii*)
Caribbean Yellowface (*Neoerythromma cultellatum*)
Coral-fronted Threadtail (*Neoneura aaroni*)
Amelia's Threadtail (*Neoneura amelia*)
Carmine Skimmer (*Orthemis discolor*)
Roseate Skimmer (*Orthemis ferruginea*)
Blue Dasher (*Pachydiplax longipennis*)
Wandering Glider (*Pantala flavescens*)
Spot-winged Glider (*Pantala hymenaea*)
Slough Amberwing (*Perithemis domitia*)
Eastern Amberwing (*Perithemis tenera*)
Ringed Forceptail (*Phyllocycla breviphylla*)
Five-striped Leaftail (*Phyllogomphoides albrighti*)
Common Whitetail (*Plathemis lydia*)
Orange-striped Threadtail (*Protoneura cara*)
Filigree Skimmer (*Pseudoleon superbus*)
Arroyo Darner (*Rhionaeschna dugesi*)
Turquoise-tipped Darner (*Rhionaeschna psilus*)
Russet-tipped Clubtail (*Stylurus plagiatus*)
Variegated Meadowhawk (*Sympetrum corruptum*)
Desert Firetail (*Telebasis salva*)
Evening Skimmer (*Tholymis citrina*)
Vermilion Saddlebags (*Tramea abdominalis*)
Striped Saddlebags (*Tramea calverti*)
Black Saddlebags (*Tramea lacerata*)
Red Saddlebags (*Tramea onusta*)
Triacanthagyna septima (*Triacanthagyna septima*)

Hill *(18)*
Blue-fronted Dancer (*Argia apicalis*)
Powdered Dancer (*Argia moesta*)
Aztec Dancer (*Argia nahuana*)
Springwater Dancer (*Argia plana*)
Blue-ringed Dancer (*Argia sedula*)
Dusky Dancer (*Argia translata*)
Flag-tailed Spinyleg (*Dromogomphus spoliatus*)
Double-striped Bluet (*Enallagma basidens*)
Familiar Bluet (*Enallagma civile*)
Stream Bluet (*Enallagma exsulans*)
Eastern Ringtail (*Erpetogomphus designatus*)
Eastern Pondhawk (*Erythemis simplicicollis*)
Sulphur-tipped Clubtail (*Gomphus militaris*)
American Rubyspot (*Hetaerina americana*)
Plateau Spreadwing (*Lestes alacer*)
Widow Skimmer (*Libellula luctuosa*)
Four-striped Leaftail (*Phyllogomphoides stigmatus*)
Desert Firetail (*Telebasis salva*)

Hockley *(7)*
Familiar Bluet (*Enallagma civile*)
Desert Forktail (*Ischnura barberi*)
Roseate Skimmer (*Orthemis ferruginea*)
Wandering Glider (*Pantala flavescens*)
Common Whitetail (*Plathemis lydia*)
Variegated Meadowhawk (*Sympetrum corruptum*)
Red Saddlebags (*Tramea onusta*)

Hood *(6)*
Blue-fronted Dancer (*Argia apicalis*)
Violet Dancer (*Argia fumipennis violacea*)
Aztec Dancer (*Argia nahuana*)
Dot-winged Baskettail (*Epitheca petechialis*)
Sulphur-tipped Clubtail (*Gomphus militaris*)

American Rubyspot (*Hetaerina americana*)

Hopkins (36)
Common Green Darner (*Anax junius*)
Comet Darner (*Anax longipes*)
Blue-fronted Dancer (*Argia apicalis*)
Violet Dancer (*Argia fumipennis violacea*)
Kiowa Dancer (*Argia immunda*)
Powdered Dancer (*Argia moesta*)
Aztec Dancer (*Argia nahuana*)
Blue-tipped Dancer (*Argia tibialis*)
Stillwater Clubtail (*Arigomphus lentulus*)
Ebony Jewelwing (*Calopteryx maculata*)
Calico Pennant (*Celithemis elisa*)
Halloween Pennant (*Celithemis eponina*)
Banded Pennant (*Celithemis fasciata*)
Flag-tailed Spinyleg (*Dromogomphus spoliatus*)
Double-striped Bluet (*Enallagma basidens*)
Familiar Bluet (*Enallagma civile*)
Stream Bluet (*Enallagma exsulans*)
Skimming Bluet (*Enallagma geminatum*)
Orange Bluet (*Enallagma signatum*)
Prince Baskettail (*Epitheca princeps*)
Mantled Baskettail (*Epitheca semiaquea*)
Eastern Pondhawk (*Erythemis simplicicollis*)
Oklahoma Clubtail (*Gomphus oklahomensis*)
Citrine Forktail (*Ischnura hastata*)
Rambur's Forktail (*Ischnura ramburii*)
Blue Corporal (*Ladona deplanata*)
Slaty Skimmer (*Libellula incesta*)
Widow Skimmer (*Libellula luctuosa*)
Roseate Skimmer (*Orthemis ferruginea*)
Blue Dasher (*Pachydiplax longipennis*)
Eastern Amberwing (*Perithemis tenera*)
Common Whitetail (*Plathemis lydia*)
Common Sanddragon (*Progomphus obscurus*)
Blue-faced Meadowhawk (*Sympetrum ambiguum*)
Variegated Meadowhawk (*Sympetrum corruptum*)
Black Saddlebags (*Tramea lacerata*)

Houston (30)
Common Green Darner (*Anax junius*)
Broad-striped Forceptail (*Aphylla angustifolia*)
Powdered Dancer (*Argia moesta*)
Blue-tipped Dancer (*Argia tibialis*)
Springtime Darner (*Basiaeschna janata*)
Fawn Darner (*Boyeria vinosa*)
Ebony Jewelwing (*Calopteryx maculata*)
Stream Cruiser (*Didymops transversa*)
Double-striped Bluet (*Enallagma basidens*)
Familiar Bluet (*Enallagma civile*)
Burgundy Bluet (*Enallagma dubium*)
Skimming Bluet (*Enallagma geminatum*)
Stripe-winged Baskettail (*Epitheca costalis*)
Common Baskettail (*Epitheca cynosura*)
Mantled Baskettail (*Epitheca semiaquea*)
Robust Baskettail (*Epitheca spinosa*)
Eastern Pondhawk (*Erythemis simplicicollis*)
Plains Clubtail (*Gomphus externus*)
Ashy Clubtail (*Gomphus lividus*)
Oklahoma Clubtail (*Gomphus oklahomensis*)
Rambur's Forktail (*Ischnura ramburii*)
Blue Corporal (*Ladona deplanata*)
Slaty Skimmer (*Libellula incesta*)
Great Blue Skimmer (*Libellula vibrans*)
Cyrano Darner (*Nasiaeschna pentacantha*)
Blue Dasher (*Pachydiplax longipennis*)
Eastern Amberwing (*Perithemis tenera*)
Common Whitetail (*Plathemis lydia*)
Common Sanddragon (*Progomphus obscurus*)
Mocha Emerald (*Somatochlora linearis*)

Howard (4)
Powdered Dancer (*Argia moesta*)
Dusky Dancer (*Argia translata*)
Familiar Bluet (*Enallagma civile*)
Plateau Spreadwing (*Lestes alacer*)

Hudspeth (24)

Common Green Darner (*Anax junius*)
Blue-fronted Dancer (*Argia apicalis*)
Lavender Dancer (*Argia hinei*)
Leonora's Dancer (*Argia leonorae*)
Powdered Dancer (*Argia moesta*)
Blue-ringed Dancer (*Argia sedula*)
Red-tailed Pennant (*Brachymesia furcata*)
Familiar Bluet (*Enallagma civile*)
White-belted Ringtail (*Erpetogomphus compositus*)
Painted Damsel (*Hesperagrion heterodoxum*)
American Rubyspot (*Hetaerina americana*)
Desert Forktail (*Ischnura barberi*)
Mexican Forktail (*Ischnura demorsa*)
Rambur's Forktail (*Ischnura ramburii*)
Plateau Spreadwing (*Lestes alacer*)
Widow Skimmer (*Libellula luctuosa*)
Twelve-spotted Skimmer (*Libellula pulchella*)
Roseate Skimmer (*Orthemis ferruginea*)
Wandering Glider (*Pantala flavescens*)
Spot-winged Glider (*Pantala hymenaea*)
Eastern Amberwing (*Perithemis tenera*)
Blue-eyed Darner (*Rhionaeschna multicolor*)
Variegated Meadowhawk (*Sympetrum corruptum*)
Black Saddlebags (*Tramea lacerata*)

Hunt (38)
Common Green Darner (*Anax junius*)
Blue-fronted Dancer (*Argia apicalis*)
Stillwater Clubtail (*Arigomphus lentulus*)
Calico Pennant (*Celithemis elisa*)
Halloween Pennant (*Celithemis eponina*)
Stream Cruiser (*Didymops transversa*)
Flag-tailed Spinyleg (*Dromogomphus spoliatus*)
Swift Setwing (*Dythemis velox*)
Double-striped Bluet (*Enallagma basidens*)
Familiar Bluet (*Enallagma civile*)
Orange Bluet (*Enallagma signatum*)
Swamp Darner (*Epiaeschna heros*)
Prince Baskettail (*Epitheca princeps*)
Mantled Baskettail (*Epitheca semiaquea*)
Eastern Ringtail (*Erpetogomphus designatus*)
Eastern Pondhawk (*Erythemis simplicicollis*)
Little Blue Dragonlet (*Erythrodiplax minuscula*)
Band-winged Dragonlet (*Erythrodiplax umbrata*)
Plains Clubtail (*Gomphus externus*)
Cocoa Clubtail (*Gomphus hybridus*)
Sulphur-tipped Clubtail (*Gomphus militaris*)
Oklahoma Clubtail (*Gomphus oklahomensis*)
Smoky Rubyspot (*Hetaerina titia*)
Citrine Forktail (*Ischnura hastata*)
Rambur's Forktail (*Ischnura ramburii*)
Slaty Skimmer (*Libellula incesta*)
Widow Skimmer (*Libellula luctuosa*)
Twelve-spotted Skimmer (*Libellula pulchella*)
Royal River Cruiser (*Macromia taeniolata*)
Blue Dasher (*Pachydiplax longipennis*)
Wandering Glider (*Pantala flavescens*)
Eastern Amberwing (*Perithemis tenera*)
Common Whitetail (*Plathemis lydia*)
Mocha Emerald (*Somatochlora linearis*)
Blue-faced Meadowhawk (*Sympetrum ambiguum*)
Variegated Meadowhawk (*Sympetrum corruptum*)
Black Saddlebags (*Tramea lacerata*)
Red Saddlebags (*Tramea onusta*)

Hutchinson (16)
Common Green Darner (*Anax junius*)
Paiute Dancer (*Argia alberta*)
Aztec Dancer (*Argia nahuana*)
Flag-tailed Spinyleg (*Dromogomphus spoliatus*)
Swift Setwing (*Dythemis velox*)
Eastern Pondhawk (*Erythemis simplicicollis*)
Sulphur-tipped Clubtail (*Gomphus militaris*)
American Rubyspot (*Hetaerina americana*)
Comanche Skimmer (*Libellula comanche*)

Bleached Skimmer (*Libellula composita*)
Twelve-spotted Skimmer (*Libellula pulchella*)
Flame Skimmer (*Libellula saturata*)
Blue Dasher (*Pachydiplax longipennis*)
Four-striped Leaftail (*Phyllogomphoides stigmatus*)
Blue-eyed Darner (*Rhionaeschna multicolor*)
Variegated Meadowhawk (*Sympetrum corruptum*)

Irion (6)
Pale-faced Clubskimmer (*Brechmorhoga mendax*)
Checkered Setwing (*Dythemis fugax*)
Swift Setwing (*Dythemis velox*)
Familiar Bluet (*Enallagma civile*)
Eastern Pondhawk (*Erythemis simplicicollis*)
American Rubyspot (*Hetaerina americana*)

Jack (35)
Common Green Darner (*Anax junius*)
Violet Dancer (*Argia fumipennis violacea*)
Kiowa Dancer (*Argia immunda*)
Springwater Dancer (*Argia plana*)
Calico Pennant (*Celithemis elisa*)
Halloween Pennant (*Celithemis eponina*)
Banded Pennant (*Celithemis fasciata*)
Checkered Setwing (*Dythemis fugax*)
Swift Setwing (*Dythemis velox*)
Double-striped Bluet (*Enallagma basidens*)
Familiar Bluet (*Enallagma civile*)
Dot-winged Baskettail (*Epitheca petechialis*)
Prince Baskettail (*Epitheca princeps*)
Eastern Ringtail (*Erpetogomphus designatus*)
Eastern Pondhawk (*Erythemis simplicicollis*)
Great Pondhawk (*Erythemis vesiculosa*)
Sulphur-tipped Clubtail (*Gomphus militaris*)
Citrine Forktail (*Ischnura hastata*)
Fragile Forktail (*Ischnura posita*)
Blue Corporal (*Ladona deplanata*)
Plateau Spreadwing (*Lestes alacer*)
Bleached Skimmer (*Libellula composita*)
Slaty Skimmer (*Libellula incesta*)
Widow Skimmer (*Libellula luctuosa*)
Twelve-spotted Skimmer (*Libellula pulchella*)
Blue Dasher (*Pachydiplax longipennis*)
Wandering Glider (*Pantala flavescens*)
Eastern Amberwing (*Perithemis tenera*)
Five-striped Leaftail (*Phyllogomphoides albrighti*)
Four-striped Leaftail (*Phyllogomphoides stigmatus*)
Common Sanddragon (*Progomphus obscurus*)
Variegated Meadowhawk (*Sympetrum corruptum*)
Autumn Meadowhawk (*Sympetrum vicinum*)
Black Saddlebags (*Tramea lacerata*)
Red Saddlebags (*Tramea onusta*)

Jackson (12)
Blue-fronted Dancer (*Argia apicalis*)
Blue-tipped Dancer (*Argia tibialis*)
Four-spotted Pennant (*Brachymesia gravida*)
Familiar Bluet (*Enallagma civile*)
Eastern Pondhawk (*Erythemis simplicicollis*)
Smoky Rubyspot (*Hetaerina titia*)
Rambur's Forktail (*Ischnura ramburii*)
Hyacinth Glider (*Miathyria marcella*)
Roseate Skimmer (*Orthemis ferruginea*)
Blue Dasher (*Pachydiplax longipennis*)
Eastern Amberwing (*Perithemis tenera*)
Russet-tipped Clubtail (*Stylurus plagiatus*)

Jasper (50)
Common Green Darner (*Anax junius*)
Violet Dancer (*Argia fumipennis violacea*)
Powdered Dancer (*Argia moesta*)
Aztec Dancer (*Argia nahuana*)
Blue-tipped Dancer (*Argia tibialis*)
Fawn Darner (*Boyeria vinosa*)
Four-spotted Pennant (*Brachymesia gravida*)

Sparkling Jewelwing (*Calopteryx dimidiata*)
Ebony Jewelwing (*Calopteryx maculata*)
Amanda's Pennant (*Celithemis amanda*)
Calico Pennant (*Celithemis elisa*)
Halloween Pennant (*Celithemis eponina*)
Banded Pennant (*Celithemis fasciata*)
Ornate Pennant (*Celithemis ornata*)
Twin-spotted Spiketail (*Cordulegaster maculata*)
Stream Cruiser (*Didymops transversa*)
Familiar Bluet (*Enallagma civile*)
Attenuated Bluet (*Enallagma daeckii*)
Turquoise Bluet (*Enallagma divagans*)
Big Bluet (*Enallagma durum*)
Orange Bluet (*Enallagma signatum*)
Slender Bluet {westfalli} (*Enallagma traviatum westfalli*)
Swamp Darner (*Epiaeschna heros*)
Common Baskettail (*Epitheca cynosura*)
Dot-winged Baskettail (*Epitheca petechialis*)
Prince Baskettail (*Epitheca princeps*)
Eastern Pondhawk (*Erythemis simplicicollis*)
Little Blue Dragonlet (*Erythrodiplax minuscula*)
Banner Clubtail (*Gomphus apomyius*)
Cocoa Clubtail (*Gomphus hybridus*)
Ashy Clubtail (*Gomphus lividus*)
Oklahoma Clubtail (*Gomphus oklahomensis*)
Smoky Rubyspot (*Hetaerina titia*)
Citrine Forktail (*Ischnura hastata*)
Rambur's Forktail (*Ischnura ramburii*)
Blue Corporal (*Ladona deplanata*)
Golden-winged Skimmer (*Libellula auripennis*)
Spangled Skimmer (*Libellula cyanea*)
Yellow-sided Skimmer (*Libellula flavida*)
Slaty Skimmer (*Libellula incesta*)
Needham's Skimmer (*Libellula needhami*)
Painted Skimmer (*Libellula semifasciata*)
Great Blue Skimmer (*Libellula vibrans*)
Hyacinth Glider (*Miathyria marcella*)
Southern Sprite (*Nehalennia integricollis*)
Blue Dasher (*Pachydiplax longipennis*)
Eastern Amberwing (*Perithemis tenera*)
Common Whitetail (*Plathemis lydia*)
Common Sanddragon (*Progomphus obscurus*)
Gray Petaltail (*Tachopteryx thoreyi*)

Jeff Davis (58)
Common Green Darner (*Anax junius*)
Giant Darner (*Anax walsinghami*)
Great Spreadwing (*Archilestes grandis*)
Blue-fronted Dancer (*Argia apicalis*)
Violet Dancer (*Argia fumipennis violacea*)
Lavender Dancer (*Argia hinei*)
Kiowa Dancer (*Argia immunda*)
Sooty Dancer (*Argia lugens*)
Powdered Dancer (*Argia moesta*)
Apache Dancer (*Argia munda*)
Aztec Dancer (*Argia nahuana*)
Springwater Dancer (*Argia plana*)
Blue-ringed Dancer (*Argia sedula*)
Pale-faced Clubskimmer (*Brechmorhoga mendax*)
Checkered Setwing (*Dythemis fugax*)
Swift Setwing (*Dythemis velox*)
Double-striped Bluet (*Enallagma basidens*)
Familiar Bluet (*Enallagma civile*)
Arroyo Bluet (*Enallagma praevarum*)
Dot-winged Baskettail (*Epitheca petechialis*)
White-belted Ringtail (*Erpetogomphus compositus*)
Eastern Ringtail (*Erpetogomphus designatus*)
Dashed Ringtail (*Erpetogomphus heterodon*)
Serpent Ringtail (*Erpetogomphus lampropeltis natrix*)
Eastern Pondhawk (*Erythemis simplicicollis*)
Plateau Dragonlet (*Erythrodiplax basifusca*)
Sulphur-tipped Clubtail (*Gomphus militaris*)
Painted Damsel (*Hesperagrion heterodoxum*)
American Rubyspot (*Hetaerina americana*)

Plains Forktail (*Ischnura damula*)
Mexican Forktail (*Ischnura demorsa*)
Black-fronted Forktail (*Ischnura denticollis*)
Citrine Forktail (*Ischnura hastata*)
Plateau Spreadwing (*Lestes alacer*)
Southern Spreadwing (*Lestes australis*)
Comanche Skimmer (*Libellula comanche*)
Neon Skimmer (*Libellula croceipennis*)
Widow Skimmer (*Libellula luctuosa*)
Flame Skimmer (*Libellula saturata*)
Roseate Skimmer (*Orthemis ferruginea*)
Blue Dasher (*Pachydiplax longipennis*)
Red Rock Skimmer (*Paltothemis lineatipes*)
Wandering Glider (*Pantala flavescens*)
Spot-winged Glider (*Pantala hymenaea*)
Eastern Amberwing (*Perithemis tenera*)
Common Whitetail (*Plathemis lydia*)
Desert Whitetail (*Plathemis subornata*)
Gray Sanddragon (*Progomphus borealis*)
Filigree Skimmer (*Pseudoleon superbus*)
Arroyo Darner (*Rhionaeschna dugesi*)
Blue-eyed Darner (*Rhionaeschna multicolor*)
Variegated Meadowhawk (*Sympetrum corruptum*)
Cardinal Meadowhawk (*Sympetrum illotum*)
Autumn Meadowhawk (*Sympetrum vicinum*)
Desert Firetail (*Telebasis salva*)
Carolina Saddlebags (*Tramea carolina*)
Black Saddlebags (*Tramea lacerata*)
Red Saddlebags (*Tramea onusta*)

Jefferson (18)
Common Green Darner (*Anax junius*)
Two-striped Forceptail (*Aphylla williamsoni*)
Blue-ringed Dancer (*Argia sedula*)
Blue-tipped Dancer (*Argia tibialis*)
Four-spotted Pennant (*Brachymesia gravida*)
Familiar Bluet (*Enallagma civile*)
Swamp Darner (*Epiaeschna heros*)
Eastern Pondhawk (*Erythemis simplicicollis*)
Great Pondhawk (*Erythemis vesiculosa*)
Little Blue Dragonlet (*Erythrodiplax minuscula*)
Rambur's Forktail (*Ischnura ramburii*)
Needham's Skimmer (*Libellula needhami*)
Roseate Skimmer (*Orthemis ferruginea*)
Blue Dasher (*Pachydiplax longipennis*)
Eastern Amberwing (*Perithemis tenera*)
Variegated Meadowhawk (*Sympetrum corruptum*)
Black Saddlebags (*Tramea lacerata*)
Red Saddlebags (*Tramea onusta*)

Jim Hogg (15)
Kiowa Dancer (*Argia immunda*)
Leonora's Dancer (*Argia leonorae*)
Blue-ringed Dancer (*Argia sedula*)
Red-tailed Pennant (*Brachymesia furcata*)
Familiar Bluet (*Enallagma civile*)
Eastern Pondhawk (*Erythemis simplicicollis*)
Citrine Forktail (*Ischnura hastata*)
Rambur's Forktail (*Ischnura ramburii*)
Comanche Skimmer (*Libellula comanche*)
Flame Skimmer (*Libellula saturata*)
Roseate Skimmer (*Orthemis ferruginea*)
Blue Dasher (*Pachydiplax longipennis*)
Common Whitetail (*Plathemis lydia*)
Variegated Meadowhawk (*Sympetrum corruptum*)
Desert Firetail (*Telebasis salva*)

Jim Wells (46)
Common Green Darner (*Anax junius*)
Broad-striped Forceptail (*Aphylla angustifolia*)
Blue-fronted Dancer (*Argia apicalis*)
Kiowa Dancer (*Argia immunda*)
Powdered Dancer (*Argia moesta*)
Blue-ringed Dancer (*Argia sedula*)
Dusky Dancer (*Argia translata*)
Four-spotted Pennant (*Brachymesia gravida*)
Halloween Pennant (*Celithemis eponina*)

Flag-tailed Spinyleg (*Dromogomphus spoliatus*)
Checkered Setwing (*Dythemis fugax*)
Black Setwing (*Dythemis nigrescens*)
Swift Setwing (*Dythemis velox*)
Double-striped Bluet (*Enallagma basidens*)
Familiar Bluet (*Enallagma civile*)
Prince Baskettail (*Epitheca princeps*)
Eastern Ringtail (*Erpetogomphus designatus*)
Eastern Pondhawk (*Erythemis simplicicollis*)
Great Pondhawk (*Erythemis vesiculosa*)
Little Blue Dragonlet (*Erythrodiplax minuscula*)
Band-winged Dragonlet (*Erythrodiplax umbrata*)
Plains Clubtail (*Gomphus externus*)
Sulphur-tipped Clubtail (*Gomphus militaris*)
Cobra Clubtail (*Gomphus vastus*)
American Rubyspot (*Hetaerina americana*)
Smoky Rubyspot (*Hetaerina titia*)
Citrine Forktail (*Ischnura hastata*)
Rambur's Forktail (*Ischnura ramburii*)
Golden-winged Skimmer (*Libellula auripennis*)
Widow Skimmer (*Libellula luctuosa*)
Needham's Skimmer (*Libellula needhami*)
Bronzed River Cruiser (*Macromia annulata*)
Royal River Cruiser (*Macromia taeniolata*)
Hyacinth Glider (*Miathyria marcella*)
Amelia's Threadtail (*Neoneura amelia*)
Roseate Skimmer (*Orthemis ferruginea*)
Blue Dasher (*Pachydiplax longipennis*)
Wandering Glider (*Pantala flavescens*)
Spot-winged Glider (*Pantala hymenaea*)
Eastern Amberwing (*Perithemis tenera*)
Common Whitetail (*Plathemis lydia*)
Common Sanddragon (*Progomphus obscurus*)
Russet-tipped Clubtail (*Stylurus plagiatus*)
Desert Firetail (*Telebasis salva*)
Black Saddlebags (*Tramea lacerata*)
Red Saddlebags (*Tramea onusta*)

Johnson (34)
Common Green Darner (*Anax junius*)
Blue-fronted Dancer (*Argia apicalis*)
Kiowa Dancer (*Argia immunda*)
Powdered Dancer (*Argia moesta*)
Blue-ringed Dancer (*Argia sedula*)
Dusky Dancer (*Argia translata*)
Jade Clubtail (*Arigomphus submedianus*)
Springtime Darner (*Basiaeschna janata*)
Four-spotted Pennant (*Brachymesia gravida*)
Flag-tailed Spinyleg (*Dromogomphus spoliatus*)
Checkered Setwing (*Dythemis fugax*)
Swift Setwing (*Dythemis velox*)
Double-striped Bluet (*Enallagma basidens*)
Familiar Bluet (*Enallagma civile*)
Turquoise Bluet (*Enallagma divagans*)
Orange Bluet (*Enallagma signatum*)
Dot-winged Baskettail (*Epitheca petechialis*)
Prince Baskettail (*Epitheca princeps*)
Eastern Pondhawk (*Erythemis simplicicollis*)
Band-winged Dragonlet (*Erythrodiplax umbrata*)
Sulphur-tipped Clubtail (*Gomphus militaris*)
Citrine Forktail (*Ischnura hastata*)
Rambur's Forktail (*Ischnura ramburii*)
Slaty Skimmer (*Libellula incesta*)
Widow Skimmer (*Libellula luctuosa*)
Orange Shadowdragon (*Neurocordulia xanthosoma*)
Roseate Skimmer (*Orthemis ferruginea*)
Blue Dasher (*Pachydiplax longipennis*)
Wandering Glider (*Pantala flavescens*)
Spot-winged Glider (*Pantala hymenaea*)
Eastern Amberwing (*Perithemis tenera*)
Common Whitetail (*Plathemis lydia*)
Variegated Meadowhawk (*Sympetrum corruptum*)
Black Saddlebags (*Tramea lacerata*)

Jones (15)
Blue-fronted Dancer (*Argia apicalis*)
Powdered Dancer (*Argia moesta*)

Blue-ringed Dancer (*Argia sedula*)
Double-striped Bluet (*Enallagma basidens*)
Familiar Bluet (*Enallagma civile*)
Eastern Ringtail (*Erpetogomphus designatus*)
Plains Clubtail (*Gomphus externus*)
Sulphur-tipped Clubtail (*Gomphus militaris*)
American Rubyspot (*Hetaerina americana*)
Plateau Spreadwing (*Lestes alacer*)
Twelve-spotted Skimmer (*Libellula pulchella*)
Bronzed River Cruiser (*Macromia annulata*)
Wandering Glider (*Pantala flavescens*)
Eastern Amberwing (*Perithemis tenera*)
Variegated Meadowhawk (*Sympetrum corruptum*)

Karnes *(20)*
Common Green Darner (*Anax junius*)
Blue-fronted Dancer (*Argia apicalis*)
Powdered Dancer (*Argia moesta*)
Blue-ringed Dancer (*Argia sedula*)
Halloween Pennant (*Celithemis eponina*)
Black Setwing (*Dythemis nigrescens*)
Eastern Pondhawk (*Erythemis simplicicollis*)
American Rubyspot (*Hetaerina americana*)
Smoky Rubyspot (*Hetaerina titia*)
Citrine Forktail (*Ischnura hastata*)
Fragile Forktail (*Ischnura posita*)
Rambur's Forktail (*Ischnura ramburii*)
Georgia River Cruiser (*Macromia illinoiensis georgina*)
Thornbush Dasher (*Micrathyria hagenii*)
Roseate Skimmer (*Orthemis ferruginea*)
Blue Dasher (*Pachydiplax longipennis*)
Eastern Amberwing (*Perithemis tenera*)
Five-striped Leaftail (*Phyllogomphoides albrighti*)
Black Saddlebags (*Tramea lacerata*)
Red Saddlebags (*Tramea onusta*)

Kaufman *(13)*
Blue-fronted Dancer (*Argia apicalis*)
Blue-ringed Dancer (*Argia sedula*)
Checkered Setwing (*Dythemis fugax*)
Prince Baskettail (*Epitheca princeps*)
Sulphur-tipped Clubtail (*Gomphus militaris*)
Fragile Forktail (*Ischnura posita*)
Rambur's Forktail (*Ischnura ramburii*)
Widow Skimmer (*Libellula luctuosa*)
Roseate Skimmer (*Orthemis ferruginea*)
Blue Dasher (*Pachydiplax longipennis*)
Eastern Amberwing (*Perithemis tenera*)
Common Whitetail (*Plathemis lydia*)
Russet-tipped Clubtail (*Stylurus plagiatus*)

Kendall *(38)*
Common Green Darner (*Anax junius*)
Blue-fronted Dancer (*Argia apicalis*)
Comanche Dancer (*Argia barretti*)
Coppery Dancer (*Argia cuprea*)
Violet Dancer (*Argia fumipennis violacea*)
Kiowa Dancer (*Argia immunda*)
Powdered Dancer (*Argia moesta*)
Aztec Dancer (*Argia nahuana*)
Springwater Dancer (*Argia plana*)
Blue-ringed Dancer (*Argia sedula*)
Dusky Dancer (*Argia translata*)
Springtime Darner (*Basiaeschna janata*)
Pale-faced Clubskimmer (*Brechmorhoga mendax*)
Black-shouldered Spinyleg (*Dromogomphus spinosus*)
Flag-tailed Spinyleg (*Dromogomphus spoliatus*)
Checkered Setwing (*Dythemis fugax*)
Black Setwing (*Dythemis nigrescens*)
Swift Setwing (*Dythemis velox*)
Double-striped Bluet (*Enallagma basidens*)
Stream Bluet (*Enallagma exsulans*)
Neotropical Bluet (*Enallagma novaehispaniae*)
Prince Baskettail (*Epitheca princeps*)
Eastern Ringtail (*Erpetogomphus designatus*)

Sulphur-tipped Clubtail (*Gomphus militaris*)
Dragonhunter (*Hagenius brevistylus*)
American Rubyspot (*Hetaerina americana*)
Smoky Rubyspot (*Hetaerina titia*)
Citrine Forktail (*Ischnura hastata*)
Rainpool Spreadwing (*Lestes forficula*)
Widow Skimmer (*Libellula luctuosa*)
Five-striped Leaftail (*Phyllogomphoides albrighti*)
Four-striped Leaftail (*Phyllogomphoides stigmatus*)
Common Sanddragon (*Progomphus obscurus*)
Orange-striped Threadtail (*Protoneura cara*)
Filigree Skimmer (*Pseudoleon superbus*)
Variegated Meadowhawk (*Sympetrum corruptum*)
Autumn Meadowhawk (*Sympetrum vicinum*)
Desert Firetail (*Telebasis salva*)

Kenedy *(24)*
Common Green Darner (*Anax junius*)
Comet Darner (*Anax longipes*)
Red-tailed Pennant (*Brachymesia furcata*)
Four-spotted Pennant (*Brachymesia gravida*)
Halloween Pennant (*Celithemis eponina*)
Familiar Bluet (*Enallagma civile*)
Eastern Pondhawk (*Erythemis simplicicollis*)
Band-winged Dragonlet (*Erythrodiplax umbrata*)
Citrine Forktail (*Ischnura hastata*)
Rambur's Forktail (*Ischnura ramburii*)
Plateau Spreadwing (*Lestes alacer*)
Southern Spreadwing (*Lestes australis*)
Needham's Skimmer (*Libellula needhami*)
Marl Pennant (*Macrodiplax balteata*)
Spot-tailed Dasher (*Micrathyria aequalis*)
Thornbush Dasher (*Micrathyria hagenii*)
Roseate Skimmer (*Orthemis ferruginea*)
Blue Dasher (*Pachydiplax longipennis*)
Spot-winged Glider (*Pantala hymenaea*)
Common Whitetail (*Plathemis lydia*)
Variegated Meadowhawk (*Sympetrum corruptum*)
Striped Saddlebags (*Tramea calverti*)
Black Saddlebags (*Tramea lacerata*)
Red Saddlebags (*Tramea onusta*)

Kent *(7)*
Powdered Dancer (*Argia moesta*)
Blue-ringed Dancer (*Argia sedula*)
Halloween Pennant (*Celithemis eponina*)
Double-striped Bluet (*Enallagma basidens*)
Familiar Bluet (*Enallagma civile*)
Desert Forktail (*Ischnura barberi*)
Red Saddlebags (*Tramea onusta*)

Kerr *(77)*
Common Green Darner (*Anax junius*)
Comet Darner (*Anax longipes*)
Great Spreadwing (*Archilestes grandis*)
Blue-fronted Dancer (*Argia apicalis*)
Comanche Dancer (*Argia barretti*)
Violet Dancer (*Argia fumipennis violacea*)
Kiowa Dancer (*Argia immunda*)
Leonora's Dancer (*Argia leonorae*)
Powdered Dancer (*Argia moesta*)
Aztec Dancer (*Argia nahuana*)
Springwater Dancer (*Argia plana*)
Blue-ringed Dancer (*Argia sedula*)
Dusky Dancer (*Argia translata*)
Springtime Darner (*Basiaeschna janata*)
Red-tailed Pennant (*Brachymesia furcata*)
Four-spotted Pennant (*Brachymesia gravida*)
Tawny Pennant (*Brachymesia herbida*)
Pale-faced Clubskimmer (*Brechmorhoga mendax*)
Halloween Pennant (*Celithemis eponina*)
Banded Pennant (*Celithemis fasciata*)
Black-shouldered Spinyleg (*Dromogomphus spinosus*)
Flag-tailed Spinyleg (*Dromogomphus spoliatus*)

Checkered Setwing (*Dythemis fugax*)
Black Setwing (*Dythemis nigrescens*)
Swift Setwing (*Dythemis velox*)
Double-striped Bluet (*Enallagma basidens*)
Familiar Bluet (*Enallagma civile*)
Stream Bluet (*Enallagma exsulans*)
Neotropical Bluet (*Enallagma novaehispaniae*)
Arroyo Bluet (*Enallagma praevarum*)
Orange Bluet (*Enallagma signatum*)
Vesper Bluet (*Enallagma vesperum*)
Dot-winged Baskettail (*Epitheca petechialis*)
Prince Baskettail (*Epitheca princeps*)
Eastern Ringtail (*Erpetogomphus designatus*)
Pin-tailed Pondhawk (*Erythemis plebeja*)
Eastern Pondhawk (*Erythemis simplicicollis*)
Great Pondhawk (*Erythemis vesiculosa*)
Little Blue Dragonlet (*Erythrodiplax minuscula*)
Band-winged Dragonlet (*Erythrodiplax umbrata*)
Pronghorn Clubtail (*Gomphus graslinellus*)
Sulphur-tipped Clubtail (*Gomphus militaris*)
Cobra Clubtail (*Gomphus vastus*)
Dragonhunter (*Hagenius brevistylus*)
American Rubyspot (*Hetaerina americana*)
Smoky Rubyspot (*Hetaerina titia*)
Citrine Forktail (*Ischnura hastata*)
Fragile Forktail (*Ischnura posita*)
Rambur's Forktail (*Ischnura ramburii*)
Plateau Spreadwing (*Lestes alacer*)
Chalky Spreadwing (*Lestes sigma*)
Comanche Skimmer (*Libellula comanche*)
Neon Skimmer (*Libellula croceipennis*)
Yellow-sided Skimmer (*Libellula flavida*)
Widow Skimmer (*Libellula luctuosa*)
Twelve-spotted Skimmer (*Libellula pulchella*)
Flame Skimmer (*Libellula saturata*)
Marl Pennant (*Macrodiplax balteata*)
Bronzed River Cruiser (*Macromia annulata*)
Thornbush Dasher (*Micrathyria hagenii*)
Coral-fronted Threadtail (*Neoneura aaroni*)
Orange Shadowdragon (*Neurocordulia xanthosoma*)
Roseate Skimmer (*Orthemis ferruginea*)
Blue Dasher (*Pachydiplax longipennis*)
Wandering Glider (*Pantala flavescens*)
Spot-winged Glider (*Pantala hymenaea*)
Eastern Amberwing (*Perithemis tenera*)
Five-striped Leaftail (*Phyllogomphoides albrighti*)
Four-striped Leaftail (*Phyllogomphoides stigmatus*)
Common Whitetail (*Plathemis lydia*)
Orange-striped Threadtail (*Protoneura cara*)
Variegated Meadowhawk (*Sympetrum corruptum*)
Autumn Meadowhawk (*Sympetrum vicinum*)
Desert Firetail (*Telebasis salva*)
Striped Saddlebags (*Tramea calverti*)
Black Saddlebags (*Tramea lacerata*)
Red Saddlebags (*Tramea onusta*)

Kimble *(57)*
Common Green Darner (*Anax junius*)
Blue-fronted Dancer (*Argia apicalis*)
Comanche Dancer (*Argia barretti*)
Violet Dancer (*Argia fumipennis violacea*)
Kiowa Dancer (*Argia immunda*)
Sooty Dancer (*Argia lugens*)
Powdered Dancer (*Argia moesta*)
Aztec Dancer (*Argia nahuana*)
Springwater Dancer (*Argia plana*)
Blue-ringed Dancer (*Argia sedula*)
Dusky Dancer (*Argia translata*)
Springtime Darner (*Basiaeschna janata*)
Pale-faced Clubskimmer (*Brechmorhoga mendax*)
Stream Cruiser (*Didymops transversa*)
Black-shouldered Spinyleg (*Dromogomphus spinosus*)

Flag-tailed Spinyleg (*Dromogomphus spoliatus*)
Checkered Setwing (*Dythemis fugax*)
Black Setwing (*Dythemis nigrescens*)
Swift Setwing (*Dythemis velox*)
Double-striped Bluet (*Enallagma basidens*)
Familiar Bluet (*Enallagma civile*)
Stream Bluet (*Enallagma exsulans*)
Stripe-winged Baskettail (*Epitheca costalis*)
Prince Baskettail (*Epitheca princeps*)
Eastern Ringtail (*Erpetogomphus designatus*)
Flame-tailed Pondhawk (*Erythemis peruviana*)
Eastern Pondhawk (*Erythemis simplicicollis*)
Sulphur-tipped Clubtail (*Gomphus militaris*)
Dragonhunter (*Hagenius brevistylus*)
American Rubyspot (*Hetaerina americana*)
Smoky Rubyspot (*Hetaerina titia*)
Citrine Forktail (*Ischnura hastata*)
Fragile Forktail (*Ischnura posita*)
Rambur's Forktail (*Ischnura ramburii*)
Plateau Spreadwing (*Lestes alacer*)
Comanche Skimmer (*Libellula comanche*)
Widow Skimmer (*Libellula luctuosa*)
Twelve-spotted Skimmer (*Libellula pulchella*)
Flame Skimmer (*Libellula saturata*)
Bronzed River Cruiser (*Macromia annulata*)
Gilded River Cruiser (*Macromia pacifica*)
Orange Shadowdragon (*Neurocordulia
    xanthosoma*)
Roseate Skimmer (*Orthemis ferruginea*)
Blue Dasher (*Pachydiplax longipennis*)
Wandering Glider (*Pantala flavescens*)
Spot-winged Glider (*Pantala hymenaea*)
Eastern Amberwing (*Perithemis tenera*)
Five-striped Leaftail (*Phyllogomphoides
    albrighti*)
Four-striped Leaftail (*Phyllogomphoides
    stigmatus*)
Common Whitetail (*Plathemis lydia*)
Orange-striped Threadtail (*Protoneura cara*)
Russet-tipped Clubtail (*Stylurus plagiatus*)
Variegated Meadowhawk (*Sympetrum
    corruptum*)
Autumn Meadowhawk (*Sympetrum vicinum*)
Desert Firetail (*Telebasis salva*)
Black Saddlebags (*Tramea lacerata*)
Red Saddlebags (*Tramea onusta*)

King (1)
Familiar Bluet (*Enallagma civile*)

Kinney (61)
Common Green Darner (*Anax junius*)
Broad-striped Forceptail (*Aphylla angustifolia*)
Blue-fronted Dancer (*Argia apicalis*)
Violet Dancer (*Argia fumipennis violacea*)
Kiowa Dancer (*Argia immunda*)
Leonora's Dancer (*Argia leonorae*)
Powdered Dancer (*Argia moesta*)
Aztec Dancer (*Argia nahuana*)
Golden-winged Dancer (*Argia rhoadsi*)
Blue-ringed Dancer (*Argia sedula*)
Dusky Dancer (*Argia translata*)
Springtime Darner (*Basiaeschna janata*)
Red-tailed Pennant (*Brachymesia furcata*)
Pale-faced Clubskimmer (*Brechmorhoga
    mendax*)
Gray-waisted Skimmer (*Cannaphila insularis
    funerea*)
Flag-tailed Spinyleg (*Dromogomphus spoliatus*)
Checkered Setwing (*Dythemis fugax*)
Black Setwing (*Dythemis nigrescens*)
Swift Setwing (*Dythemis velox*)
Double-striped Bluet (*Enallagma basidens*)
Familiar Bluet (*Enallagma civile*)
Stream Bluet (*Enallagma exsulans*)
Neotropical Bluet (*Enallagma novaehispaniae*)
Arroyo Bluet (*Enallagma praevarum*)
Orange Bluet (*Enallagma signatum*)
Dot-winged Baskettail (*Epitheca petechialis*)

Prince Baskettail (*Epitheca princeps*)
Eastern Ringtail (*Erpetogomphus designatus*)
Eastern Pondhawk (*Erythemis simplicicollis*)
Great Pondhawk (*Erythemis vesiculosa*)
Sulphur-tipped Clubtail (*Gomphus militaris*)
American Rubyspot (*Hetaerina americana*)
Smoky Rubyspot (*Hetaerina titia*)
Citrine Forktail (*Ischnura hastata*)
Fragile Forktail (*Ischnura posita*)
Rambur's Forktail (*Ischnura ramburii*)
Plateau Spreadwing (*Lestes alacer*)
Chalky Spreadwing (*Lestes sigma*)
Comanche Skimmer (*Libellula comanche*)
Neon Skimmer (*Libellula croceipennis*)
Widow Skimmer (*Libellula luctuosa*)
Flame Skimmer (*Libellula saturata*)
Bronzed River Cruiser (*Macromia annulata*)
Ivory-striped Sylph (*Macrothemis imitans
    leucozona*)
Straw-colored Sylph (*Macrothemis inacuta*)
Roseate Skimmer (*Orthemis ferruginea*)
Blue Dasher (*Pachydiplax longipennis*)
Wandering Glider (*Pantala flavescens*)
Spot-winged Glider (*Pantala hymenaea*)
Eastern Amberwing (*Perithemis tenera*)
Five-striped Leaftail (*Phyllogomphoides
    albrighti*)
Four-striped Leaftail (*Phyllogomphoides
    stigmatus*)
Common Whitetail (*Plathemis lydia*)
Filigree Skimmer (*Pseudoleon superbus*)
Russet-tipped Clubtail (*Stylurus plagiatus*)
Variegated Meadowhawk (*Sympetrum
    corruptum*)
Desert Firetail (*Telebasis salva*)
Striped Saddlebags (*Tramea calverti*)
Antillean Saddlebags (*Tramea insularis*)
Black Saddlebags (*Tramea lacerata*)
Red Saddlebags (*Tramea onusta*)

Kleberg (39)
Mexican Wedgetail (*Acanthagrion quadratum*)
Common Green Darner (*Anax junius*)
Narrow-striped Forceptail (*Aphylla protracta*)
Kiowa Dancer (*Argia immunda*)
Blue-ringed Dancer (*Argia sedula*)
Red-tailed Pennant (*Brachymesia furcata*)
Blue-faced Darner (*Coryphaeschna adnexa*)
Familiar Bluet (*Enallagma civile*)
Neotropical Bluet (*Enallagma novaehispaniae*)
Black Pondhawk (*Erythemis attala*)
Pin-tailed Pondhawk (*Erythemis plebeja*)
Eastern Pondhawk (*Erythemis simplicicollis*)
Great Pondhawk (*Erythemis vesiculosa*)
Seaside Dragonlet (*Erythrodiplax berenice*)
Band-winged Dragonlet (*Erythrodiplax umbrata*)
Smoky Rubyspot (*Hetaerina titia*)
Citrine Forktail (*Ischnura hastata*)
Rambur's Forktail (*Ischnura ramburii*)
Cream-tipped Swampdamsel (*Leptobasis
    melinogaster*)
Rainpool Spreadwing (*Lestes forficula*)
Chalky Spreadwing (*Lestes sigma*)
Twelve-spotted Skimmer (*Libellula pulchella*)
Flame Skimmer (*Libellula saturata*)
Hyacinth Glider (*Miathyria marcella*)
Spot-tailed Dasher (*Micrathyria aequalis*)
Thornbush Dasher (*Micrathyria hagenii*)
Carmine Skimmer (*Orthemis discolor*)
Roseate Skimmer (*Orthemis ferruginea*)
Blue Dasher (*Pachydiplax longipennis*)
Wandering Glider (*Pantala flavescens*)
Spot-winged Glider (*Pantala hymenaea*)
Slough Amberwing (*Perithemis domitia*)
Common Whitetail (*Plathemis lydia*)
Variegated Meadowhawk (*Sympetrum
    corruptum*)
Aztec Glider (*Tauriphila azteca*)

Desert Firetail (*Telebasis salva*)
Evening Skimmer (*Tholymis citrina*)
Black Saddlebags (*Tramea lacerata*)
Red Saddlebags (*Tramea onusta*)

Knox (10)
Blue-fronted Dancer (*Argia apicalis*)
Powdered Dancer (*Argia moesta*)
Blue-ringed Dancer (*Argia sedula*)
Flag-tailed Spinyleg (*Dromogomphus spoliatus*)
Swift Setwing (*Dythemis velox*)
Double-striped Bluet (*Enallagma basidens*)
Familiar Bluet (*Enallagma civile*)
Plateau Spreadwing (*Lestes alacer*)
Eastern Amberwing (*Perithemis tenera*)
Variegated Meadowhawk (*Sympetrum
    corruptum*)

La Salle (31)
Broad-striped Forceptail (*Aphylla angustifolia*)
Blue-fronted Dancer (*Argia apicalis*)
Powdered Dancer (*Argia moesta*)
Blue-ringed Dancer (*Argia sedula*)
Red-tailed Pennant (*Brachymesia furcata*)
Halloween Pennant (*Celithemis eponina*)
Checkered Setwing (*Dythemis fugax*)
Double-striped Bluet (*Enallagma basidens*)
Familiar Bluet (*Enallagma civile*)
Prince Baskettail (*Epitheca princeps*)
Eastern Ringtail (*Erpetogomphus designatus*)
Eastern Pondhawk (*Erythemis simplicicollis*)
Seaside Dragonlet (*Erythrodiplax berenice*)
Plains Clubtail (*Gomphus externus*)
Sulphur-tipped Clubtail (*Gomphus militaris*)
American Rubyspot (*Hetaerina americana*)
Smoky Rubyspot (*Hetaerina titia*)
Citrine Forktail (*Ischnura hastata*)
Rambur's Forktail (*Ischnura ramburii*)
Chalky Spreadwing (*Lestes sigma*)
Twelve-spotted Skimmer (*Libellula pulchella*)
Bronzed River Cruiser (*Macromia annulata*)
Thornbush Dasher (*Micrathyria hagenii*)
Roseate Skimmer (*Orthemis ferruginea*)
Blue Dasher (*Pachydiplax longipennis*)
Wandering Glider (*Pantala flavescens*)
Spot-winged Glider (*Pantala hymenaea*)
Eastern Amberwing (*Perithemis tenera*)
Common Whitetail (*Plathemis lydia*)
Black Saddlebags (*Tramea lacerata*)
Red Saddlebags (*Tramea onusta*)

Lamar (46)
Common Green Darner (*Anax junius*)
Blue-fronted Dancer (*Argia apicalis*)
Seepage Dancer (*Argia bipunctulata*)
Violet Dancer (*Argia fumipennis violacea*)
Powdered Dancer (*Argia moesta*)
Aztec Dancer (*Argia nahuana*)
Springwater Dancer (*Argia plana*)
Blue-ringed Dancer (*Argia sedula*)
Blue-tipped Dancer (*Argia tibialis*)
Springtime Darner (*Basiaeschna janata*)
Ebony Jewelwing (*Calopteryx maculata*)
Calico Pennant (*Celithemis elisa*)
Stream Cruiser (*Didymops transversa*)
Black-shouldered Spinyleg (*Dromogomphus
    spinosus*)
Azure Bluet (*Enallagma aspersum*)
Double-striped Bluet (*Enallagma basidens*)
Familiar Bluet (*Enallagma civile*)
Turquoise Bluet (*Enallagma divagans*)
Orange Bluet (*Enallagma signatum*)
Stripe-winged Baskettail (*Epitheca costalis*)
Common Baskettail (*Epitheca cynosura*)
Prince Baskettail (*Epitheca princeps*)
Mantled Baskettail (*Epitheca semiaquea*)
Eastern Pondhawk (*Erythemis simplicicollis*)
Plains Clubtail (*Gomphus externus*)
Oklahoma Clubtail (*Gomphus oklahomensis*)
Citrine Forktail (*Ischnura hastata*)

Fragile Forktail (*Ischnura posita*)
Rambur's Forktail (*Ischnura ramburii*)
Blue Corporal (*Ladona deplanata*)
Southern Spreadwing (*Lestes australis*)
Elegant Spreadwing (*Lestes inaequalis*)
Spangled Skimmer (*Libellula cyanea*)
Slaty Skimmer (*Libellula incesta*)
Widow Skimmer (*Libellula luctuosa*)
Great Blue Skimmer (*Libellula vibrans*)
Roseate Skimmer (*Orthemis ferruginea*)
Blue Dasher (*Pachydiplax longipennis*)
Spot-winged Glider (*Pantala hymenaea*)
Eastern Amberwing (*Perithemis tenera*)
Common Whitetail (*Plathemis lydia*)
Blue-faced Meadowhawk (*Sympetrum ambiguum*)
Variegated Meadowhawk (*Sympetrum corruptum*)
Autumn Meadowhawk (*Sympetrum vicinum*)
Black Saddlebags (*Tramea lacerata*)
Red Saddlebags (*Tramea onusta*)

Lampasas *(20)*
Violet Dancer (*Argia fumipennis violacea*)
Kiowa Dancer (*Argia immunda*)
Aztec Dancer (*Argia nahuana*)
Springwater Dancer (*Argia plana*)
Blue-ringed Dancer (*Argia sedula*)
Swift Setwing (*Dythemis velox*)
Double-striped Bluet (*Enallagma basidens*)
Prince Baskettail (*Epitheca princeps*)
Eastern Ringtail (*Erpetogomphus designatus*)
Sulphur-tipped Clubtail (*Gomphus militaris*)
American Rubyspot (*Hetaerina americana*)
Citrine Forktail (*Ischnura hastata*)
Rambur's Forktail (*Ischnura ramburii*)
Comanche Skimmer (*Libellula comanche*)
Widow Skimmer (*Libellula luctuosa*)
Orange Shadowdragon (*Neurocordulia xanthosoma*)
Roseate Skimmer (*Orthemis ferruginea*)
Blue Dasher (*Pachydiplax longipennis*)
Five-striped Leaftail (*Phyllogomphoides albrighti*)
Autumn Meadowhawk (*Sympetrum vicinum*)

Lavaca *(31)*
Common Green Darner (*Anax junius*)
Violet Dancer (*Argia fumipennis violacea*)
Kiowa Dancer (*Argia immunda*)
Powdered Dancer (*Argia moesta*)
Blue-ringed Dancer (*Argia sedula*)
Dusky Dancer (*Argia translata*)
Four-spotted Pennant (*Brachymesia gravida*)
Halloween Pennant (*Celithemis eponina*)
Checkered Setwing (*Dythemis fugax*)
Swift Setwing (*Dythemis velox*)
Double-striped Bluet (*Enallagma basidens*)
Familiar Bluet (*Enallagma civile*)
Eastern Ringtail (*Erpetogomphus designatus*)
Eastern Pondhawk (*Erythemis simplicicollis*)
Band-winged Dragonlet (*Erythrodiplax umbrata*)
American Rubyspot (*Hetaerina americana*)
Smoky Rubyspot (*Hetaerina titia*)
Citrine Forktail (*Ischnura hastata*)
Rambur's Forktail (*Ischnura ramburii*)
Blue Corporal (*Ladona deplanata*)
Neon Skimmer (*Libellula croceipennis*)
Widow Skimmer (*Libellula luctuosa*)
Great Blue Skimmer (*Libellula vibrans*)
Carmine Skimmer (*Orthemis discolor*)
Roseate Skimmer (*Orthemis ferruginea*)
Blue Dasher (*Pachydiplax longipennis*)
Eastern Amberwing (*Perithemis tenera*)
Common Whitetail (*Plathemis lydia*)
Common Sanddragon (*Progomphus obscurus*)
Desert Firetail (*Telebasis salva*)
Red Saddlebags (*Tramea onusta*)

Lee *(32)*

Common Green Darner (*Anax junius*)
Broad-striped Forceptail (*Aphylla angustifolia*)
Narrow-striped Forceptail (*Aphylla protracta*)
Blue-fronted Dancer (*Argia apicalis*)
Blue-ringed Dancer (*Argia sedula*)
Four-spotted Pennant (*Brachymesia gravida*)
Ebony Jewelwing (*Calopteryx maculata*)
Halloween Pennant (*Celithemis eponina*)
Stream Cruiser (*Didymops transversa*)
Familiar Bluet (*Enallagma civile*)
Turquoise Bluet (*Enallagma divagans*)
Big Bluet (*Enallagma durum*)
Stream Bluet (*Enallagma exsulans*)
Orange Bluet (*Enallagma signatum*)
Prince Baskettail (*Epitheca princeps*)
Mantled Baskettail (*Epitheca semiaquea*)
Eastern Pondhawk (*Erythemis simplicicollis*)
Oklahoma Clubtail (*Gomphus oklahomensis*)
Smoky Rubyspot (*Hetaerina titia*)
Citrine Forktail (*Ischnura hastata*)
Fragile Forktail (*Ischnura posita*)
Rambur's Forktail (*Ischnura ramburii*)
Slaty Skimmer (*Libellula incesta*)
Widow Skimmer (*Libellula luctuosa*)
Twelve-spotted Skimmer (*Libellula pulchella*)
Great Blue Skimmer (*Libellula vibrans*)
Roseate Skimmer (*Orthemis ferruginea*)
Blue Dasher (*Pachydiplax longipennis*)
Eastern Amberwing (*Perithemis tenera*)
Common Whitetail (*Plathemis lydia*)
Variegated Meadowhawk (*Sympetrum corruptum*)
Black Saddlebags (*Tramea lacerata*)

Leon *(34)*
Common Green Darner (*Anax junius*)
Blue-fronted Dancer (*Argia apicalis*)
Violet Dancer (*Argia fumipennis violacea*)
Powdered Dancer (*Argia moesta*)
Blue-ringed Dancer (*Argia sedula*)
Blue-tipped Dancer (*Argia tibialis*)
Ebony Jewelwing (*Calopteryx maculata*)
Calico Pennant (*Celithemis elisa*)
Halloween Pennant (*Celithemis eponina*)
Flag-tailed Spinyleg (*Dromogomphus spoliatus*)
Swift Setwing (*Dythemis velox*)
Familiar Bluet (*Enallagma civile*)
Swamp Darner (*Epiaeschna heros*)
Eastern Ringtail (*Erpetogomphus designatus*)
Eastern Pondhawk (*Erythemis simplicicollis*)
Little Blue Dragonlet (*Erythrodiplax minuscula*)
Cobra Clubtail (*Gomphus vastus*)
Smoky Rubyspot (*Hetaerina titia*)
Citrine Forktail (*Ischnura hastata*)
Rambur's Forktail (*Ischnura ramburii*)
Rainpool Spreadwing (*Lestes forficula*)
Slaty Skimmer (*Libellula incesta*)
Widow Skimmer (*Libellula luctuosa*)
Great Blue Skimmer (*Libellula vibrans*)
Cyrano Darner (*Nasiaeschna pentacantha*)
Blue Dasher (*Pachydiplax longipennis*)
Wandering Glider (*Pantala flavescens*)
Spot-winged Glider (*Pantala hymenaea*)
Eastern Amberwing (*Perithemis tenera*)
Common Whitetail (*Plathemis lydia*)
Common Sanddragon (*Progomphus obscurus*)
Mocha Emerald (*Somatochlora linearis*)
Russet-tipped Clubtail (*Stylurus plagiatus*)
Black Saddlebags (*Tramea lacerata*)

Liberty *(55)*
Common Green Darner (*Anax junius*)
Comet Darner (*Anax longipes*)
Two-striped Forceptail (*Aphylla williamsoni*)
Blue-fronted Dancer (*Argia apicalis*)
Powdered Dancer (*Argia moesta*)
Blue-ringed Dancer (*Argia sedula*)
Blue-tipped Dancer (*Argia tibialis*)
Bayou Clubtail (*Arigomphus maxwelli*)

Calico Pennant (*Celithemis elisa*)
Halloween Pennant (*Celithemis eponina*)
Banded Pennant (*Celithemis fasciata*)
Ornate Pennant (*Celithemis ornata*)
Double-ringed Pennant (*Celithemis verna*)
Regal Darner (*Coryphaeschna ingens*)
Stream Cruiser (*Didymops transversa*)
Black-shouldered Spinyleg (*Dromogomphus spinosus*)
Flag-tailed Spinyleg (*Dromogomphus spoliatus*)
Double-striped Bluet (*Enallagma basidens*)
Familiar Bluet (*Enallagma civile*)
Burgundy Bluet (*Enallagma dubium*)
Stream Bluet (*Enallagma exsulans*)
Orange Bluet (*Enallagma signatum*)
Swamp Darner (*Epiaeschna heros*)
Common Baskettail (*Epitheca cynosura*)
Prince Baskettail (*Epitheca princeps*)
Mantled Baskettail (*Epitheca semiaquea*)
Eastern Ringtail (*Erpetogomphus designatus*)
Eastern Pondhawk (*Erythemis simplicicollis*)
Little Blue Dragonlet (*Erythrodiplax minuscula*)
Plains Clubtail (*Gomphus externus*)
Gulf Coast Clubtail (*Gomphus modestus*)
Oklahoma Clubtail (*Gomphus oklahomensis*)
Smoky Rubyspot (*Hetaerina titia*)
Citrine Forktail (*Ischnura hastata*)
Rambur's Forktail (*Ischnura ramburii*)
Blue Corporal (*Ladona deplanata*)
Swamp Spreadwing (*Lestes vigilax*)
Golden-winged Skimmer (*Libellula auripennis*)
Slaty Skimmer (*Libellula incesta*)
Needham's Skimmer (*Libellula needhami*)
Painted Skimmer (*Libellula semifasciata*)
Great Blue Skimmer (*Libellula vibrans*)
Hyacinth Glider (*Miathyria marcella*)
Cyrano Darner (*Nasiaeschna pentacantha*)
Roseate Skimmer (*Orthemis ferruginea*)
Blue Dasher (*Pachydiplax longipennis*)
Wandering Glider (*Pantala flavescens*)
Spot-winged Glider (*Pantala hymenaea*)
Eastern Amberwing (*Perithemis tenera*)
Common Whitetail (*Plathemis lydia*)
Common Sanddragon (*Progomphus obscurus*)
Gray Petaltail (*Tachopteryx thoreyi*)
Carolina Saddlebags (*Tramea carolina*)
Black Saddlebags (*Tramea lacerata*)
Red Saddlebags (*Tramea onusta*)

Limestone *(29)*
Blue-fronted Dancer (*Argia apicalis*)
Kiowa Dancer (*Argia immunda*)
Powdered Dancer (*Argia moesta*)
Blue-ringed Dancer (*Argia sedula*)
Blue-tipped Dancer (*Argia tibialis*)
Dusky Dancer (*Argia translata*)
Four-spotted Pennant (*Brachymesia gravida*)
Flag-tailed Spinyleg (*Dromogomphus spoliatus*)
Checkered Setwing (*Dythemis fugax*)
Swift Setwing (*Dythemis velox*)
Orange Bluet (*Enallagma signatum*)
Eastern Pondhawk (*Erythemis simplicicollis*)
Sulphur-tipped Clubtail (*Gomphus militaris*)
American Rubyspot (*Hetaerina americana*)
Citrine Forktail (*Ischnura hastata*)
Rambur's Forktail (*Ischnura ramburii*)
Widow Skimmer (*Libellula luctuosa*)
Twelve-spotted Skimmer (*Libellula pulchella*)
Great Blue Skimmer (*Libellula vibrans*)
Roseate Skimmer (*Orthemis ferruginea*)
Blue Dasher (*Pachydiplax longipennis*)
Wandering Glider (*Pantala flavescens*)
Spot-winged Glider (*Pantala hymenaea*)
Eastern Amberwing (*Perithemis tenera*)
Four-striped Leaftail (*Phyllogomphoides stigmatus*)
Common Whitetail (*Plathemis lydia*)
Common Sanddragon (*Progomphus obscurus*)

Black Saddlebags (*Tramea lacerata*)
Red Saddlebags (*Tramea onusta*)

Lipscomb (8)
Blue-fronted Dancer (*Argia apicalis*)
Double-striped Bluet (*Enallagma basidens*)
Familiar Bluet (*Enallagma civile*)
Citrine Forktail (*Ischnura hastata*)
Eastern Forktail (*Ischnura verticalis*)
Lyre-tipped Spreadwing (*Lestes unguiculatus*)
Roseate Skimmer (*Orthemis ferruginea*)
Common Whitetail (*Plathemis lydia*)

Live Oak (23)
Common Green Darner (*Anax junius*)
Blue-fronted Dancer (*Argia apicalis*)
Powdered Dancer (*Argia moesta*)
Blue-ringed Dancer (*Argia sedula*)
Four-spotted Pennant (*Brachymesia gravida*)
Swift Setwing (*Dythemis velox*)
Double-striped Bluet (*Enallagma basidens*)
Familiar Bluet (*Enallagma civile*)
Big Bluet (*Enallagma durum*)
Orange Bluet (*Enallagma signatum*)
Eastern Pondhawk (*Erythemis simplicicollis*)
Band-winged Dragonlet (*Erythrodiplax umbrata*)
Plains Clubtail (*Gomphus externus*)
Sulphur-tipped Clubtail (*Gomphus militaris*)
American Rubyspot (*Hetaerina americana*)
Smoky Rubyspot (*Hetaerina titia*)
Rambur's Forktail (*Ischnura ramburii*)
Cyrano Darner (*Nasiaeschna pentacantha*)
Roseate Skimmer (*Orthemis ferruginea*)
Wandering Glider (*Pantala flavescens*)
Eastern Amberwing (*Perithemis tenera*)
Variegated Meadowhawk (*Sympetrum corruptum*)
Red Saddlebags (*Tramea onusta*)

Llano (34)
Great Spreadwing (*Archilestes grandis*)
Blue-fronted Dancer (*Argia apicalis*)
Kiowa Dancer (*Argia immunda*)
Powdered Dancer (*Argia moesta*)
Aztec Dancer (*Argia nahuana*)
Blue-ringed Dancer (*Argia sedula*)
Pale-faced Clubskimmer (*Brechmorhoga mendax*)
Halloween Pennant (*Celithemis eponina*)
Flag-tailed Spinyleg (*Dromogomphus spoliatus*)
Checkered Setwing (*Dythemis fugax*)
Swift Setwing (*Dythemis velox*)
Double-striped Bluet (*Enallagma basidens*)
Familiar Bluet (*Enallagma civile*)
Dot-winged Baskettail (*Epitheca petechialis*)
Prince Baskettail (*Epitheca princeps*)
Eastern Ringtail (*Erpetogomphus designatus*)
Eastern Pondhawk (*Erythemis simplicicollis*)
Sulphur-tipped Clubtail (*Gomphus militaris*)
American Rubyspot (*Hetaerina americana*)
Rambur's Forktail (*Ischnura ramburii*)
Comanche Skimmer (*Libellula comanche*)
Neon Skimmer (*Libellula croceipennis*)
Widow Skimmer (*Libellula luctuosa*)
Roseate Skimmer (*Orthemis ferruginea*)
Blue Dasher (*Pachydiplax longipennis*)
Wandering Glider (*Pantala flavescens*)
Spot-winged Glider (*Pantala hymenaea*)
Common Whitetail (*Plathemis lydia*)
Common Sanddragon (*Progomphus obscurus*)
Blue-faced Meadowhawk (*Sympetrum ambiguum*)
Variegated Meadowhawk (*Sympetrum corruptum*)
Desert Firetail (*Telebasis salva*)
Black Saddlebags (*Tramea lacerata*)
Red Saddlebags (*Tramea onusta*)

Loving (9)
Powdered Dancer (*Argia moesta*)
Checkered Setwing (*Dythemis fugax*)

Familiar Bluet (*Enallagma civile*)
Eastern Ringtail (*Erpetogomphus designatus*)
Desert Forktail (*Ischnura barberi*)
Bleached Skimmer (*Libellula composita*)
Common Whitetail (*Plathemis lydia*)
Common Sanddragon (*Progomphus obscurus*)
Variegated Meadowhawk (*Sympetrum corruptum*)

Lubbock (52)
Common Green Darner (*Anax junius*)
Great Spreadwing (*Archilestes grandis*)
Blue-fronted Dancer (*Argia apicalis*)
Kiowa Dancer (*Argia immunda*)
Sooty Dancer (*Argia lugens*)
Powdered Dancer (*Argia moesta*)
Springwater Dancer (*Argia plana*)
Blue-ringed Dancer (*Argia sedula*)
Dusky Dancer (*Argia translata*)
Four-spotted Pennant (*Brachymesia gravida*)
Halloween Pennant (*Celithemis eponina*)
Flag-tailed Spinyleg (*Dromogomphus spoliatus*)
Checkered Setwing (*Dythemis fugax*)
Swift Setwing (*Dythemis velox*)
Double-striped Bluet (*Enallagma basidens*)
Familiar Bluet (*Enallagma civile*)
Dot-winged Baskettail (*Epitheca petechialis*)
Prince Baskettail (*Epitheca princeps*)
Eastern Ringtail (*Erpetogomphus designatus*)
Eastern Pondhawk (*Erythemis simplicicollis*)
Great Pondhawk (*Erythemis vesiculosa*)
Band-winged Dragonlet (*Erythrodiplax umbrata*)
Sulphur-tipped Clubtail (*Gomphus militaris*)
American Rubyspot (*Hetaerina americana*)
Desert Forktail (*Ischnura barberi*)
Plains Forktail (*Ischnura damula*)
Black-fronted Forktail (*Ischnura denticollis*)
Fragile Forktail (*Ischnura posita*)
Rambur's Forktail (*Ischnura ramburii*)
Eastern Forktail (*Ischnura verticalis*)
Plateau Spreadwing (*Lestes alacer*)
Southern Spreadwing (*Lestes australis*)
Comanche Skimmer (*Libellula comanche*)
Neon Skimmer (*Libellula croceipennis*)
Widow Skimmer (*Libellula luctuosa*)
Twelve-spotted Skimmer (*Libellula pulchella*)
Flame Skimmer (*Libellula saturata*)
Roseate Skimmer (*Orthemis ferruginea*)
Blue Dasher (*Pachydiplax longipennis*)
Wandering Glider (*Pantala flavescens*)
Spot-winged Glider (*Pantala hymenaea*)
Eastern Amberwing (*Perithemis tenera*)
Four-striped Leaftail (*Phyllogomphoides stigmatus*)
Common Whitetail (*Plathemis lydia*)
Blue-eyed Darner (*Rhionaeschna multicolor*)
Turquoise-tipped Darner (*Rhionaeschna psilus*)
Variegated Meadowhawk (*Sympetrum corruptum*)
Autumn Meadowhawk (*Sympetrum vicinum*)
Desert Firetail (*Telebasis salva*)
Striped Saddlebags (*Tramea calverti*)
Black Saddlebags (*Tramea lacerata*)
Red Saddlebags (*Tramea onusta*)

Lynn (5)
Familiar Bluet (*Enallagma civile*)
Sulphur-tipped Clubtail (*Gomphus militaris*)
Plateau Spreadwing (*Lestes alacer*)
Roseate Skimmer (*Orthemis ferruginea*)
Variegated Meadowhawk (*Sympetrum corruptum*)

Madison (19)
Common Green Darner (*Anax junius*)
Blue-fronted Dancer (*Argia apicalis*)
Powdered Dancer (*Argia moesta*)
Blue-ringed Dancer (*Argia sedula*)
Blue-tipped Dancer (*Argia tibialis*)
Fawn Darner (*Boyeria vinosa*)

Ebony Jewelwing (*Calopteryx maculata*)
Halloween Pennant (*Celithemis eponina*)
Eastern Pondhawk (*Erythemis simplicicollis*)
Cobra Clubtail (*Gomphus vastus*)
Citrine Forktail (*Ischnura hastata*)
Spangled Skimmer (*Libellula cyanea*)
Yellow-sided Skimmer (*Libellula flavida*)
Slaty Skimmer (*Libellula incesta*)
Widow Skimmer (*Libellula luctuosa*)
Blue Dasher (*Pachydiplax longipennis*)
Eastern Amberwing (*Perithemis tenera*)
Common Whitetail (*Plathemis lydia*)
Black Saddlebags (*Tramea lacerata*)

Marion (43)
Common Green Darner (*Anax junius*)
Blue-fronted Dancer (*Argia apicalis*)
Violet Dancer (*Argia fumipennis violacea*)
Powdered Dancer (*Argia moesta*)
Blue-tipped Dancer (*Argia tibialis*)
Dusky Dancer (*Argia translata*)
Stillwater Clubtail (*Arigomphus lentulus*)
Bayou Clubtail (*Arigomphus maxwelli*)
Springtime Darner (*Basiaeschna janata*)
Four-spotted Pennant (*Brachymesia gravida*)
Ebony Jewelwing (*Calopteryx maculata*)
Halloween Pennant (*Celithemis eponina*)
Black-shouldered Spinyleg (*Dromogomphus spinosus*)
Flag-tailed Spinyleg (*Dromogomphus spoliatus*)
Double-striped Bluet (*Enallagma basidens*)
Burgundy Bluet (*Enallagma dubium*)
Stream Bluet (*Enallagma exsulans*)
Orange Bluet (*Enallagma signatum*)
Vesper Bluet (*Enallagma vesperum*)
Common Baskettail (*Epitheca cynosura*)
Prince Baskettail (*Epitheca princeps*)
Mantled Baskettail (*Epitheca semiaquea*)
Eastern Pondhawk (*Erythemis simplicicollis*)
Sulphur-tipped Clubtail (*Gomphus militaris*)
Citrine Forktail (*Ischnura hastata*)
Lilypad Forktail (*Ischnura kellicotti*)
Fragile Forktail (*Ischnura posita*)
Rambur's Forktail (*Ischnura ramburii*)
Chalky Spreadwing (*Lestes sigma*)
Swamp Spreadwing (*Lestes vigilax*)
Spangled Skimmer (*Libellula cyanea*)
Slaty Skimmer (*Libellula incesta*)
Widow Skimmer (*Libellula luctuosa*)
Great Blue Skimmer (*Libellula vibrans*)
Royal River Cruiser (*Macromia taeniolata*)
Orange Shadowdragon (*Neurocordulia xanthosoma*)
Blue Dasher (*Pachydiplax longipennis*)
Wandering Glider (*Pantala flavescens*)
Eastern Amberwing (*Perithemis tenera*)
Common Whitetail (*Plathemis lydia*)
Russet-tipped Clubtail (*Stylurus plagiatus*)
Black Saddlebags (*Tramea lacerata*)
Red Saddlebags (*Tramea onusta*)

Martin (2)
Widow Skimmer (*Libellula luctuosa*)
Flame Skimmer (*Libellula saturata*)

Mason (35)
Lavender Dancer (*Argia hinei*)
Kiowa Dancer (*Argia immunda*)
Powdered Dancer (*Argia moesta*)
Aztec Dancer (*Argia nahuana*)
Blue-ringed Dancer (*Argia sedula*)
Dusky Dancer (*Argia translata*)
Pale-faced Clubskimmer (*Brechmorhoga mendax*)
Ebony Jewelwing (*Calopteryx maculata*)
Flag-tailed Spinyleg (*Dromogomphus spoliatus*)
Checkered Setwing (*Dythemis fugax*)
Swift Setwing (*Dythemis velox*)
Double-striped Bluet (*Enallagma basidens*)
Familiar Bluet (*Enallagma civile*)

Stream Bluet (*Enallagma exsulans*)
Eastern Ringtail (*Erpetogomphus designatus*)
Eastern Pondhawk (*Erythemis simplicicollis*)
Dragonhunter (*Hagenius brevistylus*)
American Rubyspot (*Hetaerina americana*)
Smoky Rubyspot (*Hetaerina titia*)
Rambur's Forktail (*Ischnura ramburii*)
Comanche Skimmer (*Libellula comanche*)
Slaty Skimmer (*Libellula incesta*)
Widow Skimmer (*Libellula luctuosa*)
Twelve-spotted Skimmer (*Libellula pulchella*)
Flame Skimmer (*Libellula saturata*)
Bronzed River Cruiser (*Macromia annulata*)
Roseate Skimmer (*Orthemis ferruginea*)
Blue Dasher (*Pachydiplax longipennis*)
Five-striped Leaftail (*Phyllogomphoides
    albrighti*)
Four-striped Leaftail (*Phyllogomphoides
    stigmatus*)
Common Whitetail (*Plathemis lydia*)
Gray Sanddragon (*Progomphus borealis*)
Variegated Meadowhawk (*Sympetrum
    corruptum*)
Desert Firetail (*Telebasis salva*)
Black Saddlebags (*Tramea lacerata*)

Matagorda *(45)*
Common Green Darner (*Anax junius*)
Comet Darner (*Anax longipes*)
Broad-striped Forceptail (*Aphylla angustifolia*)
Narrow-striped Forceptail (*Aphylla protracta*)
Blue-fronted Dancer (*Argia apicalis*)
Powdered Dancer (*Argia moesta*)
Blue-ringed Dancer (*Argia sedula*)
Blue-tipped Dancer (*Argia tibialis*)
Stillwater Clubtail (*Arigomphus lentulus*)
Jade Clubtail (*Arigomphus submedianus*)
Four-spotted Pennant (*Brachymesia gravida*)
Halloween Pennant (*Celithemis eponina*)
Familiar Bluet (*Enallagma civile*)
Big Bluet (*Enallagma durum*)
Skimming Bluet (*Enallagma geminatum*)
Orange Bluet (*Enallagma signatum*)
Prince Baskettail (*Epitheca princeps*)
Eastern Ringtail (*Erpetogomphus designatus*)
Eastern Pondhawk (*Erythemis simplicicollis*)
Great Pondhawk (*Erythemis vesiculosa*)
Seaside Dragonlet (*Erythrodiplax berenice*)
Plains Clubtail (*Gomphus externus*)
Gulf Coast Clubtail (*Gomphus modestus*)
Cobra Clubtail (*Gomphus vastus*)
American Rubyspot (*Hetaerina americana*)
Smoky Rubyspot (*Hetaerina titia*)
Citrine Forktail (*Ischnura hastata*)
Rambur's Forktail (*Ischnura ramburii*)
Plateau Spreadwing (*Lestes alacer*)
Golden-winged Skimmer (*Libellula auripennis*)
Widow Skimmer (*Libellula luctuosa*)
Needham's Skimmer (*Libellula needhami*)
Great Blue Skimmer (*Libellula vibrans*)
Hyacinth Glider (*Miathyria marcella*)
Roseate Skimmer (*Orthemis ferruginea*)
Blue Dasher (*Pachydiplax longipennis*)
Wandering Glider (*Pantala flavescens*)
Spot-winged Glider (*Pantala hymenaea*)
Eastern Amberwing (*Perithemis tenera*)
Common Whitetail (*Plathemis lydia*)
Russet-tipped Clubtail (*Stylurus plagiatus*)
Variegated Meadowhawk (*Sympetrum
    corruptum*)
Striped Saddlebags (*Tramea calverti*)
Black Saddlebags (*Tramea lacerata*)
Red Saddlebags (*Tramea onusta*)

Maverick *(31)*
Common Green Darner (*Anax junius*)
Blue-fronted Dancer (*Argia apicalis*)
Kiowa Dancer (*Argia immunda*)
Powdered Dancer (*Argia moesta*)

Aztec Dancer (*Argia nahuana*)
Blue-ringed Dancer (*Argia sedula*)
Dusky Dancer (*Argia translata*)
Pale-faced Clubskimmer (*Brechmorhoga
    mendax*)
Black Setwing (*Dythemis nigrescens*)
Swift Setwing (*Dythemis velox*)
Double-striped Bluet (*Enallagma basidens*)
Familiar Bluet (*Enallagma civile*)
Neotropical Bluet (*Enallagma novaehispaniae*)
White-belted Ringtail (*Erpetogomphus
    compositus*)
Eastern Ringtail (*Erpetogomphus designatus*)
Eastern Pondhawk (*Erythemis simplicicollis*)
American Rubyspot (*Hetaerina americana*)
Smoky Rubyspot (*Hetaerina titia*)
Citrine Forktail (*Ischnura hastata*)
Fragile Forktail (*Ischnura posita*)
Comanche Skimmer (*Libellula comanche*)
Twelve-spotted Skimmer (*Libellula pulchella*)
Marl Pennant (*Macrodiplax balteata*)
Roseate Skimmer (*Orthemis ferruginea*)
Blue Dasher (*Pachydiplax longipennis*)
Slough Amberwing (*Perithemis domitia*)
Common Whitetail (*Plathemis lydia*)
Filigree Skimmer (*Pseudoleon superbus*)
Desert Firetail (*Telebasis salva*)
Black Saddlebags (*Tramea lacerata*)
Red Saddlebags (*Tramea onusta*)

McCulloch *(23)*
Common Green Darner (*Anax junius*)
Powdered Dancer (*Argia moesta*)
Blue-ringed Dancer (*Argia sedula*)
Dusky Dancer (*Argia translata*)
Flag-tailed Spinyleg (*Dromogomphus spoliatus*)
Azure Bluet (*Enallagma aspersum*)
Double-striped Bluet (*Enallagma basidens*)
Familiar Bluet (*Enallagma civile*)
Dot-winged Baskettail (*Epitheca petechialis*)
Eastern Ringtail (*Erpetogomphus designatus*)
Eastern Pondhawk (*Erythemis simplicicollis*)
American Rubyspot (*Hetaerina americana*)
Fragile Forktail (*Ischnura posita*)
Rambur's Forktail (*Ischnura ramburii*)
Comanche Skimmer (*Libellula comanche*)
Widow Skimmer (*Libellula luctuosa*)
Twelve-spotted Skimmer (*Libellula pulchella*)
Blue Dasher (*Pachydiplax longipennis*)
Eastern Amberwing (*Perithemis tenera*)
Four-striped Leaftail (*Phyllogomphoides
    stigmatus*)
Common Whitetail (*Plathemis lydia*)
Desert Firetail (*Telebasis salva*)
Black Saddlebags (*Tramea lacerata*)

McLennan *(71)*
Amazon Darner (*Anax amazili*)
Common Green Darner (*Anax junius*)
Comet Darner (*Anax longipes*)
Great Spreadwing (*Archilestes grandis*)
Blue-fronted Dancer (*Argia apicalis*)
Kiowa Dancer (*Argia immunda*)
Powdered Dancer (*Argia moesta*)
Aztec Dancer (*Argia nahuana*)
Springwater Dancer (*Argia plana*)
Blue-ringed Dancer (*Argia sedula*)
Dusky Dancer (*Argia translata*)
Stillwater Clubtail (*Arigomphus lentulus*)
Jade Clubtail (*Arigomphus submedianus*)
Springtime Darner (*Basiaeschna janata*)
Four-spotted Pennant (*Brachymesia gravida*)
Pale-faced Clubskimmer (*Brechmorhoga
    mendax*)
Calico Pennant (*Celithemis elisa*)
Halloween Pennant (*Celithemis eponina*)
Stream Cruiser (*Didymops transversa*)
Flag-tailed Spinyleg (*Dromogomphus spoliatus*)
Checkered Setwing (*Dythemis fugax*)

Swift Setwing (*Dythemis velox*)
Double-striped Bluet (*Enallagma basidens*)
Familiar Bluet (*Enallagma civile*)
Stream Bluet (*Enallagma exsulans*)
Orange Bluet (*Enallagma signatum*)
Swamp Darner (*Epiaeschna heros*)
Dot-winged Baskettail (*Epitheca petechialis*)
Prince Baskettail (*Epitheca princeps*)
Eastern Ringtail (*Erpetogomphus designatus*)
Eastern Pondhawk (*Erythemis simplicicollis*)
Great Pondhawk (*Erythemis vesiculosa*)
Little Blue Dragonlet (*Erythrodiplax minuscula*)
Band-winged Dragonlet (*Erythrodiplax umbrata*)
Plains Clubtail (*Gomphus externus*)
Sulphur-tipped Clubtail (*Gomphus militaris*)
Cobra Clubtail (*Gomphus vastus*)
American Rubyspot (*Hetaerina americana*)
Smoky Rubyspot (*Hetaerina titia*)
Citrine Forktail (*Ischnura hastata*)
Rambur's Forktail (*Ischnura ramburii*)
Blue Corporal (*Ladona deplanata*)
Comanche Skimmer (*Libellula comanche*)
Neon Skimmer (*Libellula croceipennis*)
Yellow-sided Skimmer (*Libellula flavida*)
Slaty Skimmer (*Libellula incesta*)
Widow Skimmer (*Libellula luctuosa*)
Twelve-spotted Skimmer (*Libellula pulchella*)
Flame Skimmer (*Libellula saturata*)
Great Blue Skimmer (*Libellula vibrans*)
Bronzed River Cruiser (*Macromia annulata*)
Gilded River Cruiser (*Macromia pacifica*)
Royal River Cruiser (*Macromia taeniolata*)
Cyrano Darner (*Nasiaeschna pentacantha*)
Orange Shadowdragon (*Neurocordulia
    xanthosoma*)
Roseate Skimmer (*Orthemis ferruginea*)
Blue Dasher (*Pachydiplax longipennis*)
Wandering Glider (*Pantala flavescens*)
Spot-winged Glider (*Pantala hymenaea*)
Eastern Amberwing (*Perithemis tenera*)
Five-striped Leaftail (*Phyllogomphoides
    albrighti*)
Four-striped Leaftail (*Phyllogomphoides
    stigmatus*)
Common Whitetail (*Plathemis lydia*)
Common Sanddragon (*Progomphus obscurus*)
Russet-tipped Clubtail (*Stylurus plagiatus*)
Blue-faced Meadowhawk (*Sympetrum
    ambiguum*)
Variegated Meadowhawk (*Sympetrum
    corruptum*)
Autumn Meadowhawk (*Sympetrum vicinum*)
Desert Firetail (*Telebasis salva*)
Black Saddlebags (*Tramea lacerata*)
Red Saddlebags (*Tramea onusta*)

McMullen *(28)*
Common Green Darner (*Anax junius*)
Blue-fronted Dancer (*Argia apicalis*)
Powdered Dancer (*Argia moesta*)
Blue-ringed Dancer (*Argia sedula*)
Dusky Dancer (*Argia translata*)
Jade Clubtail (*Arigomphus submedianus*)
Red-tailed Pennant (*Brachymesia furcata*)
Pale-faced Clubskimmer (*Brechmorhoga
    mendax*)
Swift Setwing (*Dythemis velox*)
Double-striped Bluet (*Enallagma basidens*)
Familiar Bluet (*Enallagma civile*)
Prince Baskettail (*Epitheca princeps*)
Eastern Ringtail (*Erpetogomphus designatus*)
Eastern Pondhawk (*Erythemis simplicicollis*)
Great Pondhawk (*Erythemis vesiculosa*)
Little Blue Dragonlet (*Erythrodiplax minuscula*)
Band-winged Dragonlet (*Erythrodiplax umbrata*)
Plains Clubtail (*Gomphus externus*)
Sulphur-tipped Clubtail (*Gomphus militaris*)
American Rubyspot (*Hetaerina americana*)

Smoky Rubyspot (*Hetaerina titia*)
Rambur's Forktail (*Ischnura ramburii*)
Twelve-spotted Skimmer (*Libellula pulchella*)
Roseate Skimmer (*Orthemis ferruginea*)
Eastern Amberwing (*Perithemis tenera*)
Common Whitetail (*Plathemis lydia*)
Common Sanddragon (*Progomphus obscurus*)
Red Saddlebags (*Tramea onusta*)

Medina *(45)*
Common Green Darner (*Anax junius*)
Blue-fronted Dancer (*Argia apicalis*)
Comanche Dancer (*Argia barretti*)
Violet Dancer (*Argia fumipennis violacea*)
Kiowa Dancer (*Argia immunda*)
Leonora's Dancer (*Argia leonorae*)
Powdered Dancer (*Argia moesta*)
Aztec Dancer (*Argia nahuana*)
Blue-ringed Dancer (*Argia sedula*)
Dusky Dancer (*Argia translata*)
Springtime Darner (*Basiaeschna janata*)
Pale-faced Clubskimmer (*Brechmorhoga
    mendax*)
Calico Pennant (*Celithemis elisa*)
Flag-tailed Spinyleg (*Dromogomphus spoliatus*)
Checkered Setwing (*Dythemis fugax*)
Black Setwing (*Dythemis nigrescens*)
Swift Setwing (*Dythemis velox*)
Double-striped Bluet (*Enallagma basidens*)
Familiar Bluet (*Enallagma civile*)
Prince Baskettail (*Epitheca princeps*)
Eastern Ringtail (*Erpetogomphus designatus*)
Plateau Dragonlet (*Erythrodiplax basifusca*)
Seaside Dragonlet (*Erythrodiplax berenice*)
Red-faced Dragonlet (*Erythrodiplax fusca*)
Dragonhunter (*Hagenius brevistylus*)
American Rubyspot (*Hetaerina americana*)
Smoky Rubyspot (*Hetaerina titia*)
Citrine Forktail (*Ischnura hastata*)
Comanche Skimmer (*Libellula comanche*)
Slaty Skimmer (*Libellula incesta*)
Widow Skimmer (*Libellula luctuosa*)
Flame Skimmer (*Libellula saturata*)
Cyrano Darner (*Nasiaeschna pentacantha*)
Coral-fronted Threadtail (*Neoneura aaroni*)
Carmine Skimmer (*Orthemis discolor*)
Roseate Skimmer (*Orthemis ferruginea*)
Blue Dasher (*Pachydiplax longipennis*)
Wandering Glider (*Pantala flavescens*)
Eastern Amberwing (*Perithemis tenera*)
Five-striped Leaftail (*Phyllogomphoides
    albrighti*)
Common Whitetail (*Plathemis lydia*)
Orange-striped Threadtail (*Protoneura cara*)
Variegated Meadowhawk (*Sympetrum
    corruptum*)
Autumn Meadowhawk (*Sympetrum vicinum*)
Black Saddlebags (*Tramea lacerata*)

Menard *(47)*
Common Green Darner (*Anax junius*)
Blue-fronted Dancer (*Argia apicalis*)
Violet Dancer (*Argia fumipennis violacea*)
Kiowa Dancer (*Argia immunda*)
Sooty Dancer (*Argia lugens*)
Powdered Dancer (*Argia moesta*)
Aztec Dancer (*Argia nahuana*)
Springwater Dancer (*Argia plana*)
Blue-ringed Dancer (*Argia sedula*)
Dusky Dancer (*Argia translata*)
Springtime Darner (*Basiaeschna janata*)
Halloween Pennant (*Celithemis eponina*)
Banded Pennant (*Celithemis fasciata*)
Flag-tailed Spinyleg (*Dromogomphus spoliatus*)
Checkered Setwing (*Dythemis fugax*)
Black Setwing (*Dythemis nigrescens*)
Swift Setwing (*Dythemis velox*)
Double-striped Bluet (*Enallagma basidens*)
Stream Bluet (*Enallagma exsulans*)

Neotropical Bluet (*Enallagma novaehispaniae*)
Arroyo Bluet (*Enallagma praevarum*)
Vesper Bluet (*Enallagma vesperum*)
Prince Baskettail (*Epitheca princeps*)
Eastern Ringtail (*Erpetogomphus designatus*)
Eastern Pondhawk (*Erythemis simplicicollis*)
Band-winged Dragonlet (*Erythrodiplax umbrata*)
Sulphur-tipped Clubtail (*Gomphus militaris*)
American Rubyspot (*Hetaerina americana*)
Smoky Rubyspot (*Hetaerina titia*)
Rambur's Forktail (*Ischnura ramburii*)
Comanche Skimmer (*Libellula comanche*)
Widow Skimmer (*Libellula luctuosa*)
Flame Skimmer (*Libellula saturata*)
Bronzed River Cruiser (*Macromia annulata*)
Roseate Skimmer (*Orthemis ferruginea*)
Blue Dasher (*Pachydiplax longipennis*)
Wandering Glider (*Pantala flavescens*)
Eastern Amberwing (*Perithemis tenera*)
Five-striped Leaftail (*Phyllogomphoides
    albrighti*)
Four-striped Leaftail (*Phyllogomphoides
    stigmatus*)
Common Whitetail (*Plathemis lydia*)
Orange-striped Threadtail (*Protoneura cara*)
Filigree Skimmer (*Pseudoleon superbus*)
Autumn Meadowhawk (*Sympetrum vicinum*)
Desert Firetail (*Telebasis salva*)
Black Saddlebags (*Tramea lacerata*)
Red Saddlebags (*Tramea onusta*)

Midland *(47)*
Common Green Darner (*Anax junius*)
Blue-fronted Dancer (*Argia apicalis*)
Violet Dancer (*Argia fumipennis violacea*)
Kiowa Dancer (*Argia immunda*)
Powdered Dancer (*Argia moesta*)
Blue-ringed Dancer (*Argia sedula*)
Red-tailed Pennant (*Brachymesia furcata*)
Four-spotted Pennant (*Brachymesia gravida*)
Halloween Pennant (*Celithemis eponina*)
Checkered Setwing (*Dythemis fugax*)
Swift Setwing (*Dythemis velox*)
Double-striped Bluet (*Enallagma basidens*)
Familiar Bluet (*Enallagma civile*)
Dot-winged Baskettail (*Epitheca petechialis*)
Eastern Pondhawk (*Erythemis simplicicollis*)
Great Pondhawk (*Erythemis vesiculosa*)
Seaside Dragonlet (*Erythrodiplax berenice*)
Band-winged Dragonlet (*Erythrodiplax umbrata*)
Smoky Rubyspot (*Hetaerina titia*)
Desert Forktail (*Ischnura barberi*)
Plains Forktail (*Ischnura damula*)
Mexican Forktail (*Ischnura demorsa*)
Black-fronted Forktail (*Ischnura denticollis*)
Citrine Forktail (*Ischnura hastata*)
Rambur's Forktail (*Ischnura ramburii*)
Southern Spreadwing (*Lestes australis*)
Comanche Skimmer (*Libellula comanche*)
Bleached Skimmer (*Libellula composita*)
Widow Skimmer (*Libellula luctuosa*)
Twelve-spotted Skimmer (*Libellula pulchella*)
Flame Skimmer (*Libellula saturata*)
Marl Pennant (*Macrodiplax balteata*)
Thornbush Dasher (*Micrathyria hagenii*)
Roseate Skimmer (*Orthemis ferruginea*)
Blue Dasher (*Pachydiplax longipennis*)
Wandering Glider (*Pantala flavescens*)
Spot-winged Glider (*Pantala hymenaea*)
Eastern Amberwing (*Perithemis tenera*)
Four-striped Leaftail (*Phyllogomphoides
    stigmatus*)
Common Whitetail (*Plathemis lydia*)
Desert Whitetail (*Plathemis subornata*)
Blue-eyed Darner (*Rhionaeschna multicolor*)
Variegated Meadowhawk (*Sympetrum
    corruptum*)
Autumn Meadowhawk (*Sympetrum vicinum*)

Desert Firetail (*Telebasis salva*)
Black Saddlebags (*Tramea lacerata*)
Red Saddlebags (*Tramea onusta*)

Milam *(18)*
Common Green Darner (*Anax junius*)
Blue-fronted Dancer (*Argia apicalis*)
Powdered Dancer (*Argia moesta*)
Blue-ringed Dancer (*Argia sedula*)
Stillwater Clubtail (*Arigomphus lentulus*)
Fawn Darner (*Boyeria vinosa*)
Flag-tailed Spinyleg (*Dromogomphus spoliatus*)
Familiar Bluet (*Enallagma civile*)
Common Baskettail (*Epitheca cynosura*)
Prince Baskettail (*Epitheca princeps*)
Eastern Ringtail (*Erpetogomphus designatus*)
Band-winged Dragonlet (*Erythrodiplax umbrata*)
Plains Clubtail (*Gomphus externus*)
Smoky Rubyspot (*Hetaerina titia*)
Blue Dasher (*Pachydiplax longipennis*)
Wandering Glider (*Pantala flavescens*)
Common Whitetail (*Plathemis lydia*)
Mocha Emerald (*Somatochlora linearis*)

Mills *(12)*
Halloween Pennant (*Celithemis eponina*)
Checkered Setwing (*Dythemis fugax*)
Eastern Ringtail (*Erpetogomphus designatus*)
Comanche Skimmer (*Libellula comanche*)
Neon Skimmer (*Libellula croceipennis*)
Widow Skimmer (*Libellula luctuosa*)
Twelve-spotted Skimmer (*Libellula pulchella*)
Eastern Amberwing (*Perithemis tenera*)
Five-striped Leaftail (*Phyllogomphoides
    albrighti*)
Four-striped Leaftail (*Phyllogomphoides
    stigmatus*)
Russet-tipped Clubtail (*Stylurus plagiatus*)
Black Saddlebags (*Tramea lacerata*)

Mitchell *(11)*
Blue-fronted Dancer (*Argia apicalis*)
Springwater Dancer (*Argia plana*)
Checkered Setwing (*Dythemis fugax*)
Familiar Bluet (*Enallagma civile*)
Eastern Ringtail (*Erpetogomphus designatus*)
Eastern Pondhawk (*Erythemis simplicicollis*)
Desert Forktail (*Ischnura barberi*)
Comanche Skimmer (*Libellula comanche*)
Widow Skimmer (*Libellula luctuosa*)
Eastern Amberwing (*Perithemis tenera*)
Red Saddlebags (*Tramea onusta*)

Montague *(10)*
Great Spreadwing (*Archilestes grandis*)
Blue-fronted Dancer (*Argia apicalis*)
Halloween Pennant (*Celithemis eponina*)
Double-striped Bluet (*Enallagma basidens*)
Familiar Bluet (*Enallagma civile*)
Vesper Bluet (*Enallagma vesperum*)
Eastern Pondhawk (*Erythemis simplicicollis*)
Widow Skimmer (*Libellula luctuosa*)
Eastern Amberwing (*Perithemis tenera*)
Variegated Meadowhawk (*Sympetrum
    corruptum*)

Montgomery *(74)*
Common Green Darner (*Anax junius*)
Comet Darner (*Anax longipes*)
Broad-striped Forceptail (*Aphylla angustifolia*)
Narrow-striped Forceptail (*Aphylla protracta*)
Two-striped Forceptail (*Aphylla williamsoni*)
Blue-fronted Dancer (*Argia apicalis*)
Violet Dancer (*Argia fumipennis violacea*)
Powdered Dancer (*Argia moesta*)
Blue-ringed Dancer (*Argia sedula*)
Blue-tipped Dancer (*Argia tibialis*)
Stillwater Clubtail (*Arigomphus lentulus*)
Jade Clubtail (*Arigomphus submedianus*)
Four-spotted Pennant (*Brachymesia gravida*)
Ebony Jewelwing (*Calopteryx maculata*)
Amanda's Pennant (*Celithemis amanda*)

Calico Pennant (*Celithemis elisa*)
Halloween Pennant (*Celithemis eponina*)
Banded Pennant (*Celithemis fasciata*)
Ornate Pennant (*Celithemis ornata*)
Double-ringed Pennant (*Celithemis verna*)
Twin-spotted Spiketail (*Cordulegaster maculata*)
Regal Darner (*Coryphaeschna ingens*)
Stream Cruiser (*Didymops transversa*)
Black-shouldered Spinyleg (*Dromogomphus spinosus*)
Swift Setwing (*Dythemis velox*)
Double-striped Bluet (*Enallagma basidens*)
Familiar Bluet (*Enallagma civile*)
Attenuated Bluet (*Enallagma daeckii*)
Burgundy Bluet (*Enallagma dubium*)
Skimming Bluet (*Enallagma geminatum*)
Orange Bluet (*Enallagma signatum*)
Vesper Bluet (*Enallagma vesperum*)
Swamp Darner (*Epiaeschna heros*)
Common Basketail (*Epitheca cynosura*)
Prince Baskettail (*Epitheca princeps*)
Eastern Pondhawk (*Erythemis simplicicollis*)
Great Pondhawk (*Erythemis vesiculosa*)
Little Blue Dragonlet (*Erythrodiplax minuscula*)
Band-winged Dragonlet (*Erythrodiplax umbrata*)
Plains Clubtail (*Gomphus externus*)
Oklahoma Clubtail (*Gomphus oklahomensis*)
Dragonhunter (*Hagenius brevistylus*)
American Rubyspot (*Hetaerina americana*)
Smoky Rubyspot (*Hetaerina titia*)
Citrine Forktail (*Ischnura hastata*)
Fragile Forktail (*Ischnura posita*)
Rambur's Forktail (*Ischnura ramburii*)
Blue Corporal (*Ladona deplanata*)
Swamp Spreadwing (*Lestes vigilax*)
Golden-winged Skimmer (*Libellula auripennis*)
Spangled Skimmer (*Libellula cyanea*)
Slaty Skimmer (*Libellula incesta*)
Widow Skimmer (*Libellula luctuosa*)
Needham's Skimmer (*Libellula needhami*)
Painted Skimmer (*Libellula semifasciata*)
Great Blue Skimmer (*Libellula vibrans*)
Hyacinth Glider (*Miathyria marcella*)
Cyrano Darner (*Nasiaeschna pentacantha*)
Southern Sprite (*Nehalennia integricollis*)
Roseate Skimmer (*Orthemis ferruginea*)
Blue Dasher (*Pachydiplax longipennis*)
Wandering Glider (*Pantala flavescens*)
Spot-winged Glider (*Pantala hymenaea*)
Eastern Amberwing (*Perithemis tenera*)
Common Whitetail (*Plathemis lydia*)
Common Sanddragon (*Progomphus obscurus*)
Mocha Emerald (*Somatochlora linearis*)
Texas Emerald (*Somatochlora margarita*)
Blue-faced Meadowhawk (*Sympetrum ambiguum*)
Variegated Meadowhawk (*Sympetrum corruptum*)
Gray Petaltail (*Tachopteryx thoreyi*)
Carolina Saddlebags (*Tramea carolina*)
Black Saddlebags (*Tramea lacerata*)
Red Saddlebags (*Tramea onusta*)

Moore (3)
Common Green Darner (*Anax junius*)
Familiar Bluet (*Enallagma civile*)
American Rubyspot (*Hetaerina americana*)

Morris (43)
Common Green Darner (*Anax junius*)
Two-striped Forceptail (*Aphylla williamsoni*)
Blue-fronted Dancer (*Argia apicalis*)
Violet Dancer (*Argia fumipennis violacea*)
Aztec Dancer (*Argia nahuana*)
Springwater Dancer (*Argia plana*)
Blue-tipped Dancer (*Argia tibialis*)
Ebony Jewelwing (*Calopteryx maculata*)
Calico Pennant (*Celithemis elisa*)
Halloween Pennant (*Celithemis eponina*)

Banded Pennant (*Celithemis fasciata*)
Twin-spotted Spiketail (*Cordulegaster maculata*)
Flag-tailed Spinyleg (*Dromogomphus spoliatus*)
Double-striped Bluet (*Enallagma basidens*)
Attenuated Bluet (*Enallagma daeckii*)
Turquoise Bluet (*Enallagma divagans*)
Skimming Bluet (*Enallagma geminatum*)
Vesper Bluet (*Enallagma vesperum*)
Swamp Darner (*Epiaeschna heros*)
Stripe-winged Baskettail (*Epitheca costalis*)
Common Baskettail (*Epitheca cynosura*)
Dot-winged Baskettail (*Epitheca petechialis*)
Prince Baskettail (*Epitheca princeps*)
Eastern Pondhawk (*Erythemis simplicicollis*)
Ashy Clubtail (*Gomphus lividus*)
Oklahoma Clubtail (*Gomphus oklahomensis*)
Citrine Forktail (*Ischnura hastata*)
Lilypad Forktail (*Ischnura kellicotti*)
Blue Corporal (*Ladona deplanata*)
Elegant Spreadwing (*Lestes inaequalis*)
Swamp Spreadwing (*Lestes vigilax*)
Spangled Skimmer (*Libellula cyanea*)
Yellow-sided Skimmer (*Libellula flavida*)
Slaty Skimmer (*Libellula incesta*)
Widow Skimmer (*Libellula luctuosa*)
Great Blue Skimmer (*Libellula vibrans*)
Blue Dasher (*Pachydiplax longipennis*)
Eastern Amberwing (*Perithemis tenera*)
Common Whitetail (*Plathemis lydia*)
Common Sanddragon (*Progomphus obscurus*)
Variegated Meadowhawk (*Sympetrum corruptum*)
Black Saddlebags (*Tramea lacerata*)
Red Saddlebags (*Tramea onusta*)

Motley (12)
Common Green Darner (*Anax junius*)
Violet Dancer (*Argia fumipennis violacea*)
Blue-ringed Dancer (*Argia sedula*)
Double-striped Bluet (*Enallagma basidens*)
Familiar Bluet (*Enallagma civile*)
Eastern Ringtail (*Erpetogomphus designatus*)
American Rubyspot (*Hetaerina americana*)
Flame Skimmer (*Libellula saturata*)
Blue Dasher (*Pachydiplax longipennis*)
Eastern Amberwing (*Perithemis tenera*)
Common Whitetail (*Plathemis lydia*)
Blue-eyed Darner (*Rhionaeschna multicolor*)

Nacogdoches (31)
Common Green Darner (*Anax junius*)
Great Spreadwing (*Archilestes grandis*)
Seepage Dancer (*Argia bipunctulata*)
Powdered Dancer (*Argia moesta*)
Blue-tipped Dancer (*Argia tibialis*)
Ebony Jewelwing (*Calopteryx maculata*)
Twin-spotted Spiketail (*Cordulegaster maculata*)
Black-shouldered Spinyleg (*Dromogomphus spinosus*)
Familiar Bluet (*Enallagma civile*)
Stream Bluet (*Enallagma exsulans*)
Swamp Darner (*Epiaeschna heros*)
Common Baskettail (*Epitheca cynosura*)
Mantled Baskettail (*Epitheca semiaquea*)
Eastern Pondhawk (*Erythemis simplicicollis*)
Selys' Sundragon (*Helocordulia selysii*)
Citrine Forktail (*Ischnura hastata*)
Rambur's Forktail (*Ischnura ramburii*)
Slaty Skimmer (*Libellula incesta*)
Widow Skimmer (*Libellula luctuosa*)
Needham's Skimmer (*Libellula needhami*)
Great Blue Skimmer (*Libellula vibrans*)
Hyacinth Glider (*Miathyria marcella*)
Roseate Skimmer (*Orthemis ferruginea*)
Blue Dasher (*Pachydiplax longipennis*)
Wandering Glider (*Pantala flavescens*)
Spot-winged Glider (*Pantala hymenaea*)
Eastern Amberwing (*Perithemis tenera*)
Common Whitetail (*Plathemis lydia*)

Common Sanddragon (*Progomphus obscurus*)
Blue-faced Meadowhawk (*Sympetrum ambiguum*)
Carolina Saddlebags (*Tramea carolina*)

Navarro (19)
Two-striped Forceptail (*Aphylla williamsoni*)
Blue-fronted Dancer (*Argia apicalis*)
Powdered Dancer (*Argia moesta*)
Jade Clubtail (*Arigomphus submedianus*)
Four-spotted Pennant (*Brachymesia gravida*)
Familiar Bluet (*Enallagma civile*)
Eastern Pondhawk (*Erythemis simplicicollis*)
Plains Clubtail (*Gomphus externus*)
American Rubyspot (*Hetaerina americana*)
Smoky Rubyspot (*Hetaerina titia*)
Fragile Forktail (*Ischnura posita*)
Rambur's Forktail (*Ischnura ramburii*)
Slaty Skimmer (*Libellula incesta*)
Widow Skimmer (*Libellula luctuosa*)
Blue Dasher (*Pachydiplax longipennis*)
Common Whitetail (*Plathemis lydia*)
Variegated Meadowhawk (*Sympetrum corruptum*)
Black Saddlebags (*Tramea lacerata*)
Red Saddlebags (*Tramea onusta*)

Newton (42)
Common Green Darner (*Anax junius*)
Narrow-striped Forceptail (*Aphylla protracta*)
Two-striped Forceptail (*Aphylla williamsoni*)
Powdered Dancer (*Argia moesta*)
Blue-tipped Dancer (*Argia tibialis*)
Bayou Clubtail (*Arigomphus maxwelli*)
Four-spotted Pennant (*Brachymesia gravida*)
Ebony Jewelwing (*Calopteryx maculata*)
Calico Pennant (*Celithemis elisa*)
Halloween Pennant (*Celithemis eponina*)
Banded Pennant (*Celithemis fasciata*)
Regal Darner (*Coryphaeschna ingens*)
Burgundy Bluet (*Enallagma dubium*)
Skimming Bluet (*Enallagma geminatum*)
Swamp Darner (*Epiaeschna heros*)
Eastern Pondhawk (*Erythemis simplicicollis*)
Little Blue Dragonlet (*Erythrodiplax minuscula*)
Band-winged Dragonlet (*Erythrodiplax umbrata*)
Gulf Coast Clubtail (*Gomphus modestus*)
Oklahoma Clubtail (*Gomphus oklahomensis*)
Citrine Forktail (*Ischnura hastata*)
Fragile Forktail (*Ischnura posita*)
Rambur's Forktail (*Ischnura ramburii*)
Blue Corporal (*Ladona deplanata*)
Swamp Spreadwing (*Lestes vigilax*)
Golden-winged Skimmer (*Libellula auripennis*)
Spangled Skimmer (*Libellula cyanea*)
Slaty Skimmer (*Libellula incesta*)
Widow Skimmer (*Libellula luctuosa*)
Painted Skimmer (*Libellula semifasciata*)
Great Blue Skimmer (*Libellula vibrans*)
Royal River Cruiser (*Macromia taeniolata*)
Roseate Skimmer (*Orthemis ferruginea*)
Blue Dasher (*Pachydiplax longipennis*)
Wandering Glider (*Pantala flavescens*)
Spot-winged Glider (*Pantala hymenaea*)
Eastern Amberwing (*Perithemis tenera*)
Common Whitetail (*Plathemis lydia*)
Common Sanddragon (*Progomphus obscurus*)
Variegated Meadowhawk (*Sympetrum corruptum*)
Carolina Saddlebags (*Tramea carolina*)
Black Saddlebags (*Tramea lacerata*)

Nueces (42)
Common Green Darner (*Anax junius*)
Blue-fronted Dancer (*Argia apicalis*)
Kiowa Dancer (*Argia immunda*)
Powdered Dancer (*Argia moesta*)
Blue-ringed Dancer (*Argia sedula*)
Red-tailed Pennant (*Brachymesia furcata*)
Four-spotted Pennant (*Brachymesia gravida*)

Pale-faced Clubskimmer (*Brechmorhoga mendax*)
Black Setwing (*Dythemis nigrescens*)
Swift Setwing (*Dythemis velox*)
Double-striped Bluet (*Enallagma basidens*)
Familiar Bluet (*Enallagma civile*)
Prince Baskettail (*Epitheca princeps*)
Pin-tailed Pondhawk (*Erythemis plebeja*)
Eastern Pondhawk (*Erythemis simplicicollis*)
Great Pondhawk (*Erythemis vesiculosa*)
Seaside Dragonlet (*Erythrodiplax berenice*)
Band-winged Dragonlet (*Erythrodiplax umbrata*)
Smoky Rubyspot (*Hetaerina titia*)
Citrine Forktail (*Ischnura hastata*)
Rambur's Forktail (*Ischnura ramburii*)
Rainpool Spreadwing (*Lestes forficula*)
Chalky Spreadwing (*Lestes sigma*)
Neon Skimmer (*Libellula croceipennis*)
Needham's Skimmer (*Libellula needhami*)
Marl Pennant (*Macrodiplax balteata*)
Hyacinth Glider (*Miathyria marcella*)
Three-striped Dasher (*Micrathyria didyma*)
Thornbush Dasher (*Micrathyria hagenii*)
Coral-fronted Threadtail (*Neoneura aaroni*)
Roseate Skimmer (*Orthemis ferruginea*)
Blue Dasher (*Pachydiplax longipennis*)
Wandering Glider (*Pantala flavescens*)
Spot-winged Glider (*Pantala hymenaea*)
Slough Amberwing (*Perithemis domitia*)
Eastern Amberwing (*Perithemis tenera*)
Common Whitetail (*Plathemis lydia*)
Turquoise-tipped Darner (*Rhionaeschna psilus*)
Variegated Meadowhawk (*Sympetrum corruptum*)
Striped Saddlebags (*Tramea calverti*)
Black Saddlebags (*Tramea lacerata*)
Red Saddlebags (*Tramea onusta*)

Ochiltree (13)
Blue-fronted Dancer (*Argia apicalis*)
Violet Dancer (*Argia fumipennis violacea*)
Blue-ringed Dancer (*Argia sedula*)
Double-striped Bluet (*Enallagma basidens*)
Familiar Bluet (*Enallagma civile*)
Eastern Pondhawk (*Erythemis simplicicollis*)
American Rubyspot (*Hetaerina americana*)
Black-fronted Forktail (*Ischnura denticollis*)
Eastern Forktail (*Ischnura verticalis*)
Lyre-tipped Spreadwing (*Lestes unguiculatus*)
Widow Skimmer (*Libellula luctuosa*)
Blue-eyed Darner (*Rhionaeschna multicolor*)
Desert Firetail (*Telebasis salva*)

Oldham (6)
Familiar Bluet (*Enallagma civile*)
American Rubyspot (*Hetaerina americana*)
Widow Skimmer (*Libellula luctuosa*)
Common Sanddragon (*Progomphus obscurus*)
Variegated Meadowhawk (*Sympetrum corruptum*)
Red Saddlebags (*Tramea onusta*)

Orange (17)
Blue-fronted Dancer (*Argia apicalis*)
Blue-tipped Dancer (*Argia tibialis*)
Bayou Clubtail (*Arigomphus maxwelli*)
Calico Pennant (*Celithemis elisa*)
Familiar Bluet (*Enallagma civile*)
Big Bluet (*Enallagma durum*)
Orange Bluet (*Enallagma signatum*)
Band-winged Dragonlet (*Erythrodiplax umbrata*)
Ashy Clubtail (*Gomphus lividus*)
Citrine Forktail (*Ischnura hastata*)
Rambur's Forktail (*Ischnura ramburii*)
Golden-winged Skimmer (*Libellula auripennis*)
Needham's Skimmer (*Libellula needhami*)
Hyacinth Glider (*Miathyria marcella*)
Blue Dasher (*Pachydiplax longipennis*)
Eastern Amberwing (*Perithemis tenera*)
Russet-tipped Clubtail (*Stylurus plagiatus*)

Palo Pinto (37)
Blue-fronted Dancer (*Argia apicalis*)
Violet Dancer (*Argia fumipennis violacea*)
Powdered Dancer (*Argia moesta*)
Springwater Dancer (*Argia plana*)
Blue-ringed Dancer (*Argia sedula*)
Dusky Dancer (*Argia translata*)
Pale-faced Clubskimmer (*Brechmorhoga mendax*)
Halloween Pennant (*Celithemis eponina*)
Stream Cruiser (*Didymops transversa*)
Flag-tailed Spinyleg (*Dromogomphus spoliatus*)
Checkered Setwing (*Dythemis fugax*)
Swift Setwing (*Dythemis velox*)
Double-striped Bluet (*Enallagma basidens*)
Familiar Bluet (*Enallagma civile*)
Stream Bluet (*Enallagma exsulans*)
Dot-winged Baskettail (*Epitheca petechialis*)
Prince Baskettail (*Epitheca princeps*)
Eastern Ringtail (*Erpetogomphus designatus*)
Eastern Pondhawk (*Erythemis simplicicollis*)
Plains Clubtail (*Gomphus externus*)
Sulphur-tipped Clubtail (*Gomphus militaris*)
American Rubyspot (*Hetaerina americana*)
Smoky Rubyspot (*Hetaerina titia*)
Citrine Forktail (*Ischnura hastata*)
Plateau Spreadwing (*Lestes alacer*)
Widow Skimmer (*Libellula luctuosa*)
Painted Skimmer (*Libellula semifasciata*)
Bronzed River Cruiser (*Macromia annulata*)
Orange Shadowdragon (*Neurocordulia xanthosoma*)
Spot-winged Glider (*Pantala hymenaea*)
Eastern Amberwing (*Perithemis tenera*)
Five-striped Leaftail (*Phyllogomphoides albrighti*)
Four-striped Leaftail (*Phyllogomphoides stigmatus*)
Common Whitetail (*Plathemis lydia*)
Variegated Meadowhawk (*Sympetrum corruptum*)
Desert Firetail (*Telebasis salva*)
Black Saddlebags (*Tramea lacerata*)

Panola (16)
Blue-fronted Dancer (*Argia apicalis*)
Seepage Dancer (*Argia bipunctulata*)
Powdered Dancer (*Argia moesta*)
Four-spotted Pennant (*Brachymesia gravida*)
Halloween Pennant (*Celithemis eponina*)
Stream Cruiser (*Didymops transversa*)
Skimming Bluet (*Enallagma geminatum*)
Orange Bluet (*Enallagma signatum*)
Common Baskettail (*Epitheca cynosura*)
Prince Baskettail (*Epitheca princeps*)
Eastern Pondhawk (*Erythemis simplicicollis*)
Citrine Forktail (*Ischnura hastata*)
Rambur's Forktail (*Ischnura ramburii*)
Blue Dasher (*Pachydiplax longipennis*)
Eastern Amberwing (*Perithemis tenera*)
Common Whitetail (*Plathemis lydia*)

Parker (31)
Blue-fronted Dancer (*Argia apicalis*)
Violet Dancer (*Argia fumipennis violacea*)
Powdered Dancer (*Argia moesta*)
Springwater Dancer (*Argia plana*)
Blue-ringed Dancer (*Argia sedula*)
Dusky Dancer (*Argia translata*)
Stillwater Clubtail (*Arigomphus lentulus*)
Ebony Jewelwing (*Calopteryx maculata*)
Banded Pennant (*Celithemis fasciata*)
Flag-tailed Spinyleg (*Dromogomphus spoliatus*)
Checkered Setwing (*Dythemis fugax*)
Swift Setwing (*Dythemis velox*)
Double-striped Bluet (*Enallagma basidens*)
Familiar Bluet (*Enallagma civile*)
Turquoise Bluet (*Enallagma divagans*)
Prince Baskettail (*Epitheca princeps*)

Eastern Pondhawk (*Erythemis simplicicollis*)
Sulphur-tipped Clubtail (*Gomphus militaris*)
Slaty Skimmer (*Libellula incesta*)
Widow Skimmer (*Libellula luctuosa*)
Cyrano Darner (*Nasiaeschna pentacantha*)
Orange Shadowdragon (*Neurocordulia xanthosoma*)
Blue Dasher (*Pachydiplax longipennis*)
Wandering Glider (*Pantala flavescens*)
Eastern Amberwing (*Perithemis tenera*)
Five-striped Leaftail (*Phyllogomphoides albrighti*)
Four-striped Leaftail (*Phyllogomphoides stigmatus*)
Common Whitetail (*Plathemis lydia*)
Mocha Emerald (*Somatochlora linearis*)
Variegated Meadowhawk (*Sympetrum corruptum*)
Desert Firetail (*Telebasis salva*)

Pecos (28)
Paiute Dancer (*Argia alberta*)
Kiowa Dancer (*Argia immunda*)
Powdered Dancer (*Argia moesta*)
Aztec Dancer (*Argia nahuana*)
Blue-ringed Dancer (*Argia sedula*)
Checkered Setwing (*Dythemis fugax*)
Swift Setwing (*Dythemis velox*)
Double-striped Bluet (*Enallagma basidens*)
Familiar Bluet (*Enallagma civile*)
Stream Bluet (*Enallagma exsulans*)
Arroyo Bluet (*Enallagma praevarum*)
Eastern Ringtail (*Erpetogomphus designatus*)
Eastern Pondhawk (*Erythemis simplicicollis*)
Sulphur-tipped Clubtail (*Gomphus militaris*)
American Rubyspot (*Hetaerina americana*)
Desert Forktail (*Ischnura barberi*)
Mexican Forktail (*Ischnura demorsa*)
Comanche Skimmer (*Libellula comanche*)
Bleached Skimmer (*Libellula composita*)
Widow Skimmer (*Libellula luctuosa*)
Flame Skimmer (*Libellula saturata*)
Roseate Skimmer (*Orthemis ferruginea*)
Blue Dasher (*Pachydiplax longipennis*)
Spot-winged Glider (*Pantala hymenaea*)
Common Whitetail (*Plathemis lydia*)
Desert Whitetail (*Plathemis subornata*)
Variegated Meadowhawk (*Sympetrum corruptum*)
Desert Firetail (*Telebasis salva*)

Polk (35)
Common Green Darner (*Anax junius*)
Two-striped Forceptail (*Aphylla williamsoni*)
Violet Dancer (*Argia fumipennis violacea*)
Kiowa Dancer (*Argia immunda*)
Powdered Dancer (*Argia moesta*)
Blue-ringed Dancer (*Argia sedula*)
Blue-tipped Dancer (*Argia tibialis*)
Bayou Clubtail (*Arigomphus maxwelli*)
Springtime Darner (*Basiaeschna janata*)
Fawn Darner (*Boyeria vinosa*)
Sparkling Jewelwing (*Calopteryx dimidiata*)
Ebony Jewelwing (*Calopteryx maculata*)
Stream Cruiser (*Didymops transversa*)
Black-shouldered Spinyleg (*Dromogomphus spinosus*)
Familiar Bluet (*Enallagma civile*)
Turquoise Bluet (*Enallagma divagans*)
Vesper Bluet (*Enallagma vesperum*)
Swamp Darner (*Epiaeschna heros*)
Common Baskettail (*Epitheca cynosura*)
Mantled Baskettail (*Epitheca semiaquea*)
Eastern Pondhawk (*Erythemis simplicicollis*)
Little Blue Dragonlet (*Erythrodiplax minuscula*)
Ashy Clubtail (*Gomphus lividus*)
Oklahoma Clubtail (*Gomphus oklahomensis*)
Selys' Sundragon (*Helocordulia selysii*)
American Rubyspot (*Hetaerina americana*)

Citrine Forktail (*Ischnura hastata*)
Rambur's Forktail (*Ischnura ramburii*)
Blue Corporal (*Ladona deplanata*)
Neon Skimmer (*Libellula croceipennis*)
Great Blue Skimmer (*Libellula vibrans*)
Blue Dasher (*Pachydiplax longipennis*)
Wandering Glider (*Pantala flavescens*)
Common Whitetail (*Plathemis lydia*)
Common Sanddragon (*Progomphus obscurus*)

Potter *(5)*
Common Green Darner (*Anax junius*)
Familiar Bluet (*Enallagma civile*)
American Rubyspot (*Hetaerina americana*)
Blue-eyed Darner (*Rhionaeschna multicolor*)
Variegated Meadowhawk (*Sympetrum corruptum*)

Presidio *(65)*
Common Green Darner (*Anax junius*)
Giant Darner (*Anax walsinghami*)
Great Spreadwing (*Archilestes grandis*)
Blue-fronted Dancer (*Argia apicalis*)
Violet Dancer (*Argia fumipennis violacea*)
Lavender Dancer (*Argia hinei*)
Kiowa Dancer (*Argia immunda*)
Leonora's Dancer (*Argia leonorae*)
Sooty Dancer (*Argia lugens*)
Powdered Dancer (*Argia moesta*)
Aztec Dancer (*Argia nahuana*)
Fiery-eyed Dancer (*Argia oenea*)
Amethyst Dancer (*Argia pallens*)
Springwater Dancer (*Argia plana*)
Blue-ringed Dancer (*Argia sedula*)
Tezpi Dancer (*Argia tezpi*)
Red-tailed Pennant (*Brachymesia furcata*)
Tawny Pennant (*Brachymesia herbida*)
Pale-faced Clubskimmer (*Brechmorhoga mendax*)
Checkered Setwing (*Dythemis fugax*)
Mayan Setwing (*Dythemis maya*)
Swift Setwing (*Dythemis velox*)
Familiar Bluet (*Enallagma civile*)
Arroyo Bluet (*Enallagma praevarum*)
White-belted Ringtail (*Erpetogomphus compositus*)
Eastern Ringtail (*Erpetogomphus designatus*)
Serpent Ringtail (*Erpetogomphus lampropeltis natrix*)
Western Pondhawk (*Erythemis collocata*)
Eastern Pondhawk (*Erythemis simplicicollis*)
Plateau Dragonlet (*Erythrodiplax basifusca*)
Plains Clubtail (*Gomphus externus*)
Sulphur-tipped Clubtail (*Gomphus militaris*)
Painted Damsel (*Hesperagrion heterodoxum*)
American Rubyspot (*Hetaerina americana*)
Smoky Rubyspot (*Hetaerina titia*)
Desert Forktail (*Ischnura barberi*)
Mexican Forktail (*Ischnura demorsa*)
Black-fronted Forktail (*Ischnura denticollis*)
Citrine Forktail (*Ischnura hastata*)
Fragile Forktail (*Ischnura posita*)
Rambur's Forktail (*Ischnura ramburii*)
Plateau Spreadwing (*Lestes alacer*)
Comanche Skimmer (*Libellula comanche*)
Neon Skimmer (*Libellula croceipennis*)
Widow Skimmer (*Libellula luctuosa*)
Flame Skimmer (*Libellula saturata*)
Bronzed River Cruiser (*Macromia annulata*)
Roseate Skimmer (*Orthemis ferruginea*)
Blue Dasher (*Pachydiplax longipennis*)
Red Rock Skimmer (*Paltothemis lineatipes*)
Wandering Glider (*Pantala flavescens*)
Spot-winged Glider (*Pantala hymenaea*)
Eastern Amberwing (*Perithemis tenera*)
Common Whitetail (*Plathemis lydia*)
Desert Whitetail (*Plathemis subornata*)
Gray Sanddragon (*Progomphus borealis*)
Filigree Skimmer (*Pseudoleon superbus*)

Arroyo Darner (*Rhionaeschna dugesi*)
Blue-eyed Darner (*Rhionaeschna multicolor*)
Brimstone Clubtail (*Stylurus intricatus*)
Russet-tipped Clubtail (*Stylurus plagiatus*)
Variegated Meadowhawk (*Sympetrum corruptum*)
Desert Firetail (*Telebasis salva*)
Black Saddlebags (*Tramea lacerata*)
Red Saddlebags (*Tramea onusta*)

Rains *(13)*
Common Green Darner (*Anax junius*)
Jade Clubtail (*Arigomphus submedianus*)
Halloween Pennant (*Celithemis eponina*)
Familiar Bluet (*Enallagma civile*)
Orange Bluet (*Enallagma signatum*)
Swamp Darner (*Epiaeschna heros*)
Eastern Pondhawk (*Erythemis simplicicollis*)
Plains Clubtail (*Gomphus externus*)
Slaty Skimmer (*Libellula incesta*)
Great Blue Skimmer (*Libellula vibrans*)
Blue Dasher (*Pachydiplax longipennis*)
Spot-winged Glider (*Pantala hymenaea*)
Common Whitetail (*Plathemis lydia*)

Randall *(34)*
Shadow Darner (*Aeshna umbrosa*)
Common Green Darner (*Anax junius*)
Blue-fronted Dancer (*Argia apicalis*)
Lavender Dancer (*Argia hinei*)
Sooty Dancer (*Argia lugens*)
Powdered Dancer (*Argia moesta*)
Aztec Dancer (*Argia nahuana*)
Blue-ringed Dancer (*Argia sedula*)
Pale-faced Clubskimmer (*Brechmorhoga mendax*)
Familiar Bluet (*Enallagma civile*)
Eastern Ringtail (*Erpetogomphus designatus*)
Eastern Pondhawk (*Erythemis simplicicollis*)
Pronghorn Clubtail (*Gomphus graslinellus*)
Sulphur-tipped Clubtail (*Gomphus militaris*)
American Rubyspot (*Hetaerina americana*)
Plains Forktail (*Ischnura damula*)
Eastern Forktail (*Ischnura verticalis*)
Comanche Skimmer (*Libellula comanche*)
Widow Skimmer (*Libellula luctuosa*)
Twelve-spotted Skimmer (*Libellula pulchella*)
Flame Skimmer (*Libellula saturata*)
Blue Dasher (*Pachydiplax longipennis*)
Red Rock Skimmer (*Paltothemis lineatipes*)
Wandering Glider (*Pantala flavescens*)
Spot-winged Glider (*Pantala hymenaea*)
Eastern Amberwing (*Perithemis tenera*)
Common Whitetail (*Plathemis lydia*)
Gray Sanddragon (*Progomphus borealis*)
Common Sanddragon (*Progomphus obscurus*)
Blue-eyed Darner (*Rhionaeschna multicolor*)
Variegated Meadowhawk (*Sympetrum corruptum*)
Autumn Meadowhawk (*Sympetrum vicinum*)
Black Saddlebags (*Tramea lacerata*)
Red Saddlebags (*Tramea onusta*)

Reagan *(11)*
Familiar Bluet (*Enallagma civile*)
Eastern Pondhawk (*Erythemis simplicicollis*)
Widow Skimmer (*Libellula luctuosa*)
Roseate Skimmer (*Orthemis ferruginea*)
Blue Dasher (*Pachydiplax longipennis*)
Wandering Glider (*Pantala flavescens*)
Spot-winged Glider (*Pantala hymenaea*)
Eastern Amberwing (*Perithemis tenera*)
Variegated Meadowhawk (*Sympetrum corruptum*)
Black Saddlebags (*Tramea lacerata*)
Red Saddlebags (*Tramea onusta*)

Real *(63)*
Common Green Darner (*Anax junius*)
Great Spreadwing (*Archilestes grandis*)
Comanche Dancer (*Argia barretti*)

Coppery Dancer (*Argia cuprea*)
Violet Dancer (*Argia fumipennis violacea*)
Lavender Dancer (*Argia hinei*)
Kiowa Dancer (*Argia immunda*)
Powdered Dancer (*Argia moesta*)
Aztec Dancer (*Argia nahuana*)
Springwater Dancer (*Argia plana*)
Blue-ringed Dancer (*Argia sedula*)
Dusky Dancer (*Argia translata*)
Springtime Darner (*Basiaeschna janata*)
Pale-faced Clubskimmer (*Brechmorhoga mendax*)
Halloween Pennant (*Celithemis eponina*)
Banded Pennant (*Celithemis fasciata*)
Black-shouldered Spinyleg (*Dromogomphus spinosus*)
Flag-tailed Spinyleg (*Dromogomphus spoliatus*)
Checkered Setwing (*Dythemis fugax*)
Black Setwing (*Dythemis nigrescens*)
Swift Setwing (*Dythemis velox*)
Double-striped Bluet (*Enallagma basidens*)
Familiar Bluet (*Enallagma civile*)
Stream Bluet (*Enallagma exsulans*)
Neotropical Bluet (*Enallagma novaehispaniae*)
Arroyo Bluet (*Enallagma praevarum*)
Orange Bluet (*Enallagma signatum*)
Dot-winged Baskettail (*Epitheca petechialis*)
Prince Baskettail (*Epitheca princeps*)
Eastern Ringtail (*Erpetogomphus designatus*)
Eastern Pondhawk (*Erythemis simplicicollis*)
Pronghorn Clubtail (*Gomphus graslinellus*)
Sulphur-tipped Clubtail (*Gomphus militaris*)
Dragonhunter (*Hagenius brevistylus*)
American Rubyspot (*Hetaerina americana*)
Citrine Forktail (*Ischnura hastata*)
Rambur's Forktail (*Ischnura ramburii*)
Comanche Skimmer (*Libellula comanche*)
Neon Skimmer (*Libellula croceipennis*)
Widow Skimmer (*Libellula luctuosa*)
Twelve-spotted Skimmer (*Libellula pulchella*)
Flame Skimmer (*Libellula saturata*)
Bronzed River Cruiser (*Macromia annulata*)
Ivory-striped Sylph (*Macrothemis imitans leucozona*)
Jade-striped Sylph (*Macrothemis inequiunguis*)
Cyrano Darner (*Nasiaeschna pentacantha*)
Coral-fronted Threadtail (*Neoneura aaroni*)
Roseate Skimmer (*Orthemis ferruginea*)
Blue Dasher (*Pachydiplax longipennis*)
Red Rock Skimmer (*Paltothemis lineatipes*)
Wandering Glider (*Pantala flavescens*)
Spot-winged Glider (*Pantala hymenaea*)
Slough Amberwing (*Perithemis domitia*)
Eastern Amberwing (*Perithemis tenera*)
Five-striped Leaftail (*Phyllogomphoides albrighti*)
Four-striped Leaftail (*Phyllogomphoides stigmatus*)
Common Whitetail (*Plathemis lydia*)
Orange-striped Threadtail (*Protoneura cara*)
Variegated Meadowhawk (*Sympetrum corruptum*)
Autumn Meadowhawk (*Sympetrum vicinum*)
Desert Firetail (*Telebasis salva*)
Black Saddlebags (*Tramea lacerata*)
Red Saddlebags (*Tramea onusta*)

Red River *(13)*
Common Green Darner (*Anax junius*)
Blue-fronted Dancer (*Argia apicalis*)
Powdered Dancer (*Argia moesta*)
Jade Clubtail (*Arigomphus submedianus*)
Orange Bluet (*Enallagma signatum*)
Eastern Pondhawk (*Erythemis simplicicollis*)
Oklahoma Clubtail (*Gomphus oklahomensis*)
Citrine Forktail (*Ischnura hastata*)
Fragile Forktail (*Ischnura posita*)
Blue Dasher (*Pachydiplax longipennis*)

Wandering Glider (*Pantala flavescens*)
Variegated Meadowhawk (*Sympetrum corruptum*)
Black Saddlebags (*Tramea lacerata*)

Reeves (65)
Amazon Darner (*Anax amazili*)
Common Green Darner (*Anax junius*)
Paiute Dancer (*Argia alberta*)
Blue-fronted Dancer (*Argia apicalis*)
Violet Dancer (*Argia fumipennis violacea*)
Kiowa Dancer (*Argia immunda*)
Leonora's Dancer (*Argia leonorae*)
Sooty Dancer (*Argia lugens*)
Powdered Dancer (*Argia moesta*)
Apache Dancer (*Argia munda*)
Aztec Dancer (*Argia nahuana*)
Amethyst Dancer (*Argia pallens*)
Springwater Dancer (*Argia plana*)
Blue-ringed Dancer (*Argia sedula*)
Dusky Dancer (*Argia translata*)
Red-tailed Pennant (*Brachymesia furcata*)
Four-spotted Pennant (*Brachymesia gravida*)
Pale-faced Clubskimmer (*Brechmorhoga mendax*)
Checkered Setwing (*Dythemis fugax*)
Black Setwing (*Dythemis nigrescens*)
Swift Setwing (*Dythemis velox*)
Double-striped Bluet (*Enallagma basidens*)
Familiar Bluet (*Enallagma civile*)
Arroyo Bluet (*Enallagma praevarum*)
Eastern Ringtail (*Erpetogomphus designatus*)
Dashed Ringtail (*Erpetogomphus heterodon*)
Western Pondhawk (*Erythemis collocata*)
Eastern Pondhawk (*Erythemis simplicicollis*)
Plateau Dragonlet (*Erythrodiplax basifusca*)
Seaside Dragonlet (*Erythrodiplax berenice*)
Band-winged Dragonlet (*Erythrodiplax umbrata*)
Sulphur-tipped Clubtail (*Gomphus militaris*)
American Rubyspot (*Hetaerina americana*)
Desert Forktail (*Ischnura barberi*)
Plains Forktail (*Ischnura damula*)
Mexican Forktail (*Ischnura demorsa*)
Black-fronted Forktail (*Ischnura denticollis*)
Citrine Forktail (*Ischnura hastata*)
Rambur's Forktail (*Ischnura ramburii*)
Plateau Spreadwing (*Lestes alacer*)
Comanche Skimmer (*Libellula comanche*)
Bleached Skimmer (*Libellula composita*)
Widow Skimmer (*Libellula luctuosa*)
Twelve-spotted Skimmer (*Libellula pulchella*)
Flame Skimmer (*Libellula saturata*)
Marl Pennant (*Macrodiplax balteata*)
Bronzed River Cruiser (*Macromia annulata*)
Roseate Skimmer (*Orthemis ferruginea*)
Blue Dasher (*Pachydiplax longipennis*)
Red Rock Skimmer (*Paltothemis lineatipes*)
Wandering Glider (*Pantala flavescens*)
Spot-winged Glider (*Pantala hymenaea*)
Eastern Amberwing (*Perithemis tenera*)
Four-striped Leaftail (*Phyllogomphoides stigmatus*)
Common Whitetail (*Plathemis lydia*)
Desert Whitetail (*Plathemis subornata*)
Gray Sanddragon (*Progomphus borealis*)
Filigree Skimmer (*Pseudoleon superbus*)
Arroyo Darner (*Rhionaeschna dugesi*)
Blue-eyed Darner (*Rhionaeschna multicolor*)
Russet-tipped Clubtail (*Stylurus plagiatus*)
Variegated Meadowhawk (*Sympetrum corruptum*)
Desert Firetail (*Telebasis salva*)
Black Saddlebags (*Tramea lacerata*)
Red Saddlebags (*Tramea onusta*)

Refugio (16)
Common Green Darner (*Anax junius*)
Blue-fronted Dancer (*Argia apicalis*)
Kiowa Dancer (*Argia immunda*)

Familiar Bluet (*Enallagma civile*)
Eastern Pondhawk (*Erythemis simplicicollis*)
Seaside Dragonlet (*Erythrodiplax berenice*)
Band-winged Dragonlet (*Erythrodiplax umbrata*)
Smoky Rubyspot (*Hetaerina titia*)
Rambur's Forktail (*Ischnura ramburii*)
Hyacinth Glider (*Miathyria marcella*)
Roseate Skimmer (*Orthemis ferruginea*)
Blue Dasher (*Pachydiplax longipennis*)
Wandering Glider (*Pantala flavescens*)
Eastern Amberwing (*Perithemis tenera*)
Black Saddlebags (*Tramea lacerata*)
Red Saddlebags (*Tramea onusta*)

Roberts (1)
Common Green Darner (*Anax junius*)

Robertson (51)
Common Green Darner (*Anax junius*)
Blue-fronted Dancer (*Argia apicalis*)
Violet Dancer (*Argia fumipennis violacea*)
Kiowa Dancer (*Argia immunda*)
Powdered Dancer (*Argia moesta*)
Aztec Dancer (*Argia nahuana*)
Blue-ringed Dancer (*Argia sedula*)
Blue-tipped Dancer (*Argia tibialis*)
Fawn Darner (*Boyeria vinosa*)
Ebony Jewelwing (*Calopteryx maculata*)
Calico Pennant (*Celithemis elisa*)
Halloween Pennant (*Celithemis eponina*)
Black-shouldered Spinyleg (*Dromogomphus spinosus*)
Flag-tailed Spinyleg (*Dromogomphus spoliatus*)
Swift Setwing (*Dythemis velox*)
Double-striped Bluet (*Enallagma basidens*)
Familiar Bluet (*Enallagma civile*)
Stream Bluet (*Enallagma exsulans*)
Swamp Darner (*Epiaeschna heros*)
Prince Baskettail (*Epitheca princeps*)
Eastern Ringtail (*Erpetogomphus designatus*)
Eastern Pondhawk (*Erythemis simplicicollis*)
Little Blue Dragonlet (*Erythrodiplax minuscula*)
Plains Clubtail (*Gomphus externus*)
Sulphur-tipped Clubtail (*Gomphus militaris*)
Cobra Clubtail (*Gomphus vastus*)
Dragonhunter (*Hagenius brevistylus*)
American Rubyspot (*Hetaerina americana*)
Smoky Rubyspot (*Hetaerina titia*)
Citrine Forktail (*Ischnura hastata*)
Rambur's Forktail (*Ischnura ramburii*)
Comanche Skimmer (*Libellula comanche*)
Neon Skimmer (*Libellula croceipennis*)
Yellow-sided Skimmer (*Libellula flavida*)
Slaty Skimmer (*Libellula incesta*)
Widow Skimmer (*Libellula luctuosa*)
Great Blue Skimmer (*Libellula vibrans*)
Georgia River Cruiser (*Macromia illinoiensis georgina*)
Cyrano Darner (*Nasiaeschna pentacantha*)
Roseate Skimmer (*Orthemis ferruginea*)
Blue Dasher (*Pachydiplax longipennis*)
Wandering Glider (*Pantala flavescens*)
Spot-winged Glider (*Pantala hymenaea*)
Eastern Amberwing (*Perithemis tenera*)
Four-striped Leaftail (*Phyllogomphoides stigmatus*)
Common Whitetail (*Plathemis lydia*)
Common Sanddragon (*Progomphus obscurus*)
Russet-tipped Clubtail (*Stylurus plagiatus*)
Blue-faced Meadowhawk (*Sympetrum ambiguum*)
Autumn Meadowhawk (*Sympetrum vicinum*)
Desert Firetail (*Telebasis salva*)

Rockwall (17)
Blue-fronted Dancer (*Argia apicalis*)
Powdered Dancer (*Argia moesta*)
Blue-ringed Dancer (*Argia sedula*)
Flag-tailed Spinyleg (*Dromogomphus spoliatus*)
Familiar Bluet (*Enallagma civile*)

Eastern Pondhawk (*Erythemis simplicicollis*)
Citrine Forktail (*Ischnura hastata*)
Fragile Forktail (*Ischnura posita*)
Rambur's Forktail (*Ischnura ramburii*)
Widow Skimmer (*Libellula luctuosa*)
Roseate Skimmer (*Orthemis ferruginea*)
Blue Dasher (*Pachydiplax longipennis*)
Wandering Glider (*Pantala flavescens*)
Spot-winged Glider (*Pantala hymenaea*)
Eastern Amberwing (*Perithemis tenera*)
Common Whitetail (*Plathemis lydia*)
Desert Firetail (*Telebasis salva*)

Runnels (1)
Familiar Bluet (*Enallagma civile*)

Rusk (30)
Blue-fronted Dancer (*Argia apicalis*)
Violet Dancer (*Argia fumipennis violacea*)
Kiowa Dancer (*Argia immunda*)
Blue-ringed Dancer (*Argia sedula*)
Blue-tipped Dancer (*Argia tibialis*)
Fawn Darner (*Boyeria vinosa*)
Ebony Jewelwing (*Calopteryx maculata*)
Halloween Pennant (*Celithemis eponina*)
Banded Pennant (*Celithemis fasciata*)
Familiar Bluet (*Enallagma civile*)
Turquoise Bluet (*Enallagma divagans*)
Stream Bluet (*Enallagma exsulans*)
Prince Baskettail (*Epitheca princeps*)
Eastern Pondhawk (*Erythemis simplicicollis*)
Sulphur-tipped Clubtail (*Gomphus militaris*)
Oklahoma Clubtail (*Gomphus oklahomensis*)
Cobra Clubtail (*Gomphus vastus*)
Dragonhunter (*Hagenius brevistylus*)
Citrine Forktail (*Ischnura hastata*)
Rambur's Forktail (*Ischnura ramburii*)
Blue Corporal (*Ladona deplanata*)
Comanche Skimmer (*Libellula comanche*)
Slaty Skimmer (*Libellula incesta*)
Widow Skimmer (*Libellula luctuosa*)
Great Blue Skimmer (*Libellula vibrans*)
Blue Dasher (*Pachydiplax longipennis*)
Eastern Amberwing (*Perithemis tenera*)
Common Whitetail (*Plathemis lydia*)
Common Sanddragon (*Progomphus obscurus*)
Gray Petaltail (*Tachopteryx thoreyi*)

Sabine (20)
Common Green Darner (*Anax junius*)
Narrow-striped Forceptail (*Aphylla protracta*)
Two-striped Forceptail (*Aphylla williamsoni*)
Four-spotted Pennant (*Brachymesia gravida*)
Ebony Jewelwing (*Calopteryx maculata*)
Halloween Pennant (*Celithemis eponina*)
Regal Darner (*Coryphaeschna ingens*)
Swamp Darner (*Epiaeschna heros*)
Common Baskettail (*Epitheca cynosura*)
Prince Baskettail (*Epitheca princeps*)
Eastern Pondhawk (*Erythemis simplicicollis*)
Plains Clubtail (*Gomphus externus*)
Slaty Skimmer (*Libellula incesta*)
Blue Dasher (*Pachydiplax longipennis*)
Wandering Glider (*Pantala flavescens*)
Spot-winged Glider (*Pantala hymenaea*)
Eastern Amberwing (*Perithemis tenera*)
Mocha Emerald (*Somatochlora linearis*)
Black Saddlebags (*Tramea lacerata*)
Red Saddlebags (*Tramea onusta*)

San Augustine (22)
Common Green Darner (*Anax junius*)
Ebony Jewelwing (*Calopteryx maculata*)
Halloween Pennant (*Celithemis eponina*)
Swamp Darner (*Epiaeschna heros*)
Stripe-winged Baskettail (*Epitheca costalis*)
Common Baskettail (*Epitheca cynosura*)
Prince Baskettail (*Epitheca princeps*)
Eastern Pondhawk (*Erythemis simplicicollis*)
Slaty Skimmer (*Libellula incesta*)
Great Blue Skimmer (*Libellula vibrans*)

Roseate Skimmer (*Orthemis ferruginea*)
Blue Dasher (*Pachydiplax longipennis*)
Wandering Glider (*Pantala flavescens*)
Spot-winged Glider (*Pantala hymenaea*)
Eastern Amberwing (*Perithemis tenera*)
Common Whitetail (*Plathemis lydia*)
Mocha Emerald (*Somatochlora linearis*)
Texas Emerald (*Somatochlora margarita*)
Clamp-tipped Emerald (*Somatochlora tenebrosa*)
Gray Petaltail (*Tachopteryx thoreyi*)
Carolina Saddlebags (*Tramea carolina*)
Black Saddlebags (*Tramea lacerata*)

San Jacinto (97)
Common Green Darner (*Anax junius*)
Comet Darner (*Anax longipes*)
Two-striped Forceptail (*Aphylla williamsoni*)
Blue-fronted Dancer (*Argia apicalis*)
Seepage Dancer (*Argia bipunctulata*)
Violet Dancer (*Argia fumipennis violacea*)
Powdered Dancer (*Argia moesta*)
Blue-ringed Dancer (*Argia sedula*)
Blue-tipped Dancer (*Argia tibialis*)
Stillwater Clubtail (*Arigomphus lentulus*)
Bayou Clubtail (*Arigomphus maxwelli*)
Jade Clubtail (*Arigomphus submedianus*)
Springtime Darner (*Basiaeschna janata*)
Fawn Darner (*Boyeria vinosa*)
Four-spotted Pennant (*Brachymesia gravida*)
Sparkling Jewelwing (*Calopteryx dimidiata*)
Ebony Jewelwing (*Calopteryx maculata*)
Calico Pennant (*Celithemis elisa*)
Halloween Pennant (*Celithemis eponina*)
Banded Pennant (*Celithemis fasciata*)
Ornate Pennant (*Celithemis ornata*)
Double-ringed Pennant (*Celithemis verna*)
Twin-spotted Spiketail (*Cordulegaster maculata*)
Arrowhead Spiketail (*Cordulegaster obliqua*)
Regal Darner (*Coryphaeschna ingens*)
Stream Cruiser (*Didymops transversa*)
Black-shouldered Spinyleg (*Dromogomphus spinosus*)
Flag-tailed Spinyleg (*Dromogomphus spoliatus*)
Swift Setwing (*Dythemis velox*)
Double-striped Bluet (*Enallagma basidens*)
Familiar Bluet (*Enallagma civile*)
Attenuated Bluet (*Enallagma daeckii*)
Turquoise Bluet (*Enallagma divagans*)
Burgundy Bluet (*Enallagma dubium*)
Stream Bluet (*Enallagma exsulans*)
Skimming Bluet (*Enallagma geminatum*)
Orange Bluet (*Enallagma signatum*)
Vesper Bluet (*Enallagma vesperum*)
Swamp Darner (*Epiaeschna heros*)
Stripe-winged Baskettail (*Epitheca costalis*)
Common Baskettail (*Epitheca cynosura*)
Prince Baskettail (*Epitheca princeps*)
Mantled Baskettail (*Epitheca semiaquea*)
Eastern Ringtail (*Erpetogomphus designatus*)
Eastern Pondhawk (*Erythemis simplicicollis*)
Great Pondhawk (*Erythemis vesiculosa*)
Little Blue Dragonlet (*Erythrodiplax minuscula*)
Harlequin Darner (*Gomphaeschna furcillata*)
Banner Clubtail (*Gomphus apomyius*)
Plains Clubtail (*Gomphus externus*)
Ashy Clubtail (*Gomphus lividus*)
Sulphur-tipped Clubtail (*Gomphus militaris*)
Gulf Coast Clubtail (*Gomphus modestus*)
Oklahoma Clubtail (*Gomphus oklahomensis*)
Dragonhunter (*Hagenius brevistylus*)
Selys' Sundragon (*Helocordulia selysii*)
American Rubyspot (*Hetaerina americana*)
Smoky Rubyspot (*Hetaerina titia*)
Citrine Forktail (*Ischnura hastata*)
Lilypad Forktail (*Ischnura kellicotti*)
Fragile Forktail (*Ischnura posita*)
Furtive Forktail (*Ischnura prognata*)
Rambur's Forktail (*Ischnura ramburii*)

Blue Corporal (*Ladona deplanata*)
Elegant Spreadwing (*Lestes inaequalis*)
Swamp Spreadwing (*Lestes vigilax*)
Golden-winged Skimmer (*Libellula auripennis*)
Yellow-sided Skimmer (*Libellula flavida*)
Slaty Skimmer (*Libellula incesta*)
Widow Skimmer (*Libellula luctuosa*)
Needham's Skimmer (*Libellula needhami*)
Painted Skimmer (*Libellula semifasciata*)
Great Blue Skimmer (*Libellula vibrans*)
Georgia River Cruiser (*Macromia illinoiensis georgina*)
Royal River Cruiser (*Macromia taeniolata*)
Hyacinth Glider (*Miathyria marcella*)
Cyrano Darner (*Nasiaeschna pentacantha*)
Southern Sprite (*Nehalennia integricollis*)
Alabama Shadowdragon (*Neurocordulia alabamensis*)
Carmine Skimmer (*Orthemis discolor*)
Roseate Skimmer (*Orthemis ferruginea*)
Blue Dasher (*Pachydiplax longipennis*)
Wandering Glider (*Pantala flavescens*)
Spot-winged Glider (*Pantala hymenaea*)
Eastern Amberwing (*Perithemis tenera*)
Common Whitetail (*Plathemis lydia*)
Common Sanddragon (*Progomphus obscurus*)
Mocha Emerald (*Somatochlora linearis*)
Texas Emerald (*Somatochlora margarita*)
Laura's Clubtail (*Stylurus laurae*)
Blue-faced Meadowhawk (*Sympetrum ambiguum*)
Variegated Meadowhawk (*Sympetrum corruptum*)
Gray Petaltail (*Tachopteryx thoreyi*)
Duckweed Firetail (*Telebasis byersi*)
Carolina Saddlebags (*Tramea carolina*)
Black Saddlebags (*Tramea lacerata*)
Red Saddlebags (*Tramea onusta*)

San Patricio (59)
Common Green Darner (*Anax junius*)
Narrow-striped Forceptail (*Aphylla protracta*)
Blue-fronted Dancer (*Argia apicalis*)
Powdered Dancer (*Argia moesta*)
Blue-ringed Dancer (*Argia sedula*)
Dusky Dancer (*Argia translata*)
Red-tailed Pennant (*Brachymesia furcata*)
Four-spotted Pennant (*Brachymesia gravida*)
Halloween Pennant (*Celithemis eponina*)
Regal Darner (*Coryphaeschna ingens*)
Flag-tailed Spinyleg (*Dromogomphus spoliatus*)
Checkered Setwing (*Dythemis fugax*)
Black Setwing (*Dythemis nigrescens*)
Swift Setwing (*Dythemis velox*)
Double-striped Bluet (*Enallagma basidens*)
Familiar Bluet (*Enallagma civile*)
Big Bluet (*Enallagma durum*)
Orange Bluet (*Enallagma signatum*)
Swamp Darner (*Epiaeschna heros*)
Prince Baskettail (*Epitheca princeps*)
Eastern Ringtail (*Erpetogomphus designatus*)
Eastern Pondhawk (*Erythemis simplicicollis*)
Great Pondhawk (*Erythemis vesiculosa*)
Seaside Dragonlet (*Erythrodiplax berenice*)
Little Blue Dragonlet (*Erythrodiplax minuscula*)
Band-winged Dragonlet (*Erythrodiplax umbrata*)
Plains Clubtail (*Gomphus externus*)
Sulphur-tipped Clubtail (*Gomphus militaris*)
American Rubyspot (*Hetaerina americana*)
Smoky Rubyspot (*Hetaerina titia*)
Citrine Forktail (*Ischnura hastata*)
Rambur's Forktail (*Ischnura ramburii*)
Plateau Spreadwing (*Lestes alacer*)
Southern Spreadwing (*Lestes australis*)
Rainpool Spreadwing (*Lestes forficula*)
Chalky Spreadwing (*Lestes sigma*)
Golden-winged Skimmer (*Libellula auripennis*)
Widow Skimmer (*Libellula luctuosa*)

Needham's Skimmer (*Libellula needhami*)
Twelve-spotted Skimmer (*Libellula pulchella*)
Marl Pennant (*Macrodiplax balteata*)
Bronzed River Cruiser (*Macromia annulata*)
Gilded River Cruiser (*Macromia pacifica*)
Straw-colored Sylph (*Macrothemis inacuta*)
Hyacinth Glider (*Miathyria marcella*)
Coral-fronted Threadtail (*Neoneura aaroni*)
Roseate Skimmer (*Orthemis ferruginea*)
Blue Dasher (*Pachydiplax longipennis*)
Wandering Glider (*Pantala flavescens*)
Spot-winged Glider (*Pantala hymenaea*)
Eastern Amberwing (*Perithemis tenera*)
Five-striped Leaftail (*Phyllogomphoides albrighti*)
Common Whitetail (*Plathemis lydia*)
Common Sanddragon (*Progomphus obscurus*)
Russet-tipped Clubtail (*Stylurus plagiatus*)
Variegated Meadowhawk (*Sympetrum corruptum*)
Striped Saddlebags (*Tramea calverti*)
Black Saddlebags (*Tramea lacerata*)
Red Saddlebags (*Tramea onusta*)

San Saba (35)
Comet Darner (*Anax longipes*)
Great Spreadwing (*Archilestes grandis*)
Blue-fronted Dancer (*Argia apicalis*)
Violet Dancer (*Argia fumipennis violacea*)
Kiowa Dancer (*Argia immunda*)
Powdered Dancer (*Argia moesta*)
Aztec Dancer (*Argia nahuana*)
Springwater Dancer (*Argia plana*)
Blue-ringed Dancer (*Argia sedula*)
Dusky Dancer (*Argia translata*)
Halloween Pennant (*Celithemis eponina*)
Flag-tailed Spinyleg (*Dromogomphus spoliatus*)
Checkered Setwing (*Dythemis fugax*)
Double-striped Bluet (*Enallagma basidens*)
Familiar Bluet (*Enallagma civile*)
Eastern Ringtail (*Erpetogomphus designatus*)
Plains Clubtail (*Gomphus externus*)
Cobra Clubtail (*Gomphus vastus*)
American Rubyspot (*Hetaerina americana*)
Citrine Forktail (*Ischnura hastata*)
Rambur's Forktail (*Ischnura ramburii*)
Plateau Spreadwing (*Lestes alacer*)
Chalky Spreadwing (*Lestes sigma*)
Neon Skimmer (*Libellula croceipennis*)
Bronzed River Cruiser (*Macromia annulata*)
Orange Shadowdragon (*Neurocordulia xanthosoma*)
Roseate Skimmer (*Orthemis ferruginea*)
Eastern Amberwing (*Perithemis tenera*)
Five-striped Leaftail (*Phyllogomphoides albrighti*)
Four-striped Leaftail (*Phyllogomphoides stigmatus*)
Common Whitetail (*Plathemis lydia*)
Common Sanddragon (*Progomphus obscurus*)
Russet-tipped Clubtail (*Stylurus plagiatus*)
Desert Firetail (*Telebasis salva*)
Red Saddlebags (*Tramea onusta*)

Schleicher (5)
Familiar Bluet (*Enallagma civile*)
Eastern Pondhawk (*Erythemis simplicicollis*)
Citrine Forktail (*Ischnura hastata*)
Plateau Spreadwing (*Lestes alacer*)
Southern Spreadwing (*Lestes australis*)

Scurry (1)
Plateau Spreadwing (*Lestes alacer*)

Shackelford (4)
Blue-fronted Dancer (*Argia apicalis*)
Powdered Dancer (*Argia moesta*)
Halloween Pennant (*Celithemis eponina*)
Widow Skimmer (*Libellula luctuosa*)

Shelby (6)
Four-spotted Pennant (*Brachymesia gravida*)

Ebony Jewelwing (*Calopteryx maculata*)
Halloween Pennant (*Celithemis eponina*)
Twin-spotted Spiketail (*Cordulegaster maculata*)
Arrowhead Spiketail (*Cordulegaster obliqua*)
Eastern Pondhawk (*Erythemis simplicicollis*)

Sherman (5)
Familiar Bluet (*Enallagma civile*)
Eastern Forktail (*Ischnura verticalis*)
Southern Spreadwing (*Lestes australis*)
Lyre-tipped Spreadwing (*Lestes unguiculatus*)
Variegated Meadowhawk (*Sympetrum corruptum*)

Smith (28)
Blue-fronted Dancer (*Argia apicalis*)
Powdered Dancer (*Argia moesta*)
Blue-tipped Dancer (*Argia tibialis*)
Springtime Darner (*Basiaeschna janata*)
Ebony Jewelwing (*Calopteryx maculata*)
Black-shouldered Spinyleg (*Dromogomphus spinosus*)
Skimming Bluet (*Enallagma geminatum*)
Orange Bluet (*Enallagma signatum*)
Swamp Darner (*Epiaeschna heros*)
Stripe-winged Baskettail (*Epitheca costalis*)
Common Baskettail (*Epitheca cynosura*)
Dot-winged Baskettail (*Epitheca petechialis*)
Mantled Baskettail (*Epitheca semiaquea*)
Eastern Pondhawk (*Erythemis simplicicollis*)
Band-winged Dragonlet (*Erythrodiplax umbrata*)
Cocoa Clubtail (*Gomphus hybridus*)
Oklahoma Clubtail (*Gomphus oklahomensis*)
Smoky Rubyspot (*Hetaerina titia*)
Spangled Skimmer (*Libellula cyanea*)
Slaty Skimmer (*Libellula incesta*)
Widow Skimmer (*Libellula luctuosa*)
Great Blue Skimmer (*Libellula vibrans*)
Cyrano Darner (*Nasiaeschna pentacantha*)
Blue Dasher (*Pachydiplax longipennis*)
Eastern Amberwing (*Perithemis tenera*)
Common Whitetail (*Plathemis lydia*)
Variegated Meadowhawk (*Sympetrum corruptum*)
Black Saddlebags (*Tramea lacerata*)

Somervell (34)
Common Green Darner (*Anax junius*)
Great Spreadwing (*Archilestes grandis*)
Blue-fronted Dancer (*Argia apicalis*)
Violet Dancer (*Argia fumipennis violacea*)
Kiowa Dancer (*Argia immunda*)
Powdered Dancer (*Argia moesta*)
Aztec Dancer (*Argia nahuana*)
Blue-ringed Dancer (*Argia sedula*)
Dusky Dancer (*Argia translata*)
Four-spotted Pennant (*Brachymesia gravida*)
Pale-faced Clubskimmer (*Brechmorhoga mendax*)
Halloween Pennant (*Celithemis eponina*)
Flag-tailed Spinyleg (*Dromogomphus spoliatus*)
Checkered Setwing (*Dythemis fugax*)
Swift Setwing (*Dythemis velox*)
Eastern Ringtail (*Erpetogomphus designatus*)
Eastern Pondhawk (*Erythemis simplicicollis*)
American Rubyspot (*Hetaerina americana*)
Smoky Rubyspot (*Hetaerina titia*)
Citrine Forktail (*Ischnura hastata*)
Widow Skimmer (*Libellula luctuosa*)
Twelve-spotted Skimmer (*Libellula pulchella*)
Roseate Skimmer (*Orthemis ferruginea*)
Blue Dasher (*Pachydiplax longipennis*)
Wandering Glider (*Pantala flavescens*)
Spot-winged Glider (*Pantala hymenaea*)
Eastern Amberwing (*Perithemis tenera*)
Five-striped Leaftail (*Phyllogomphoides albrighti*)
Four-striped Leaftail (*Phyllogomphoides stigmatus*)
Common Whitetail (*Plathemis lydia*)

Common Sanddragon (*Progomphus obscurus*)
Autumn Meadowhawk (*Sympetrum vicinum*)
Black Saddlebags (*Tramea lacerata*)
Red Saddlebags (*Tramea onusta*)

Starr (55)
Common Green Darner (*Anax junius*)
Broad-striped Forceptail (*Aphylla angustifolia*)
Narrow-striped Forceptail (*Aphylla protracta*)
Blue-fronted Dancer (*Argia apicalis*)
Kiowa Dancer (*Argia immunda*)
Powdered Dancer (*Argia moesta*)
Blue-ringed Dancer (*Argia sedula*)
Red-tailed Pennant (*Brachymesia furcata*)
Four-spotted Pennant (*Brachymesia gravida*)
Tawny Pennant (*Brachymesia herbida*)
Flag-tailed Spinyleg (*Dromogomphus spoliatus*)
Black Setwing (*Dythemis nigrescens*)
Swift Setwing (*Dythemis velox*)
Double-striped Bluet (*Enallagma basidens*)
Familiar Bluet (*Enallagma civile*)
Big Bluet (*Enallagma durum*)
Neotropical Bluet (*Enallagma novaehispaniae*)
Orange Bluet (*Enallagma signatum*)
Prince Baskettail (*Epitheca princeps*)
Eastern Ringtail (*Erpetogomphus designatus*)
Pin-tailed Pondhawk (*Erythemis plebeja*)
Eastern Pondhawk (*Erythemis simplicicollis*)
Great Pondhawk (*Erythemis vesiculosa*)
Band-winged Dragonlet (*Erythrodiplax umbrata*)
Plains Clubtail (*Gomphus externus*)
Tamaulipan Clubtail (*Gomphus gonzalezi*)
Sulphur-tipped Clubtail (*Gomphus militaris*)
American Rubyspot (*Hetaerina americana*)
Smoky Rubyspot (*Hetaerina titia*)
Citrine Forktail (*Ischnura hastata*)
Rambur's Forktail (*Ischnura ramburii*)
Rainpool Spreadwing (*Lestes forficula*)
Chalky Spreadwing (*Lestes sigma*)
Neon Skimmer (*Libellula croceipennis*)
Needham's Skimmer (*Libellula needhami*)
Flame Skimmer (*Libellula saturata*)
Marl Pennant (*Macrodiplax balteata*)
Straw-colored Sylph (*Macrothemis inacuta*)
Hyacinth Glider (*Miathyria marcella*)
Thornbush Dasher (*Micrathyria hagenii*)
Carmine Skimmer (*Orthemis discolor*)
Roseate Skimmer (*Orthemis ferruginea*)
Blue Dasher (*Pachydiplax longipennis*)
Wandering Glider (*Pantala flavescens*)
Spot-winged Glider (*Pantala hymenaea*)
Eastern Amberwing (*Perithemis tenera*)
Ringed Forceptail (*Phyllocycla breviphylla*)
Five-striped Leaftail (*Phyllogomphoides albrighti*)
Filigree Skimmer (*Pseudoleon superbus*)
Turquoise-tipped Darner (*Rhionaeschna psilus*)
Russet-tipped Clubtail (*Stylurus plagiatus*)
Variegated Meadowhawk (*Sympetrum corruptum*)
Striped Saddlebags (*Tramea calverti*)
Black Saddlebags (*Tramea lacerata*)
Red Saddlebags (*Tramea onusta*)

Stephens (7)
Common Green Darner (*Anax junius*)
Double-striped Bluet (*Enallagma basidens*)
Familiar Bluet (*Enallagma civile*)
Sulphur-tipped Clubtail (*Gomphus militaris*)
Citrine Forktail (*Ischnura hastata*)
Eastern Amberwing (*Perithemis tenera*)
Variegated Meadowhawk (*Sympetrum corruptum*)

Stonewall (1)
Eastern Ringtail (*Erpetogomphus designatus*)

Sutton (25)
Common Green Darner (*Anax junius*)
Great Spreadwing (*Archilestes grandis*)
Blue-fronted Dancer (*Argia apicalis*)

Violet Dancer (*Argia fumipennis violacea*)
Kiowa Dancer (*Argia immunda*)
Powdered Dancer (*Argia moesta*)
Blue-ringed Dancer (*Argia sedula*)
Dusky Dancer (*Argia translata*)
Halloween Pennant (*Celithemis eponina*)
Checkered Setwing (*Dythemis fugax*)
Swift Setwing (*Dythemis velox*)
Double-striped Bluet (*Enallagma basidens*)
Familiar Bluet (*Enallagma civile*)
Stream Bluet (*Enallagma exsulans*)
Arroyo Bluet (*Enallagma praevarum*)
Eastern Pondhawk (*Erythemis simplicicollis*)
Band-winged Dragonlet (*Erythrodiplax umbrata*)
American Rubyspot (*Hetaerina americana*)
Citrine Forktail (*Ischnura hastata*)
Comanche Skimmer (*Libellula comanche*)
Widow Skimmer (*Libellula luctuosa*)
Roseate Skimmer (*Orthemis ferruginea*)
Blue Dasher (*Pachydiplax longipennis*)
Desert Firetail (*Telebasis salva*)
Red Saddlebags (*Tramea onusta*)

Swisher (1)
Common Green Darner (*Anax junius*)

Tarrant (71)
Common Green Darner (*Anax junius*)
Great Spreadwing (*Archilestes grandis*)
Blue-fronted Dancer (*Argia apicalis*)
Violet Dancer (*Argia fumipennis violacea*)
Kiowa Dancer (*Argia immunda*)
Powdered Dancer (*Argia moesta*)
Aztec Dancer (*Argia nahuana*)
Springwater Dancer (*Argia plana*)
Blue-ringed Dancer (*Argia sedula*)
Dusky Dancer (*Argia translata*)
Jade Clubtail (*Arigomphus submedianus*)
Springtime Darner (*Basiaeschna janata*)
Four-spotted Pennant (*Brachymesia gravida*)
Pale-faced Clubskimmer (*Brechmorhoga mendax*)
Ebony Jewelwing (*Calopteryx maculata*)
Halloween Pennant (*Celithemis eponina*)
Banded Pennant (*Celithemis fasciata*)
Regal Darner (*Coryphaeschna ingens*)
Stream Cruiser (*Didymops transversa*)
Flag-tailed Spinyleg (*Dromogomphus spoliatus*)
Checkered Setwing (*Dythemis fugax*)
Swift Setwing (*Dythemis velox*)
Double-striped Bluet (*Enallagma basidens*)
Familiar Bluet (*Enallagma civile*)
Turquoise Bluet (*Enallagma divagans*)
Stream Bluet (*Enallagma exsulans*)
Skimming Bluet (*Enallagma geminatum*)
Orange Bluet (*Enallagma signatum*)
Swamp Darner (*Epiaeschna heros*)
Dot-winged Baskettail (*Epitheca petechialis*)
Prince Baskettail (*Epitheca princeps*)
Mantled Baskettail (*Epitheca semiaquea*)
Eastern Ringtail (*Erpetogomphus designatus*)
Pin-tailed Pondhawk (*Erythemis plebeja*)
Eastern Pondhawk (*Erythemis simplicicollis*)
Great Pondhawk (*Erythemis vesiculosa*)
Band-winged Dragonlet (*Erythrodiplax umbrata*)
Plains Clubtail (*Gomphus externus*)
Sulphur-tipped Clubtail (*Gomphus militaris*)
Cobra Clubtail (*Gomphus vastus*)
American Rubyspot (*Hetaerina americana*)
Citrine Forktail (*Ischnura hastata*)
Lilypad Forktail (*Ischnura kellicotti*)
Rambur's Forktail (*Ischnura ramburii*)
Southern Spreadwing (*Lestes australis*)
Comanche Skimmer (*Libellula comanche*)
Neon Skimmer (*Libellula croceipennis*)
Slaty Skimmer (*Libellula incesta*)
Widow Skimmer (*Libellula luctuosa*)
Twelve-spotted Skimmer (*Libellula pulchella*)
Great Blue Skimmer (*Libellula vibrans*)

Georgia River Cruiser (*Macromia illinoiensis georgina*)
Royal River Cruiser (*Macromia taeniolata*)
Cyrano Darner (*Nasiaeschna pentacantha*)
Roseate Skimmer (*Orthemis ferruginea*)
Blue Dasher (*Pachydiplax longipennis*)
Wandering Glider (*Pantala flavescens*)
Spot-winged Glider (*Pantala hymenaea*)
Eastern Amberwing (*Perithemis tenera*)
Four-striped Leaftail (*Phyllogomphoides stigmatus*)
Common Whitetail (*Plathemis lydia*)
Common Sanddragon (*Progomphus obscurus*)
Mocha Emerald (*Somatochlora linearis*)
Blue-faced Meadowhawk (*Sympetrum ambiguum*)
Variegated Meadowhawk (*Sympetrum corruptum*)
Autumn Meadowhawk (*Sympetrum vicinum*)
Desert Firetail (*Telebasis salva*)
Striped Saddlebags (*Tramea calverti*)
Carolina Saddlebags (*Tramea carolina*)
Black Saddlebags (*Tramea lacerata*)
Red Saddlebags (*Tramea onusta*)

Taylor *(29)*
Common Green Darner (*Anax junius*)
Blue-fronted Dancer (*Argia apicalis*)
Kiowa Dancer (*Argia immunda*)
Powdered Dancer (*Argia moesta*)
Blue-ringed Dancer (*Argia sedula*)
Checkered Setwing (*Dythemis fugax*)
Swift Setwing (*Dythemis velox*)
Double-striped Bluet (*Enallagma basidens*)
Familiar Bluet (*Enallagma civile*)
Eastern Ringtail (*Erpetogomphus designatus*)
Eastern Pondhawk (*Erythemis simplicicollis*)
Great Pondhawk (*Erythemis vesiculosa*)
Sulphur-tipped Clubtail (*Gomphus militaris*)
American Rubyspot (*Hetaerina americana*)
Citrine Forktail (*Ischnura hastata*)
Comanche Skimmer (*Libellula comanche*)
Neon Skimmer (*Libellula croceipennis*)
Widow Skimmer (*Libellula luctuosa*)
Twelve-spotted Skimmer (*Libellula pulchella*)
Thornbush Dasher (*Micrathyria hagenii*)
Roseate Skimmer (*Orthemis ferruginea*)
Blue Dasher (*Pachydiplax longipennis*)
Wandering Glider (*Pantala flavescens*)
Spot-winged Glider (*Pantala hymenaea*)
Eastern Amberwing (*Perithemis tenera*)
Common Whitetail (*Plathemis lydia*)
Russet-tipped Clubtail (*Stylurus plagiatus*)
Variegated Meadowhawk (*Sympetrum corruptum*)
Red Saddlebags (*Tramea onusta*)

Terrell *(51)*
Common Green Darner (*Anax junius*)
Violet Dancer (*Argia fumipennis violacea*)
Kiowa Dancer (*Argia immunda*)
Leonora's Dancer (*Argia leonorae*)
Powdered Dancer (*Argia moesta*)
Aztec Dancer (*Argia nahuana*)
Blue-ringed Dancer (*Argia sedula*)
Dusky Dancer (*Argia translata*)
Red-tailed Pennant (*Brachymesia furcata*)
Four-spotted Pennant (*Brachymesia gravida*)
Pale-faced Clubskimmer (*Brechmorhoga mendax*)
Halloween Pennant (*Celithemis eponina*)
Flag-tailed Spinyleg (*Dromogomphus spoliatus*)
Checkered Setwing (*Dythemis fugax*)
Black Setwing (*Dythemis nigrescens*)
Swift Setwing (*Dythemis velox*)
Double-striped Bluet (*Enallagma basidens*)
Familiar Bluet (*Enallagma civile*)
Neotropical Bluet (*Enallagma novaehispaniae*)
Stripe-winged Baskettail (*Epitheca costalis*)

Dot-winged Baskettail (*Epitheca petechialis*)
Prince Baskettail (*Epitheca princeps*)
Eastern Ringtail (*Erpetogomphus designatus*)
Eastern Pondhawk (*Erythemis simplicicollis*)
Plateau Dragonlet (*Erythrodiplax basifusca*)
Sulphur-tipped Clubtail (*Gomphus militaris*)
American Rubyspot (*Hetaerina americana*)
Smoky Rubyspot (*Hetaerina titia*)
Mexican Forktail (*Ischnura demorsa*)
Black-fronted Forktail (*Ischnura denticollis*)
Citrine Forktail (*Ischnura hastata*)
Rambur's Forktail (*Ischnura ramburii*)
Comanche Skimmer (*Libellula comanche*)
Neon Skimmer (*Libellula croceipennis*)
Widow Skimmer (*Libellula luctuosa*)
Flame Skimmer (*Libellula saturata*)
Marl Pennant (*Macrodiplax balteata*)
Bronzed River Cruiser (*Macromia annulata*)
Roseate Skimmer (*Orthemis ferruginea*)
Blue Dasher (*Pachydiplax longipennis*)
Wandering Glider (*Pantala flavescens*)
Spot-winged Glider (*Pantala hymenaea*)
Five-striped Leaftail (*Phyllogomphoides albrighti*)
Four-striped Leaftail (*Phyllogomphoides stigmatus*)
Common Whitetail (*Plathemis lydia*)
Orange-striped Threadtail (*Protoneura cara*)
Filigree Skimmer (*Pseudoleon superbus*)
Desert Firetail (*Telebasis salva*)
Antillean Saddlebags (*Tramea insularis*)
Black Saddlebags (*Tramea lacerata*)
Red Saddlebags (*Tramea onusta*)

Throckmorton *(4)*
Blue-fronted Dancer (*Argia apicalis*)
Powdered Dancer (*Argia moesta*)
Flag-tailed Spinyleg (*Dromogomphus spoliatus*)
Familiar Bluet (*Enallagma civile*)

Titus *(28)*
Common Green Darner (*Anax junius*)
Violet Dancer (*Argia fumipennis violacea*)
Aztec Dancer (*Argia nahuana*)
Blue-tipped Dancer (*Argia tibialis*)
Fawn Darner (*Boyeria vinosa*)
Ebony Jewelwing (*Calopteryx maculata*)
Calico Pennant (*Celithemis elisa*)
Halloween Pennant (*Celithemis eponina*)
Banded Pennant (*Celithemis fasciata*)
Black-shouldered Spinyleg (*Dromogomphus spinosus*)
Double-striped Bluet (*Enallagma basidens*)
Familiar Bluet (*Enallagma civile*)
Eastern Pondhawk (*Erythemis simplicicollis*)
Little Blue Dragonlet (*Erythrodiplax minuscula*)
Citrine Forktail (*Ischnura hastata*)
Rambur's Forktail (*Ischnura ramburii*)
Slaty Skimmer (*Libellula incesta*)
Widow Skimmer (*Libellula luctuosa*)
Twelve-spotted Skimmer (*Libellula pulchella*)
Great Blue Skimmer (*Libellula vibrans*)
Blue Dasher (*Pachydiplax longipennis*)
Eastern Amberwing (*Perithemis tenera*)
Common Sanddragon (*Progomphus obscurus*)
Coppery Emerald (*Somatochlora georgiana*)
Variegated Meadowhawk (*Sympetrum corruptum*)
Gray Petaltail (*Tachopteryx thoreyi*)
Desert Firetail (*Telebasis salva*)
Black Saddlebags (*Tramea lacerata*)

Tom Green *(17)*
Common Green Darner (*Anax junius*)
Kiowa Dancer (*Argia immunda*)
Powdered Dancer (*Argia moesta*)
Aztec Dancer (*Argia nahuana*)
Blue-ringed Dancer (*Argia sedula*)
Double-striped Bluet (*Enallagma basidens*)
Stream Bluet (*Enallagma exsulans*)

Eastern Pondhawk (*Erythemis simplicicollis*)
Rambur's Forktail (*Ischnura ramburii*)
Comanche Skimmer (*Libellula comanche*)
Widow Skimmer (*Libellula luctuosa*)
Marl Pennant (*Macrodiplax balteata*)
Roseate Skimmer (*Orthemis ferruginea*)
Blue Dasher (*Pachydiplax longipennis*)
Eastern Amberwing (*Perithemis tenera*)
Common Whitetail (*Plathemis lydia*)
Desert Firetail (*Telebasis salva*)

Travis *(102)*
Amazon Darner (*Anax amazili*)
Common Green Darner (*Anax junius*)
Comet Darner (*Anax longipes*)
Giant Darner (*Anax walsinghami*)
Broad-striped Forceptail (*Aphylla angustifolia*)
Great Spreadwing (*Archilestes grandis*)
Blue-fronted Dancer (*Argia apicalis*)
Comanche Dancer (*Argia barretti*)
Violet Dancer (*Argia fumipennis violacea*)
Lavender Dancer (*Argia hinei*)
Kiowa Dancer (*Argia immunda*)
Leonora's Dancer (*Argia leonorae*)
Powdered Dancer (*Argia moesta*)
Aztec Dancer (*Argia nahuana*)
Springwater Dancer (*Argia plana*)
Blue-ringed Dancer (*Argia sedula*)
Dusky Dancer (*Argia translata*)
Jade Clubtail (*Arigomphus submedianus*)
Springtime Darner (*Basiaeschna janata*)
Fawn Darner (*Boyeria vinosa*)
Red-tailed Pennant (*Brachymesia furcata*)
Four-spotted Pennant (*Brachymesia gravida*)
Pale-faced Clubskimmer (*Brechmorhoga mendax*)
Ebony Jewelwing (*Calopteryx maculata*)
Gray-waisted Skimmer (*Cannaphila insularis funerea*)
Calico Pennant (*Celithemis elisa*)
Halloween Pennant (*Celithemis eponina*)
Banded Pennant (*Celithemis fasciata*)
Regal Darner (*Coryphaeschna ingens*)
Stream Cruiser (*Didymops transversa*)
Flag-tailed Spinyleg (*Dromogomphus spoliatus*)
Checkered Setwing (*Dythemis fugax*)
Black Setwing (*Dythemis nigrescens*)
Swift Setwing (*Dythemis velox*)
Double-striped Bluet (*Enallagma basidens*)
Familiar Bluet (*Enallagma civile*)
Stream Bluet (*Enallagma exsulans*)
Skimming Bluet (*Enallagma geminatum*)
Neotropical Bluet (*Enallagma novaehispaniae*)
Orange Bluet (*Enallagma signatum*)
Vesper Bluet (*Enallagma vesperum*)
Swamp Darner (*Epiaeschna heros*)
Dot-winged Baskettail (*Epitheca petechialis*)
Prince Baskettail (*Epitheca princeps*)
Mantled Baskettail (*Epitheca semiaquea*)
Eastern Ringtail (*Erpetogomphus designatus*)
Pin-tailed Pondhawk (*Erythemis plebeja*)
Eastern Pondhawk (*Erythemis simplicicollis*)
Great Pondhawk (*Erythemis vesiculosa*)
Little Blue Dragonlet (*Erythrodiplax minuscula*)
Band-winged Dragonlet (*Erythrodiplax umbrata*)
Plains Clubtail (*Gomphus externus*)
Pronghorn Clubtail (*Gomphus graslinellus*)
Sulphur-tipped Clubtail (*Gomphus militaris*)
Cobra Clubtail (*Gomphus vastus*)
Dragonhunter (*Hagenius brevistylus*)
American Rubyspot (*Hetaerina americana*)
Smoky Rubyspot (*Hetaerina titia*)
Citrine Forktail (*Ischnura hastata*)
Fragile Forktail (*Ischnura posita*)
Rambur's Forktail (*Ischnura ramburii*)
Blue Corporal (*Ladona deplanata*)
Plateau Spreadwing (*Lestes alacer*)
Southern Spreadwing (*Lestes australis*)

Comanche Skimmer (*Libellula comanche*)
Neon Skimmer (*Libellula croceipennis*)
Slaty Skimmer (*Libellula incesta*)
Widow Skimmer (*Libellula luctuosa*)
Needham's Skimmer (*Libellula needhami*)
Twelve-spotted Skimmer (*Libellula pulchella*)
Flame Skimmer (*Libellula saturata*)
Great Blue Skimmer (*Libellula vibrans*)
Marl Pennant (*Macrodiplax balteata*)
Bronzed River Cruiser (*Macromia annulata*)
Georgia River Cruiser (*Macromia illinoiensis georgina*)
Royal River Cruiser (*Macromia taeniolata*)
Straw-colored Sylph (*Macrothemis inacuta*)
Jade-striped Sylph (*Macrothemis inequiunguis*)
Hyacinth Glider (*Miathyria marcella*)
Thornbush Dasher (*Micrathyria hagenii*)
Orange Shadowdragon (*Neurocordulia xanthosoma*)
Carmine Skimmer (*Orthemis discolor*)
Roseate Skimmer (*Orthemis ferruginea*)
Blue Dasher (*Pachydiplax longipennis*)
Wandering Glider (*Pantala flavescens*)
Spot-winged Glider (*Pantala hymenaea*)
Eastern Amberwing (*Perithemis tenera*)
Five-striped Leaftail (*Phyllogomphoides albrighti*)
Four-striped Leaftail (*Phyllogomphoides stigmatus*)
Common Whitetail (*Plathemis lydia*)
Common Sanddragon (*Progomphus obscurus*)
Filigree Skimmer (*Pseudoleon superbus*)
Blue-eyed Darner (*Rhionaeschna multicolor*)
Turquoise-tipped Darner (*Rhionaeschna psilus*)
Russet-tipped Clubtail (*Stylurus plagiatus*)
Blue-faced Meadowhawk (*Sympetrum ambiguum*)
Variegated Meadowhawk (*Sympetrum corruptum*)
Autumn Meadowhawk (*Sympetrum vicinum*)
Desert Firetail (*Telebasis salva*)
Striped Saddlebags (*Tramea calverti*)
Black Saddlebags (*Tramea lacerata*)
Red Saddlebags (*Tramea onusta*)

Trinity *(21)*
Jade Clubtail (*Arigomphus submedianus*)
Calico Pennant (*Celithemis elisa*)
Banded Pennant (*Celithemis fasciata*)
Swamp Darner (*Epiaeschna heros*)
Prince Baskettail (*Epitheca princeps*)
Mantled Baskettail (*Epitheca semiaquea*)
Eastern Pondhawk (*Erythemis simplicicollis*)
Little Blue Dragonlet (*Erythrodiplax minuscula*)
Oklahoma Clubtail (*Gomphus oklahomensis*)
Citrine Forktail (*Ischnura hastata*)
Rambur's Forktail (*Ischnura ramburii*)
Blue Corporal (*Ladona deplanata*)
Golden-winged Skimmer (*Libellula auripennis*)
Widow Skimmer (*Libellula luctuosa*)
Great Blue Skimmer (*Libellula vibrans*)
Royal River Cruiser (*Macromia taeniolata*)
Blue Dasher (*Pachydiplax longipennis*)
Eastern Amberwing (*Perithemis tenera*)
Common Whitetail (*Plathemis lydia*)
Fine-lined Emerald (*Somatochlora filosa*)
Black Saddlebags (*Tramea lacerata*)

Tyler *(50)*
Seepage Dancer (*Argia bipunctulata*)
Violet Dancer (*Argia fumipennis violacea*)
Kiowa Dancer (*Argia immunda*)
Powdered Dancer (*Argia moesta*)
Blue-tipped Dancer (*Argia tibialis*)
Fawn Darner (*Boyeria vinosa*)
Four-spotted Pennant (*Brachymesia gravida*)
Sparkling Jewelwing (*Calopteryx dimidiata*)
Ebony Jewelwing (*Calopteryx maculata*)
Amanda's Pennant (*Celithemis amanda*)

Halloween Pennant (*Celithemis eponina*)
Turquoise Bluet (*Enallagma divagans*)
Burgundy Bluet (*Enallagma dubium*)
Orange Bluet (*Enallagma signatum*)
Vesper Bluet (*Enallagma vesperum*)
Swamp Darner (*Epiaeschna heros*)
Stripe-winged Baskettail (*Epitheca costalis*)
Common Baskettail (*Epitheca cynosura*)
Prince Baskettail (*Epitheca princeps*)
Eastern Pondhawk (*Erythemis simplicicollis*)
Little Blue Dragonlet (*Erythrodiplax minuscula*)
Cocoa Clubtail (*Gomphus hybridus*)
Oklahoma Clubtail (*Gomphus oklahomensis*)
Smoky Rubyspot (*Hetaerina titia*)
Citrine Forktail (*Ischnura hastata*)
Fragile Forktail (*Ischnura posita*)
Rambur's Forktail (*Ischnura ramburii*)
Blue Corporal (*Ladona deplanata*)
Swamp Spreadwing (*Lestes vigilax*)
Golden-winged Skimmer (*Libellula auripennis*)
Bar-winged Skimmer (*Libellula axilena*)
Yellow-sided Skimmer (*Libellula flavida*)
Slaty Skimmer (*Libellula incesta*)
Painted Skimmer (*Libellula semifasciata*)
Great Blue Skimmer (*Libellula vibrans*)
Georgia River Cruiser (*Macromia illinoiensis georgina*)
Hyacinth Glider (*Miathyria marcella*)
Cyrano Darner (*Nasiaeschna pentacantha*)
Roseate Skimmer (*Orthemis ferruginea*)
Blue Dasher (*Pachydiplax longipennis*)
Eastern Amberwing (*Perithemis tenera*)
Common Whitetail (*Plathemis lydia*)
Common Sanddragon (*Progomphus obscurus*)
Coppery Emerald (*Somatochlora georgiana*)
Mocha Emerald (*Somatochlora linearis*)
Texas Emerald (*Somatochlora margarita*)
Gray Petaltail (*Tachopteryx thoreyi*)
Carolina Saddlebags (*Tramea carolina*)
Black Saddlebags (*Tramea lacerata*)
Red Saddlebags (*Tramea onusta*)

Upshur *(16)*
Blue-fronted Dancer (*Argia apicalis*)
Fawn Darner (*Boyeria vinosa*)
Ebony Jewelwing (*Calopteryx maculata*)
Halloween Pennant (*Celithemis eponina*)
Twin-spotted Spiketail (*Cordulegaster maculata*)
Stream Bluet (*Enallagma exsulans*)
Dot-winged Baskettail (*Epitheca petechialis*)
Eastern Pondhawk (*Erythemis simplicicollis*)
Citrine Forktail (*Ischnura hastata*)
Spangled Skimmer (*Libellula cyanea*)
Slaty Skimmer (*Libellula incesta*)
Blue Dasher (*Pachydiplax longipennis*)
Wandering Glider (*Pantala flavescens*)
Eastern Amberwing (*Perithemis tenera*)
Variegated Meadowhawk (*Sympetrum corruptum*)
Black Saddlebags (*Tramea lacerata*)

Upton *(1)*
Red Saddlebags (*Tramea onusta*)

Uvalde *(76)*
Amazon Darner (*Anax amazili*)
Common Green Darner (*Anax junius*)
Comet Darner (*Anax longipes*)
Broad-striped Forceptail (*Aphylla angustifolia*)
Narrow-striped Forceptail (*Aphylla protracta*)
Great Spreadwing (*Archilestes grandis*)
Blue-fronted Dancer (*Argia apicalis*)
Comanche Dancer (*Argia barretti*)
Coppery Dancer (*Argia cuprea*)
Violet Dancer (*Argia fumipennis violacea*)
Kiowa Dancer (*Argia immunda*)
Leonora's Dancer (*Argia leonorae*)
Powdered Dancer (*Argia moesta*)
Aztec Dancer (*Argia nahuana*)
Springwater Dancer (*Argia plana*)

Blue-ringed Dancer (*Argia sedula*)
Dusky Dancer (*Argia translata*)
Jade Clubtail (*Arigomphus submedianus*)
Springtime Darner (*Basiaeschna janata*)
Red-tailed Pennant (*Brachymesia furcata*)
Four-spotted Pennant (*Brachymesia gravida*)
Pale-faced Clubskimmer (*Brechmorhoga mendax*)
Halloween Pennant (*Celithemis eponina*)
Banded Pennant (*Celithemis fasciata*)
Stream Cruiser (*Didymops transversa*)
Black-shouldered Spinyleg (*Dromogomphus spinosus*)
Flag-tailed Spinyleg (*Dromogomphus spoliatus*)
Checkered Setwing (*Dythemis fugax*)
Black Setwing (*Dythemis nigrescens*)
Swift Setwing (*Dythemis velox*)
Double-striped Bluet (*Enallagma basidens*)
Familiar Bluet (*Enallagma civile*)
Stream Bluet (*Enallagma exsulans*)
Neotropical Bluet (*Enallagma novaehispaniae*)
Arroyo Bluet (*Enallagma praevarum*)
Orange Bluet (*Enallagma signatum*)
Stripe-winged Baskettail (*Epitheca costalis*)
Dot-winged Baskettail (*Epitheca petechialis*)
Prince Baskettail (*Epitheca princeps*)
Eastern Ringtail (*Erpetogomphus designatus*)
Pin-tailed Pondhawk (*Erythemis plebeja*)
Eastern Pondhawk (*Erythemis simplicicollis*)
Great Pondhawk (*Erythemis vesiculosa*)
Little Blue Dragonlet (*Erythrodiplax minuscula*)
Band-winged Dragonlet (*Erythrodiplax umbrata*)
Sulphur-tipped Clubtail (*Gomphus militaris*)
Dragonhunter (*Hagenius brevistylus*)
American Rubyspot (*Hetaerina americana*)
Smoky Rubyspot (*Hetaerina titia*)
Citrine Forktail (*Ischnura hastata*)
Rambur's Forktail (*Ischnura ramburii*)
Plateau Spreadwing (*Lestes alacer*)
Comanche Skimmer (*Libellula comanche*)
Neon Skimmer (*Libellula croceipennis*)
Widow Skimmer (*Libellula luctuosa*)
Twelve-spotted Skimmer (*Libellula pulchella*)
Bronzed River Cruiser (*Macromia annulata*)
Ivory-striped Sylph (*Macrothemis imitans leucozona*)
Thornbush Dasher (*Micrathyria hagenii*)
Cyrano Darner (*Nasiaeschna pentacantha*)
Roseate Skimmer (*Orthemis ferruginea*)
Blue Dasher (*Pachydiplax longipennis*)
Wandering Glider (*Pantala flavescens*)
Spot-winged Glider (*Pantala hymenaea*)
Eastern Amberwing (*Perithemis tenera*)
Five-striped Leaftail (*Phyllogomphoides albrighti*)
Four-striped Leaftail (*Phyllogomphoides stigmatus*)
Common Whitetail (*Plathemis lydia*)
Orange-striped Threadtail (*Protoneura cara*)
Filigree Skimmer (*Pseudoleon superbus*)
Variegated Meadowhawk (*Sympetrum corruptum*)
Autumn Meadowhawk (*Sympetrum vicinum*)
Desert Firetail (*Telebasis salva*)
Striped Saddlebags (*Tramea calverti*)
Black Saddlebags (*Tramea lacerata*)
Red Saddlebags (*Tramea onusta*)

Val Verde *(79)*
Mexican Wedgetail (*Acanthagrion quadratum*)
Common Green Darner (*Anax junius*)
Broad-striped Forceptail (*Aphylla angustifolia*)
Great Spreadwing (*Archilestes grandis*)
Paiute Dancer (*Argia alberta*)
Blue-fronted Dancer (*Argia apicalis*)
Comanche Dancer (*Argia barretti*)
Violet Dancer (*Argia fumipennis violacea*)
Kiowa Dancer (*Argia immunda*)

Leonora's Dancer (*Argia leonorae*)
Powdered Dancer (*Argia moesta*)
Aztec Dancer (*Argia nahuana*)
Springwater Dancer (*Argia plana*)
Golden-winged Dancer (*Argia rhoadsi*)
Blue-ringed Dancer (*Argia sedula*)
Dusky Dancer (*Argia translata*)
Springtime Darner (*Basiaeschna janata*)
Red-tailed Pennant (*Brachymesia furcata*)
Four-spotted Pennant (*Brachymesia gravida*)
Pale-faced Clubskimmer (*Brechmorhoga mendax*)
Gray-waisted Skimmer (*Cannaphila insularis funerea*)
Halloween Pennant (*Celithemis eponina*)
Banded Pennant (*Celithemis fasciata*)
Stream Cruiser (*Didymops transversa*)
Flag-tailed Spinyleg (*Dromogomphus spoliatus*)
Checkered Setwing (*Dythemis fugax*)
Black Setwing (*Dythemis nigrescens*)
Swift Setwing (*Dythemis velox*)
Double-striped Bluet (*Enallagma basidens*)
Familiar Bluet (*Enallagma civile*)
Stream Bluet (*Enallagma exsulans*)
Neotropical Bluet (*Enallagma novaehispaniae*)
Arroyo Bluet (*Enallagma praevarum*)
Stripe-winged Baskettail (*Epitheca costalis*)
Dot-winged Baskettail (*Epitheca petechialis*)
Prince Baskettail (*Epitheca princeps*)
White-belted Ringtail (*Erpetogomphus compositus*)
Eastern Ringtail (*Erpetogomphus designatus*)
Eastern Pondhawk (*Erythemis simplicicollis*)
Great Pondhawk (*Erythemis vesiculosa*)
Band-winged Dragonlet (*Erythrodiplax umbrata*)
Pronghorn Clubtail (*Gomphus graslinellus*)
Sulphur-tipped Clubtail (*Gomphus militaris*)
American Rubyspot (*Hetaerina americana*)
Citrine Forktail (*Ischnura hastata*)
Fragile Forktail (*Ischnura posita*)
Rambur's Forktail (*Ischnura ramburii*)
Plateau Spreadwing (*Lestes alacer*)
Chalky Spreadwing (*Lestes sigma*)
Comanche Skimmer (*Libellula comanche*)
Neon Skimmer (*Libellula croceipennis*)
Widow Skimmer (*Libellula luctuosa*)
Twelve-spotted Skimmer (*Libellula pulchella*)
Flame Skimmer (*Libellula saturata*)
Marl Pennant (*Macrodiplax balteata*)
Bronzed River Cruiser (*Macromia annulata*)
Ivory-striped Sylph (*Macrothemis imitans leucozona*)
Hyacinth Glider (*Miathyria marcella*)
Thornbush Dasher (*Micrathyria hagenii*)
Coral-fronted Threadtail (*Neoneura aaroni*)
Orange Shadowdragon (*Neurocordulia xanthosoma*)
Roseate Skimmer (*Orthemis ferruginea*)
Blue Dasher (*Pachydiplax longipennis*)
Wandering Glider (*Pantala flavescens*)
Spot-winged Glider (*Pantala hymenaea*)
Slough Amberwing (*Perithemis domitia*)
Eastern Amberwing (*Perithemis tenera*)
Five-striped Leaftail (*Phyllogomphoides albrighti*)
Four-striped Leaftail (*Phyllogomphoides stigmatus*)
Common Whitetail (*Plathemis lydia*)
Orange-striped Threadtail (*Protoneura cara*)
Filigree Skimmer (*Pseudoleon superbus*)
Turquoise-tipped Darner (*Rhionaeschna psilus*)
Variegated Meadowhawk (*Sympetrum corruptum*)
Desert Firetail (*Telebasis salva*)
Striped Saddlebags (*Tramea calverti*)
Antillean Saddlebags (*Tramea insularis*)
Black Saddlebags (*Tramea lacerata*)

Red Saddlebags (*Tramea onusta*)

Van Zandt *(16)*
Common Green Darner (*Anax junius*)
Blue-fronted Dancer (*Argia apicalis*)
Blue-tipped Dancer (*Argia tibialis*)
Black-shouldered Spinyleg (*Dromogomphus spinosus*)
Familiar Bluet (*Enallagma civile*)
Orange Bluet (*Enallagma signatum*)
Swamp Darner (*Epiaeschna heros*)
Eastern Pondhawk (*Erythemis simplicicollis*)
Plains Clubtail (*Gomphus externus*)
Yellow-sided Skimmer (*Libellula flavida*)
Slaty Skimmer (*Libellula incesta*)
Flame Skimmer (*Libellula saturata*)
Blue Dasher (*Pachydiplax longipennis*)
Spot-winged Glider (*Pantala hymenaea*)
Eastern Amberwing (*Perithemis tenera*)
Common Whitetail (*Plathemis lydia*)

Victoria *(43)*
Common Green Darner (*Anax junius*)
Blue-fronted Dancer (*Argia apicalis*)
Powdered Dancer (*Argia moesta*)
Blue-ringed Dancer (*Argia sedula*)
Blue-tipped Dancer (*Argia tibialis*)
Stillwater Clubtail (*Arigomphus lentulus*)
Jade Clubtail (*Arigomphus submedianus*)
Red-tailed Pennant (*Brachymesia furcata*)
Four-spotted Pennant (*Brachymesia gravida*)
Tawny Pennant (*Brachymesia herbida*)
Halloween Pennant (*Celithemis eponina*)
Flag-tailed Spinyleg (*Dromogomphus spoliatus*)
Checkered Setwing (*Dythemis fugax*)
Swift Setwing (*Dythemis velox*)
Double-striped Bluet (*Enallagma basidens*)
Familiar Bluet (*Enallagma civile*)
Orange Bluet (*Enallagma signatum*)
Prince Baskettail (*Epitheca princeps*)
Pin-tailed Pondhawk (*Erythemis plebeja*)
Eastern Pondhawk (*Erythemis simplicicollis*)
Great Pondhawk (*Erythemis vesiculosa*)
Little Blue Dragonlet (*Erythrodiplax minuscula*)
Band-winged Dragonlet (*Erythrodiplax umbrata*)
Sulphur-tipped Clubtail (*Gomphus militaris*)
Cobra Clubtail (*Gomphus vastus*)
American Rubyspot (*Hetaerina americana*)
Smoky Rubyspot (*Hetaerina titia*)
Citrine Forktail (*Ischnura hastata*)
Rambur's Forktail (*Ischnura ramburii*)
Chalky Spreadwing (*Lestes sigma*)
Widow Skimmer (*Libellula luctuosa*)
Coral-fronted Threadtail (*Neoneura aaroni*)
Roseate Skimmer (*Orthemis ferruginea*)
Blue Dasher (*Pachydiplax longipennis*)
Wandering Glider (*Pantala flavescens*)
Spot-winged Glider (*Pantala hymenaea*)
Eastern Amberwing (*Perithemis tenera*)
Common Sanddragon (*Progomphus obscurus*)
Russet-tipped Clubtail (*Stylurus plagiatus*)
Desert Firetail (*Telebasis salva*)
Striped Saddlebags (*Tramea calverti*)
Black Saddlebags (*Tramea lacerata*)
Red Saddlebags (*Tramea onusta*)

Walker *(47)*
Common Green Darner (*Anax junius*)
Comet Darner (*Anax longipes*)
Blue-fronted Dancer (*Argia apicalis*)
Powdered Dancer (*Argia moesta*)
Blue-tipped Dancer (*Argia tibialis*)
Stillwater Clubtail (*Arigomphus lentulus*)
Jade Clubtail (*Arigomphus submedianus*)
Springtime Darner (*Basiaeschna janata*)
Four-spotted Pennant (*Brachymesia gravida*)
Ebony Jewelwing (*Calopteryx maculata*)
Calico Pennant (*Celithemis elisa*)
Halloween Pennant (*Celithemis eponina*)
Stream Cruiser (*Didymops transversa*)

Black-shouldered Spinyleg (*Dromogomphus spinosus*)
Flag-tailed Spinyleg (*Dromogomphus spoliatus*)
Turquoise Bluet (*Enallagma divagans*)
Burgundy Bluet (*Enallagma dubium*)
Skimming Bluet (*Enallagma geminatum*)
Orange Bluet (*Enallagma signatum*)
Swamp Darner (*Epiaeschna heros*)
Stripe-winged Baskettail (*Epitheca costalis*)
Common Baskettail (*Epitheca cynosura*)
Prince Baskettail (*Epitheca princeps*)
Mantled Baskettail (*Epitheca semiaquea*)
Eastern Pondhawk (*Erythemis simplicicollis*)
Great Pondhawk (*Erythemis vesiculosa*)
Ashy Clubtail (*Gomphus lividus*)
Oklahoma Clubtail (*Gomphus oklahomensis*)
Citrine Forktail (*Ischnura hastata*)
Fragile Forktail (*Ischnura posita*)
Rambur's Forktail (*Ischnura ramburii*)
Blue Corporal (*Ladona deplanata*)
Swamp Spreadwing (*Lestes vigilax*)
Bar-winged Skimmer (*Libellula axilena*)
Slaty Skimmer (*Libellula incesta*)
Widow Skimmer (*Libellula luctuosa*)
Painted Skimmer (*Libellula semifasciata*)
Great Blue Skimmer (*Libellula vibrans*)
Royal River Cruiser (*Macromia taeniolata*)
Hyacinth Glider (*Miathyria marcella*)
Cyrano Darner (*Nasiaeschna pentacantha*)
Blue Dasher (*Pachydiplax longipennis*)
Eastern Amberwing (*Perithemis tenera*)
Common Whitetail (*Plathemis lydia*)
Common Sanddragon (*Progomphus obscurus*)
Variegated Meadowhawk (*Sympetrum corruptum*)
Black Saddlebags (*Tramea lacerata*)

Waller *(21)*
Common Green Darner (*Anax junius*)
Jade Clubtail (*Arigomphus submedianus*)
Four-spotted Pennant (*Brachymesia gravida*)
Calico Pennant (*Celithemis elisa*)
Halloween Pennant (*Celithemis eponina*)
Regal Darner (*Coryphaeschna ingens*)
Familiar Bluet (*Enallagma civile*)
Swamp Darner (*Epiaeschna heros*)
Prince Baskettail (*Epitheca princeps*)
Eastern Pondhawk (*Erythemis simplicicollis*)
Rambur's Forktail (*Ischnura ramburii*)
Widow Skimmer (*Libellula luctuosa*)
Needham's Skimmer (*Libellula needhami*)
Blue Dasher (*Pachydiplax longipennis*)
Wandering Glider (*Pantala flavescens*)
Spot-winged Glider (*Pantala hymenaea*)
Eastern Amberwing (*Perithemis tenera*)
Common Whitetail (*Plathemis lydia*)
Carolina Saddlebags (*Tramea carolina*)
Black Saddlebags (*Tramea lacerata*)
Red Saddlebags (*Tramea onusta*)

Ward *(16)*
Blue-fronted Dancer (*Argia apicalis*)
Powdered Dancer (*Argia moesta*)
Blue-ringed Dancer (*Argia sedula*)
Checkered Setwing (*Dythemis fugax*)
Familiar Bluet (*Enallagma civile*)
Eastern Ringtail (*Erpetogomphus designatus*)
Seaside Dragonlet (*Erythrodiplax berenice*)
American Rubyspot (*Hetaerina americana*)
Desert Forktail (*Ischnura barberi*)
Rambur's Forktail (*Ischnura ramburii*)
Plateau Spreadwing (*Lestes alacer*)
Bleached Skimmer (*Libellula composita*)
Flame Skimmer (*Libellula saturata*)
Marl Pennant (*Macrodiplax balteata*)
Wandering Glider (*Pantala flavescens*)
Variegated Meadowhawk (*Sympetrum corruptum*)

Washington *(25)*

Blue-fronted Dancer (*Argia apicalis*)
Violet Dancer (*Argia fumipennis violacea*)
Kiowa Dancer (*Argia immunda*)
Powdered Dancer (*Argia moesta*)
Blue-ringed Dancer (*Argia sedula*)
Blue-tipped Dancer (*Argia tibialis*)
Ebony Jewelwing (*Calopteryx maculata*)
Checkered Setwing (*Dythemis fugax*)
Swift Setwing (*Dythemis velox*)
Double-striped Bluet (*Enallagma basidens*)
Eastern Ringtail (*Erpetogomphus designatus*)
Eastern Pondhawk (*Erythemis simplicicollis*)
Sulphur-tipped Clubtail (*Gomphus militaris*)
American Rubyspot (*Hetaerina americana*)
Smoky Rubyspot (*Hetaerina titia*)
Slaty Skimmer (*Libellula incesta*)
Widow Skimmer (*Libellula luctuosa*)
Needham's Skimmer (*Libellula needhami*)
Roseate Skimmer (*Orthemis ferruginea*)
Blue Dasher (*Pachydiplax longipennis*)
Wandering Glider (*Pantala flavescens*)
Four-striped Leaftail (*Phyllogomphoides stigmatus*)
Common Whitetail (*Plathemis lydia*)
Common Sanddragon (*Progomphus obscurus*)
Black Saddlebags (*Tramea lacerata*)

Webb (39)
Common Green Darner (*Anax junius*)
Broad-striped Forceptail (*Aphylla angustifolia*)
Blue-fronted Dancer (*Argia apicalis*)
Kiowa Dancer (*Argia immunda*)
Powdered Dancer (*Argia moesta*)
Blue-ringed Dancer (*Argia sedula*)
Dusky Dancer (*Argia translata*)
Pale-faced Clubskimmer (*Brechmorhoga mendax*)
Halloween Pennant (*Celithemis eponina*)
Flag-tailed Spinyleg (*Dromogomphus spoliatus*)
Checkered Setwing (*Dythemis fugax*)
Black Setwing (*Dythemis nigrescens*)
Swift Setwing (*Dythemis velox*)
Double-striped Bluet (*Enallagma basidens*)
Familiar Bluet (*Enallagma civile*)
Prince Baskettail (*Epitheca princeps*)
Eastern Ringtail (*Erpetogomphus designatus*)
Blue-faced Ringtail (*Erpetogomphus eutainia*)
Eastern Pondhawk (*Erythemis simplicicollis*)
Little Blue Dragonlet (*Erythrodiplax minuscula*)
Band-winged Dragonlet (*Erythrodiplax umbrata*)
Sulphur-tipped Clubtail (*Gomphus militaris*)
Smoky Rubyspot (*Hetaerina titia*)
Citrine Forktail (*Ischnura hastata*)
Rambur's Forktail (*Ischnura ramburii*)
Neon Skimmer (*Libellula croceipennis*)
Bronzed River Cruiser (*Macromia annulata*)
Straw-colored Sylph (*Macrothemis inacuta*)
Thornbush Dasher (*Micrathyria hagenii*)
Carmine Skimmer (*Orthemis discolor*)
Roseate Skimmer (*Orthemis ferruginea*)
Eastern Amberwing (*Perithemis tenera*)
Five-striped Leaftail (*Phyllogomphoides albrighti*)
Common Whitetail (*Plathemis lydia*)
Filigree Skimmer (*Pseudoleon superbus*)
Variegated Meadowhawk (*Sympetrum corruptum*)
Desert Firetail (*Telebasis salva*)
Black Saddlebags (*Tramea lacerata*)
Red Saddlebags (*Tramea onusta*)

Wharton (15)
Broad-striped Forceptail (*Aphylla angustifolia*)
Blue-fronted Dancer (*Argia apicalis*)
Powdered Dancer (*Argia moesta*)
Blue-ringed Dancer (*Argia sedula*)
Eastern Pondhawk (*Erythemis simplicicollis*)
Great Pondhawk (*Erythemis vesiculosa*)
American Rubyspot (*Hetaerina americana*)

Smoky Rubyspot (*Hetaerina titia*)
Citrine Forktail (*Ischnura hastata*)
Rambur's Forktail (*Ischnura ramburii*)
Roseate Skimmer (*Orthemis ferruginea*)
Blue Dasher (*Pachydiplax longipennis*)
Wandering Glider (*Pantala flavescens*)
Common Sanddragon (*Progomphus obscurus*)
Russet-tipped Clubtail (*Stylurus plagiatus*)

Wheeler (12)
Common Green Darner (*Anax junius*)
Kiowa Dancer (*Argia immunda*)
Banded Pennant (*Celithemis fasciata*)
Double-striped Bluet (*Enallagma basidens*)
Dot-winged Baskettail (*Epitheca petechialis*)
American Rubyspot (*Hetaerina americana*)
Plateau Spreadwing (*Lestes alacer*)
Widow Skimmer (*Libellula luctuosa*)
Twelve-spotted Skimmer (*Libellula pulchella*)
Flame Skimmer (*Libellula saturata*)
Common Whitetail (*Plathemis lydia*)
Black Saddlebags (*Tramea lacerata*)

Wichita (13)
Blue-fronted Dancer (*Argia apicalis*)
Blue-ringed Dancer (*Argia sedula*)
Familiar Bluet (*Enallagma civile*)
American Rubyspot (*Hetaerina americana*)
Rambur's Forktail (*Ischnura ramburii*)
Twelve-spotted Skimmer (*Libellula pulchella*)
Roseate Skimmer (*Orthemis ferruginea*)
Blue Dasher (*Pachydiplax longipennis*)
Wandering Glider (*Pantala flavescens*)
Eastern Amberwing (*Perithemis tenera*)
Common Whitetail (*Plathemis lydia*)
Variegated Meadowhawk (*Sympetrum corruptum*)
Black Saddlebags (*Tramea lacerata*)

Wilbarger (15)
Blue-fronted Dancer (*Argia apicalis*)
Powdered Dancer (*Argia moesta*)
Blue-ringed Dancer (*Argia sedula*)
Blue-tipped Dancer (*Argia tibialis*)
Flag-tailed Spinyleg (*Dromogomphus spoliatus*)
Double-striped Bluet (*Enallagma basidens*)
Familiar Bluet (*Enallagma civile*)
Eastern Pondhawk (*Erythemis simplicicollis*)
American Rubyspot (*Hetaerina americana*)
Twelve-spotted Skimmer (*Libellula pulchella*)
Roseate Skimmer (*Orthemis ferruginea*)
Blue Dasher (*Pachydiplax longipennis*)
Spot-winged Glider (*Pantala hymenaea*)
Common Sanddragon (*Progomphus obscurus*)
Variegated Meadowhawk (*Sympetrum corruptum*)

Willacy (32)
Amazon Darner (*Anax amazili*)
Common Green Darner (*Anax junius*)
Blue-fronted Dancer (*Argia apicalis*)
Blue-ringed Dancer (*Argia sedula*)
Red-tailed Pennant (*Brachymesia furcata*)
Four-spotted Pennant (*Brachymesia gravida*)
Checkered Setwing (*Dythemis fugax*)
Black Setwing (*Dythemis nigrescens*)
Double-striped Bluet (*Enallagma basidens*)
Familiar Bluet (*Enallagma civile*)
Big Bluet (*Enallagma durum*)
Prince Baskettail (*Epitheca princeps*)
Pin-tailed Pondhawk (*Erythemis plebeja*)
Eastern Pondhawk (*Erythemis simplicicollis*)
Great Pondhawk (*Erythemis vesiculosa*)
Band-winged Dragonlet (*Erythrodiplax umbrata*)
Citrine Forktail (*Ischnura hastata*)
Rambur's Forktail (*Ischnura ramburii*)
Needham's Skimmer (*Libellula needhami*)
Marl Pennant (*Macrodiplax balteata*)
Hyacinth Glider (*Miathyria marcella*)
Spot-tailed Dasher (*Micrathyria aequalis*)
Three-striped Dasher (*Micrathyria didyma*)

Thornbush Dasher (*Micrathyria hagenii*)
Roseate Skimmer (*Orthemis ferruginea*)
Wandering Glider (*Pantala flavescens*)
Spot-winged Glider (*Pantala hymenaea*)
Eastern Amberwing (*Perithemis tenera*)
Variegated Meadowhawk (*Sympetrum corruptum*)
Desert Firetail (*Telebasis salva*)
Striped Saddlebags (*Tramea calverti*)
Red Saddlebags (*Tramea onusta*)

Williamson (58)
Common Green Darner (*Anax junius*)
Broad-striped Forceptail (*Aphylla angustifolia*)
Great Spreadwing (*Archilestes grandis*)
Blue-fronted Dancer (*Argia apicalis*)
Violet Dancer (*Argia fumipennis violacea*)
Kiowa Dancer (*Argia immunda*)
Leonora's Dancer (*Argia leonorae*)
Powdered Dancer (*Argia moesta*)
Aztec Dancer (*Argia nahuana*)
Blue-ringed Dancer (*Argia sedula*)
Dusky Dancer (*Argia translata*)
Fawn Darner (*Boyeria vinosa*)
Pale-faced Clubskimmer (*Brechmorhoga mendax*)
Calico Pennant (*Celithemis elisa*)
Halloween Pennant (*Celithemis eponina*)
Checkered Setwing (*Dythemis fugax*)
Swift Setwing (*Dythemis velox*)
Double-striped Bluet (*Enallagma basidens*)
Familiar Bluet (*Enallagma civile*)
Stream Bluet (*Enallagma exsulans*)
Orange Bluet (*Enallagma signatum*)
Dot-winged Baskettail (*Epitheca petechialis*)
Prince Baskettail (*Epitheca princeps*)
Eastern Ringtail (*Erpetogomphus designatus*)
Pin-tailed Pondhawk (*Erythemis plebeja*)
Eastern Pondhawk (*Erythemis simplicicollis*)
Great Pondhawk (*Erythemis vesiculosa*)
Band-winged Dragonlet (*Erythrodiplax umbrata*)
Sulphur-tipped Clubtail (*Gomphus militaris*)
Dragonhunter (*Hagenius brevistylus*)
American Rubyspot (*Hetaerina americana*)
Smoky Rubyspot (*Hetaerina titia*)
Citrine Forktail (*Ischnura hastata*)
Fragile Forktail (*Ischnura posita*)
Rambur's Forktail (*Ischnura ramburii*)
Southern Spreadwing (*Lestes australis*)
Comanche Skimmer (*Libellula comanche*)
Neon Skimmer (*Libellula croceipennis*)
Slaty Skimmer (*Libellula incesta*)
Widow Skimmer (*Libellula luctuosa*)
Twelve-spotted Skimmer (*Libellula pulchella*)
Great Blue Skimmer (*Libellula vibrans*)
Thornbush Dasher (*Micrathyria hagenii*)
Cyrano Darner (*Nasiaeschna pentacantha*)
Roseate Skimmer (*Orthemis ferruginea*)
Blue Dasher (*Pachydiplax longipennis*)
Wandering Glider (*Pantala flavescens*)
Spot-winged Glider (*Pantala hymenaea*)
Eastern Amberwing (*Perithemis tenera*)
Four-striped Leaftail (*Phyllogomphoides stigmatus*)
Common Whitetail (*Plathemis lydia*)
Turquoise-tipped Darner (*Rhionaeschna psilus*)
Variegated Meadowhawk (*Sympetrum corruptum*)
Autumn Meadowhawk (*Sympetrum vicinum*)
Desert Firetail (*Telebasis salva*)
Striped Saddlebags (*Tramea calverti*)
Black Saddlebags (*Tramea lacerata*)
Red Saddlebags (*Tramea onusta*)

Wilson (38)
Common Green Darner (*Anax junius*)
Narrow-striped Forceptail (*Aphylla protracta*)
Blue-fronted Dancer (*Argia apicalis*)
Kiowa Dancer (*Argia immunda*)

Powdered Dancer (*Argia moesta*)
Blue-ringed Dancer (*Argia sedula*)
Dusky Dancer (*Argia translata*)
Pale-faced Clubskimmer (*Brechmorhoga mendax*)
Flag-tailed Spinyleg (*Dromogomphus spoliatus*)
Black Setwing (*Dythemis nigrescens*)
Swift Setwing (*Dythemis velox*)
Double-striped Bluet (*Enallagma basidens*)
Familiar Bluet (*Enallagma civile*)
Stream Bluet (*Enallagma exsulans*)
Neotropical Bluet (*Enallagma novaehispaniae*)
Orange Bluet (*Enallagma signatum*)
Prince Baskettail (*Epitheca princeps*)
Eastern Ringtail (*Erpetogomphus designatus*)
Eastern Pondhawk (*Erythemis simplicicollis*)
Great Pondhawk (*Erythemis vesiculosa*)
Plains Clubtail (*Gomphus externus*)
Sulphur-tipped Clubtail (*Gomphus militaris*)
American Rubyspot (*Hetaerina americana*)
Smoky Rubyspot (*Hetaerina titia*)
Citrine Forktail (*Ischnura hastata*)
Twelve-spotted Skimmer (*Libellula pulchella*)
Bronzed River Cruiser (*Macromia annulata*)
Cyrano Darner (*Nasiaeschna pentacantha*)
Carmine Skimmer (*Orthemis discolor*)
Roseate Skimmer (*Orthemis ferruginea*)
Blue Dasher (*Pachydiplax longipennis*)
Wandering Glider (*Pantala flavescens*)
Spot-winged Glider (*Pantala hymenaea*)
Eastern Amberwing (*Perithemis tenera*)
Five-striped Leaftail (*Phyllogomphoides albrighti*)
Common Sanddragon (*Progomphus obscurus*)
Russet-tipped Clubtail (*Stylurus plagiatus*)
Desert Firetail (*Telebasis salva*)

Winkler (6)
Familiar Bluet (*Enallagma civile*)
Widow Skimmer (*Libellula luctuosa*)
Flame Skimmer (*Libellula saturata*)
Blue Dasher (*Pachydiplax longipennis*)
Blue-eyed Darner (*Rhionaeschna multicolor*)
Black Saddlebags (*Tramea lacerata*)

Wise (48)
Common Green Darner (*Anax junius*)
Great Spreadwing (*Archilestes grandis*)
Blue-fronted Dancer (*Argia apicalis*)
Violet Dancer (*Argia fumipennis violacea*)
Powdered Dancer (*Argia moesta*)
Aztec Dancer (*Argia nahuana*)
Blue-ringed Dancer (*Argia sedula*)
Four-spotted Pennant (*Brachymesia gravida*)
Ebony Jewelwing (*Calopteryx maculata*)
Calico Pennant (*Celithemis elisa*)
Halloween Pennant (*Celithemis eponina*)
Banded Pennant (*Celithemis fasciata*)
Flag-tailed Spinyleg (*Dromogomphus spoliatus*)
Checkered Setwing (*Dythemis fugax*)
Swift Setwing (*Dythemis velox*)
Azure Bluet (*Enallagma aspersum*)
Double-striped Bluet (*Enallagma basidens*)
Familiar Bluet (*Enallagma civile*)
Orange Bluet (*Enallagma signatum*)
Prince Baskettail (*Epitheca princeps*)
Eastern Ringtail (*Erpetogomphus designatus*)
Eastern Pondhawk (*Erythemis simplicicollis*)
Band-winged Dragonlet (*Erythrodiplax umbrata*)
Sulphur-tipped Clubtail (*Gomphus militaris*)
American Rubyspot (*Hetaerina americana*)
Citrine Forktail (*Ischnura hastata*)
Rambur's Forktail (*Ischnura ramburii*)
Golden-winged Skimmer (*Libellula auripennis*)
Comanche Skimmer (*Libellula comanche*)
Neon Skimmer (*Libellula croceipennis*)
Spangled Skimmer (*Libellula cyanea*)
Slaty Skimmer (*Libellula incesta*)
Widow Skimmer (*Libellula luctuosa*)

Twelve-spotted Skimmer (*Libellula pulchella*)
Great Blue Skimmer (*Libellula vibrans*)
Roseate Skimmer (*Orthemis ferruginea*)
Blue Dasher (*Pachydiplax longipennis*)
Wandering Glider (*Pantala flavescens*)
Spot-winged Glider (*Pantala hymenaea*)
Eastern Amberwing (*Perithemis tenera*)
Four-striped Leaftail (*Phyllogomphoides stigmatus*)
Common Whitetail (*Plathemis lydia*)
Common Sanddragon (*Progomphus obscurus*)
Blue-faced Meadowhawk (*Sympetrum ambiguum*)
Variegated Meadowhawk (*Sympetrum corruptum*)
Carolina Saddlebags (*Tramea carolina*)
Black Saddlebags (*Tramea lacerata*)
Red Saddlebags (*Tramea onusta*)

Wood (47)
Common Green Darner (*Anax junius*)
Blue-fronted Dancer (*Argia apicalis*)
Seepage Dancer (*Argia bipunctulata*)
Violet Dancer (*Argia fumipennis violacea*)
Kiowa Dancer (*Argia immunda*)
Powdered Dancer (*Argia moesta*)
Springwater Dancer (*Argia plana*)
Blue-tipped Dancer (*Argia tibialis*)
Springtime Darner (*Basiaeschna janata*)
Fawn Darner (*Boyeria vinosa*)
Ebony Jewelwing (*Calopteryx maculata*)
Calico Pennant (*Celithemis elisa*)
Halloween Pennant (*Celithemis eponina*)
Banded Pennant (*Celithemis fasciata*)
Twin-spotted Spiketail (*Cordulegaster maculata*)
Stream Cruiser (*Didymops transversa*)
Black-shouldered Spinyleg (*Dromogomphus spinosus*)
Swift Setwing (*Dythemis velox*)
Double-striped Bluet (*Enallagma basidens*)
Familiar Bluet (*Enallagma civile*)
Vesper Bluet (*Enallagma vesperum*)
Swamp Darner (*Epiaeschna heros*)
Prince Baskettail (*Epitheca princeps*)
Eastern Pondhawk (*Erythemis simplicicollis*)
Little Blue Dragonlet (*Erythrodiplax minuscula*)
Sulphur-tipped Clubtail (*Gomphus militaris*)
Oklahoma Clubtail (*Gomphus oklahomensis*)
Cobra Clubtail (*Gomphus vastus*)
Dragonhunter (*Hagenius brevistylus*)
Smoky Rubyspot (*Hetaerina titia*)
Citrine Forktail (*Ischnura hastata*)
Rambur's Forktail (*Ischnura ramburii*)
Blue Corporal (*Ladona deplanata*)
Swamp Spreadwing (*Lestes vigilax*)
Spangled Skimmer (*Libellula cyanea*)
Slaty Skimmer (*Libellula incesta*)
Widow Skimmer (*Libellula luctuosa*)
Twelve-spotted Skimmer (*Libellula pulchella*)
Great Blue Skimmer (*Libellula vibrans*)
Roseate Skimmer (*Orthemis ferruginea*)
Blue Dasher (*Pachydiplax longipennis*)
Wandering Glider (*Pantala flavescens*)
Eastern Amberwing (*Perithemis tenera*)
Common Whitetail (*Plathemis lydia*)
Common Sanddragon (*Progomphus obscurus*)
Blue-faced Meadowhawk (*Sympetrum ambiguum*)
Variegated Meadowhawk (*Sympetrum corruptum*)

Yoakum (1)
Familiar Bluet (*Enallagma civile*)

Young (6)
Familiar Bluet (*Enallagma civile*)
Dot-winged Baskettail (*Epitheca petechialis*)
Plains Clubtail (*Gomphus externus*)
Blue Corporal (*Ladona deplanata*)
Blue Dasher (*Pachydiplax longipennis*)

Variegated Meadowhawk (*Sympetrum corruptum*)

Zapata (30)
Common Green Darner (*Anax junius*)
Broad-striped Forceptail (*Aphylla angustifolia*)
Blue-fronted Dancer (*Argia apicalis*)
Powdered Dancer (*Argia moesta*)
Blue-ringed Dancer (*Argia sedula*)
Red-tailed Pennant (*Brachymesia furcata*)
Tawny Pennant (*Brachymesia herbida*)
Pale-faced Clubskimmer (*Brechmorhoga mendax*)
Flag-tailed Spinyleg (*Dromogomphus spoliatus*)
Checkered Setwing (*Dythemis fugax*)
Black Setwing (*Dythemis nigrescens*)
Familiar Bluet (*Enallagma civile*)
Neotropical Bluet (*Enallagma novaehispaniae*)
Little Blue Dragonlet (*Erythrodiplax minuscula*)
Band-winged Dragonlet (*Erythrodiplax umbrata*)
Sulphur-tipped Clubtail (*Gomphus militaris*)
Smoky Rubyspot (*Hetaerina titia*)
Rambur's Forktail (*Ischnura ramburii*)
Straw-colored Sylph (*Macrothemis inacuta*)
Coral-fronted Threadtail (*Neoneura aaroni*)
Amelia's Threadtail (*Neoneura amelia*)
Carmine Skimmer (*Orthemis discolor*)
Roseate Skimmer (*Orthemis ferruginea*)
Blue Dasher (*Pachydiplax longipennis*)
Eastern Amberwing (*Perithemis tenera*)
Russet-tipped Clubtail (*Stylurus plagiatus*)
Variegated Meadowhawk (*Sympetrum corruptum*)
Evening Skimmer (*Tholymis citrina*)
Black Saddlebags (*Tramea lacerata*)
Red Saddlebags (*Tramea onusta*)

Zavala (34)
Comanche Dancer (*Argia barretti*)
Kiowa Dancer (*Argia immunda*)
Powdered Dancer (*Argia moesta*)
Blue-ringed Dancer (*Argia sedula*)
Dusky Dancer (*Argia translata*)
Red-tailed Pennant (*Brachymesia furcata*)
Four-spotted Pennant (*Brachymesia gravida*)
Pale-faced Clubskimmer (*Brechmorhoga mendax*)
Flag-tailed Spinyleg (*Dromogomphus spoliatus*)
Checkered Setwing (*Dythemis fugax*)
Black Setwing (*Dythemis nigrescens*)
Swift Setwing (*Dythemis velox*)
Double-striped Bluet (*Enallagma basidens*)
Familiar Bluet (*Enallagma civile*)
Eastern Pondhawk (*Erythemis simplicicollis*)
Sulphur-tipped Clubtail (*Gomphus militaris*)
American Rubyspot (*Hetaerina americana*)
Smoky Rubyspot (*Hetaerina titia*)
Citrine Forktail (*Ischnura hastata*)
Widow Skimmer (*Libellula luctuosa*)
Twelve-spotted Skimmer (*Libellula pulchella*)
Thornbush Dasher (*Micrathyria hagenii*)
Coral-fronted Threadtail (*Neoneura aaroni*)
Roseate Skimmer (*Orthemis ferruginea*)
Blue Dasher (*Pachydiplax longipennis*)
Eastern Amberwing (*Perithemis tenera*)
Five-striped Leaftail (*Phyllogomphoides albrighti*)
Four-striped Leaftail (*Phyllogomphoides stigmatus*)
Common Whitetail (*Plathemis lydia*)
Filigree Skimmer (*Pseudoleon superbus*)
Desert Firetail (*Telebasis salva*)
Antillean Saddlebags (*Tramea insularis*)
Black Saddlebags (*Tramea lacerata*)
Red Saddlebags (*Tramea onusta*)

Calopteryx dimidiata Burmeister / Sparkling Jewelwing

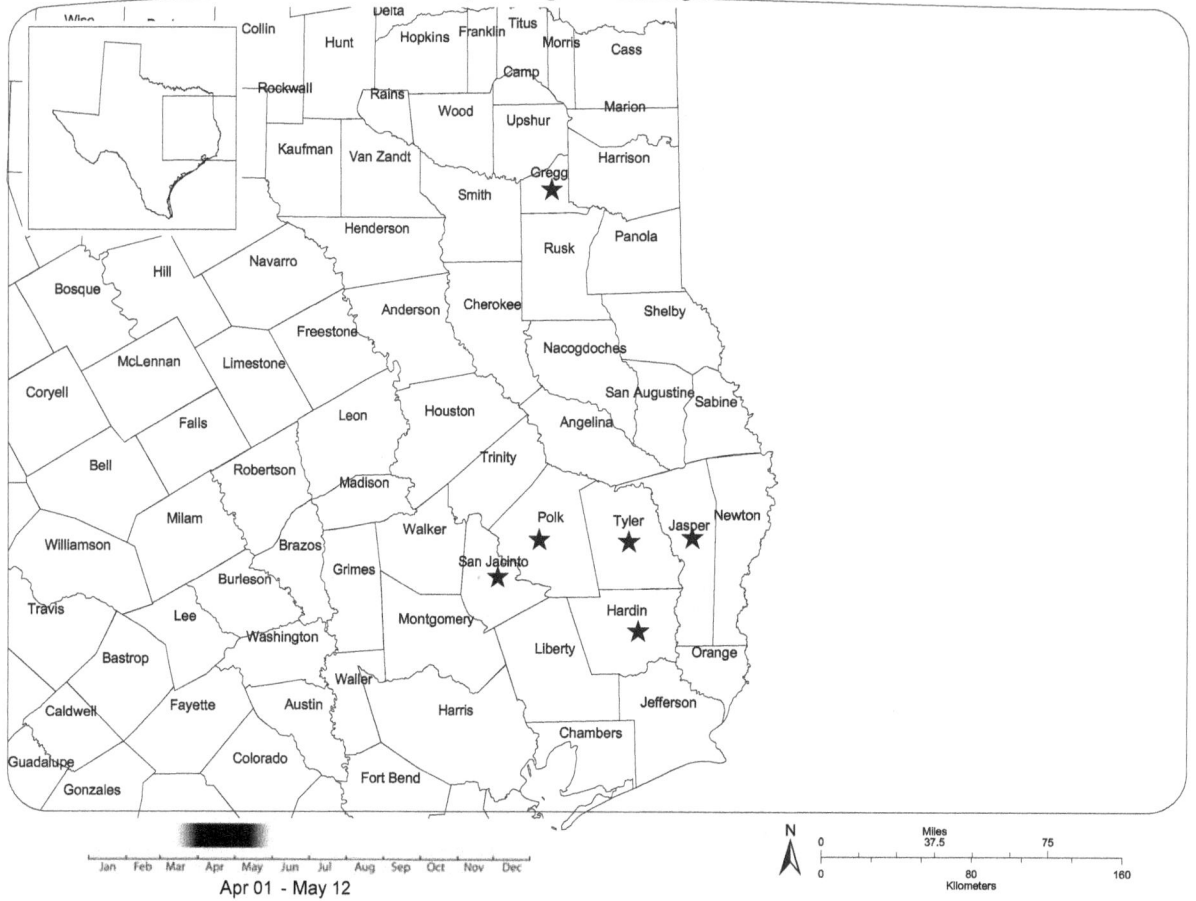

Apr 01 - May 12

HABITAT: Sandy bottomed streams; occasionally rivers with little canopy cover.

Gregg
Hardin
Jasper
Polk
San Jacinto
Tyler

Calopteryx maculata (Beauvois) / Ebony Jewelwing

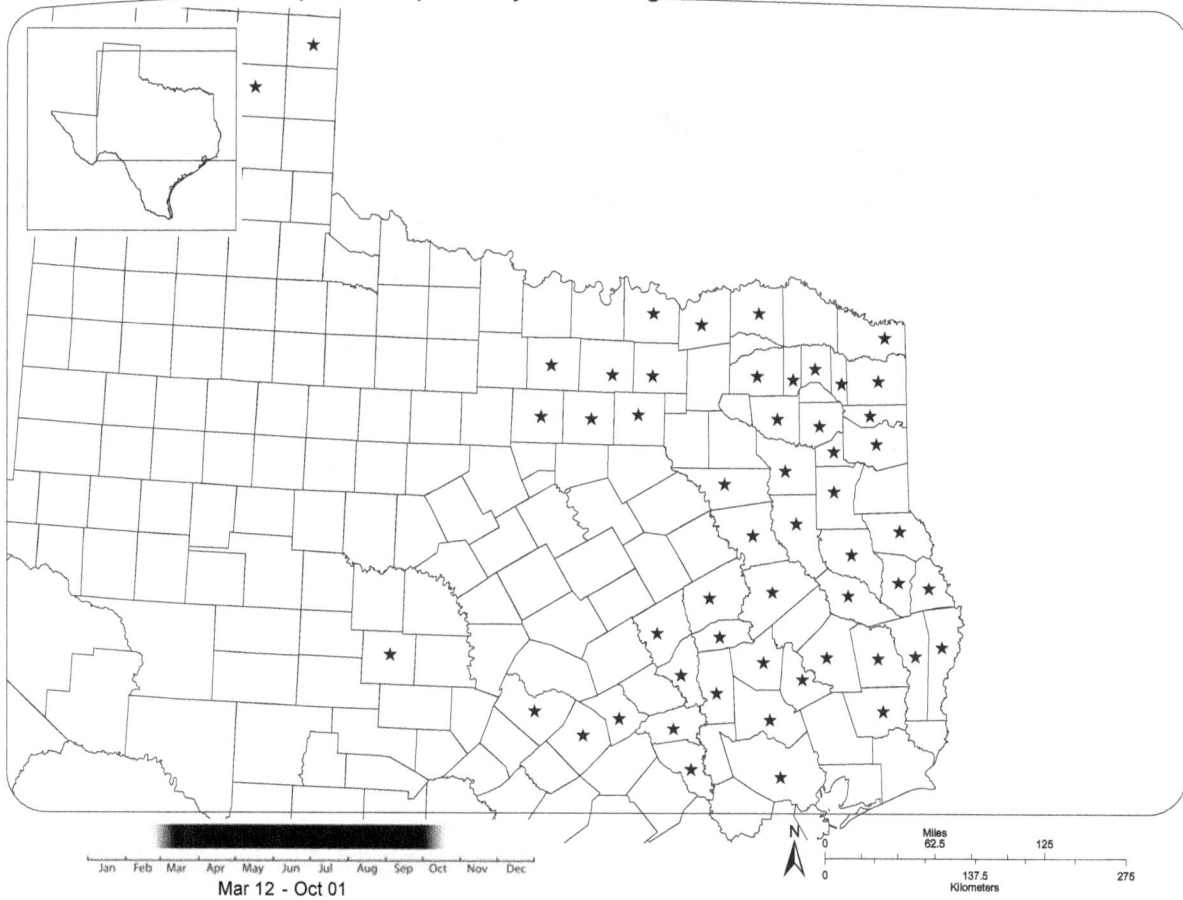

Jan Feb Mar Apr May Jun Jul Aug Sep Oct Nov Dec

Mar 12 - Oct 01

HABITAT: Small, slow moving, canopy covered streams and occasionally exposed streams and rivulets.

Anderson	Lee	Wood
Angelina	Leon	
Austin	Madison	
Bastrop	Marion	
Bowie	Mason	
Brazos	Montgomery	
Cass	Morris	
Cherokee	Nacogdoches	
Collin	Newton	
Dallas	Parker	
Denton	Polk	
Fannin	Robertson	
Franklin	Rusk	
Gray	Sabine	
Grayson	San Augustine	
Gregg	San Jacinto	
Grimes	Shelby	
Hardin	Smith	
Harris	Tarrant	
Harrison	Titus	
Hemphill	Travis	
Henderson	Tyler	
Hopkins	Upshur	
Houston	Walker	
Jasper	Washington	
Lamar	Wise	

Hetaerina americana (Fabricius) / American Rubyspot

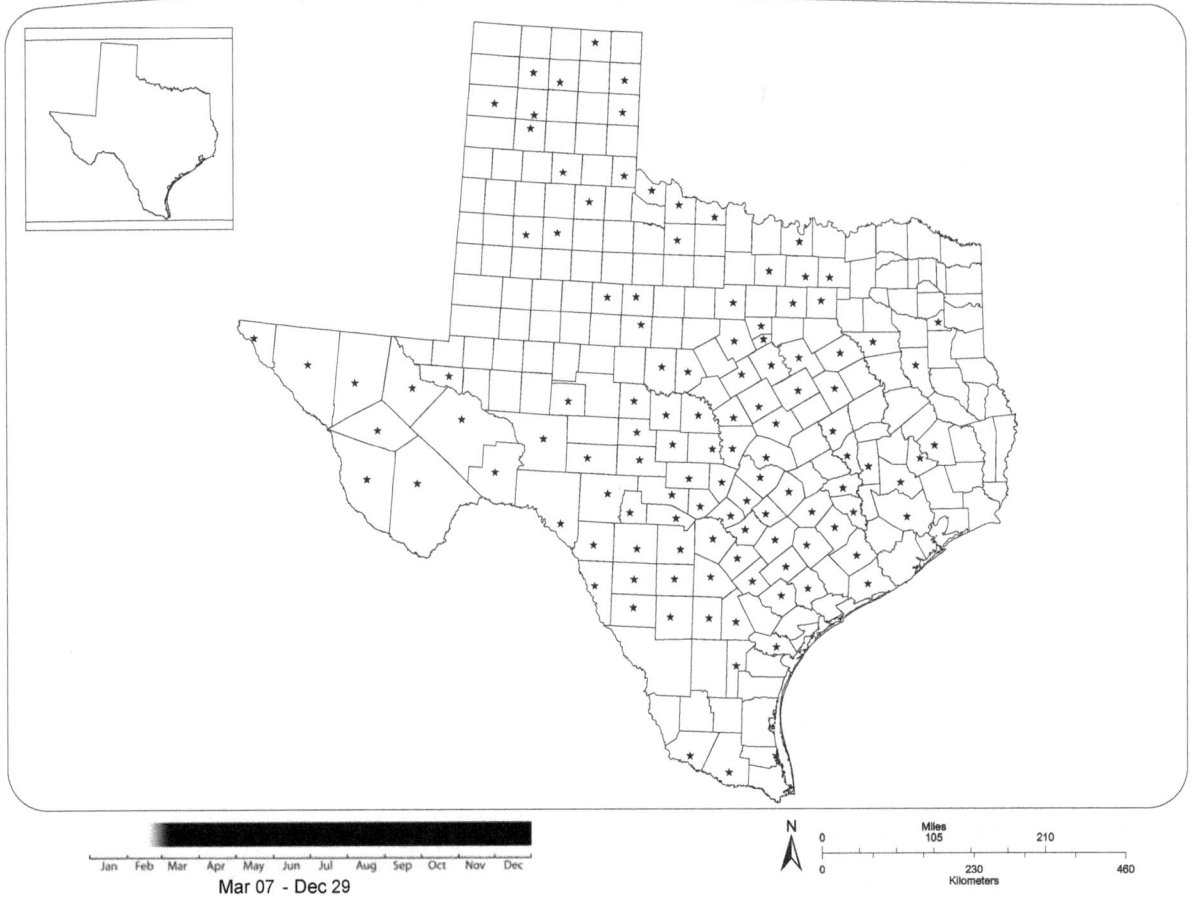

Mar 07 - Dec 29

HABITAT: Wide, open streams and rivers.

Atascosa	Culberson	Hudspeth	Montgomery	Val Verde
Austin	Dallas	Hutchinson	Moore	Victoria
Bandera	Denton	Irion	Motley	Ward
Bastrop	DeWitt	Jeff Davis	Navarro	Washington
Baylor	Dimmit	Jim Wells	Ochiltree	Wharton
Bell	Edwards	Jones	Oldham	Wheeler
Bexar	El Paso	Karnes	Palo Pinto	Wichita
Blanco	Erath	Kendall	Pecos	Wilbarger
Bosque	Fayette	Kerr	Polk	Williamson
Brazos	Fisher	Kimble	Potter	Wilson
Brewster	Frio	Kinney	Presidio	Wise
Briscoe	Gillespie	La Salle	Randall	Zavala
Brown	Goliad	Lampasas	Real	
Burnet	Gonzales	Lavaca	Reeves	
Caldwell	Gregg	Limestone	Robertson	
Cherokee	Grimes	Live Oak	San Jacinto	
Childress	Guadalupe	Llano	San Patricio	
Coleman	Hamilton	Lubbock	San Saba	
Collin	Hardeman	Mason	Somervell	
Colorado	Harris	Matagorda	Starr	
Comal	Hays	Maverick	Sutton	
Concho	Hemphill	McCulloch	Tarrant	
Cooke	Henderson	McLennan	Taylor	
Coryell	Hidalgo	McMullen	Terrell	
Crockett	Hill	Medina	Travis	
Crosby	Hood	Menard	Uvalde	

Hetaerina titia (Drury) / Smoky Rubyspot

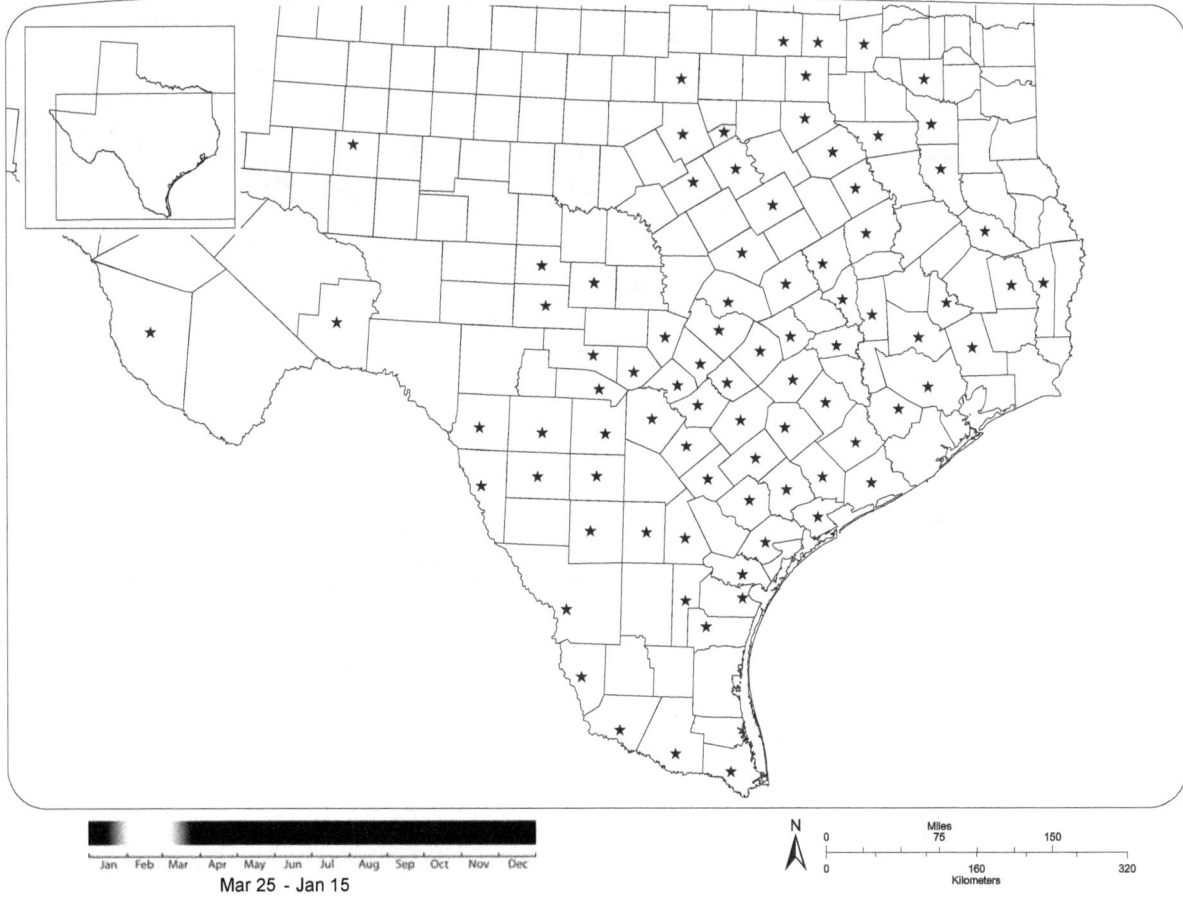

Mar 25 - Jan 15

HABITAT: Small to medium-sized streams and rivers with strong current

Angelina	Grimes	McLennan	Williamson
Bandera	Guadalupe	McMullen	Wilson
Bastrop	Hamilton	Medina	Wood
Bell	Harris	Menard	Zapata
Bexar	Hays	Midland	Zavala
Blanco	Henderson	Milam	
Bosque	Hidalgo	Montgomery	
Brazos	Hunt	Navarro	
Caldwell	Jackson	Nueces	
Calhoun	Jasper	Palo Pinto	
Cameron	Jim Wells	Presidio	
Cherokee	Karnes	Refugio	
Collin	Kendall	Robertson	
Colorado	Kerr	San Jacinto	
Comal	Kimble	San Patricio	
Dallas	Kinney	Smith	
Denton	Kleberg	Somervell	
DeWitt	La Salle	Starr	
Ellis	Lavaca	Terrell	
Erath	Lee	Travis	
Fayette	Leon	Tyler	
Fort Bend	Liberty	Uvalde	
Freestone	Live Oak	Victoria	
Frio	Mason	Washington	
Goliad	Matagorda	Webb	
Gonzales	Maverick	Wharton	

Hetaerina vulnerata Hagen in Selys / Canyon Rubyspot

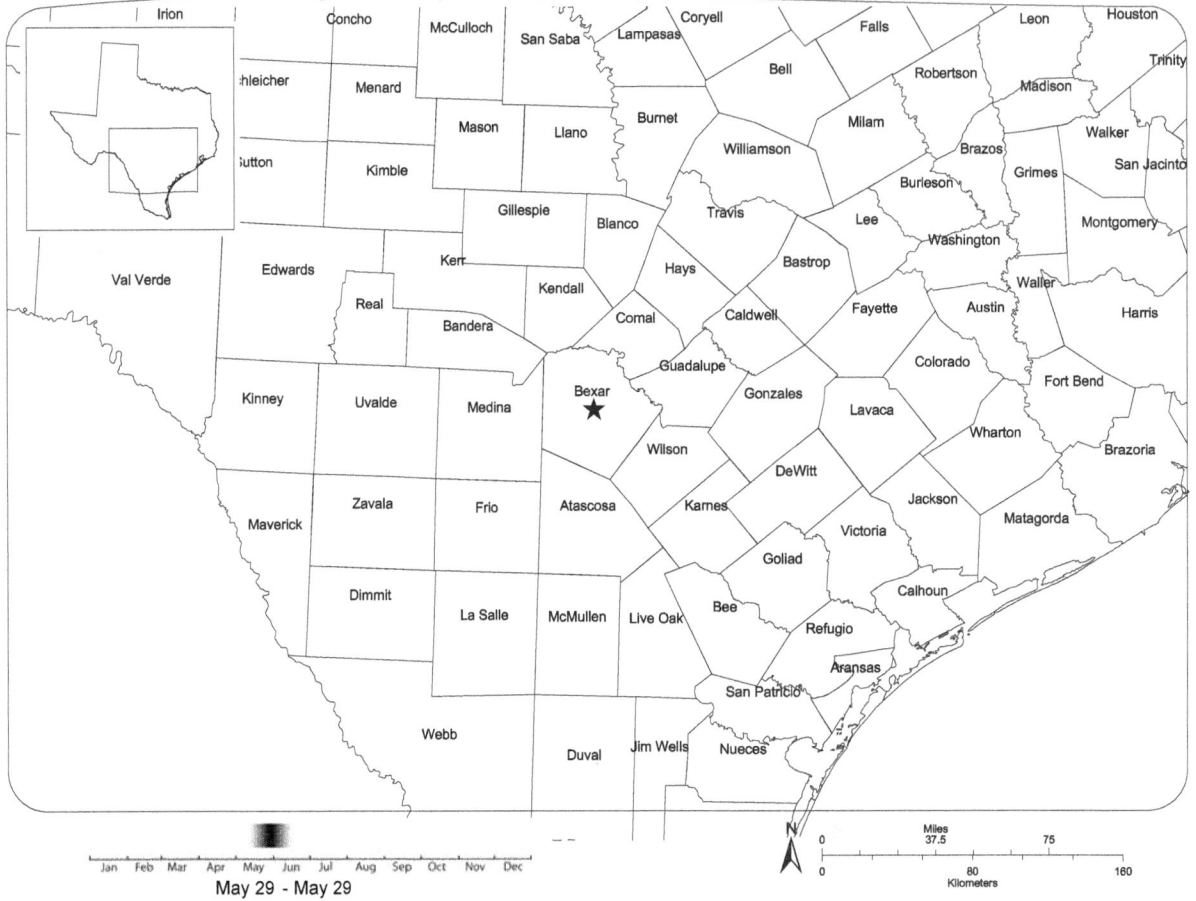

May 29 - May 29

HABITAT: Streams and rivers in open canopy woodlands

Bexar

Archilestes grandis (Rambur) / Great Spreadwing

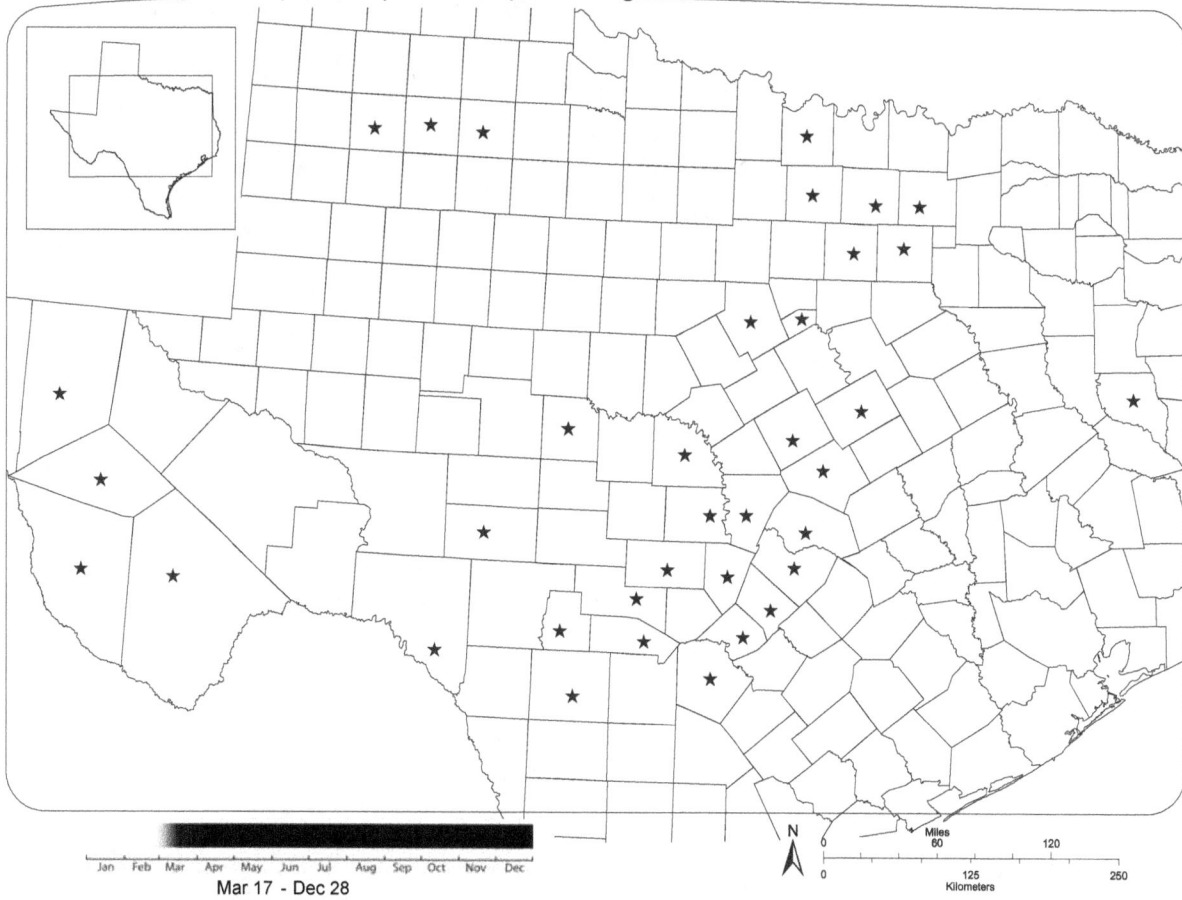

Mar 17 - Dec 28

HABITAT: Small permanent ponds or streams with slow or moderate flow

Bandera	Real
Bell	San Saba
Bexar	Somervell
Blanco	Sutton
Brewster	Tarrant
Burnet	Travis
Collin	Uvalde
Comal	Val Verde
Concho	Williamson
Coryell	Wise
Crosby	
Culberson	
Dallas	
Denton	
Dickens	
Erath	
Gillespie	
Hays	
Jeff Davis	
Kerr	
Llano	
Lubbock	
McLennan	
Montague	
Nacogdoches	
Presidio	

Lestes alacer Hagen / Plateau Spreadwing

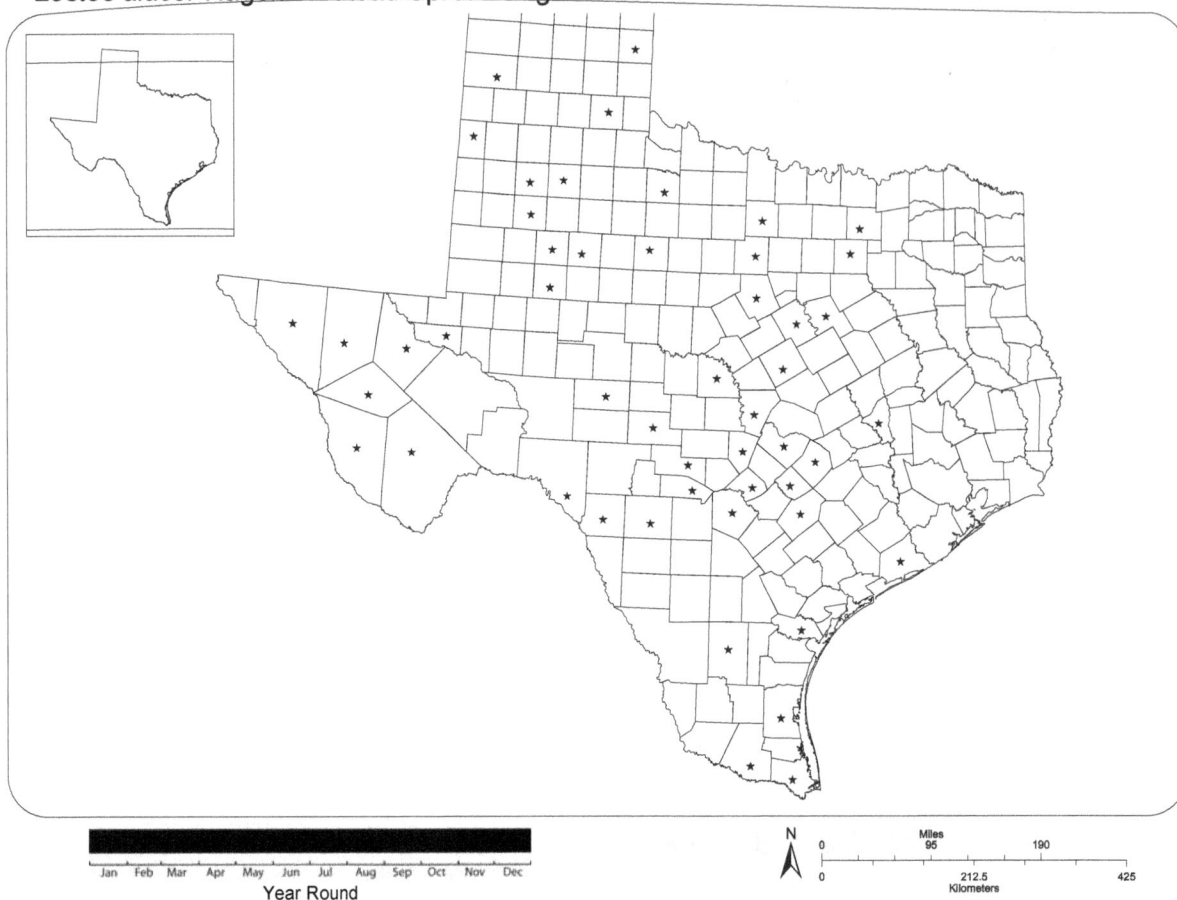

Jan Feb Mar Apr May Jun Jul Aug Sep Oct Nov Dec
Year Round

HABITAT: Still, slow moving waters

Bailey	Hudspeth
Bandera	Jack
Bastrop	Jeff Davis
Bexar	Jones
Blanco	Kenedy
Borden	Kerr
Bosque	Kimble
Brazos	Kinney
Brewster	Knox
Burnet	Lubbock
Caldwell	Lynn
Cameron	Matagorda
Collin	Palo Pinto
Comal	Presidio
Coryell	Reeves
Crosby	San Patricio
Culberson	San Saba
Dallas	Schleicher
Deaf Smith	Scurry
Duval	Travis
Erath	Uvalde
Gonzales	Val Verde
Hall	Ward
Hidalgo	Wheeler
Hill	
Howard	

Lestes australis Walker / Southern Spreadwing

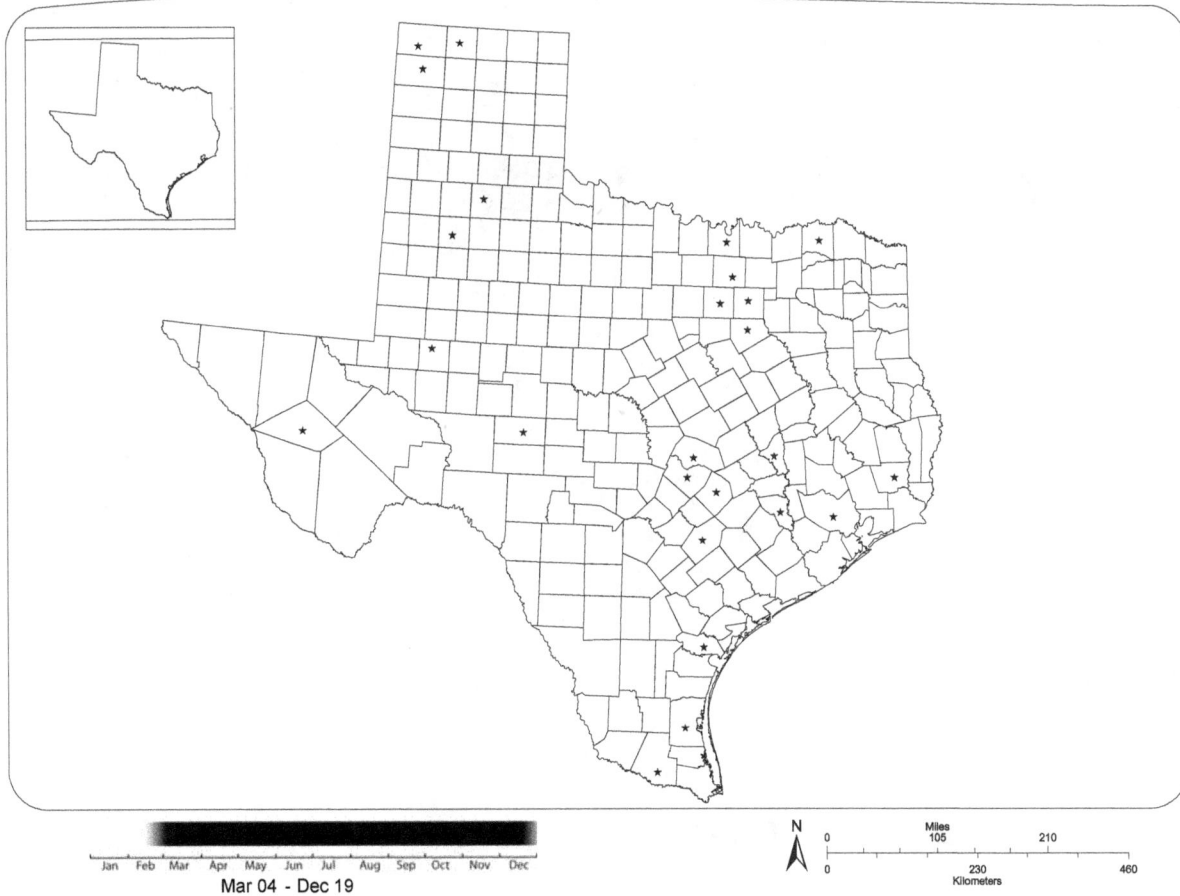

Mar 04 - Dec 19

HABITAT: Still, slow moving waters, including permanent or ephemeral ponds, marshes and lakes with moderate vegetation.

Austin
Bastrop
Brazos
Cooke
Dallam
Dallas
Denton
Ellis
Floyd
Gonzales
Hardin
Harris
Hartley
Hidalgo
Jeff Davis
Kenedy
Lamar
Lubbock
Midland
San Patricio
Schleicher
Sherman
Tarrant
Travis
Williamson

Lestes forficula Rambur / Rainpool Spreadwing

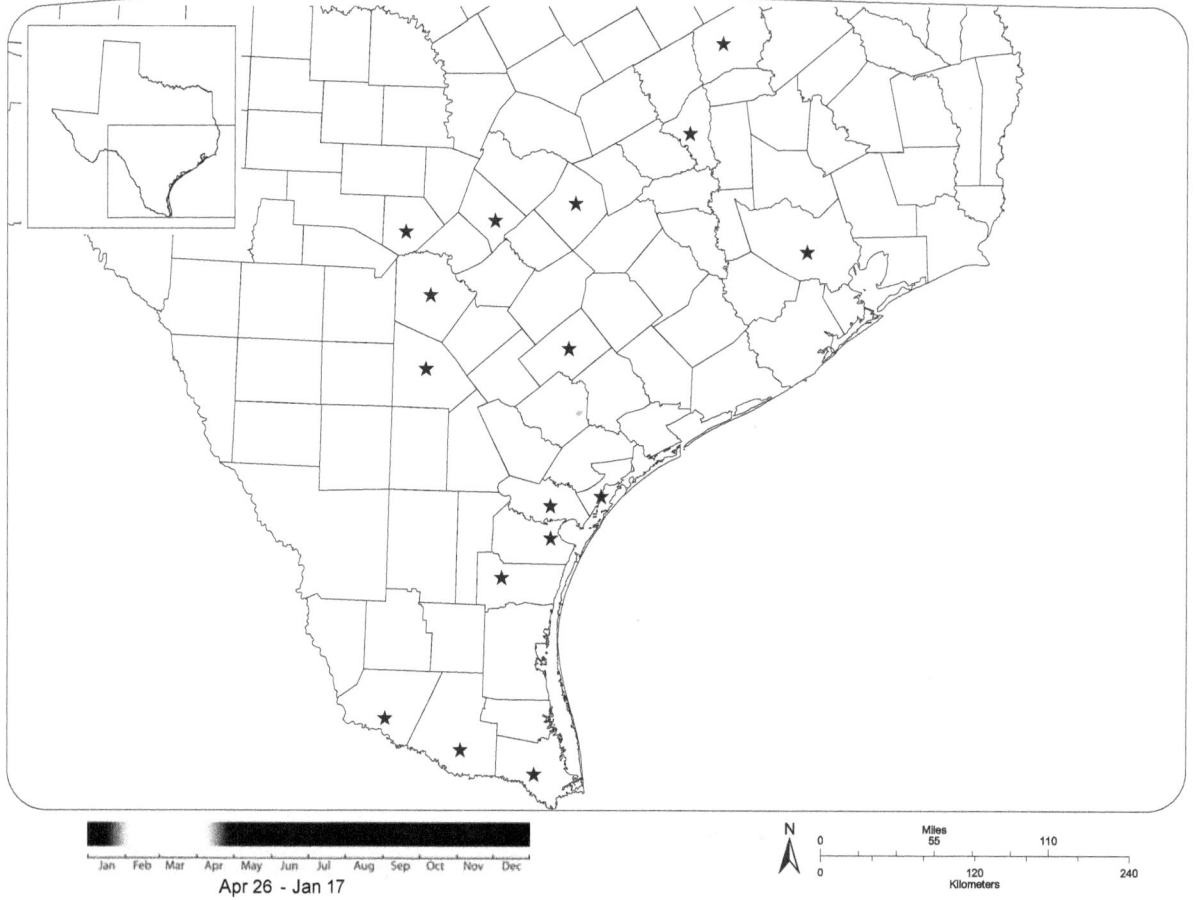

Apr 26 - Jan 17

HABITAT: Ponds, pools, other standing bodies of water; possibly slow reaches of streams, with heavy emergent vegetation.

Aransas
Atascosa
Bastrop
Bexar
Brazos
Cameron
DeWitt
Harris
Hays
Hidalgo
Kendall
Kleberg
Leon
Nueces
San Patricio
Starr

Lestes inaequalis Walsh / Elegant Spreadwing

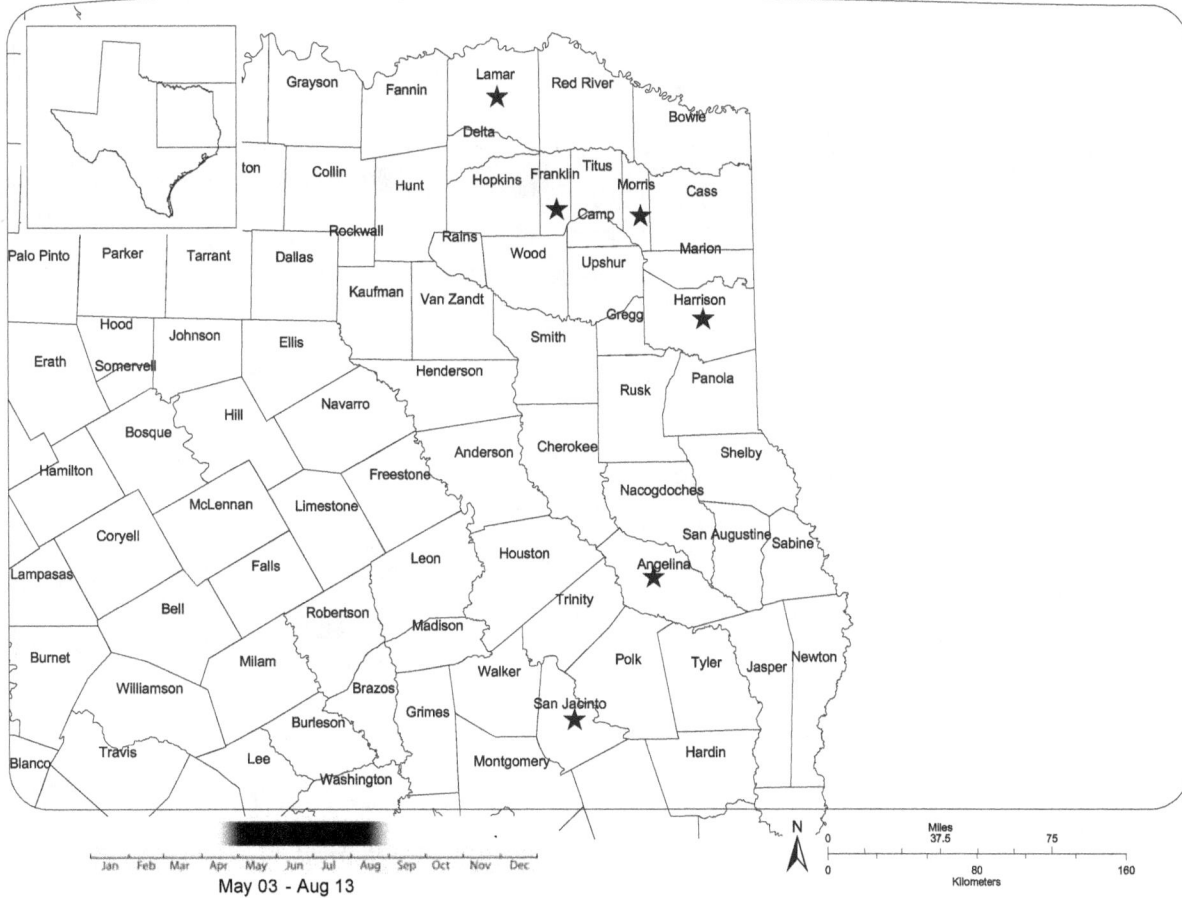

May 03 - Aug 13

HABITAT: Canopy-covered permanent ponds, lakes, slow moving streams and marshes with plenty of emergent vegetation and heavily wooded shorelines.

Angelina
Franklin
Harrison
Lamar
Morris

Lestes rectangularis Say / Slender Spreadwing

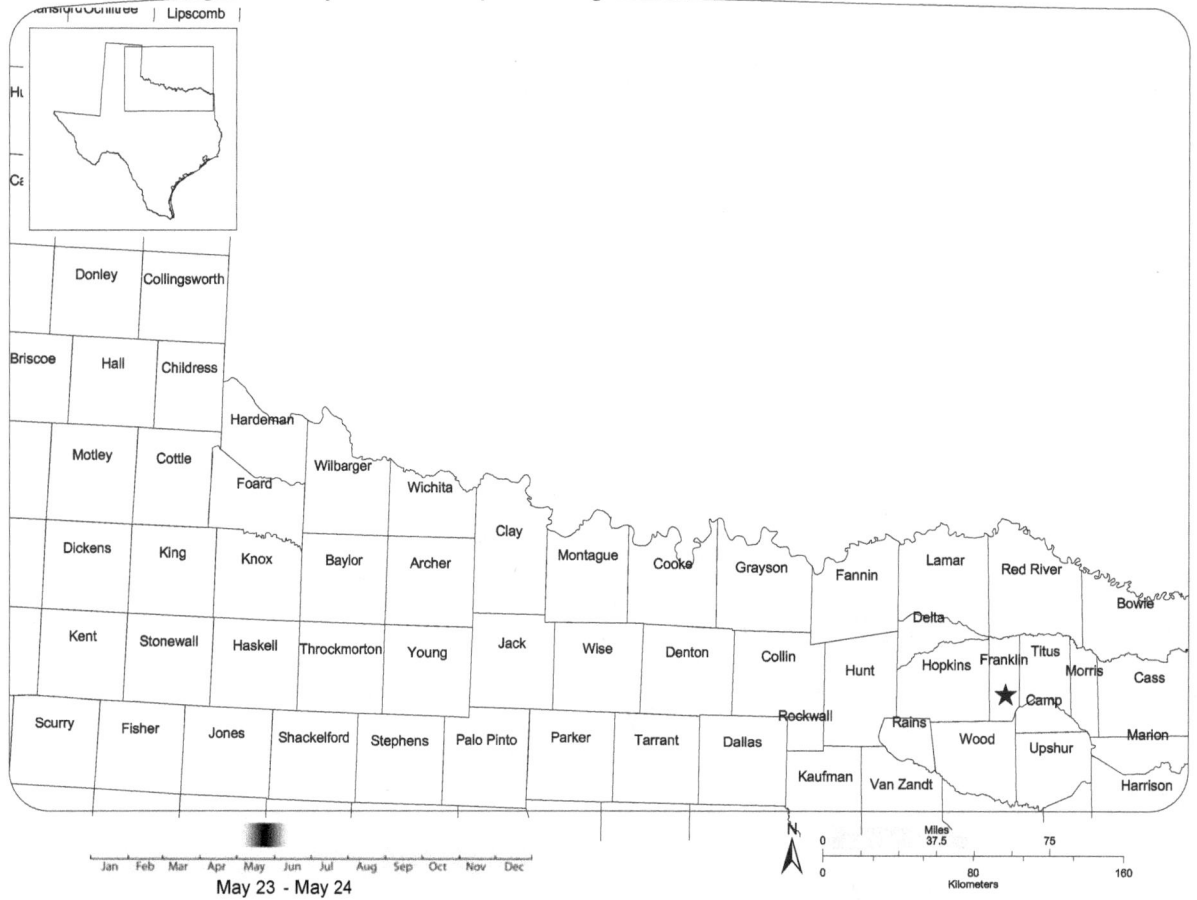

May 23 - May 24

HABITAT: Lakes or ponds with regular shade and dense emergent vegetation; often found in bays and sand-bottomed lakes.

Franklin
Hemphill

Lestes sigma Calvert / Chalky Spreadwing

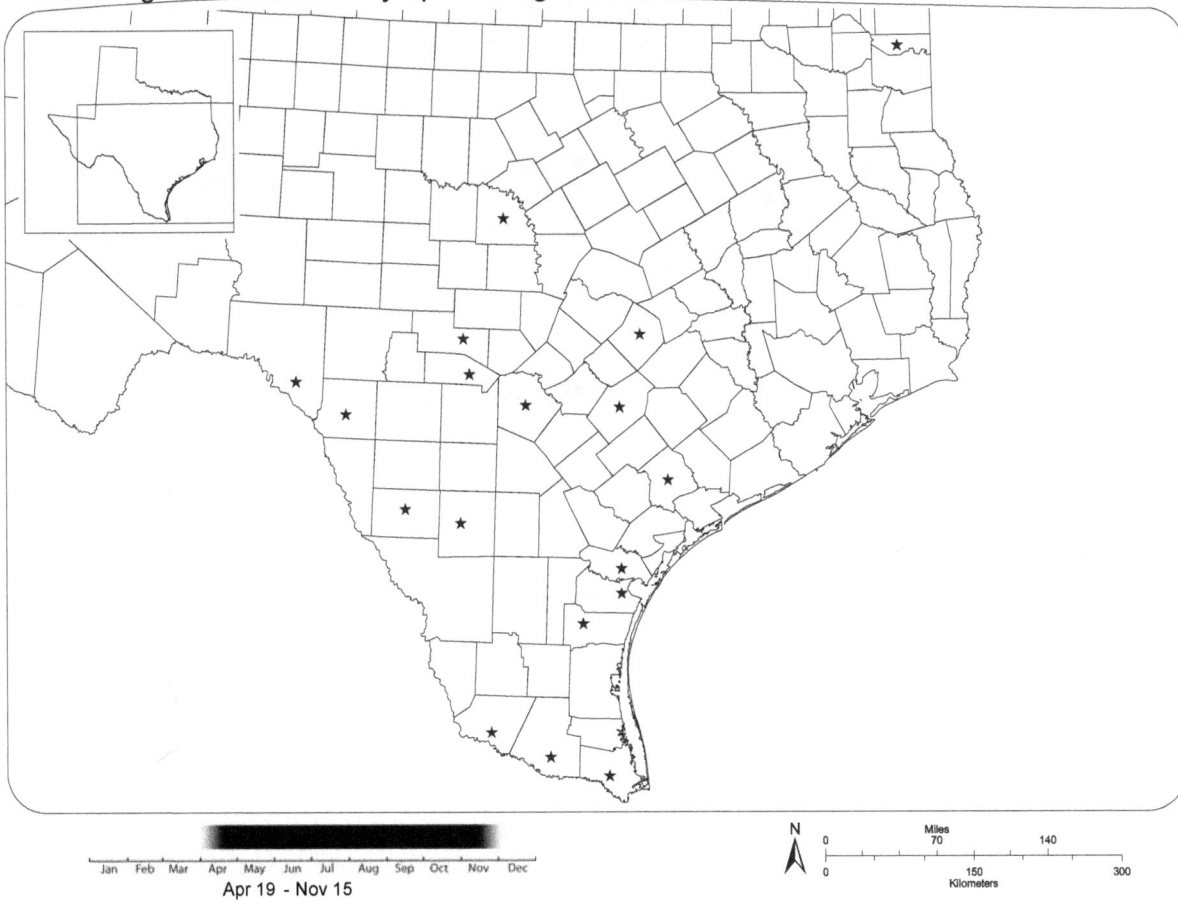

Apr 19 - Nov 15

HABITAT: Temporary pools and ponds.

Bandera
Bastrop
Bexar
Cameron
Dimmit
Gonzales
Hidalgo
Kerr
Kinney
Kleberg
La Salle
Marion
Nueces
San Patricio
San Saba
Starr
Val Verde
Victoria

Lestes unguiculatus Hagen / Lyre-tipped Spreadwing

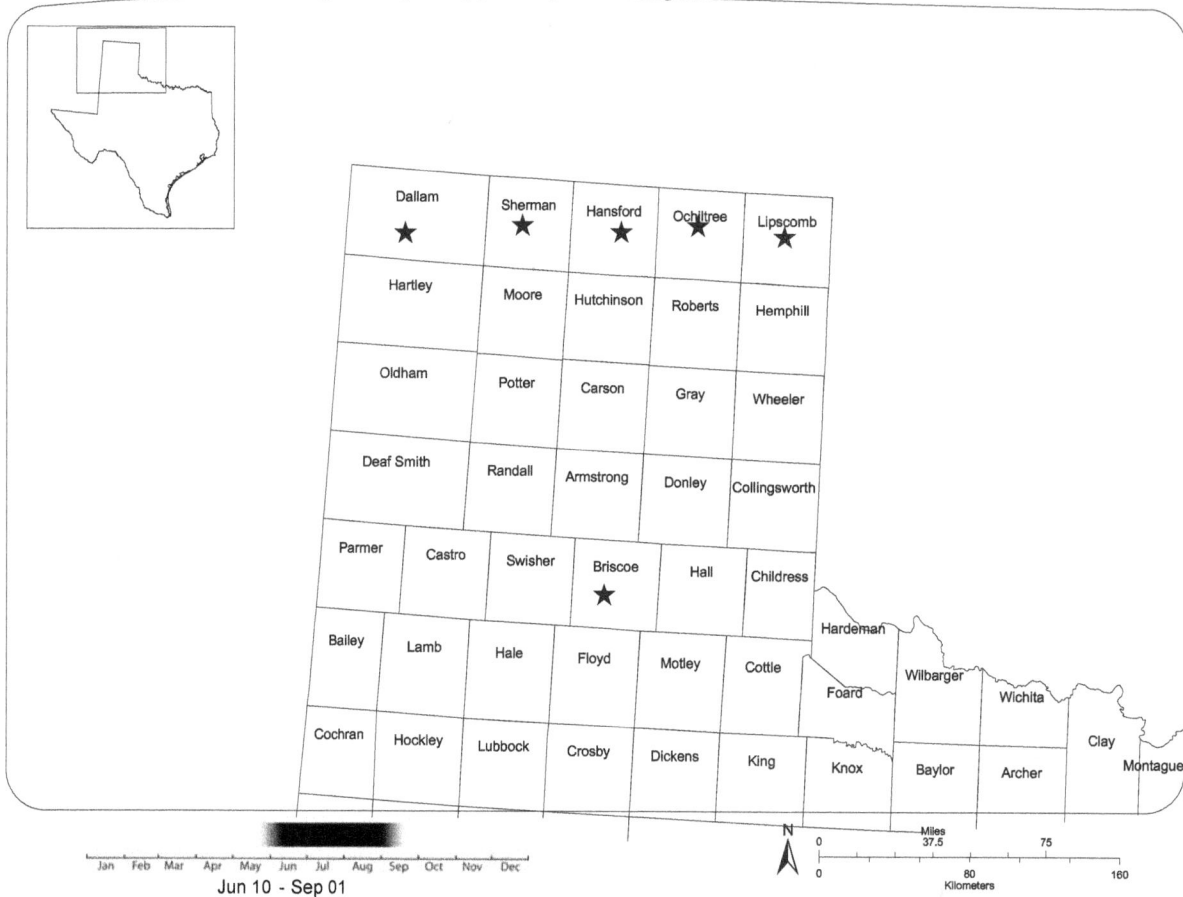

Jan Feb Mar Apr May Jun Jul Aug Sep Oct Nov Dec

Jun 10 - Sep 01

HABITAT: Open pools, ponds, sloughs and slow reaches of streams.

Briscoe
Dallam
Hansford
Lipscomb
Ochiltree
Sherman

Lestes vigilax Hagen in Selys / Swamp Spreadwing

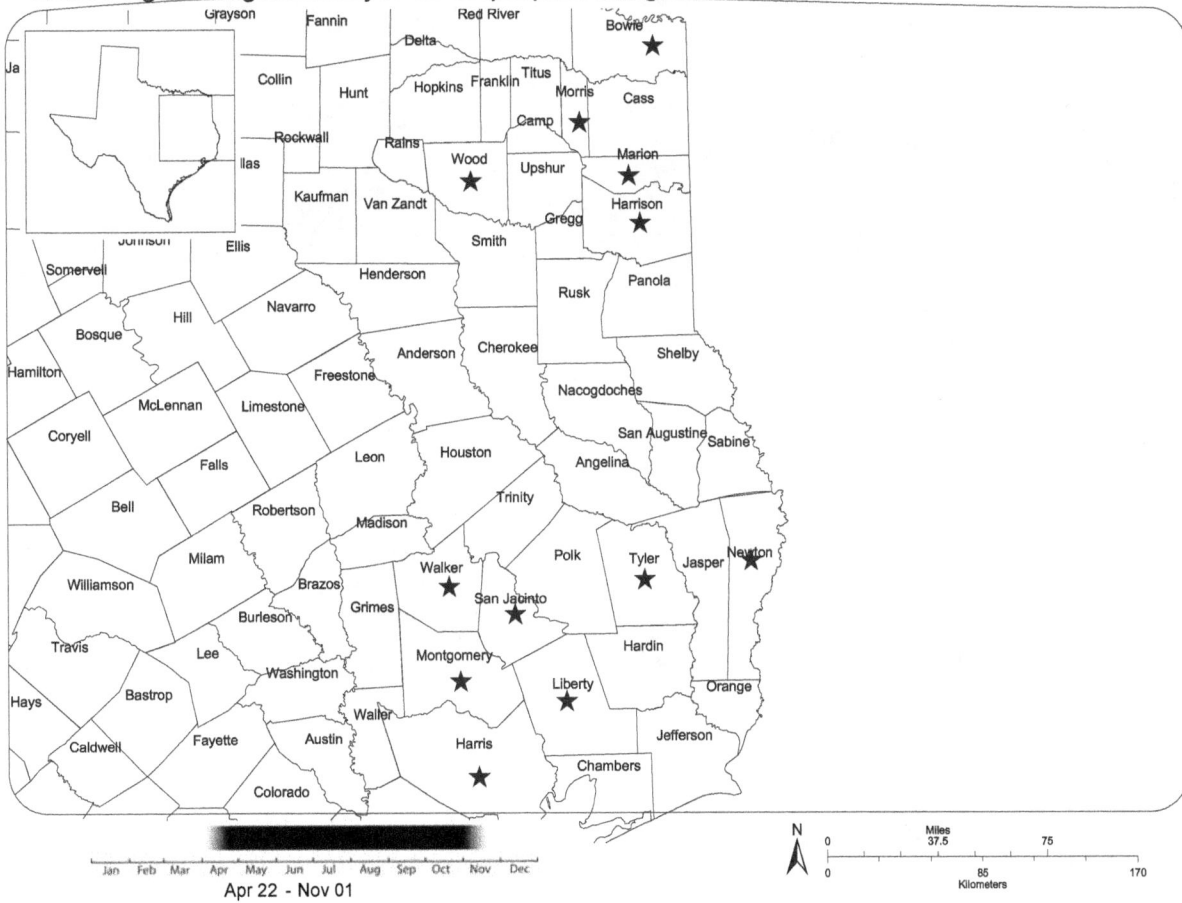

Apr 22 - Nov 01

HABITAT: Generally found in shaded acidic waters such as bogs, lakes, swamps, oxbows and slow streams.

Bowie
Harris
Harrison
Liberty
Marion
Montgomery
Morris
Newton
San Jacinto
Tyler
Walker
Wood

Neoneura aaroni Calvert / Coral-fronted Threadtail

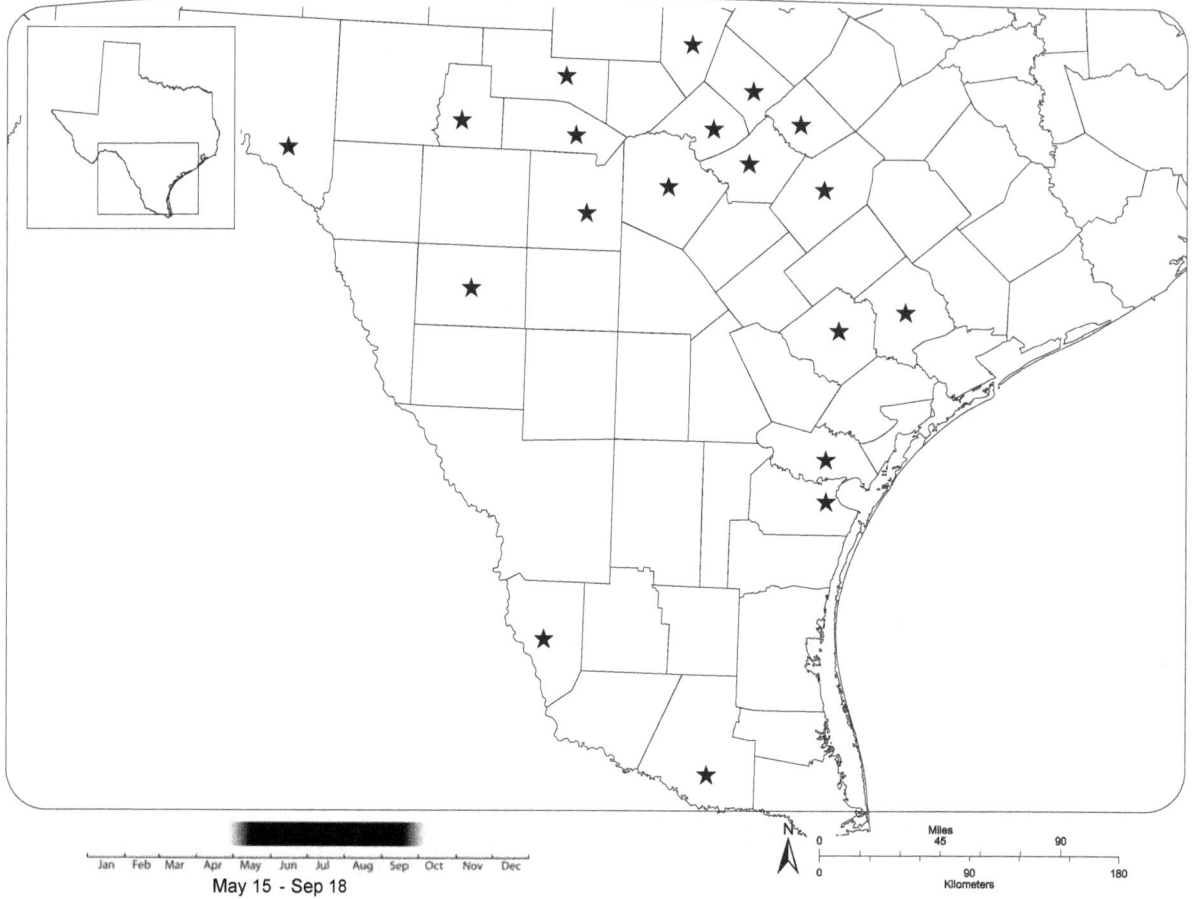

May 15 - Sep 18

HABITAT: Protected areas of slow-moving rivers and streams with emergent or floating vegetation or detritus.

Bandera
Bexar
Blanco
Caldwell
Comal
Goliad
Gonzales
Guadalupe
Hays
Hidalgo
Kerr
Medina
Nueces
Real
San Patricio
Val Verde
Victoria
Zapata

Neoneura amelia Calvert / Amelia's Threadtail

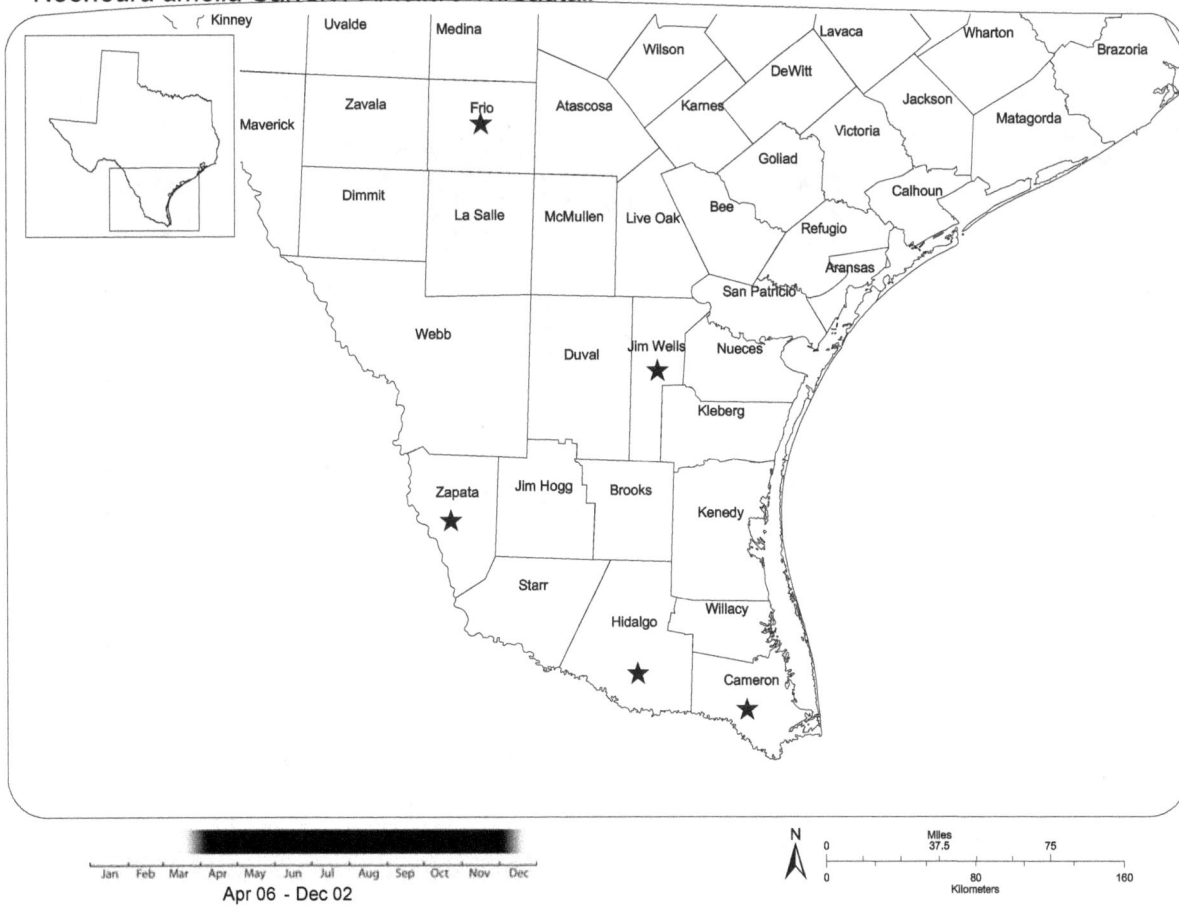

Apr 06 - Dec 02

HABITAT: Prefers protected, well-shaded areas of slow-moving rivers and streams with emergent or floating vegetation, detritus or debris.

Cameron
Frio
Hidalgo
Jim Wells
Zapata

Protoneura cara Calvert / Orange-striped Threadtail

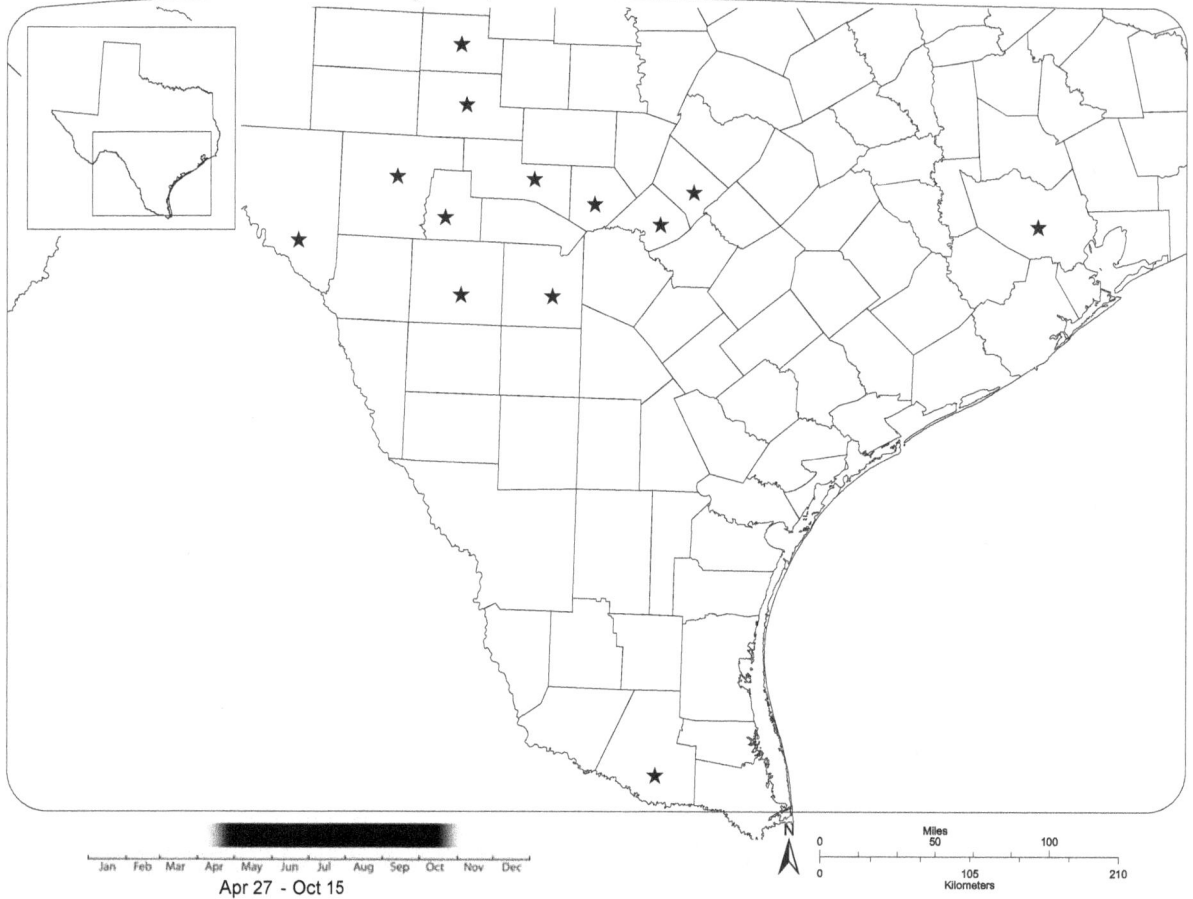

Apr 27 - Oct 15

HABITAT: Well-shaded, slow moving streams with ample leaf litter and debris.

Comal
Edwards
Harris
Hays
Hidalgo
Kendall
Kerr
Kimble
Medina
Menard
Real
Terrell
Uvalde
Val Verde

Acanthagrion quadratum Selys / Mexican Wedgetail

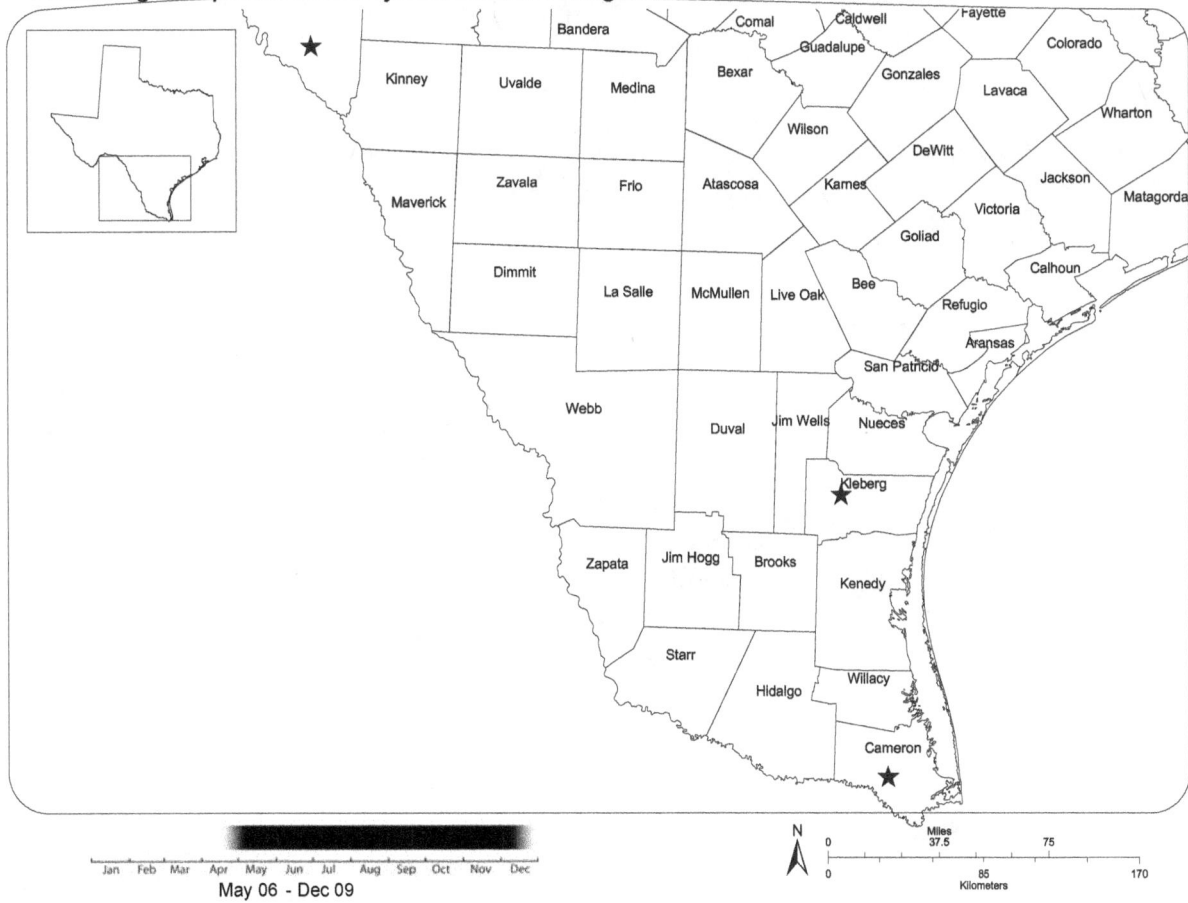

May 06 - Dec 09

HABITAT: Weedy ponds and slow backwaters.

Cameron
Kleberg
Val Verde

Argia alberta Kennedy / Paiute Dancer

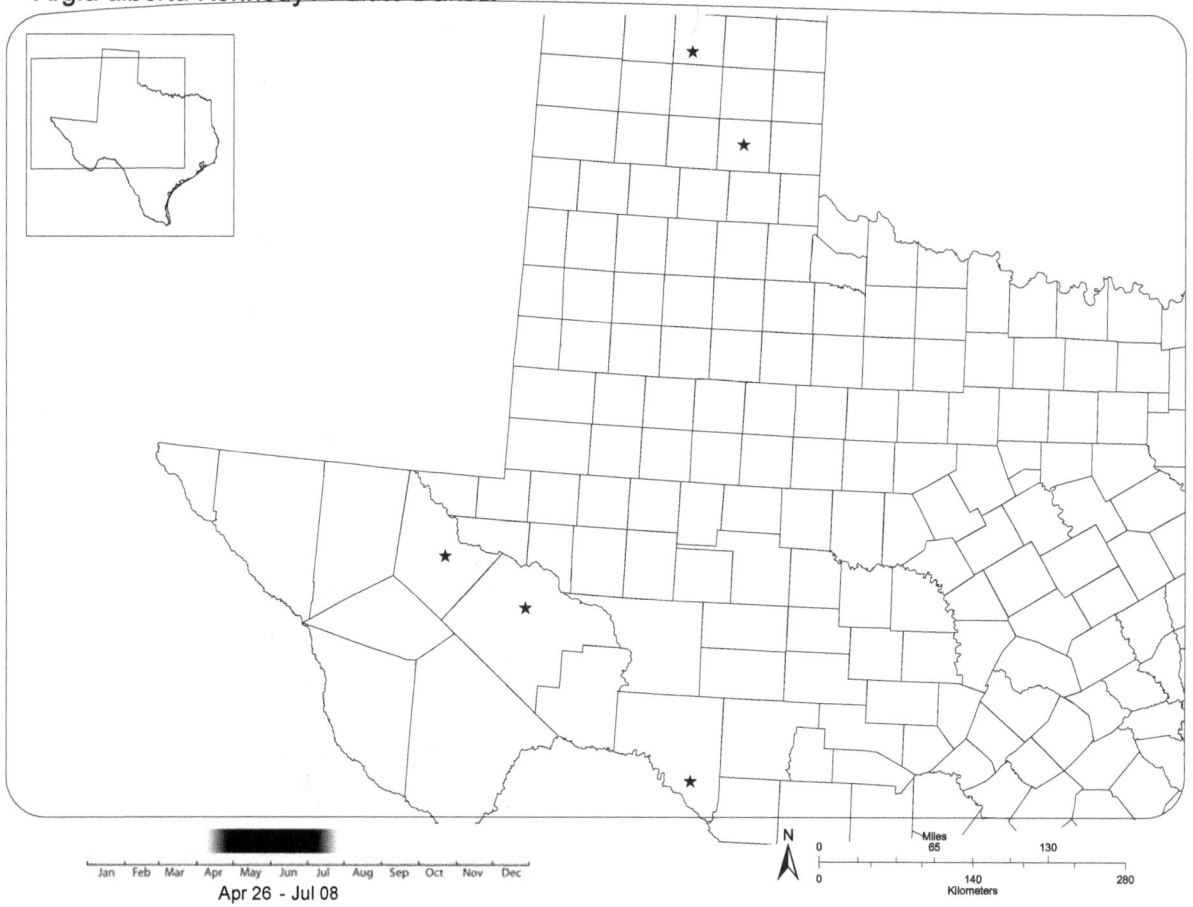

Jan Feb Mar Apr May Jun Jul Aug Sep Oct Nov Dec

Apr 26 - Jul 08

HABITAT: Small flowing streams or marshy springs.

Donley
Hutchinson
Pecos
Reeves
Val Verde

Argia apicalis (Say) / Blue-fronted Dancer

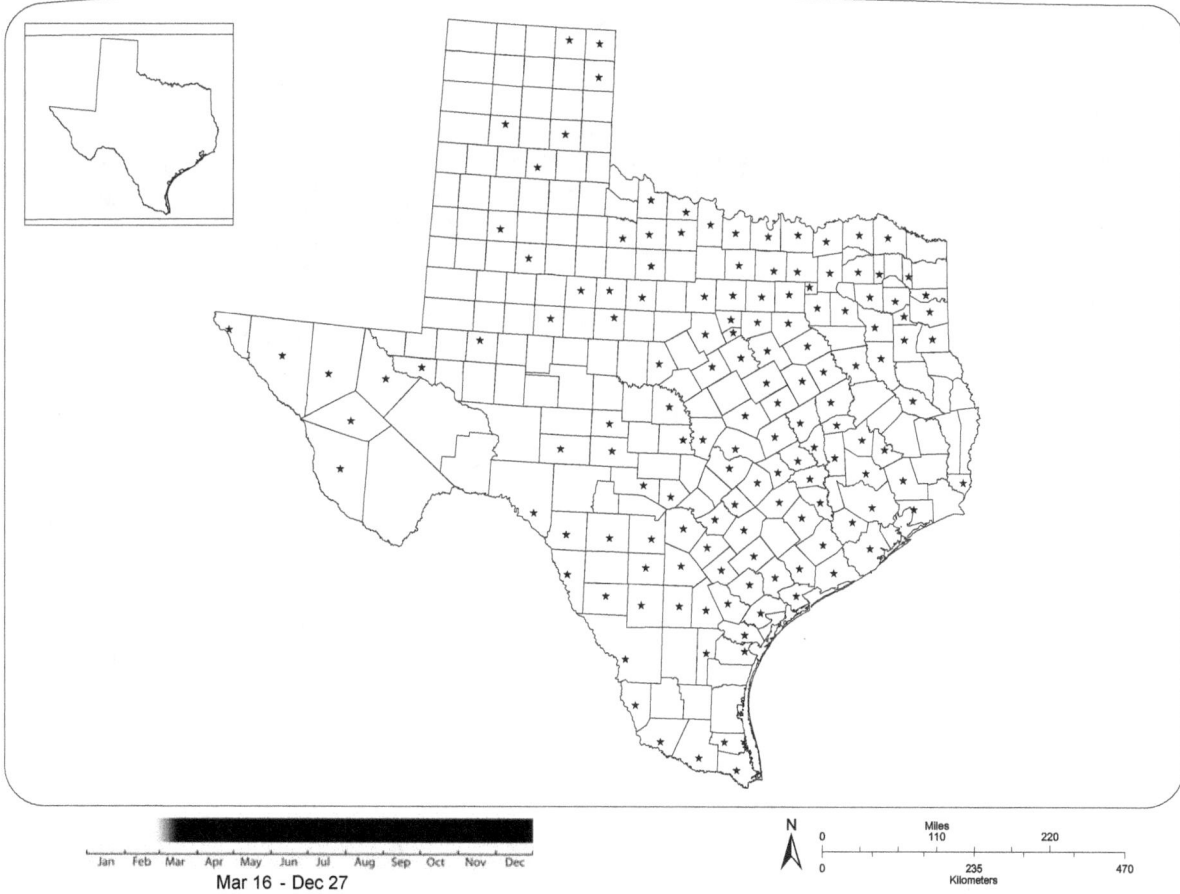

Mar 16 - Dec 27

HABITAT: Large rivers and occasionally streams, lakes or ponds.

Anderson	Culberson	Harrison	Lipscomb	Randall	Washington
Angelina	Dallas	Hemphill	Live Oak	Red River	Webb
Archer	Denton	Hidalgo	Llano	Reeves	Wharton
Atascosa	DeWitt	Hill	Lubbock	Refugio	Wichita
Austin	Dimmit	Hood	Madison	Robertson	Wilbarger
Bastrop	Donley	Hopkins	Marion	Rockwall	Willacy
Baylor	El Paso	Hudspeth	Matagorda	Rusk	Williamson
Bee	Ellis	Hunt	Maverick	San Jacinto	Wilson
Bell	Erath	Jackson	McLennan	San Patricio	Wise
Bexar	Falls	Jeff Davis	McMullen	San Saba	Wood
Bosque	Fannin	Jim Wells	Medina	Shackelford	Zapata
Brazoria	Fayette	Johnson	Menard	Smith	
Brazos	Fisher	Jones	Midland	Somervell	
Briscoe	Fort Bend	Karnes	Milam	Starr	
Brown	Franklin	Kaufman	Mitchell	Sutton	
Burleson	Freestone	Kendall	Montague	Tarrant	
Burnet	Frio	Kerr	Montgomery	Taylor	
Caldwell	Garza	Kimble	Morris	Throckmorton	
Calhoun	Goliad	Kinney	Navarro	Travis	
Cameron	Gonzales	Knox	Nueces	Upshur	
Chambers	Grayson	La Salle	Ochiltree	Uvalde	
Cherokee	Gregg	Lamar	Orange	Val Verde	
Clay	Grimes	Lee	Palo Pinto	Van Zandt	
Collin	Guadalupe	Leon	Panola	Victoria	
Colorado	Hamilton	Liberty	Parker	Walker	
Cooke	Harris	Limestone	Presidio	Ward	

Argia barretti Calvert / Comanche Dancer

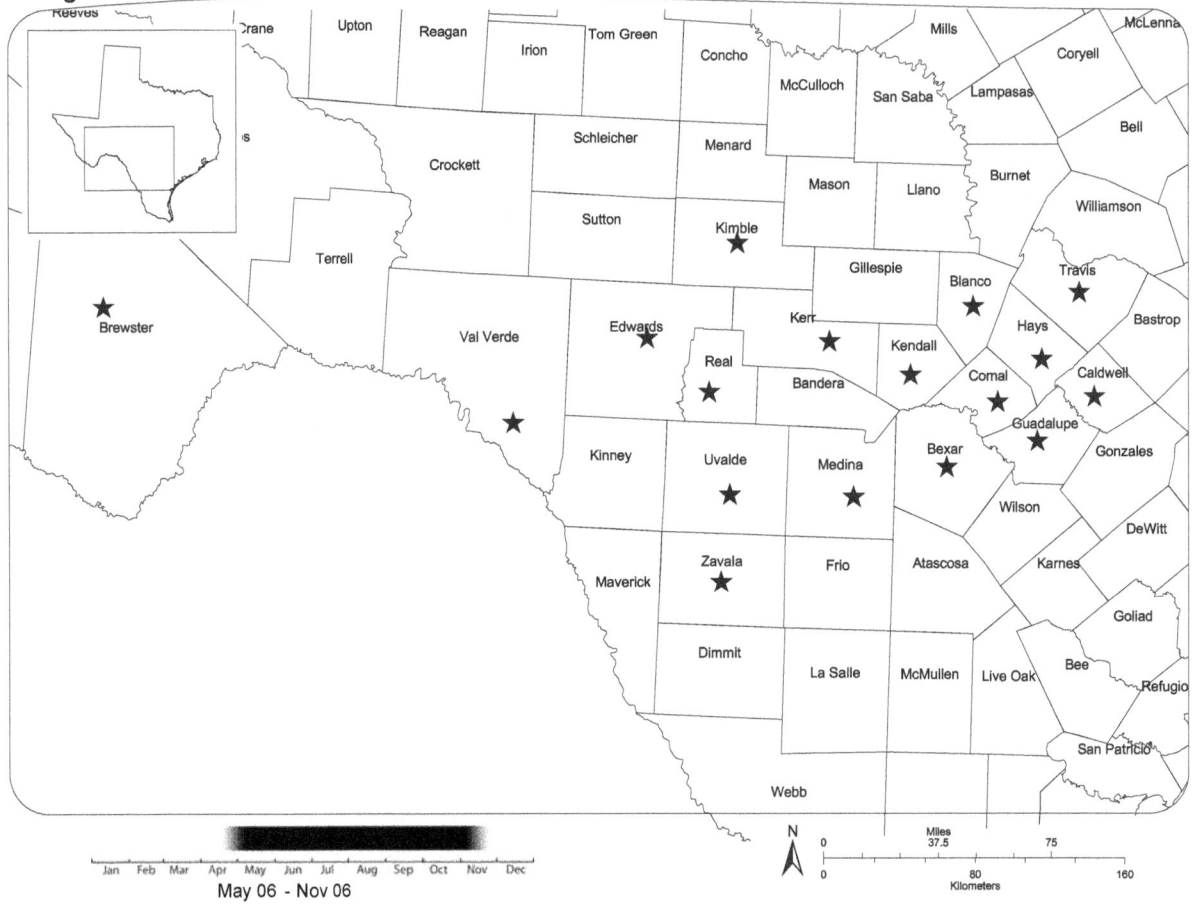

May 06 - Nov 06

HABITAT: Rivers and streams.

Bexar
Blanco
Brewster
Caldwell
Comal
Edwards
Guadalupe
Hays
Kendall
Kerr
Kimble
Medina
Real
Travis
Uvalde
Val Verde
Zavala

Argia bipunctulata (Hagen) / Seepage Dancer

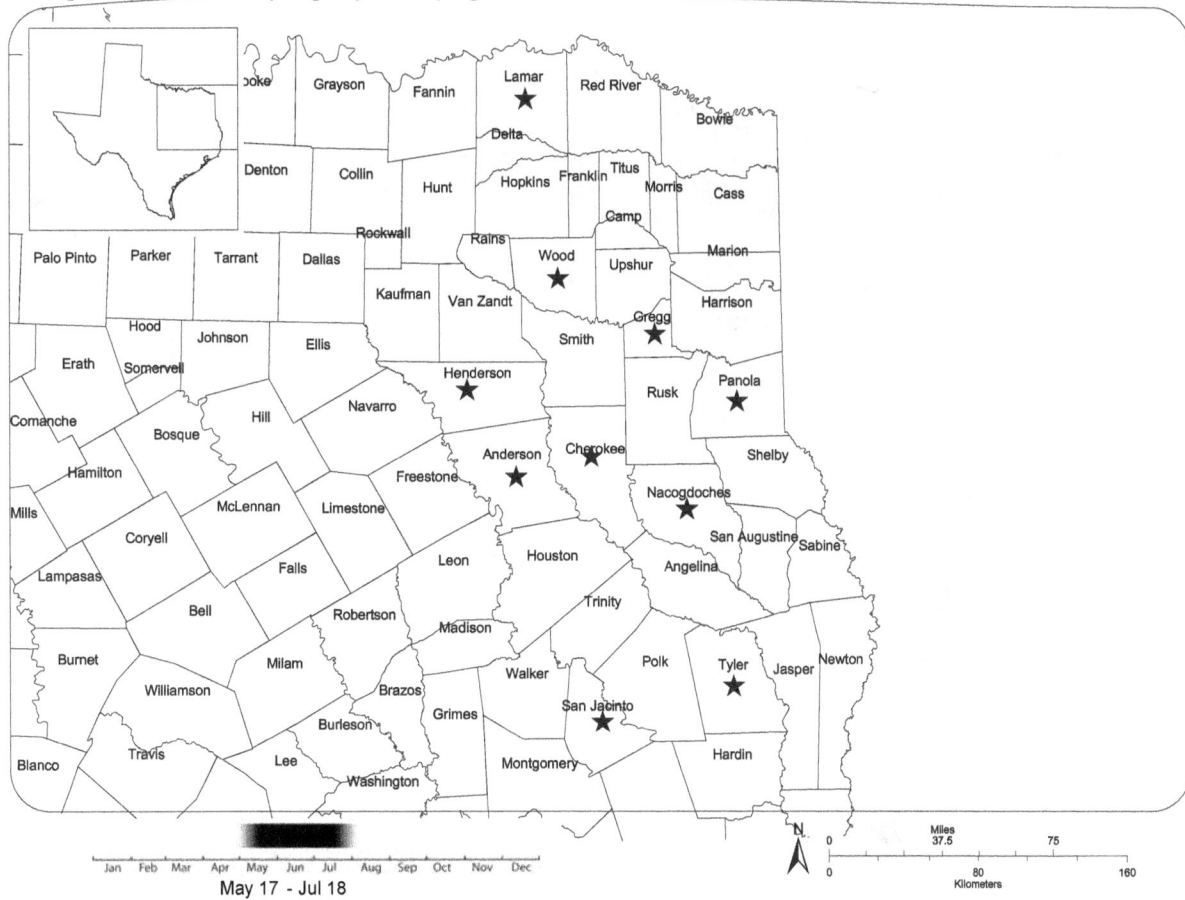

May 17 - Jul 18

HABITAT: Associated with sunny sphagnum seepages, small lakes, ponds and streams.

Anderson
Cherokee
Gregg
Henderson
Lamar
Nacogdoches
Panola
San Jacinto
Tyler
Wood

Argia cuprea (Hagen) / Coppery Dancer

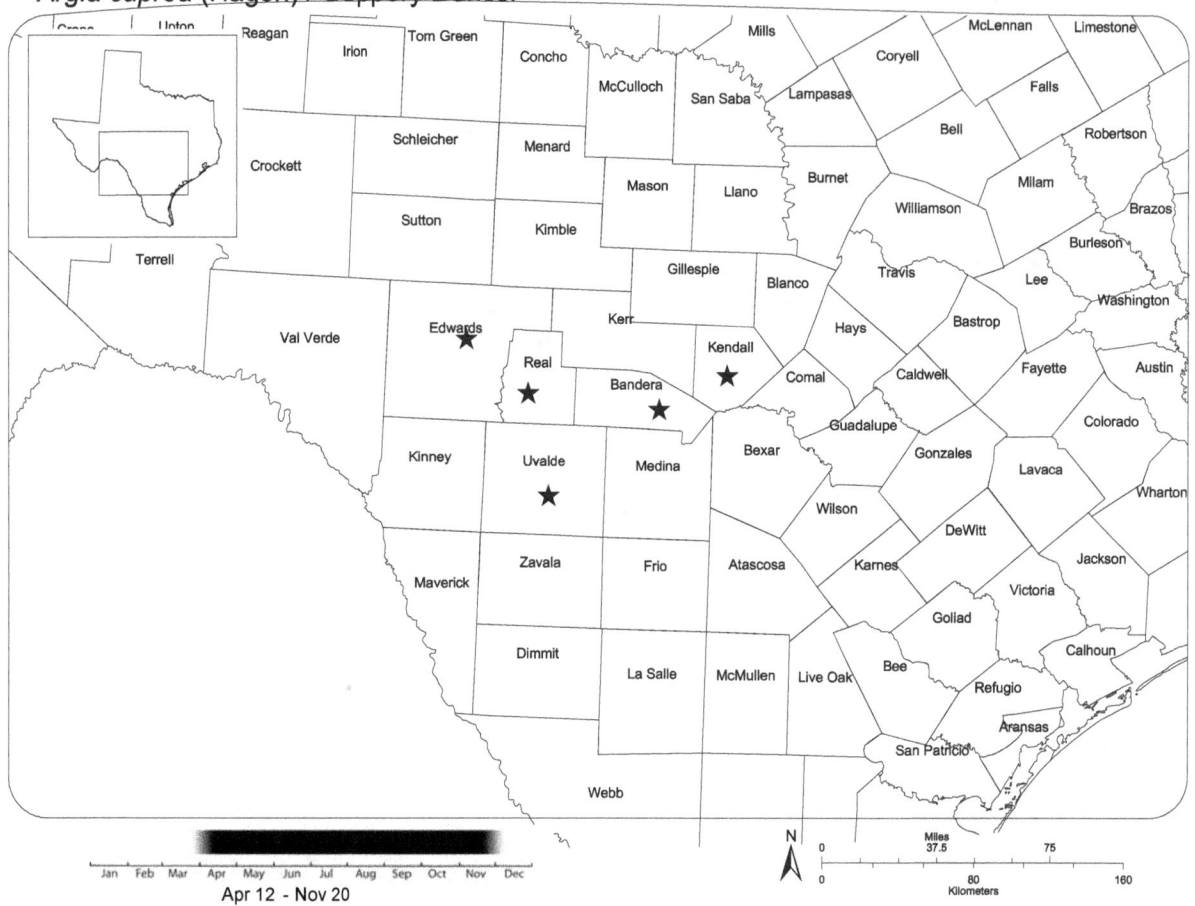

Apr 12 - Nov 20

HABITAT: Rivers and streams.

Bandera
Edwards
Kendall
Real
Uvalde

Argia fumipennis violacea (Hagen) / Violet Dancer

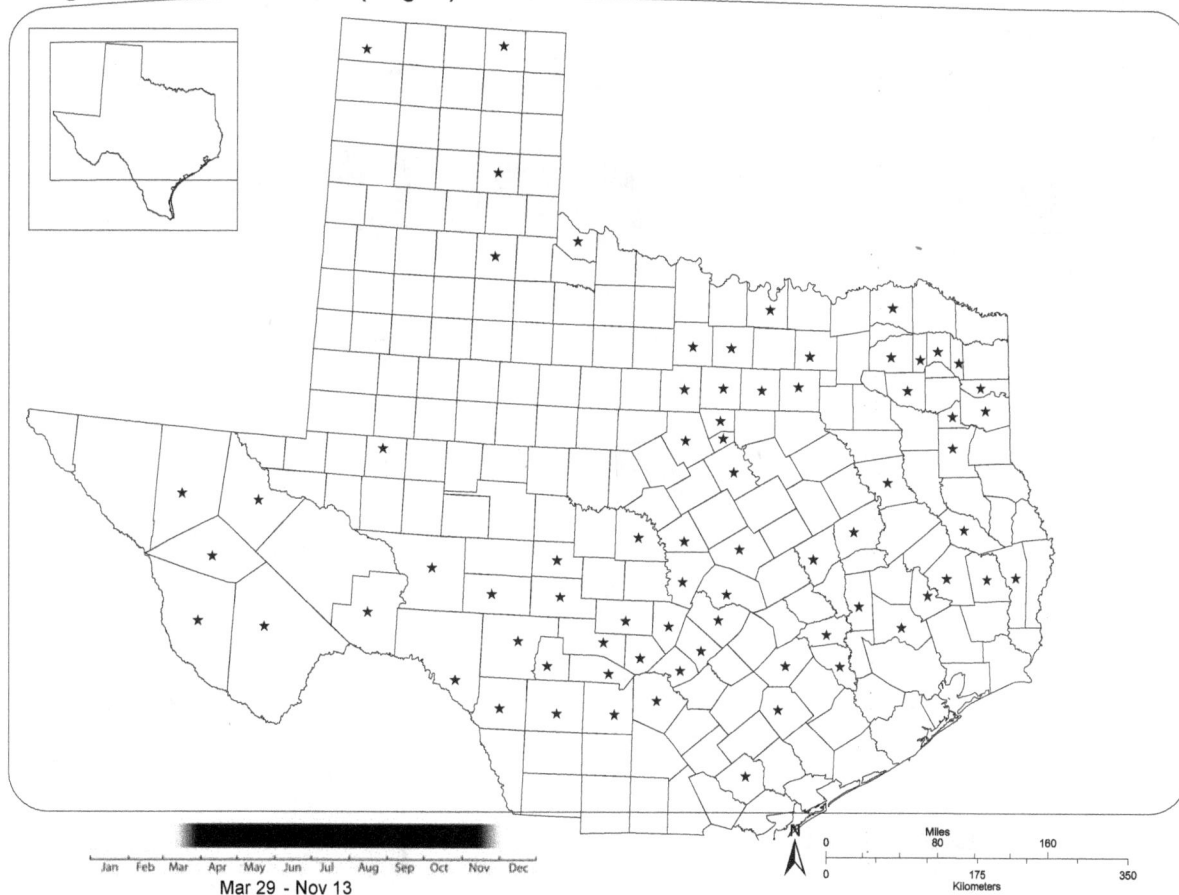

Mar 29 - Nov 13

HABITAT: Shallow streams with exposed rocks and small lakes.

Anderson	Hardeman	Polk
Angelina	Harrison	Presidio
Austin	Hays	Real
Bandera	Hood	Reeves
Bell	Hopkins	Robertson
Bexar	Jack	Rusk
Blanco	Jasper	San Jacinto
Bosque	Jeff Davis	San Saba
Brewster	Kendall	Somervell
Burnet	Kerr	Sutton
Collin	Kimble	Tarrant
Comal	Kinney	Terrell
Cooke	Lamar	Titus
Crockett	Lampasas	Travis
Culberson	Lavaca	Tyler
Dallam	Leon	Uvalde
Dallas	Marion	Val Verde
Donley	Medina	Washington
Edwards	Menard	Williamson
Erath	Midland	Wise
Fayette	Montgomery	Wood
Franklin	Morris	
Gillespie	Motley	
Goliad	Ochiltree	
Gregg	Palo Pinto	
Grimes	Parker	

Argia hinei Kennedy / Lavender Dancer

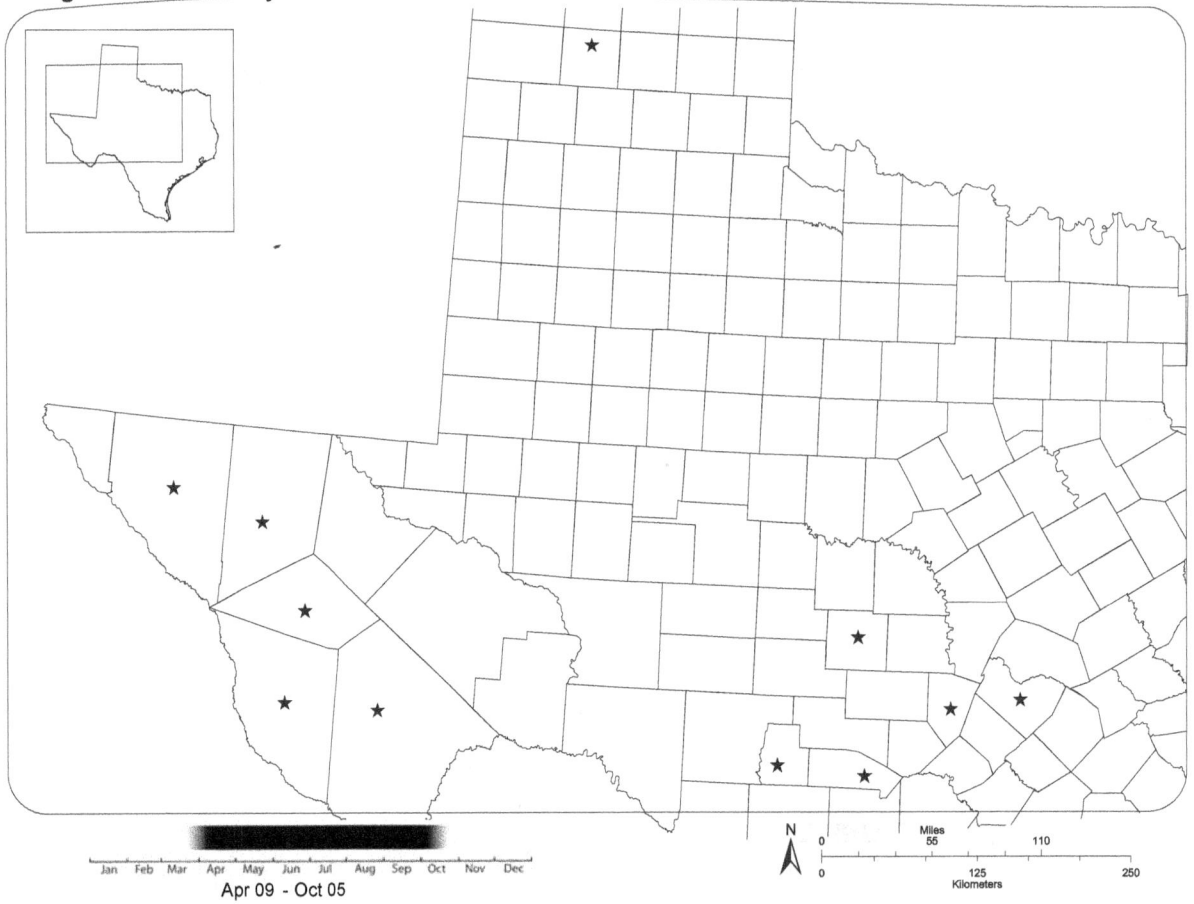

Jan Feb Mar Apr May Jun Jul Aug Sep Oct Nov Dec

Apr 09 - Oct 05

HABITAT: Semidesert creeks, streams and rivers.

Bandera
Blanco
Brewster
Culberson
Hudspeth
Jeff Davis
Mason
Presidio
Randall
Real
Travis

Argia immunda (Hagen) / Kiowa Dancer

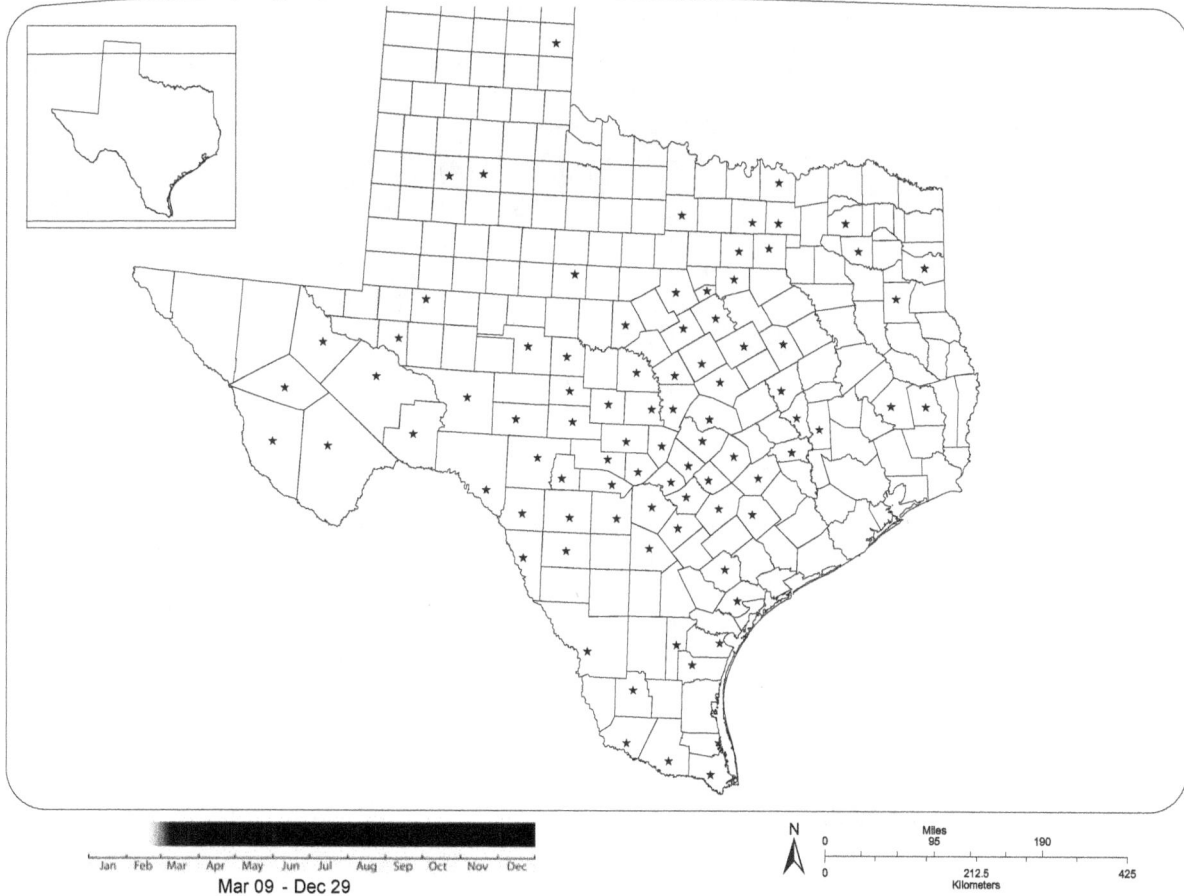

Mar 09 - Dec 29

HABITAT: Streams and rivers.

Atascosa	Goliad	Maverick	Washington
Bandera	Gonzales	McLennan	Webb
Bastrop	Grayson	Medina	Wheeler
Bell	Grimes	Menard	Williamson
Bexar	Guadalupe	Midland	Wilson
Blanco	Hamilton	Nueces	Wood
Bosque	Harrison	Pecos	Zavala
Brazos	Hays	Polk	
Brewster	Hidalgo	Presidio	
Brown	Hopkins	Real	
Burnet	Jack	Reeves	
Caldwell	Jeff Davis	Refugio	
Cameron	Jim Hogg	Robertson	
Collin	Jim Wells	Rusk	
Comal	Johnson	San Saba	
Concho	Kendall	Somervell	
Coryell	Kerr	Starr	
Crane	Kimble	Sutton	
Crockett	Kinney	Tarrant	
Crosby	Kleberg	Taylor	
Dallas	Lampasas	Terrell	
Denton	Lavaca	Tom Green	
Edwards	Limestone	Travis	
Erath	Llano	Tyler	
Fayette	Lubbock	Uvalde	
Gillespie	Mason	Val Verde	

Argia leonorae Garrison / Leonora's Dancer

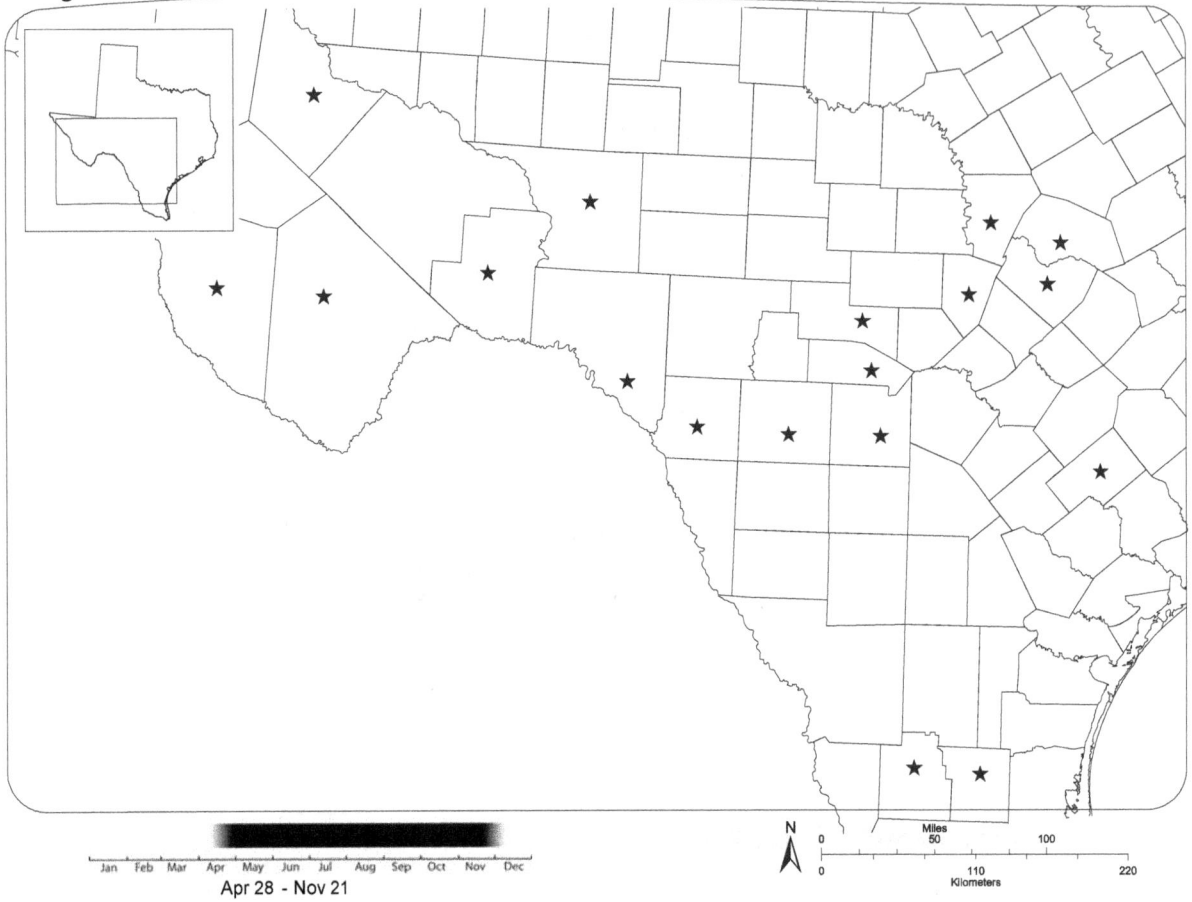

Apr 28 - Nov 21

HABITAT: Small streams and seepages.

Bandera
Blanco
Brewster
Brooks
Burnet
Crockett
Culberson
DeWitt
Hudspeth
Jim Hogg
Kerr
Kinney
Medina
Presidio
Reeves
Terrell
Travis
Uvalde
Val Verde
Williamson

Argia lugens (Hagen) / Sooty Dancer

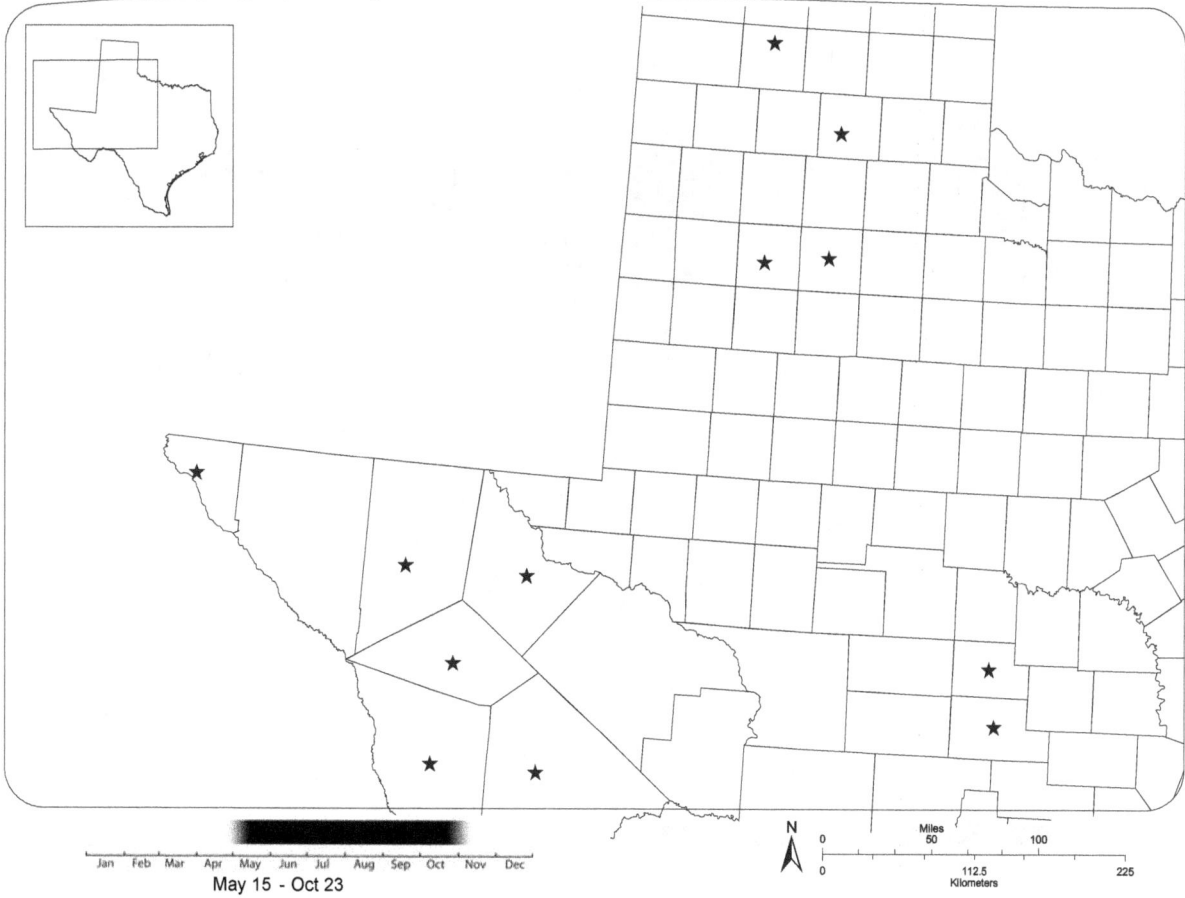

May 15 - Oct 23

HABITAT: Rocky, desert rivers and streams.

Brewster
Briscoe
Crosby
Culberson
El Paso
Jeff Davis
Kimble
Lubbock
Menard
Presidio
Randall
Reeves

Argia moesta (Hagen) / Powdered Dancer

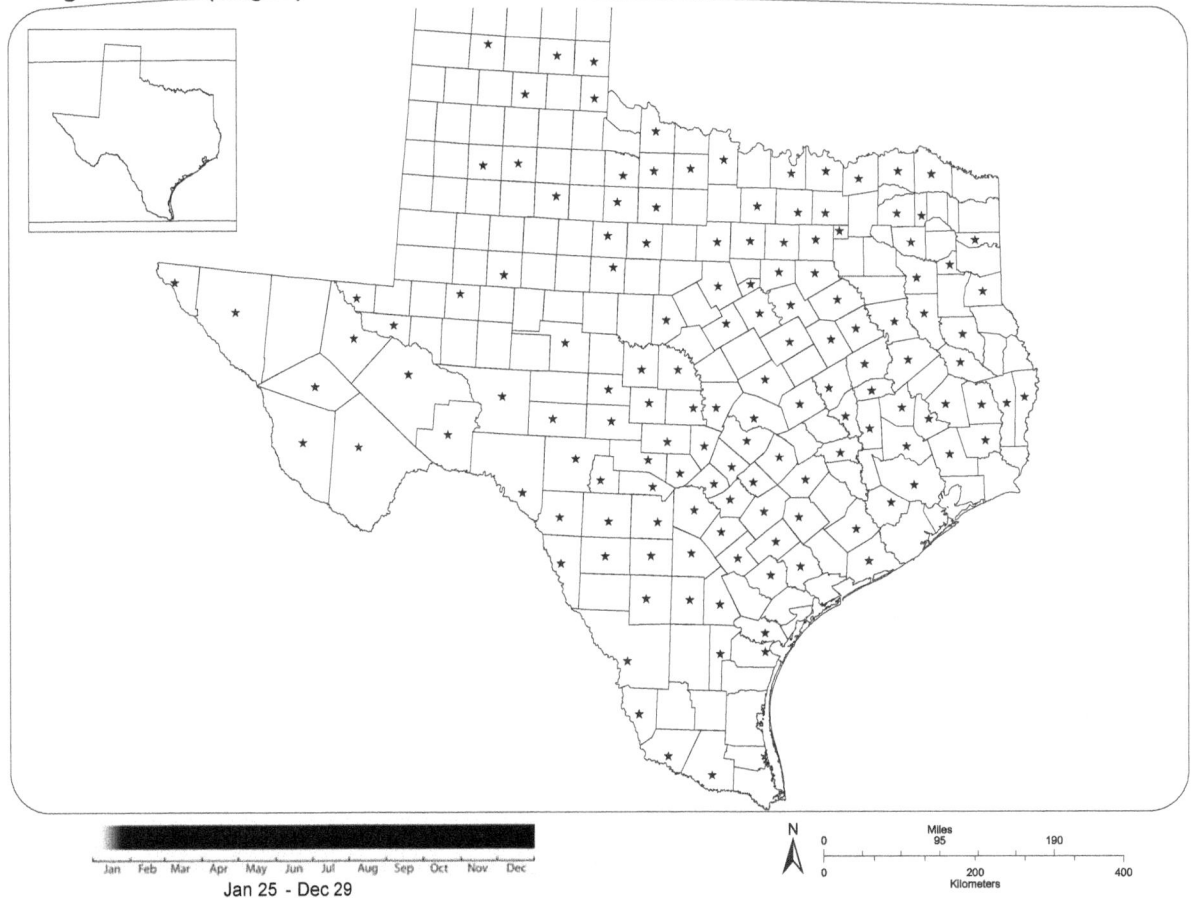

Jan 25 - Dec 29

HABITAT: Swift currents of rivers and lakes with emergent stones and rocky shores.

Anderson	Dallas	Hidalgo	Loving	Real	Webb
Angelina	Denton	Hill	Lubbock	Red River	Wharton
Archer	DeWitt	Hopkins	Madison	Reeves	Wilbarger
Atascosa	Donley	Houston	Marion	Robertson	Williamson
Bandera	Edwards	Howard	Mason	Rockwall	Wilson
Bastrop	El Paso	Hudspeth	Matagorda	San Jacinto	Wise
Baylor	Ellis	Jasper	Maverick	San Patricio	Wood
Bell	Erath	Jeff Davis	McCulloch	San Saba	Zapata
Bexar	Fannin	Jim Wells	McLennan	Shackelford	Zavala
Blanco	Fayette	Johnson	McMullen	Smith	
Bosque	Fort Bend	Jones	Medina	Somervell	
Brazos	Franklin	Karnes	Menard	Starr	
Brewster	Freestone	Kendall	Midland	Sutton	
Briscoe	Frio	Kent	Milam	Tarrant	
Brown	Gillespie	Kerr	Montgomery	Taylor	
Burnet	Goliad	Kimble	Nacogdoches	Terrell	
Caldwell	Gonzales	Kinney	Navarro	Throckmorton	
Cherokee	Grayson	Knox	Newton	Tom Green	
Childress	Gregg	La Salle	Nueces	Travis	
Clay	Grimes	Lamar	Palo Pinto	Tyler	
Collin	Guadalupe	Lavaca	Panola	Uvalde	
Collingsworth	Hamilton	Leon	Parker	Val Verde	
Comal	Hardin	Liberty	Pecos	Victoria	
Cooke	Harris	Limestone	Polk	Walker	
Crockett	Haskell	Live Oak	Presidio	Ward	
Crosby	Hays	Llano	Randall	Washington	

Argia munda Calvert / Apache Dancer

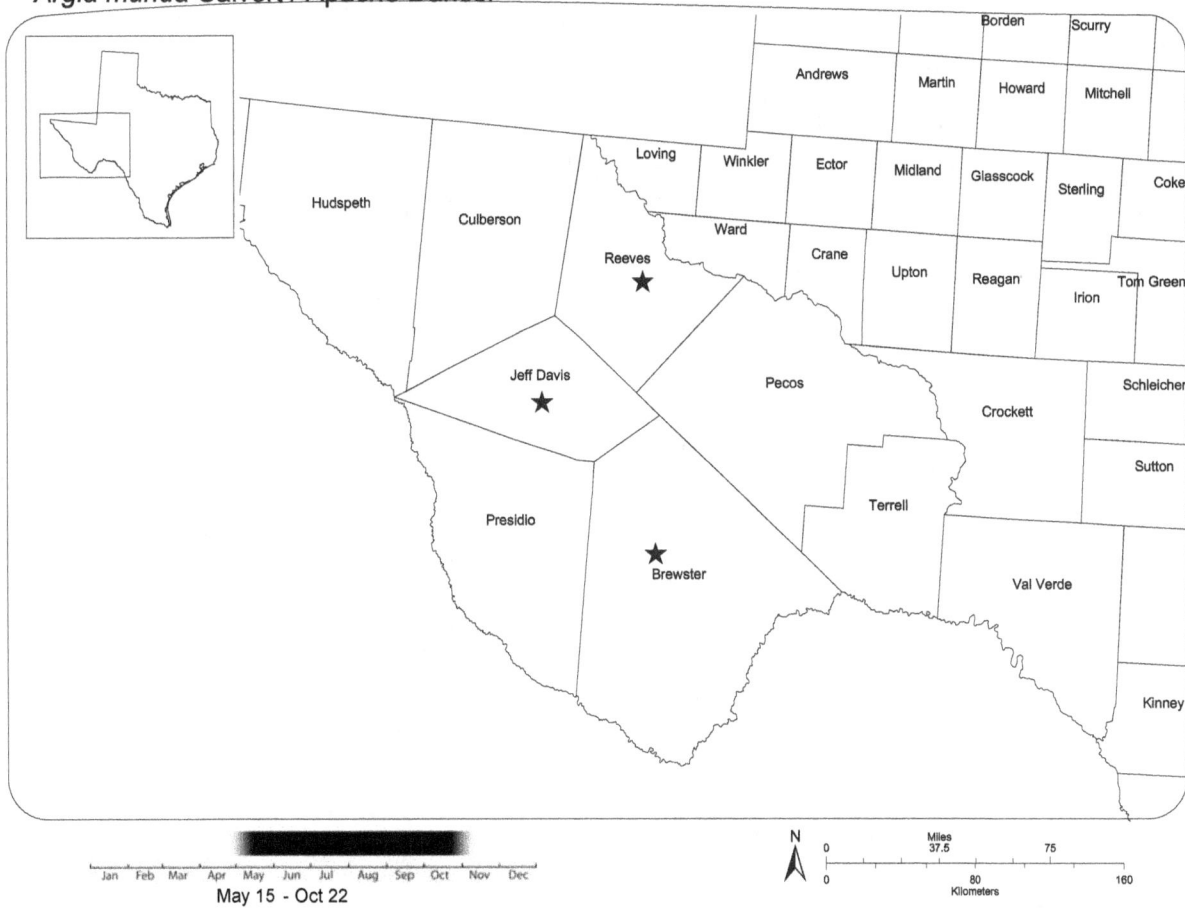

May 15 - Oct 22

HABITAT: Primarily found at desert streams.

Brewster
Jeff Davis
Reeves

Argia nahuana Calvert / Aztec Dancer

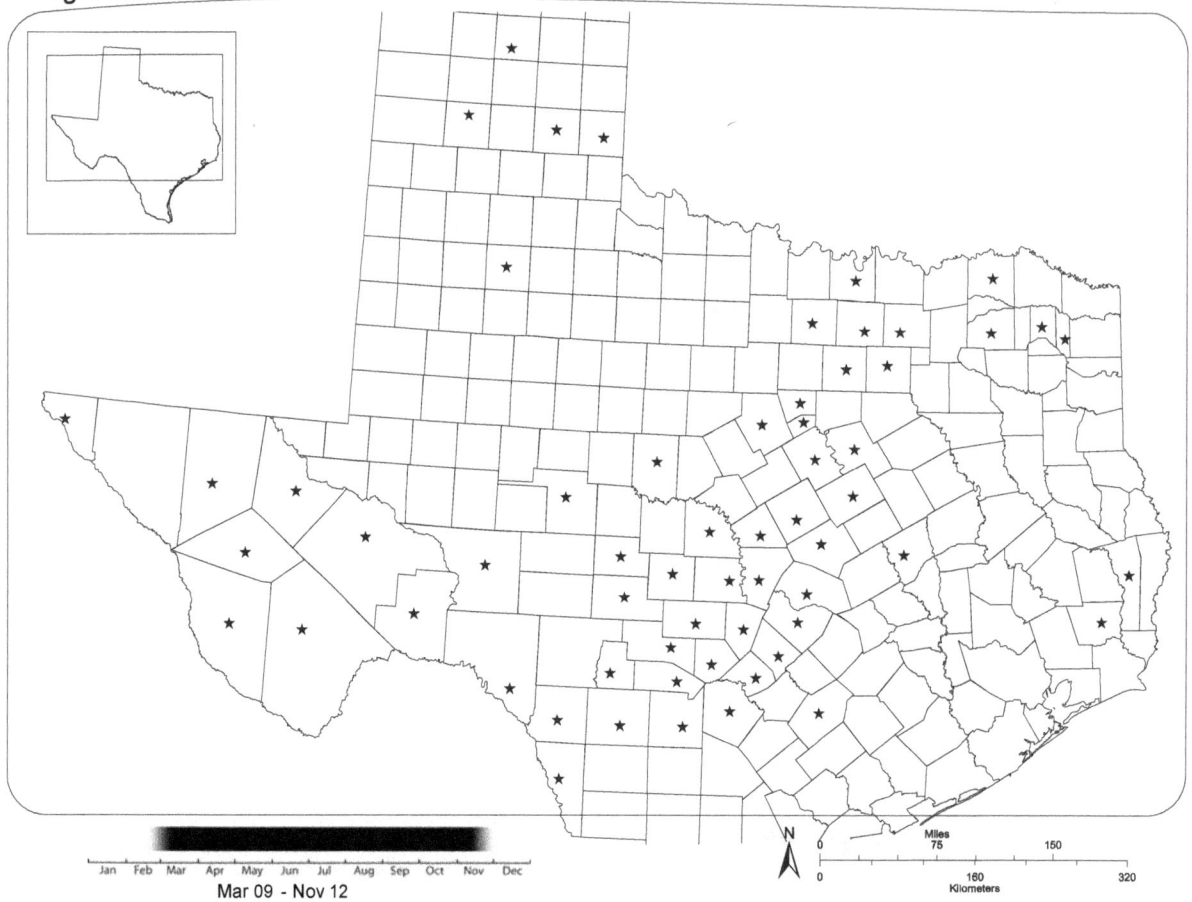

Jan Feb Mar Apr May Jun Jul Aug Sep Oct Nov Dec

Mar 09 - Nov 12

HABITAT: Small, shallow, clear water streams, fully exposed to sunlight with only moderate marginal vegetation.

Bandera	Hood	Tarrant
Bell	Hopkins	Terrell
Bexar	Hutchinson	Titus
Blanco	Jasper	Tom Green
Bosque	Jeff Davis	Travis
Brewster	Kendall	Uvalde
Burnet	Kerr	Val Verde
Coleman	Kimble	Williamson
Collin	Kinney	Wise
Collingsworth	Lamar	
Comal	Lampasas	
Cooke	Llano	
Coryell	Mason	
Crockett	Maverick	
Crosby	McLennan	
Culberson	Medina	
Dallas	Menard	
Denton	Morris	
Donley	Pecos	
El Paso	Presidio	
Erath	Randall	
Gillespie	Real	
Gonzales	Reeves	
Hardin	Robertson	
Hays	San Saba	
Hill	Somervell	

Argia oenea Hagen in Selys / Fiery-eyed Dancer

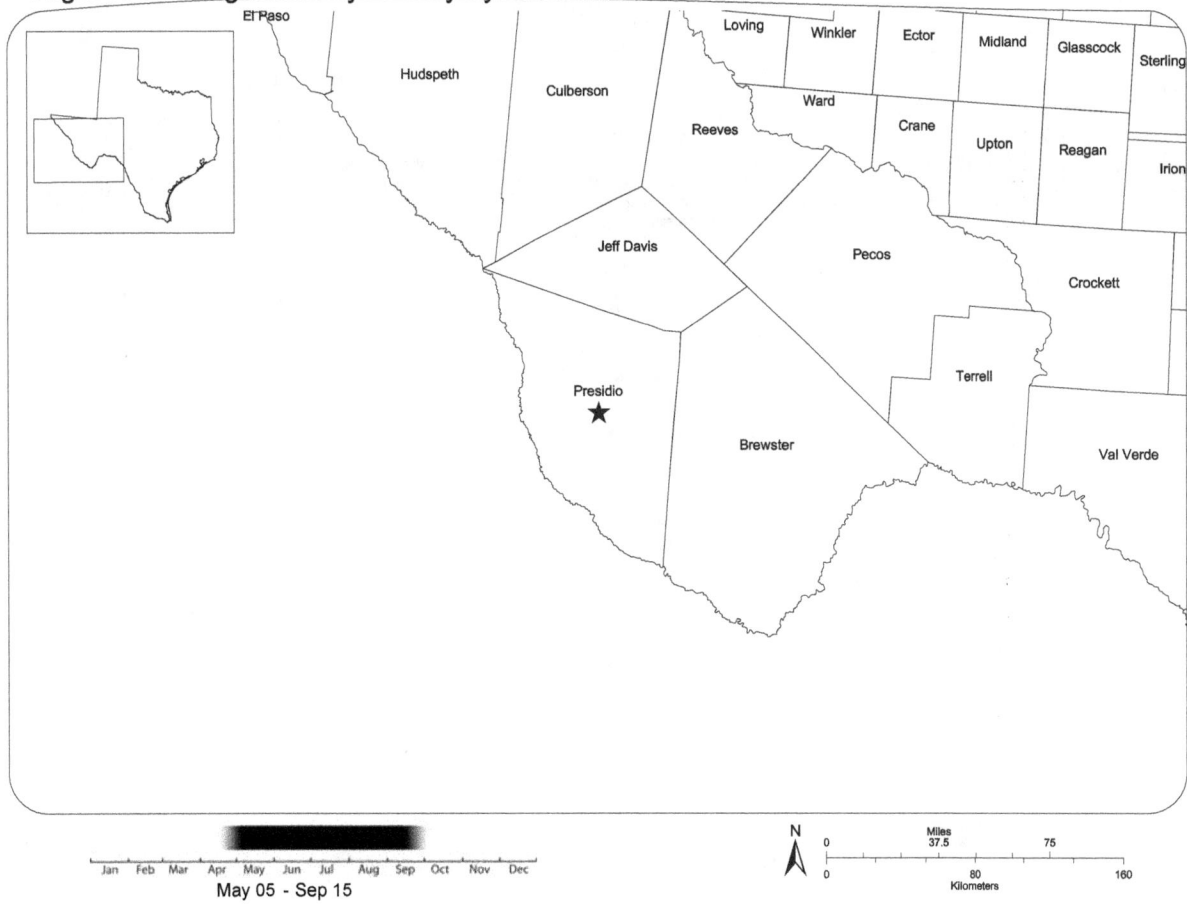

May 05 - Sep 15

HABITAT: Streams and rivers.

Presidio

Argia pallens Calvert / Amethyst Dancer

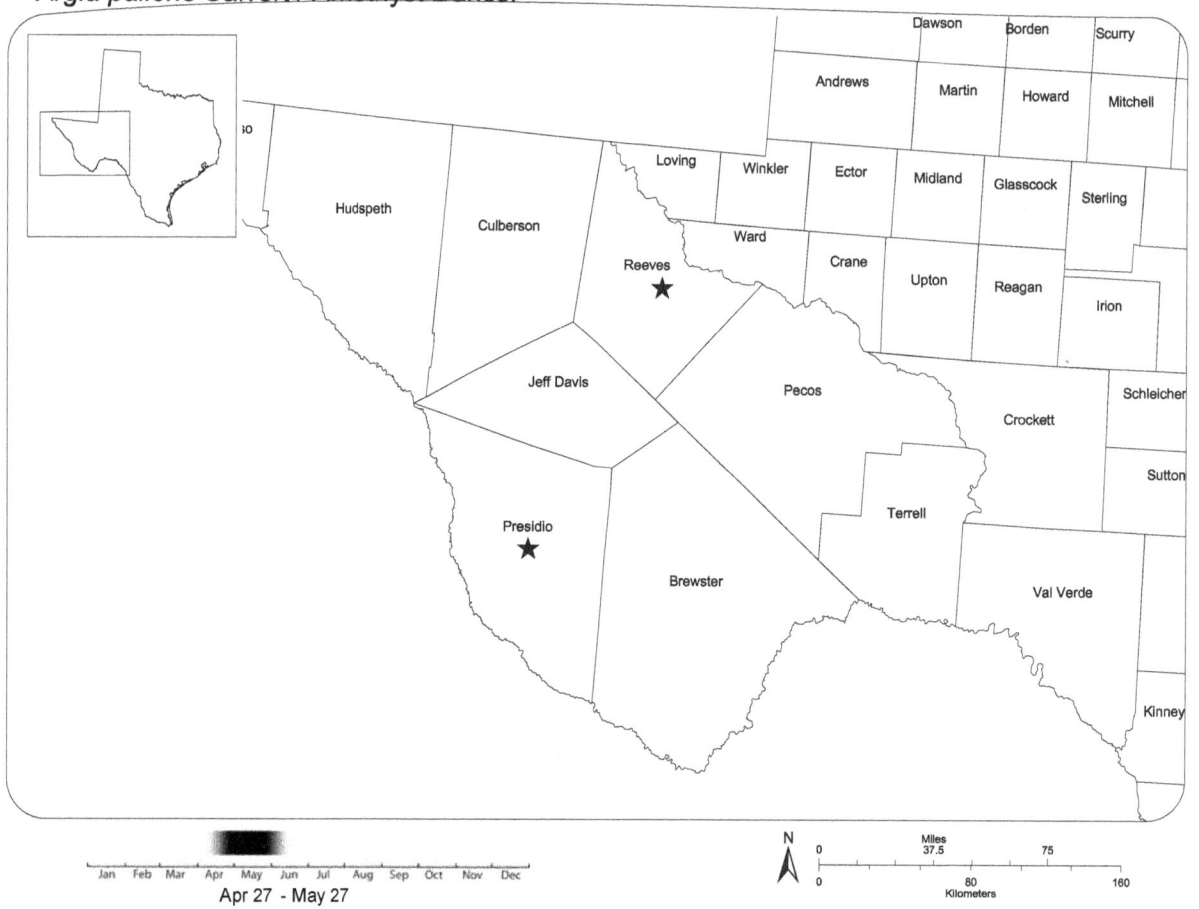

Apr 27 - May 27

HABITAT: Small desert streams.

Presidio
Reeves

Argia plana Calvert / Springwater Dancer

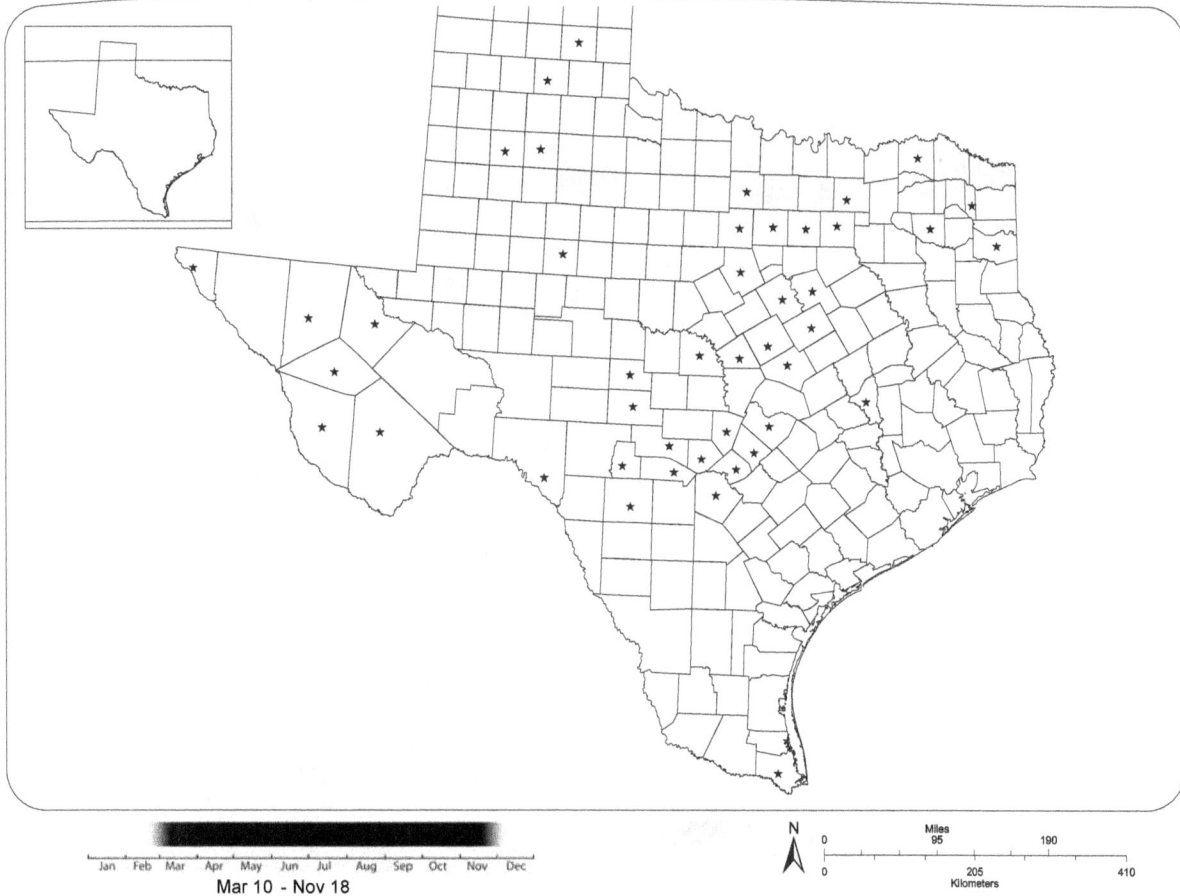

Mar 10 - Nov 18

HABITAT: Small shallow, canopied spring seepages with clay substrate.

Bandera	Lamar
Bell	Lampasas
Bexar	Lubbock
Blanco	McLennan
Bosque	Menard
Brazos	Mitchell
Brewster	Morris
Briscoe	Palo Pinto
Cameron	Parker
Collin	Presidio
Comal	Real
Coryell	Reeves
Crosby	San Saba
Culberson	Tarrant
Dallas	Travis
Donley	Uvalde
El Paso	Val Verde
Erath	Wood
Harrison	
Hays	
Hill	
Jack	
Jeff Davis	
Kendall	
Kerr	
Kimble	

Argia rhoadsi Calvert / Golden-winged Dancer

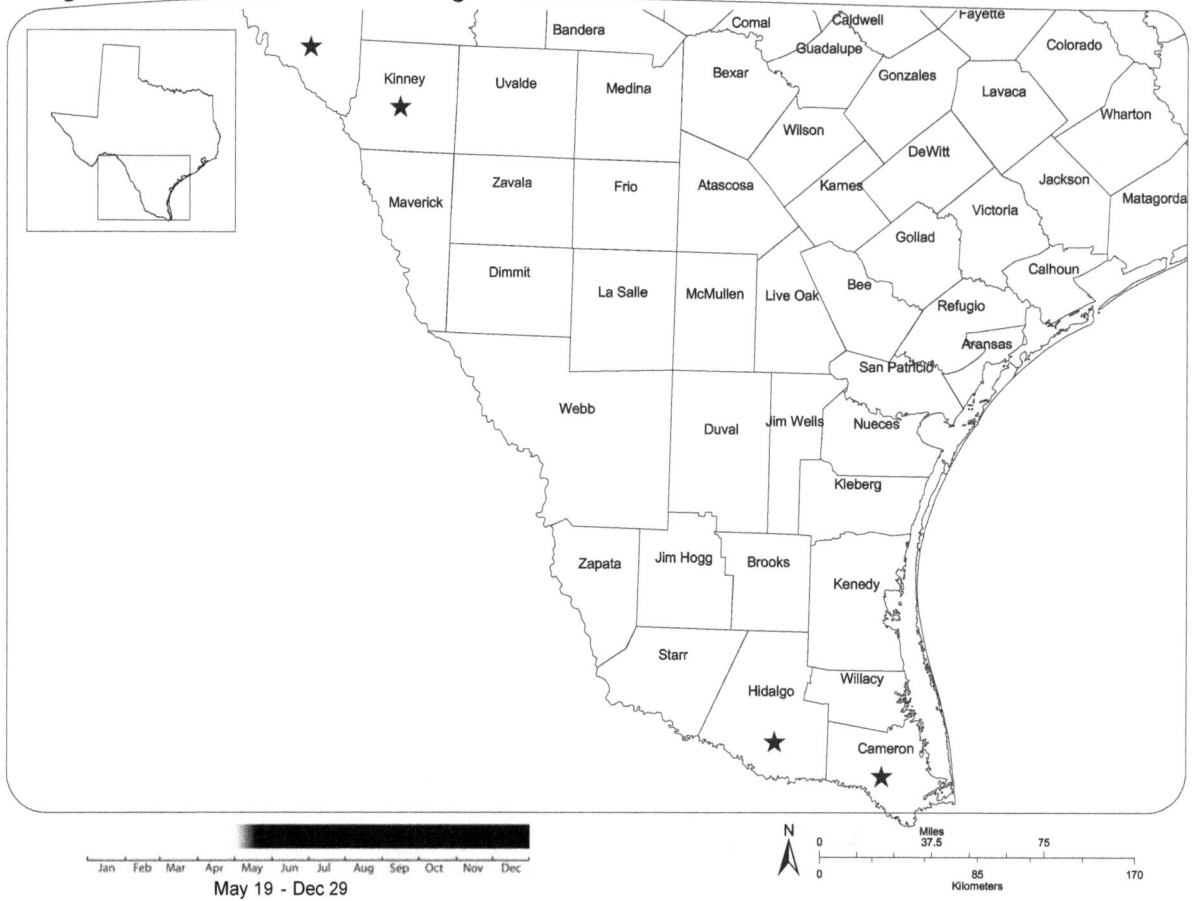

May 19 - Dec 29

HABITAT: Lagoons and pools formed at edges of streams and rivers.

Cameron
Hidalgo
Kinney
Val Verde

Argia sedula (Hagen) / Blue-ringed Dancer

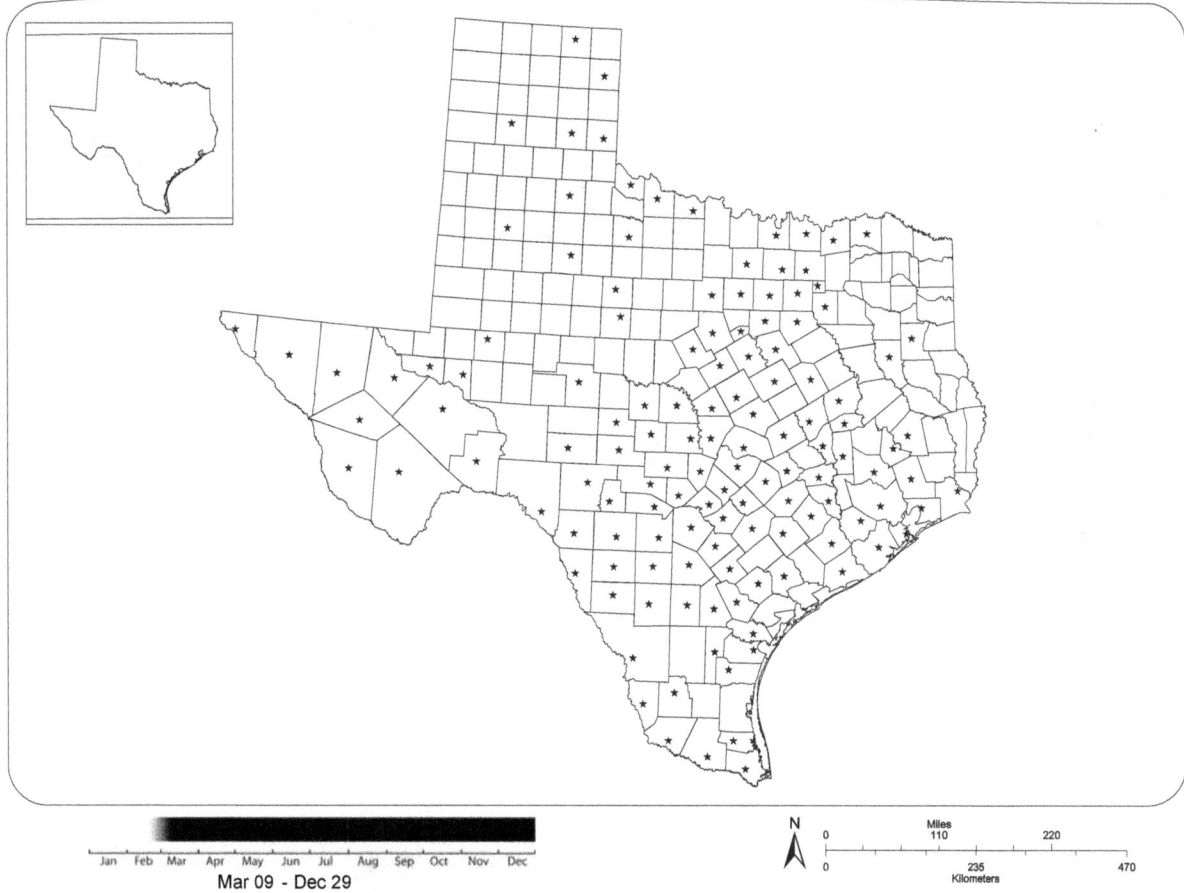

Mar 09 - Dec 29

HABITAT: Lakes, ditches, streams and rivers with gentle current and dense vegetation.

Atascosa	Dallas	Hudspeth	Lubbock	Rusk	Zavala
Austin	Denton	Jeff Davis	Madison	San Jacinto	
Bandera	Dimmit	Jefferson	Mason	San Patricio	
Bastrop	Donley	Jim Hogg	Matagorda	San Saba	
Bee	Edwards	Jim Wells	Maverick	Somervell	
Bell	El Paso	Johnson	McCulloch	Starr	
Bexar	Ellis	Jones	McLennan	Sutton	
Blanco	Erath	Karnes	McMullen	Tarrant	
Bosque	Fannin	Kaufman	Medina	Taylor	
Brazoria	Fayette	Kendall	Menard	Terrell	
Brazos	Fort Bend	Kent	Midland	Tom Green	
Brewster	Frio	Kerr	Milam	Travis	
Burnet	Galveston	Kimble	Montgomery	Uvalde	
Caldwell	Gillespie	Kinney	Motley	Val Verde	
Cameron	Goliad	Kleberg	Nueces	Victoria	
Chambers	Gonzales	Knox	Ochiltree	Ward	
Cherokee	Grayson	La Salle	Palo Pinto	Washington	
Collin	Grimes	Lamar	Parker	Webb	
Collingsworth	Guadalupe	Lampasas	Pecos	Wharton	
Colorado	Hamilton	Lavaca	Polk	Wichita	
Comal	Hardeman	Lee	Presidio	Wilbarger	
Comanche	Harris	Leon	Randall	Willacy	
Cooke	Hays	Liberty	Real	Williamson	
Coryell	Hemphill	Limestone	Reeves	Wilson	
Crane	Hidalgo	Live Oak	Robertson	Wise	
Culberson	Hill	Llano	Rockwall	Zapata	

Argia tezpi Calvert / Tezpi Dancer

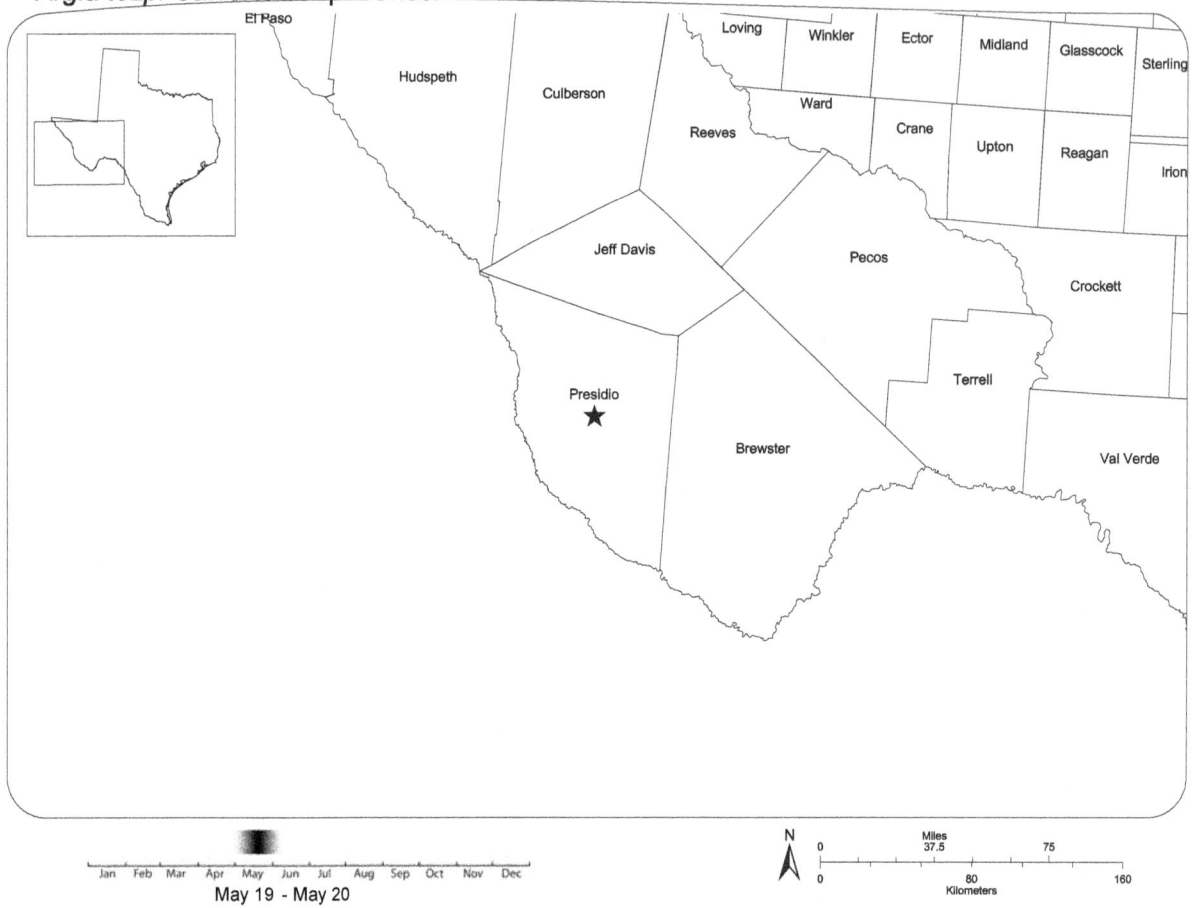

May 19 - May 20

HABITAT: Streams and rivers of the arid southwest.

Presidio

Argia tibialis (Rambur) / Blue-tipped Dancer

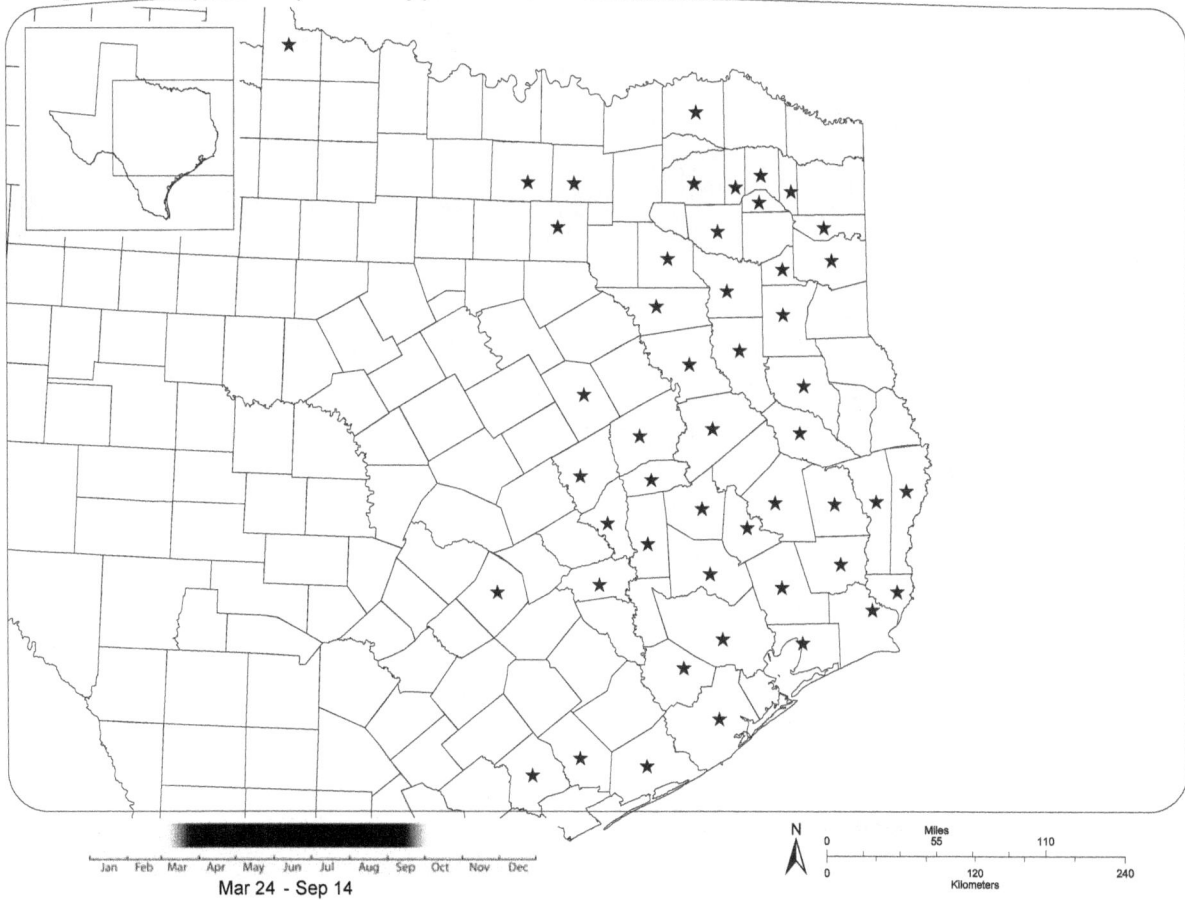

Mar 24 - Sep 14

HABITAT: Streams and rivers of various flows, also sloughs.

Anderson
Angelina
Bastrop
Brazoria
Brazos
Camp
Chambers
Cherokee
Collin
Dallas
Denton
Fort Bend
Franklin
Gregg
Grimes
Hardin
Harris
Harrison
Henderson
Hopkins
Houston
Jackson
Jasper
Jefferson
Lamar
Leon

Liberty
Limestone
Madison
Marion
Matagorda
Montgomery
Morris
Nacogdoches
Newton
Orange
Polk
Robertson
Rusk
San Jacinto
Smith
Titus
Tyler
Van Zandt
Victoria
Walker
Washington
Wilbarger
Wood

Argia translata Hagen in Selys / Dusky Dancer

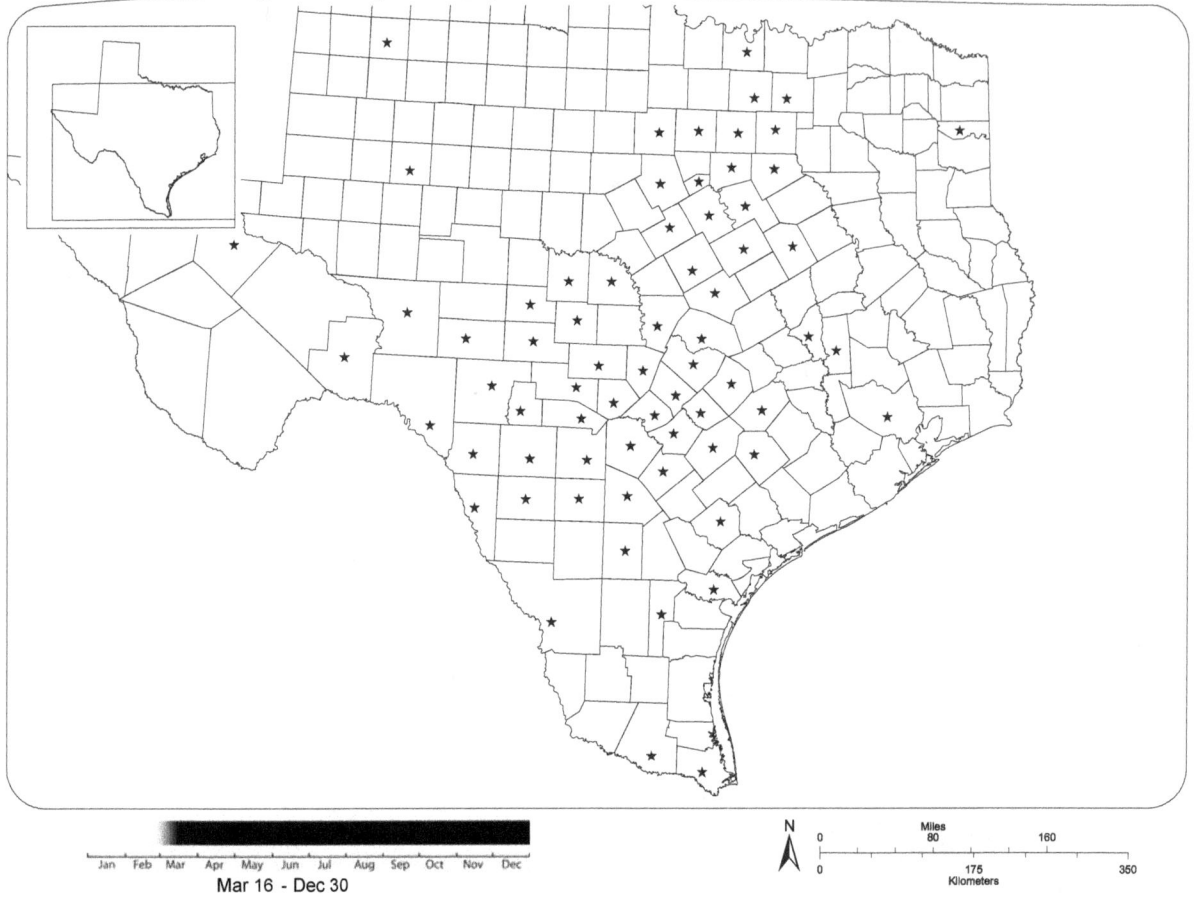

Mar 16 - Dec 30

HABITAT: Streams and rivers generally with a lot of exposure to sun and only moderate vegetation.

Atascosa	Grimes	Parker
Bandera	Guadalupe	Real
Bastrop	Hamilton	Reeves
Bell	Harris	San Patricio
Bexar	Hays	San Saba
Blanco	Hidalgo	Somervell
Bosque	Hill	Sutton
Brazos	Howard	Tarrant
Burnet	Jim Wells	Terrell
Caldwell	Johnson	Travis
Cameron	Kendall	Uvalde
Collin	Kerr	Val Verde
Comal	Kimble	Webb
Cooke	Kinney	Williamson
Coryell	Lavaca	Wilson
Crockett	Limestone	Zavala
Dallas	Lubbock	
Denton	Marion	
Edwards	Mason	
Ellis	Maverick	
Erath	McCulloch	
Fayette	McLennan	
Frio	McMullen	
Gillespie	Medina	
Goliad	Menard	
Gonzales	Palo Pinto	

Enallagma antennatum (Say) / Rainbow Bluet

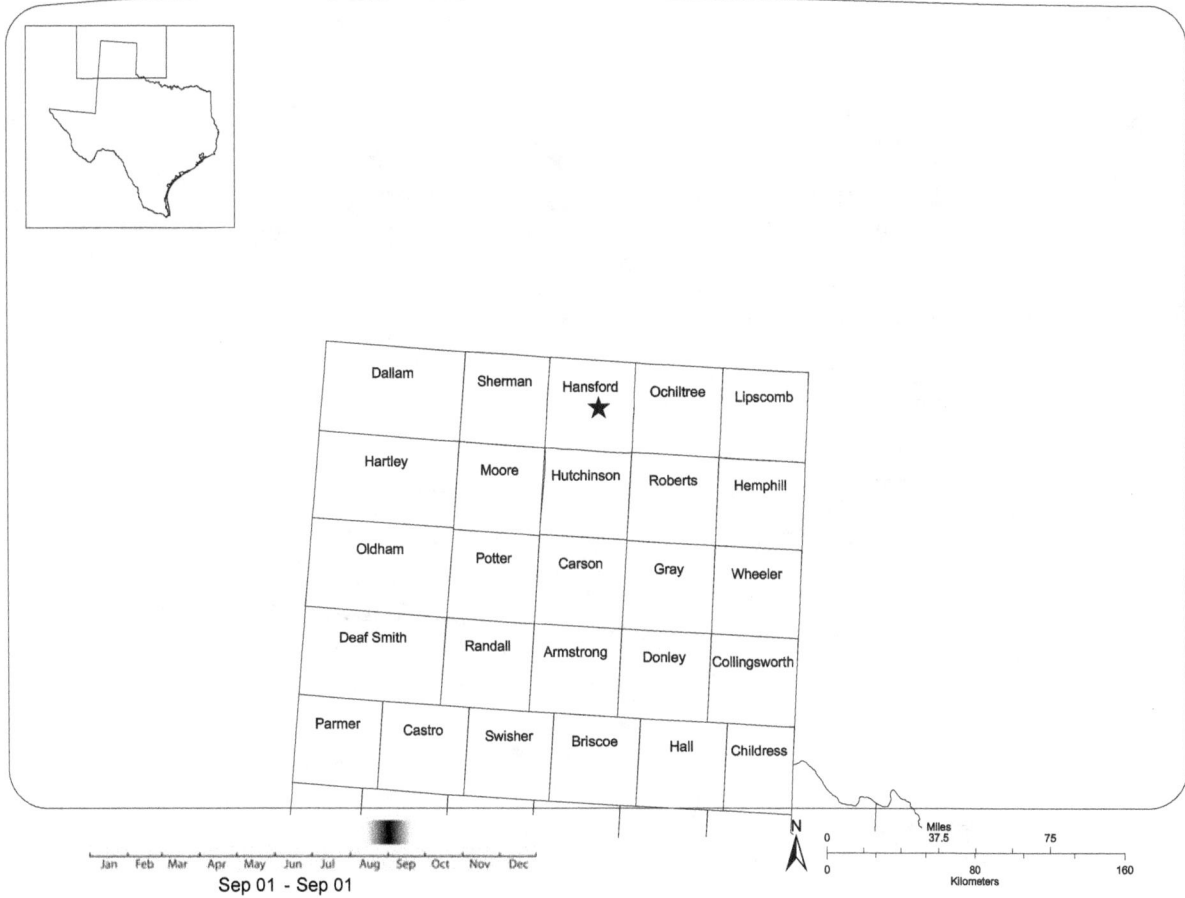

Dallam	Sherman	Hansford ★	Ochiltree	Lipscomb
Hartley	Moore	Hutchinson	Roberts	Hemphill
Oldham	Potter	Carson	Gray	Wheeler
Deaf Smith	Randall	Armstrong	Donley	Collingsworth
Parmer	Castro	Swisher	Briscoe	Hall	Childress

Jan Feb Mar Apr May Jun Jul Aug Sep Oct Nov Dec

Sep 01 - Sep 01

N

Miles
0   37.5   75

0   80   160
Kilometers

HABITAT: Slow streams, lakes, gravel and borrow pits.

Hansford

Enallagma aspersum (Hagen) / Azure Bluet

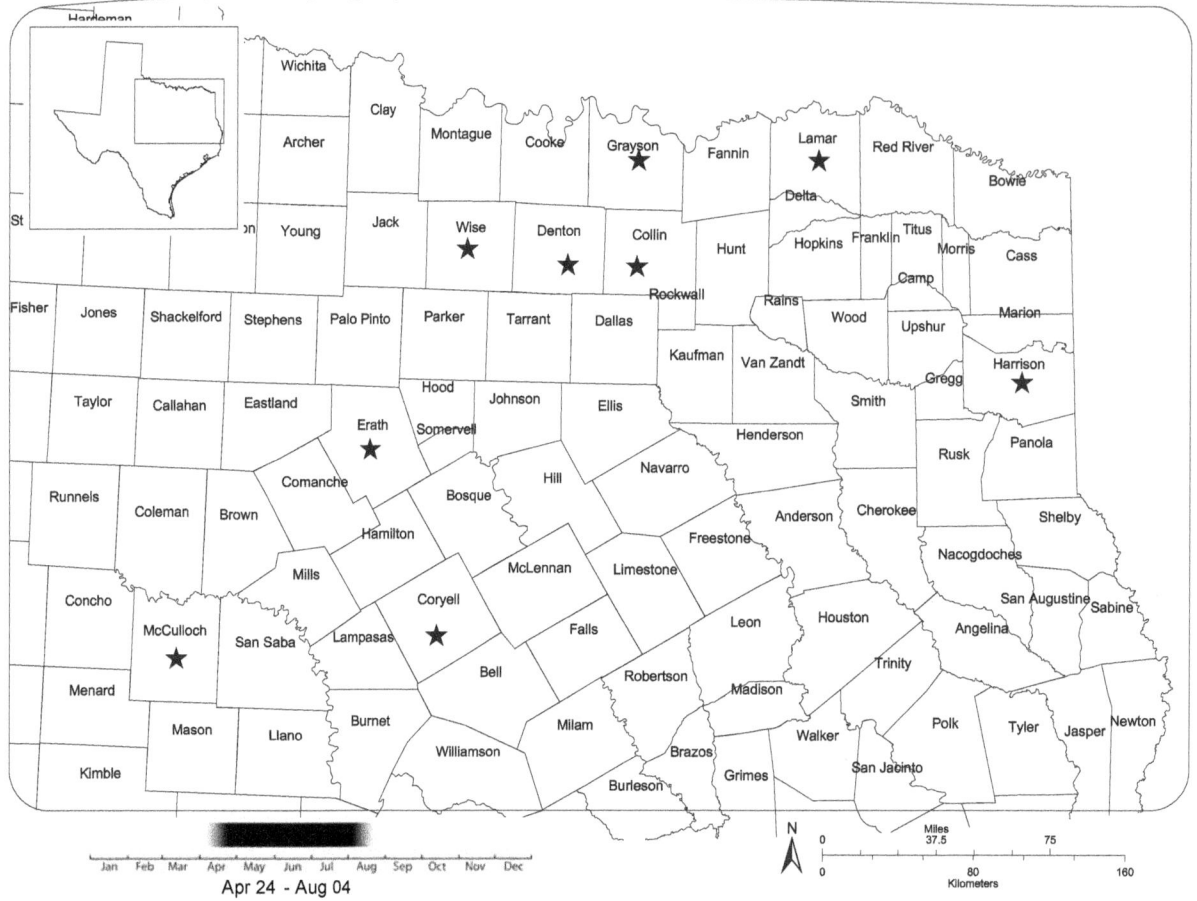

Apr 24 - Aug 04

HABITAT: Fishless lakes and semipermanent ponds and bogs.

Collin
Coryell
Denton
Erath
Grayson
Harrison
Lamar
McCulloch
Wise

Enallagma basidens Calvert / Double-striped Bluet

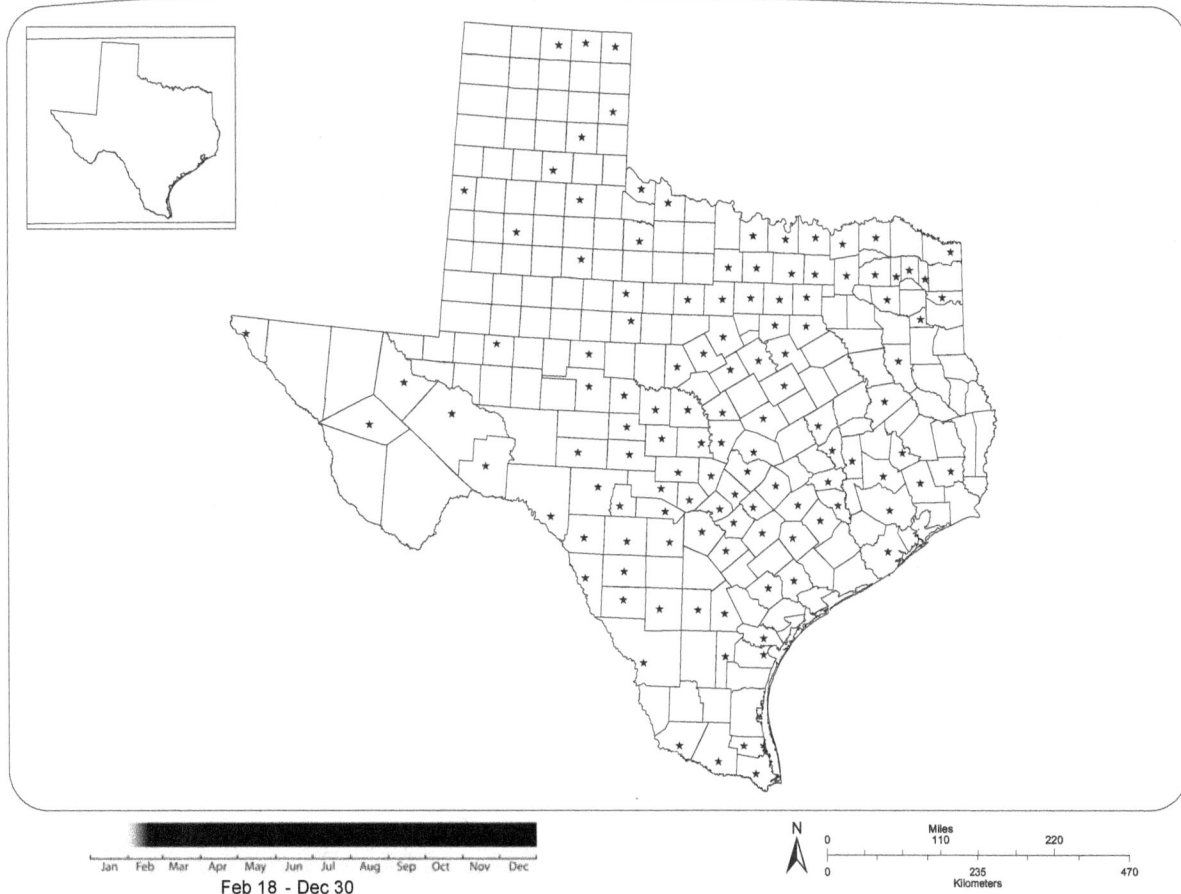

Feb 18 - Dec 30

HABITAT: Various permanent and semipermanent ponds, lakes and reservoirs as well as slow reaches of streams and rivers.

Austin	Dimmit	Hunt	McMullen	Tom Green
Bailey	Donley	Jack	Medina	Travis
Bandera	Edwards	Jeff Davis	Menard	Uvalde
Bastrop	El Paso	Jim Wells	Midland	Val Verde
Bell	Ellis	Johnson	Montague	Victoria
Bexar	Erath	Jones	Montgomery	Washington
Blanco	Fannin	Kendall	Morris	Webb
Bosque	Fayette	Kent	Motley	Wheeler
Bowie	Franklin	Kerr	Nueces	Wilbarger
Brazoria	Gillespie	Kimble	Ochiltree	Willacy
Brazos	Goliad	Kinney	Palo Pinto	Williamson
Briscoe	Gonzales	Knox	Parker	Wilson
Brown	Grayson	La Salle	Pecos	Wise
Burnet	Gregg	Lamar	Real	Wood
Caldwell	Grimes	Lampasas	Reeves	Zavala
Cameron	Guadalupe	Lavaca	Robertson	
Cherokee	Hamilton	Liberty	San Jacinto	
Coke	Hansford	Lipscomb	San Patricio	
Collin	Hardeman	Live Oak	San Saba	
Colorado	Hardin	Llano	Starr	
Comal	Harris	Lubbock	Stephens	
Comanche	Hays	Marion	Sutton	
Concho	Hidalgo	Mason	Tarrant	
Cooke	Hill	Maverick	Taylor	
Dallas	Hopkins	McCulloch	Terrell	
Denton	Houston	McLennan	Titus	

Enallagma carunculatum Morse / Tule Bluet

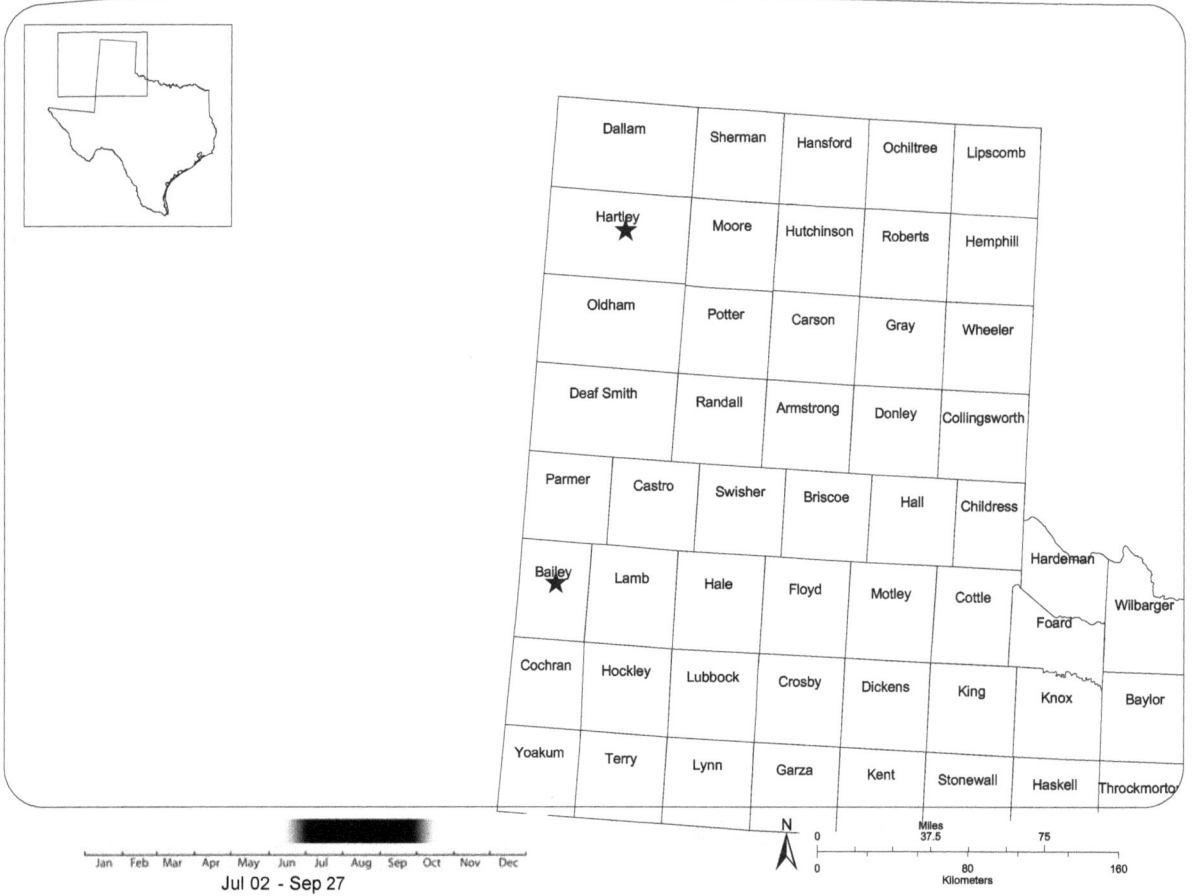

Dallam	Sherman	Hansford	Ochiltree	Lipscomb
Hartley ★	Moore	Hutchinson	Roberts	Hemphill
Oldham	Potter	Carson	Gray	Wheeler
Deaf Smith	Randall	Armstrong	Donley	Collingsworth

Parmer, Castro, Swisher, Briscoe, Hall, Childress, Bailey ★, Lamb, Hale, Floyd, Motley, Cottle, Hardeman, Foard, Wilbarger, Cochran, Hockley, Lubbock, Crosby, Dickens, King, Knox, Baylor, Yoakum, Terry, Lynn, Garza, Kent, Stonewall, Haskell, Throckmorton

Jan Feb Mar Apr May Jun Jul Aug Sep Oct Nov Dec

Jul 02 - Sep 27

Miles 37.5 75

Kilometers 80 160

HABITAT: Slow reaches of rivers and occasionally lakes and ponds.

Bailey
Hartley

Enallagma civile (Hagen) / Familiar Bluet

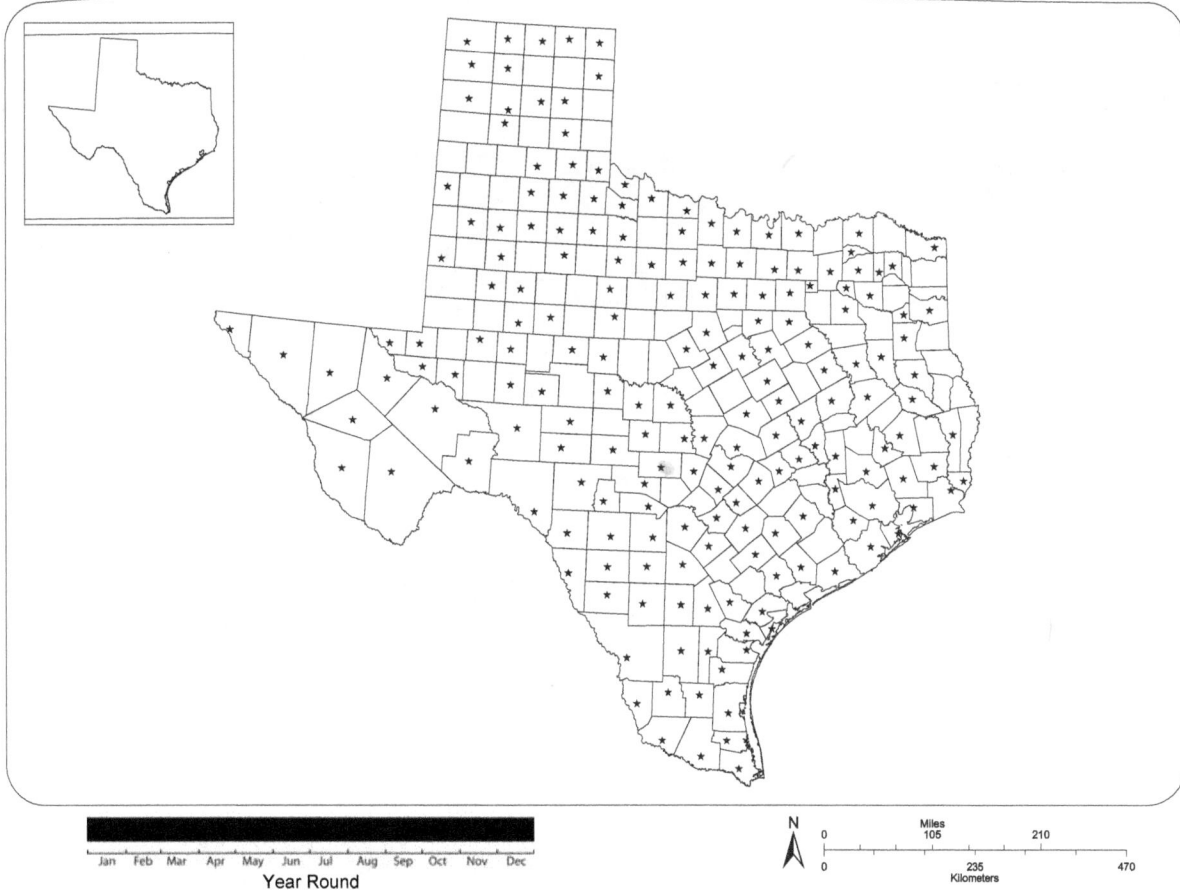

Jan Feb Mar Apr May Jun Jul Aug Sep Oct Nov Dec
Year Round

N
Miles
0   105   210
0   235   470
Kilometers

HABITAT: Ephemeral or permanent ponds and lakes. Also slow flowing streams, irregardless of salinity and vegetation.

Anderson	Cherokee	Ellis	Haskell	Kinney	Montgomery	San Patricio	Wise
Angelina	Childress	Erath	Hays	Kleberg	Moore	San Saba	Wood
Aransas	Clay	Falls	Hemphill	Knox	Motley	Schleicher	Yoakum
Archer	Coke	Floyd	Hidalgo	La Salle	Nacogdoches	Sherman	Young
Atascosa	Collin	Foard	Hill	Lamar	Navarro	Starr	Zapata
Bailey	Colorado	Fort Bend	Hockley	Lavaca	Nueces	Stephens	Zavala
Bandera	Comanche	Franklin	Hopkins	Lee	Ochiltree	Sutton	
Bastrop	Concho	Freestone	Houston	Leon	Oldham	Tarrant	
Bee	Cooke	Frio	Howard	Liberty	Orange	Taylor	
Bell	Cottle	Galveston	Hudspeth	Lipscomb	Palo Pinto	Terrell	
Bexar	Crane	Gillespie	Hunt	Live Oak	Parker	Throckmorton	
Blanco	Crockett	Glasscock	Irion	Llano	Pecos	Titus	
Borden	Crosby	Gonzales	Jack	Loving	Polk	Travis	
Bosque	Culberson	Gray	Jackson	Lubbock	Potter	Uvalde	
Bowie	Dallam	Grayson	Jasper	Lynn	Presidio	Val Verde	
Brazoria	Dallas	Gregg	Jeff Davis	Mason	Rains	Van Zandt	
Brazos	Dawson	Grimes	Jefferson	Matagorda	Randall	Victoria	
Brewster	Delta	Guadalupe	Jim Hogg	Maverick	Reagan	Waller	
Briscoe	Denton	Hall	Jim Wells	McCulloch	Real	Ward	
Brooks	DeWitt	Hamilton	Johnson	McLennan	Reeves	Webb	
Burleson	Dickens	Hansford	Jones	McMullen	Refugio	Wichita	
Burnet	Dimmit	Hardeman	Kenedy	Medina	Robertson	Wilbarger	
Caldwell	Donley	Hardin	Kent	Midland	Rockwall	Willacy	
Cameron	Duval	Harris	Kerr	Milam	Runnels	Williamson	
Carson	Edwards	Harrison	Kimble	Mitchell	Rusk	Wilson	
Chambers	El Paso	Hartley	King	Montague	San Jacinto	Winkler	

Enallagma clausum Morse / Alkali Bluet

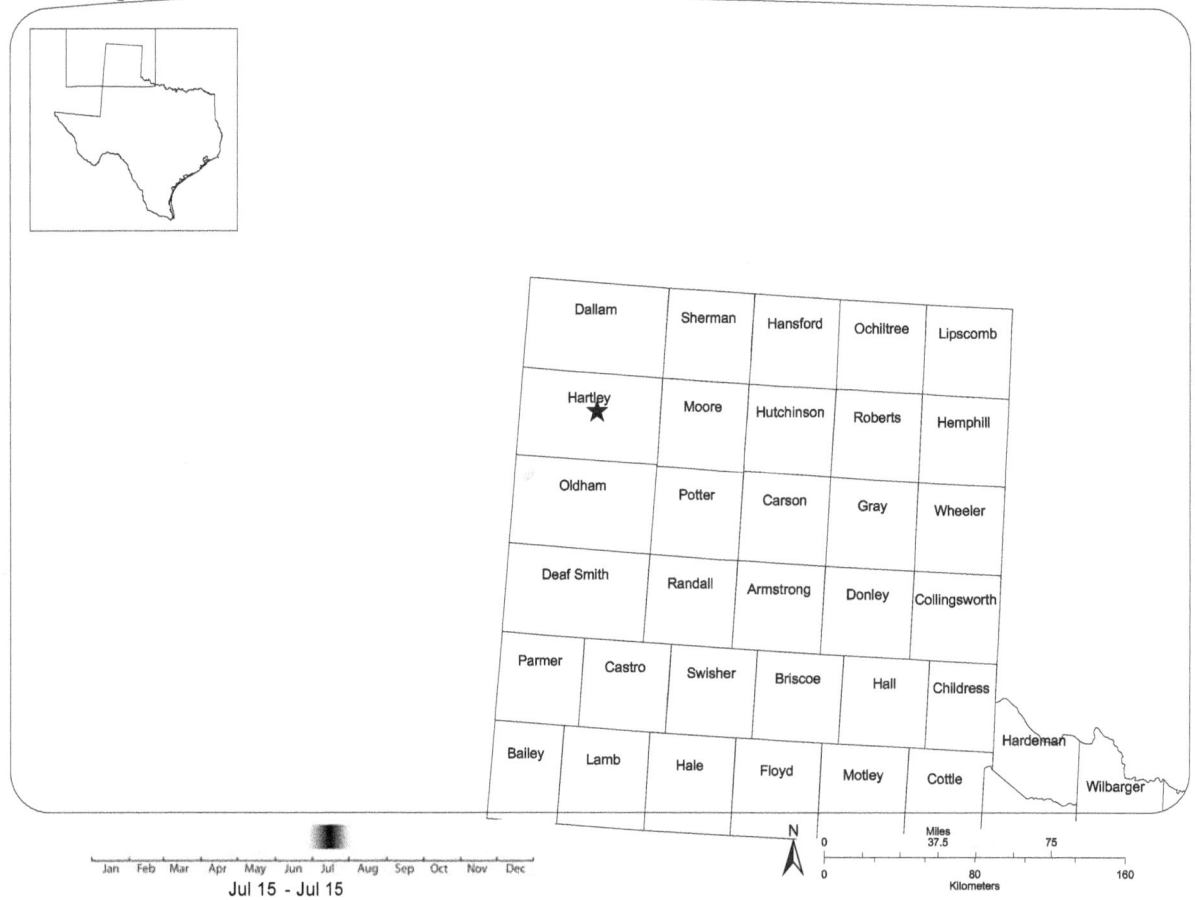

Dallam	Sherman	Hansford	Ochiltree	Lipscomb
Hartley ★	Moore	Hutchinson	Roberts	Hemphill
Oldham	Potter	Carson	Gray	Wheeler
Deaf Smith	Randall	Armstrong	Donley	Collingsworth
Parmer / Castro	Swisher	Briscoe	Hall	Childress
Bailey / Lamb	Hale	Floyd	Motley	Cottle	Hardeman / Wilbarger

Jan Feb Mar Apr May Jun Jul Aug Sep Oct Nov Dec

Jul 15 - Jul 15

N

Miles
37.5   75
0

0   80   160
Kilometers

HABITAT: Ponds and lakes, especially those with saline or alkaline water.

Hartley

Enallagma daeckii (Calvert) / Attenuated Bluet

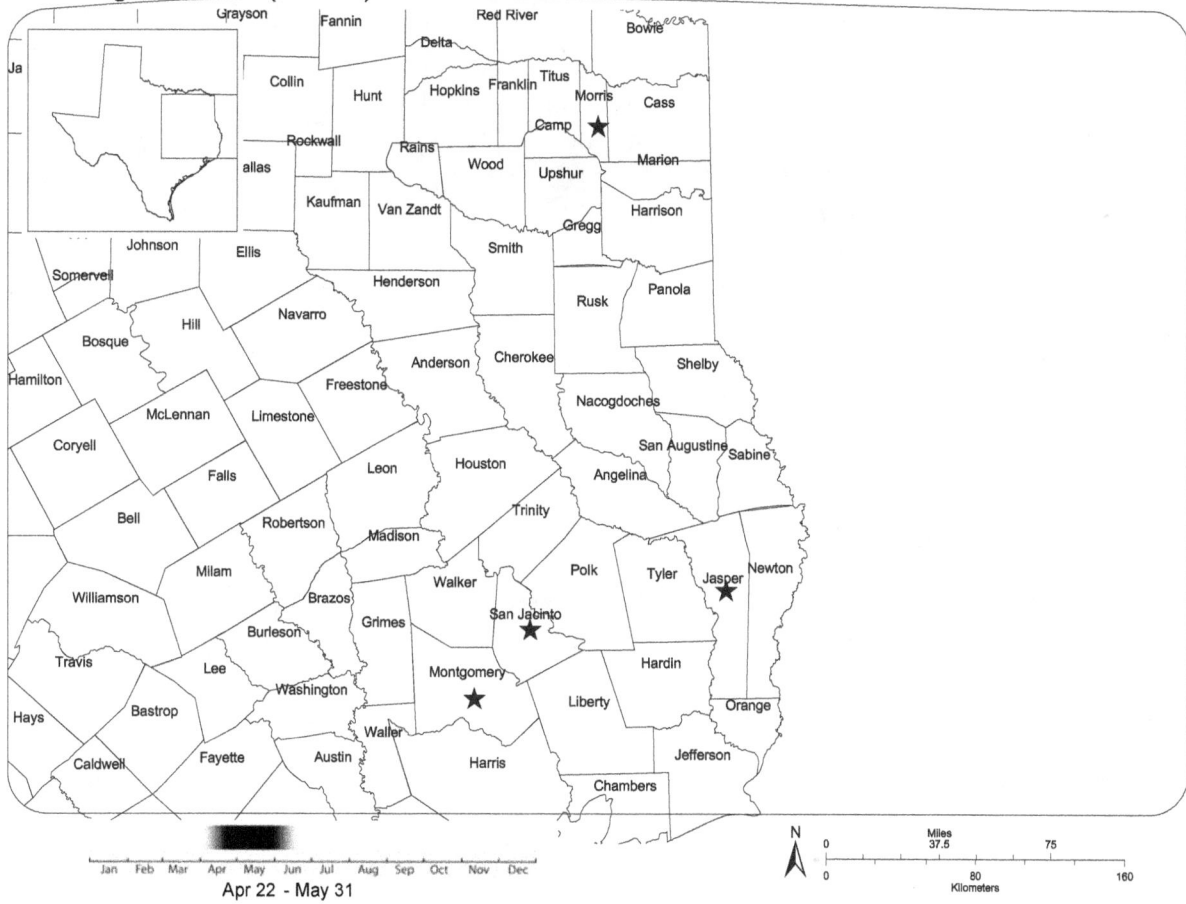

Apr 22 - May 31

HABITAT: Swamp margins and shady, often heavily vegetated pond, lakes and stream backwaters.

Jasper
Montgomery
Morris
San Jacinto

Enallagma divagans Selys / Turquoise Bluet

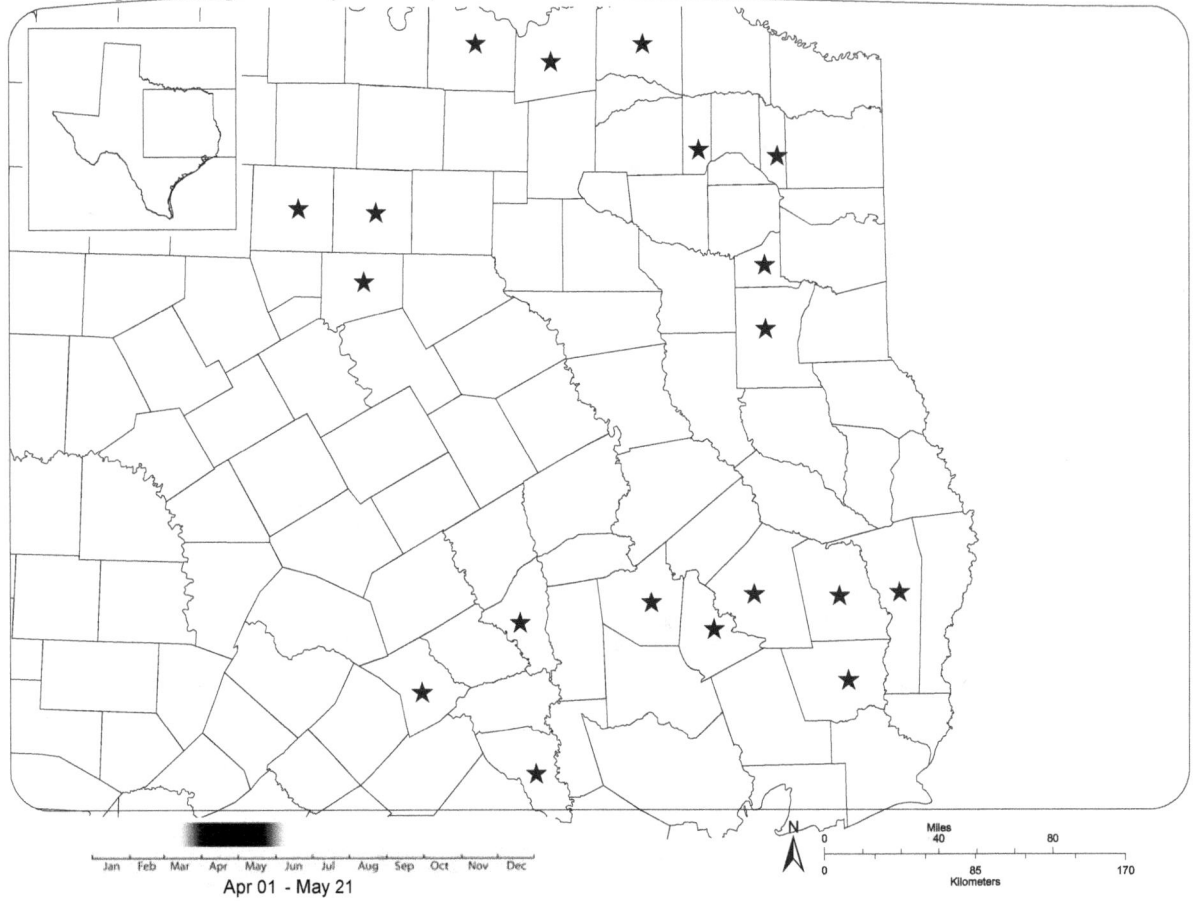

Jan Feb Mar Apr May Jun Jul Aug Sep Oct Nov Dec

Apr 01 - May 21

HABITAT: Shaded sluggish creeks and streams, sloughs or lakes.

Austin
Brazos
Fannin
Franklin
Grayson
Gregg
Hardin
Jasper
Johnson
Lamar
Lee
Morris
Parker
Polk
Rusk
San Jacinto
Tarrant
Tyler
Walker

Enallagma doubledayi (Selys) / Atlantic Bluet

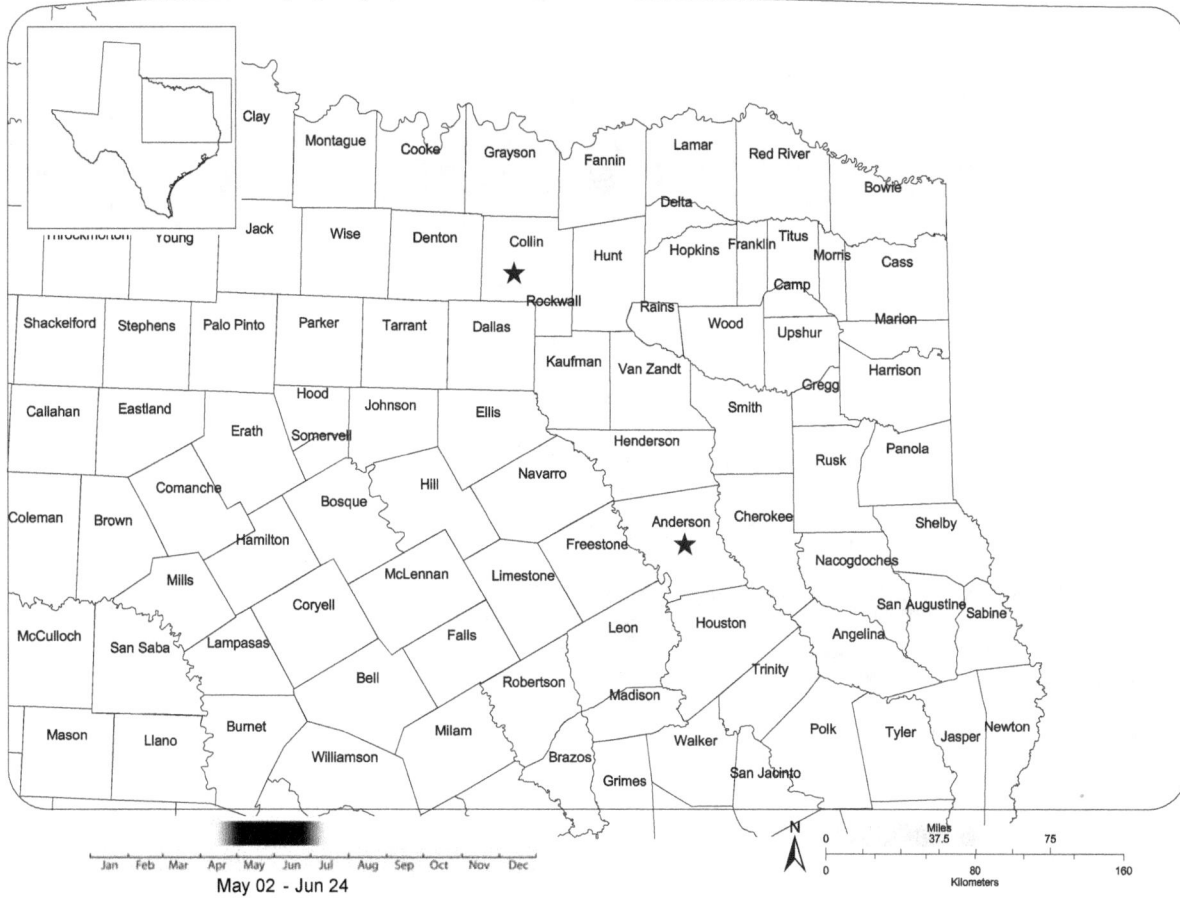

May 02 - Jun 24

HABITAT: Newly formed or ephemeral ponds and lakes, and occasionally sluggish streams.

Anderson
Collin

Enallagma dubium Root / Burgundy Bluet

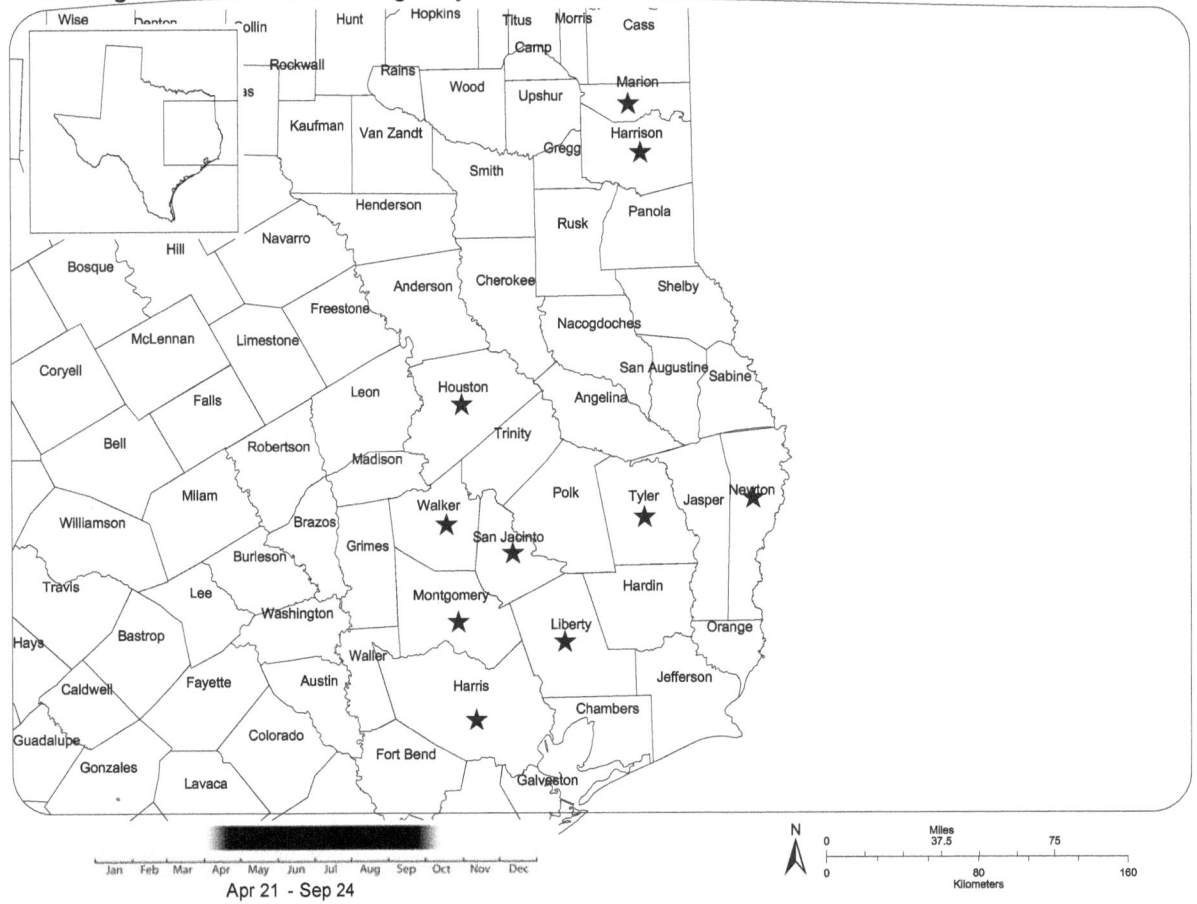

Apr 21 - Sep 24

HABITAT: Heavily vegetated black water ponds, lakes, oxbows, sloughs and slow reaches of streams, often associated with lily pads.

Harris
Harrison
Houston
Liberty
Marion
Montgomery
Newton
San Jacinto
Tyler
Walker

Enallagma durum (Hagen) / Big Bluet

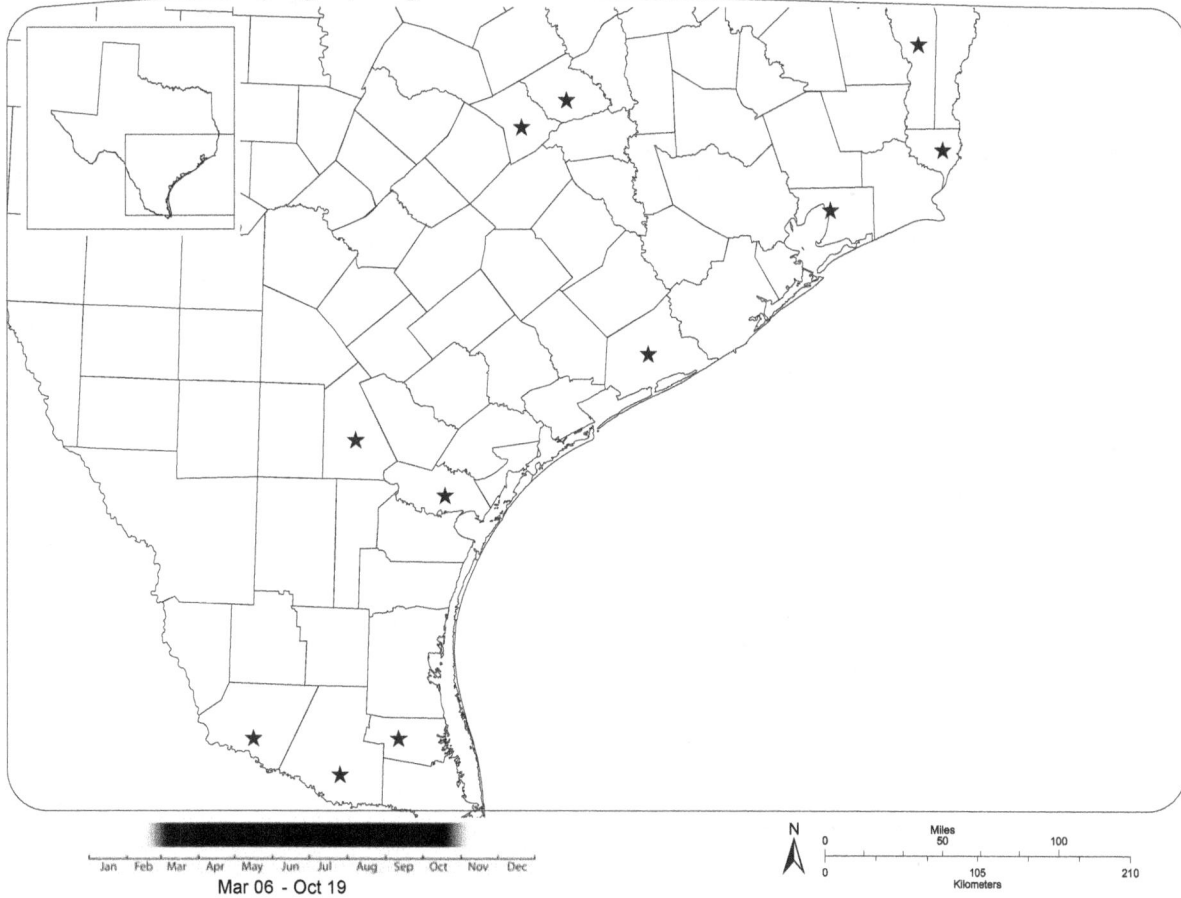

Mar 06 - Oct 19

HABITAT: Along the shores of lakes and rivers, often with brackish water and emergent vegetation.

Burleson
Chambers
Hidalgo
Jasper
Lee
Live Oak
Matagorda
Orange
San Patricio
Starr
Willacy

Enallagma exsulans (Hagen) / Stream Bluet

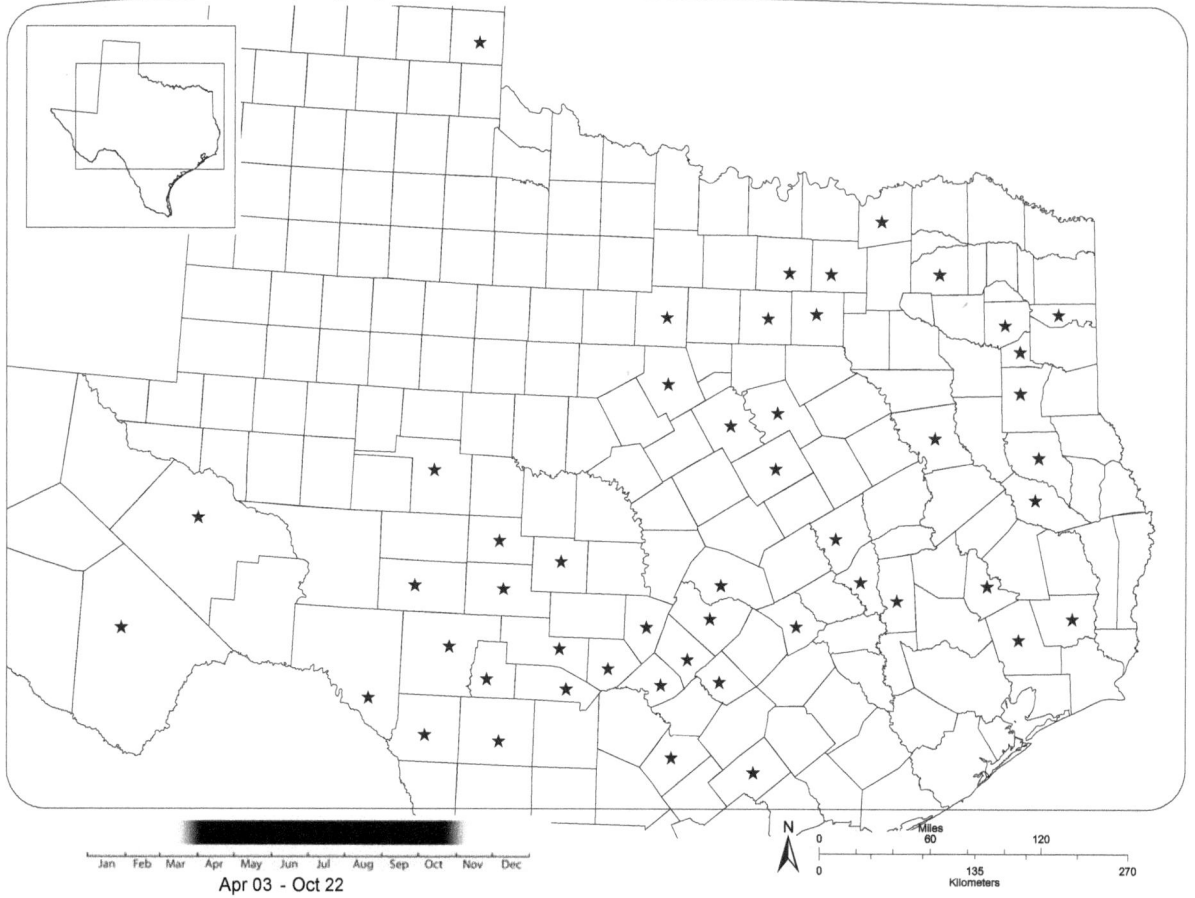

Apr 03 - Oct 22

HABITAT: Common along shores of slow moving streams, rivers and occasionally lakes.

Anderson	Kinney
Angelina	Lee
Bandera	Liberty
Blanco	Marion
Bosque	Mason
Brazos	McLennan
Brewster	Menard
Caldwell	Nacogdoches
Collin	Palo Pinto
Collingsworth	Pecos
Comal	Real
Dallas	Robertson
Denton	Rusk
DeWitt	San Jacinto
Edwards	Sutton
Erath	Tarrant
Fannin	Tom Green
Gregg	Travis
Grimes	Upshur
Hardin	Uvalde
Hays	Val Verde
Hill	Williamson
Hopkins	Wilson
Kendall	
Kerr	
Kimble	

Enallagma geminatum Kellicott / Skimming Bluet

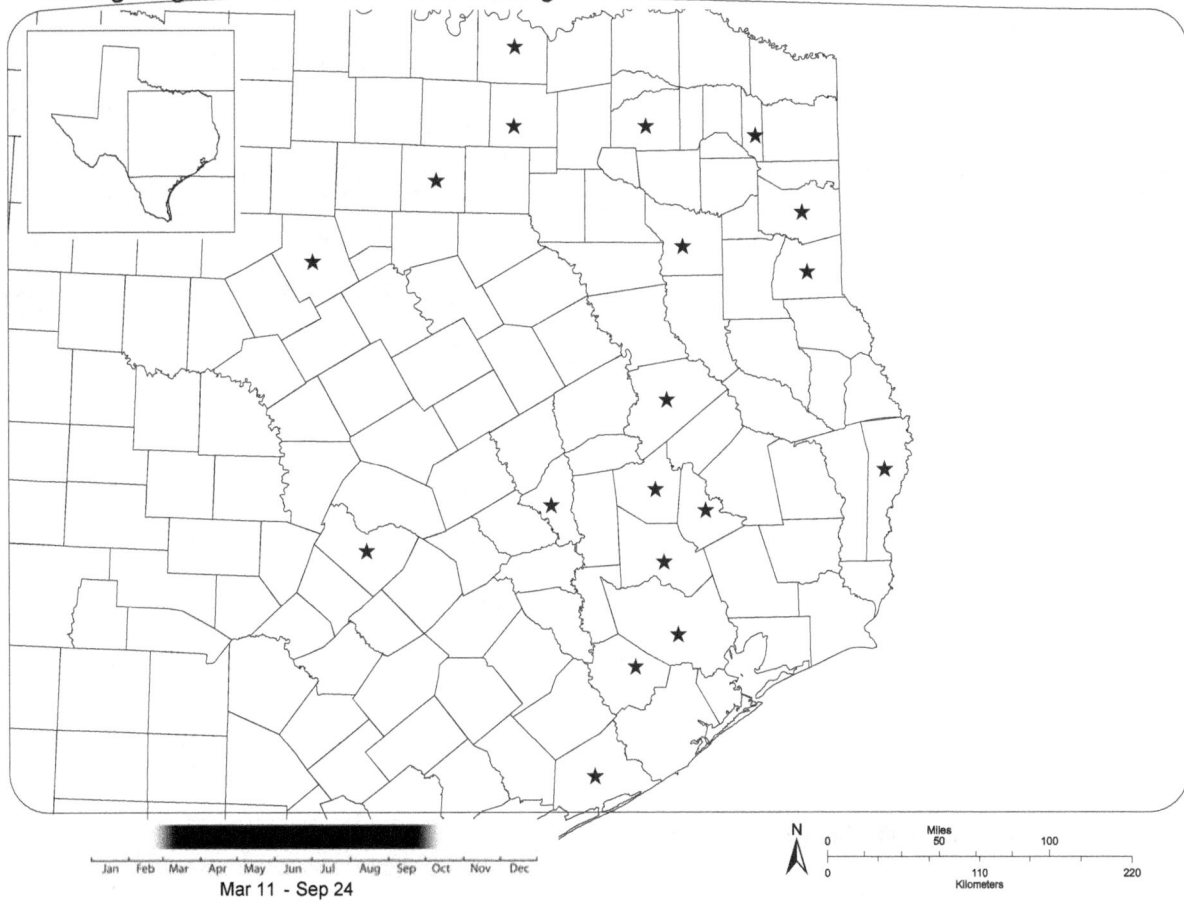

Mar 11 - Sep 24

HABITAT: Prefers open, muddy, heavily vegetated ponds and lakes with fish, and more rarely slow moving streams and swampy, small order streams.

Brazos
Collin
Erath
Fort Bend
Grayson
Harris
Harrison
Hopkins
Houston
Matagorda
Montgomery
Morris
Newton
Panola
San Jacinto
Smith
Tarrant
Travis
Walker

Enallagma novaehispaniae Calvert / Neotropical Bluet

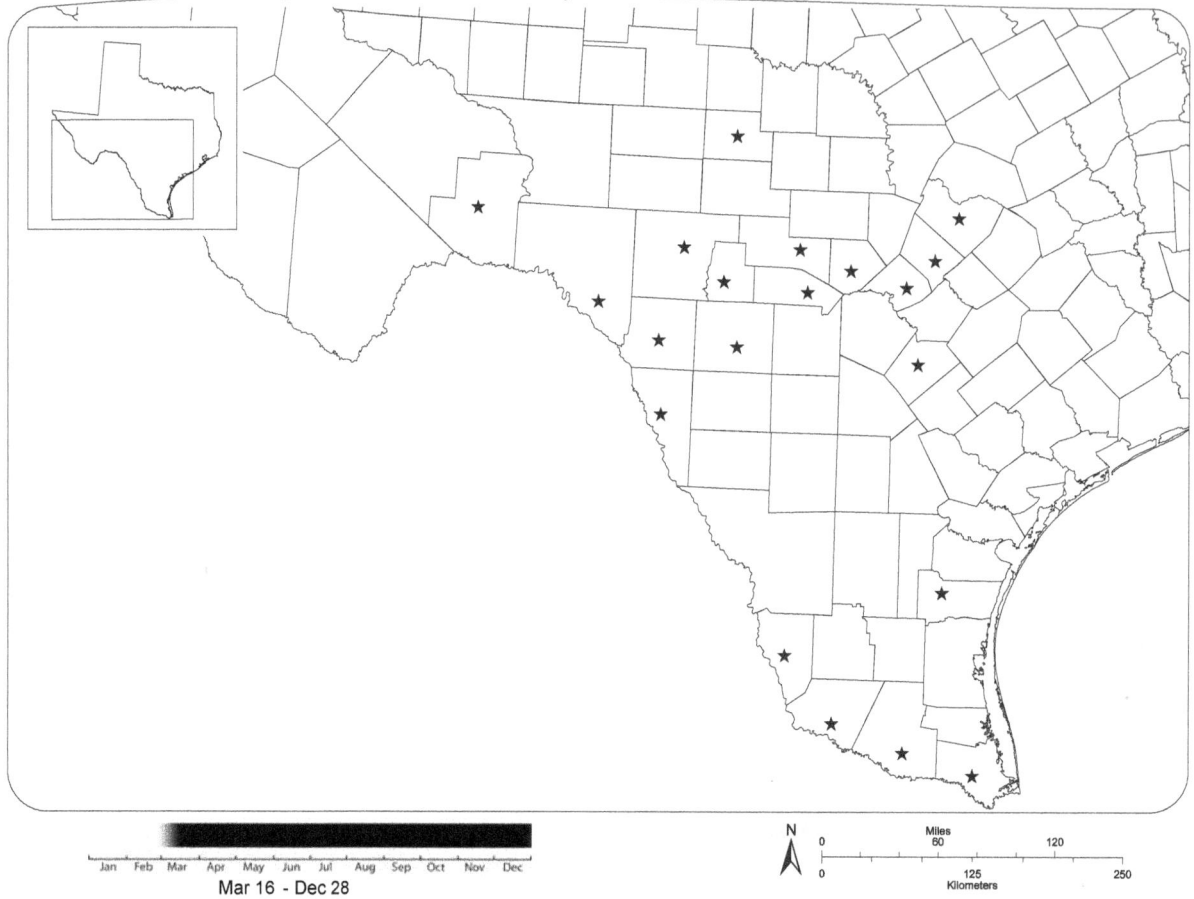

Jan Feb Mar Apr May Jun Jul Aug Sep Oct Nov Dec

Mar 16 - Dec 28

HABITAT: Clear streams and rivers with a strong current.

Bandera
Cameron
Comal
Culberson
Edwards
Hays
Hidalgo
Kendall
Kerr
Kinney
Kleberg
Maverick
Menard
Real
Starr
Terrell
Travis
Uvalde
Val Verde
Wilson
Zapata

Enallagma praevarum (Hagen) / Arroyo Bluet

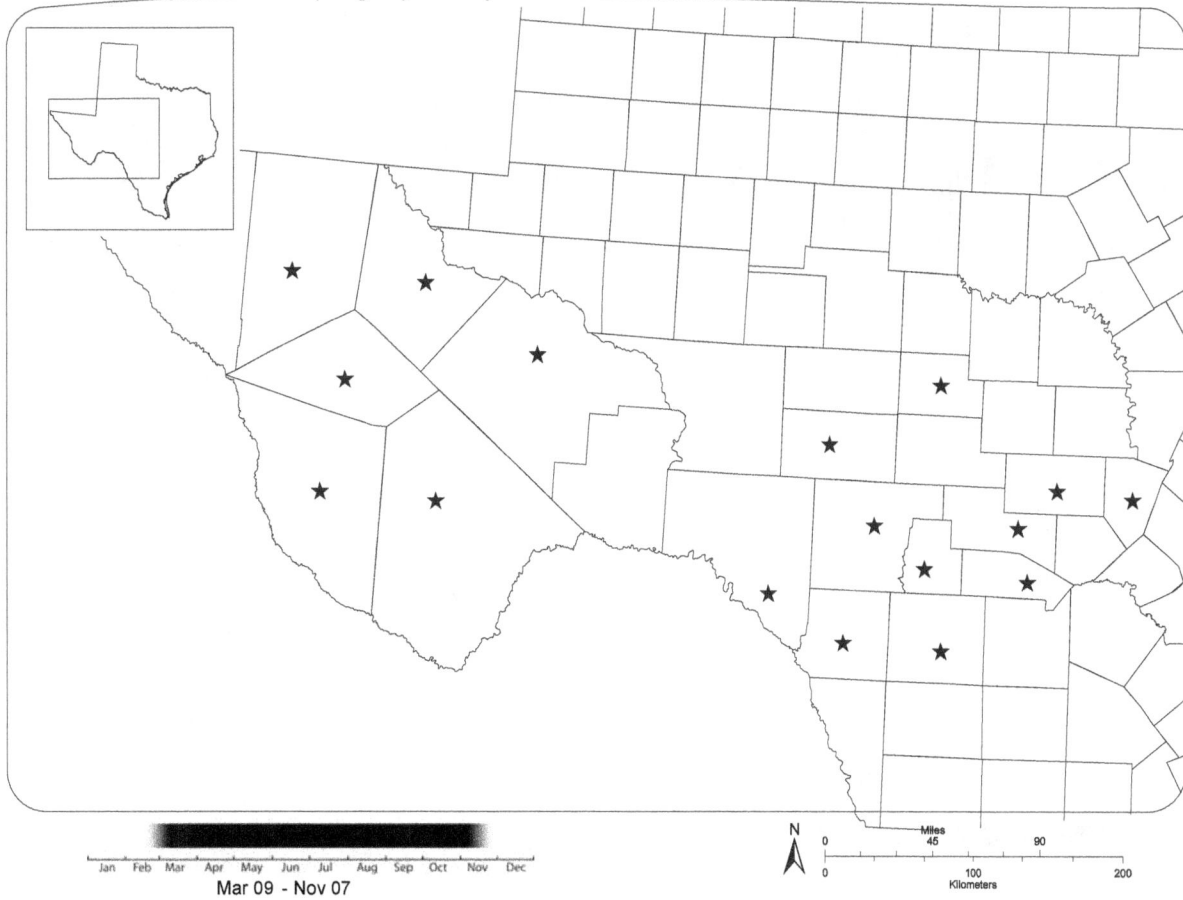

Mar 09 - Nov 07

HABITAT: Common in ponds and slow reaches of streams.

Bandera
Blanco
Brewster
Culberson
Edwards
El Paso
Gillespie
Jeff Davis
Kerr
Kinney
Menard
Pecos
Presidio
Real
Reeves
Sutton
Uvalde
Val Verde

Enallagma signatum (Hagen) / Orange Bluet

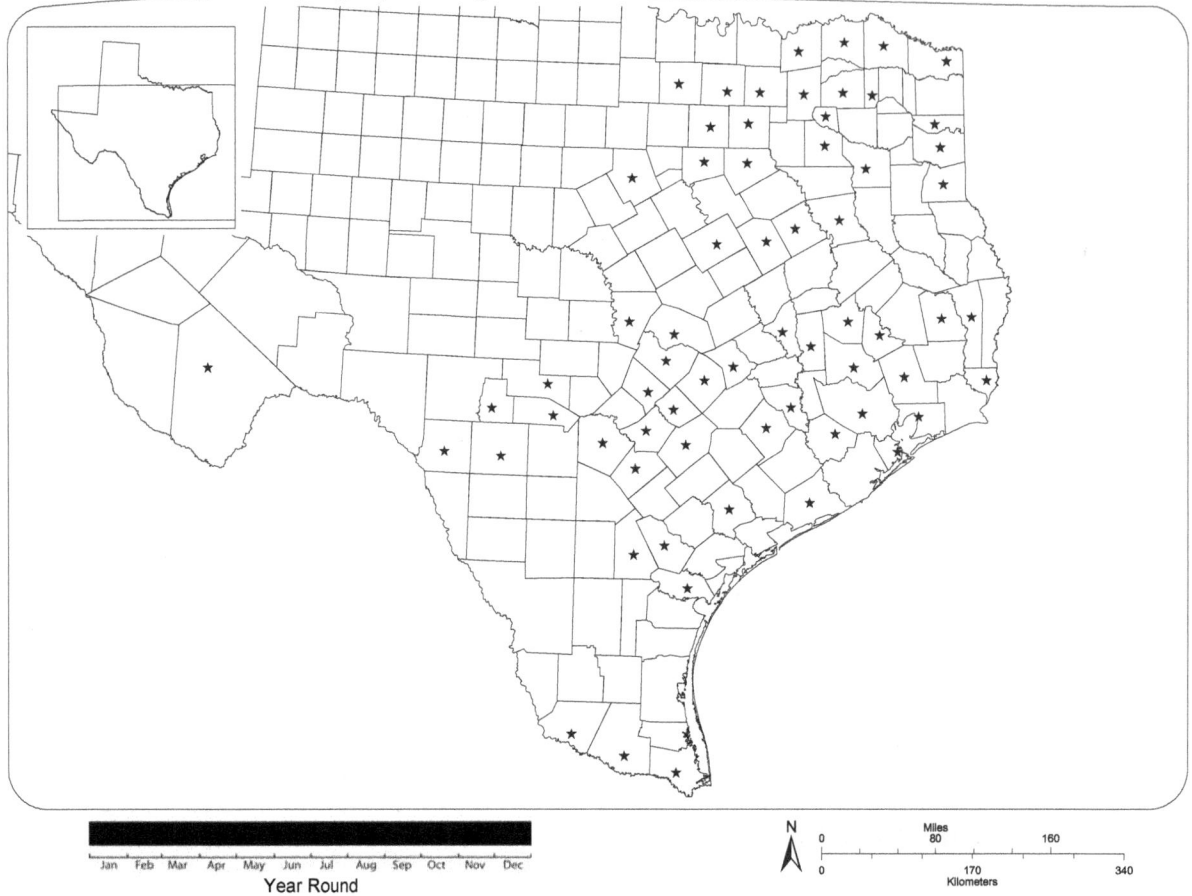

Jan Feb Mar Apr May Jun Jul Aug Sep Oct Nov Dec

Year Round

N

| Miles | | |
| 0 | 80 | 160 |

| Kilometers | | |
| 0 | 170 | 340 |

HABITAT: Various ponds and lakes as well as slow moving streams and rivers

Anderson	Guadalupe	San Patricio
Austin	Harris	Smith
Bandera	Harrison	Starr
Bastrop	Hays	Tarrant
Bee	Hidalgo	Travis
Bexar	Hopkins	Tyler
Bowie	Hunt	Uvalde
Brazos	Jasper	Van Zandt
Brewster	Johnson	Victoria
Burnet	Kerr	Walker
Caldwell	Kinney	Williamson
Cameron	Lamar	Wilson
Chambers	Lee	Wise
Collin	Liberty	
Colorado	Limestone	
Dallas	Live Oak	
Denton	Marion	
Ellis	Matagorda	
Erath	McLennan	
Fannin	Montgomery	
Fort Bend	Orange	
Franklin	Panola	
Freestone	Rains	
Galveston	Real	
Gonzales	Red River	
Grimes	San Jacinto	

Enallagma traviatum westfalli Donnelly / Slender Bluet

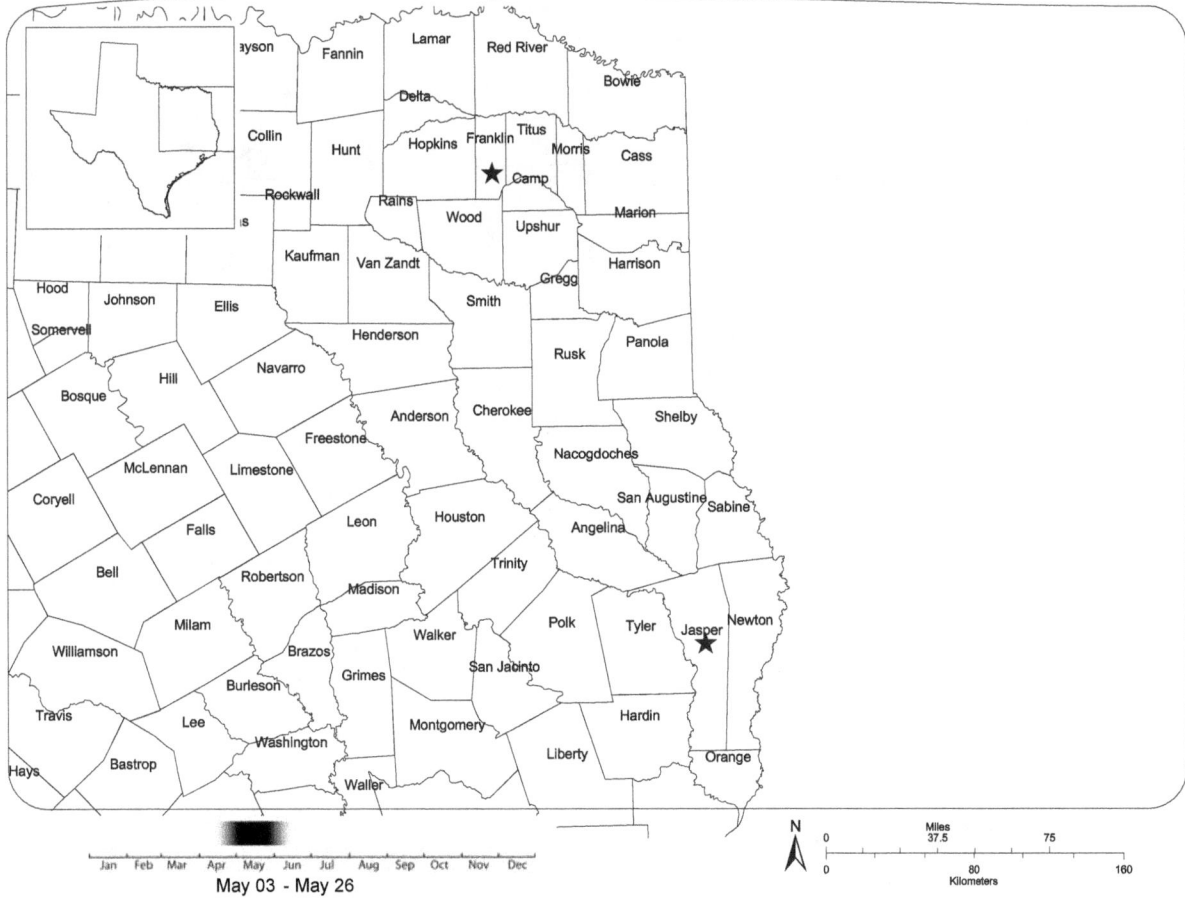

Jan Feb Mar Apr May Jun Jul Aug Sep Oct Nov Dec

May 03 - May 26

HABITAT: Permanent ponds and lakes with from sparse to abundant emergent vegetation.

Franklin
Jasper

Enallagma vesperum Calvert / Vesper Bluet

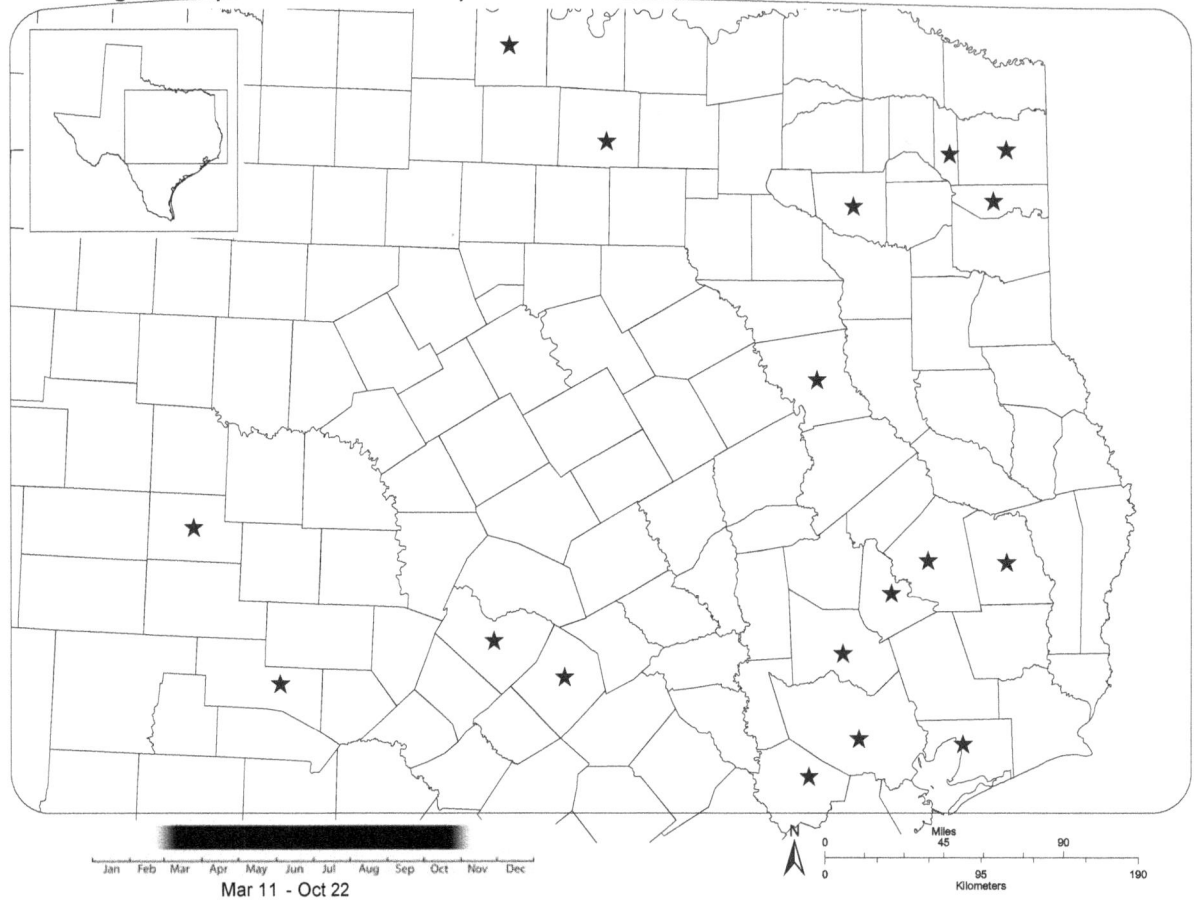

Mar 11 - Oct 22

HABITAT: Most commonly found in heavily vegetated ponds and lakes, but occasionally in slow reaches of streams.

Anderson
Bastrop
Cass
Chambers
Denton
Fort Bend
Harris
Kerr
Marion
Menard
Montague
Montgomery
Morris
Polk
San Jacinto
Travis
Tyler
Wood

Hesperagrion heterodoxum (Selys) / Painted Damsel

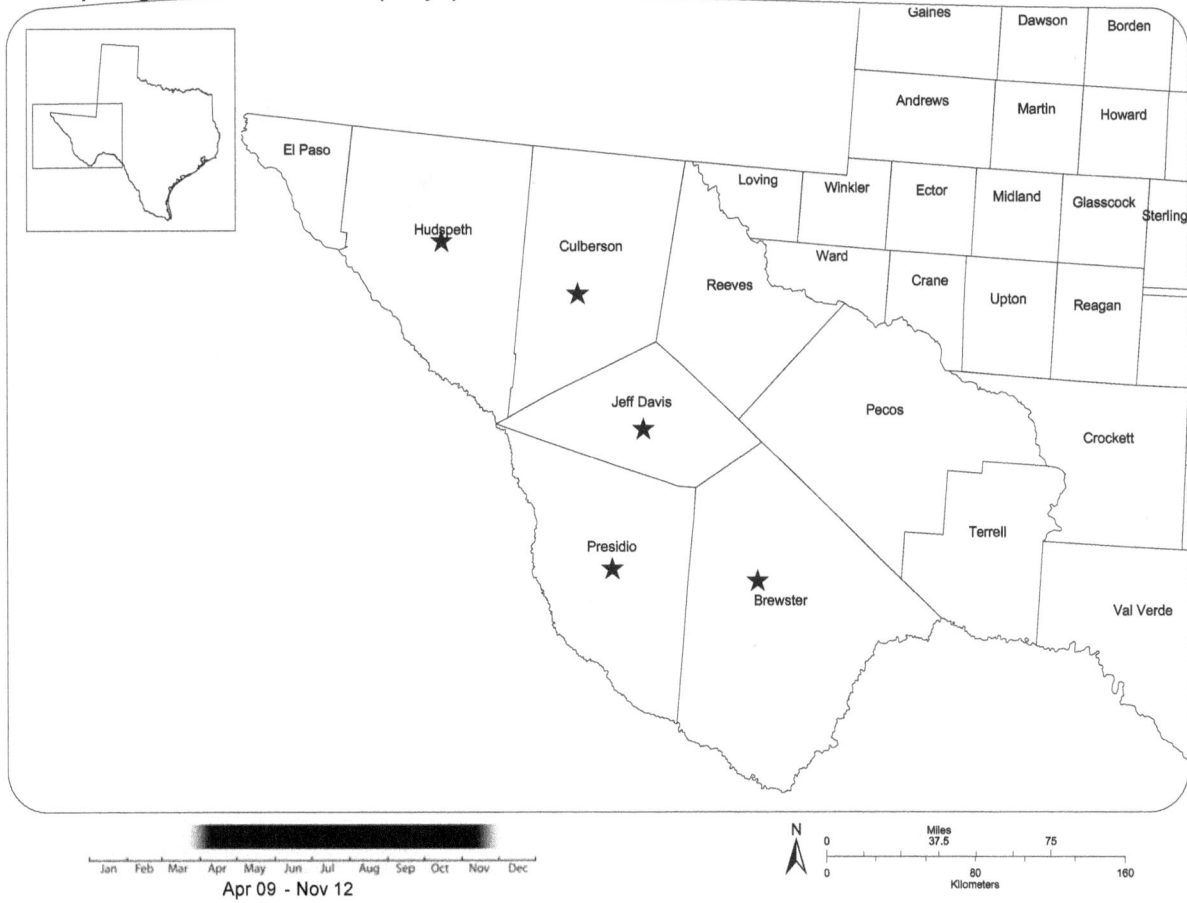

Jan Feb Mar Apr May Jun Jul Aug Sep Oct Nov Dec

Apr 09 - Nov 12

HABITAT: Permanent and ephemeral creeks and streams with moderate emergent vegetation.

Brewster
Culberson
Hudspeth
Jeff Davis
Presidio

Ischnura barberi Currie / Desert Forktail

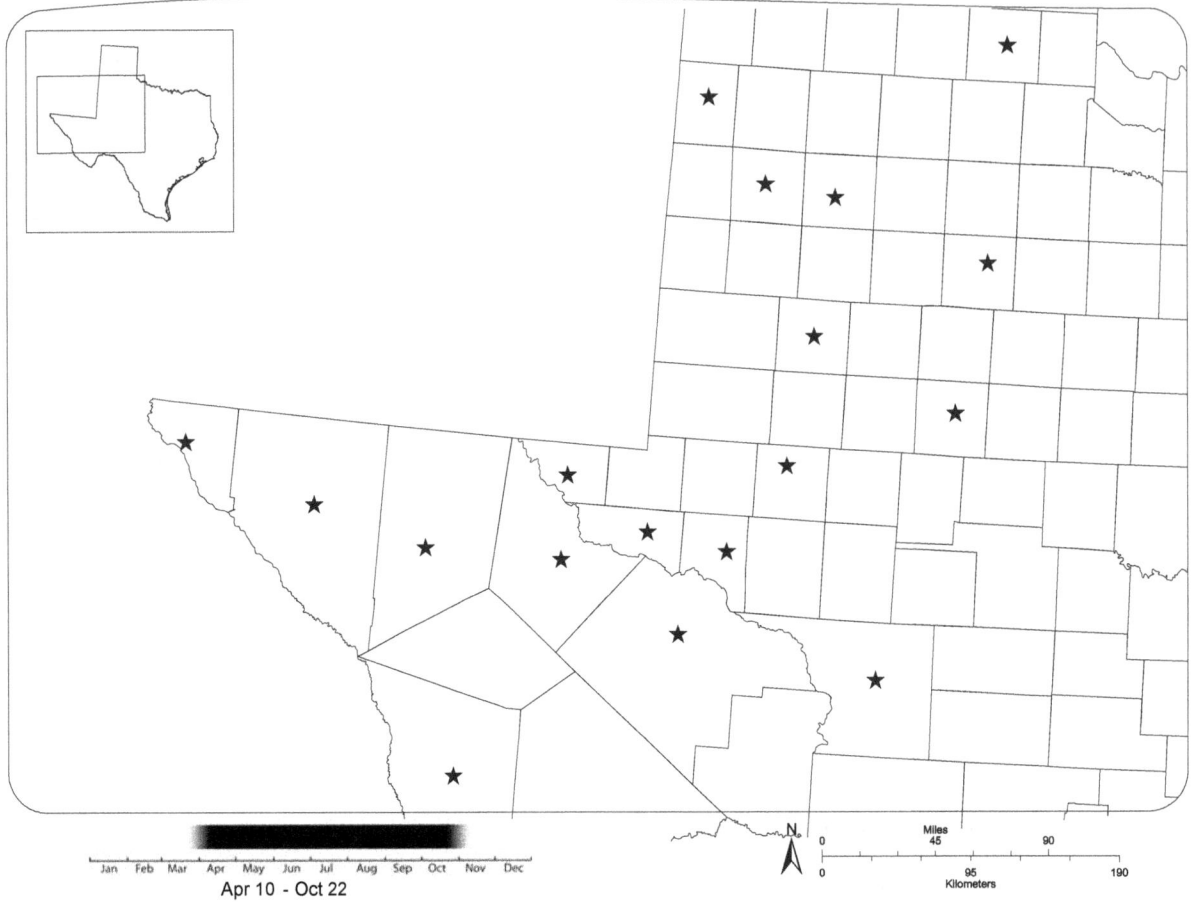

Apr 10 - Oct 22

HABITAT: Alkaline and saline, desert springs, pools, irrigation ditches and canals.

Bailey
Crane
Crockett
Culberson
Dawson
El Paso
Hall
Hockley
Hudspeth
Kent
Loving
Lubbock
Midland
Mitchell
Pecos
Presidio
Reeves
Ward

Ischnura damula Calvert / Plains Forktail

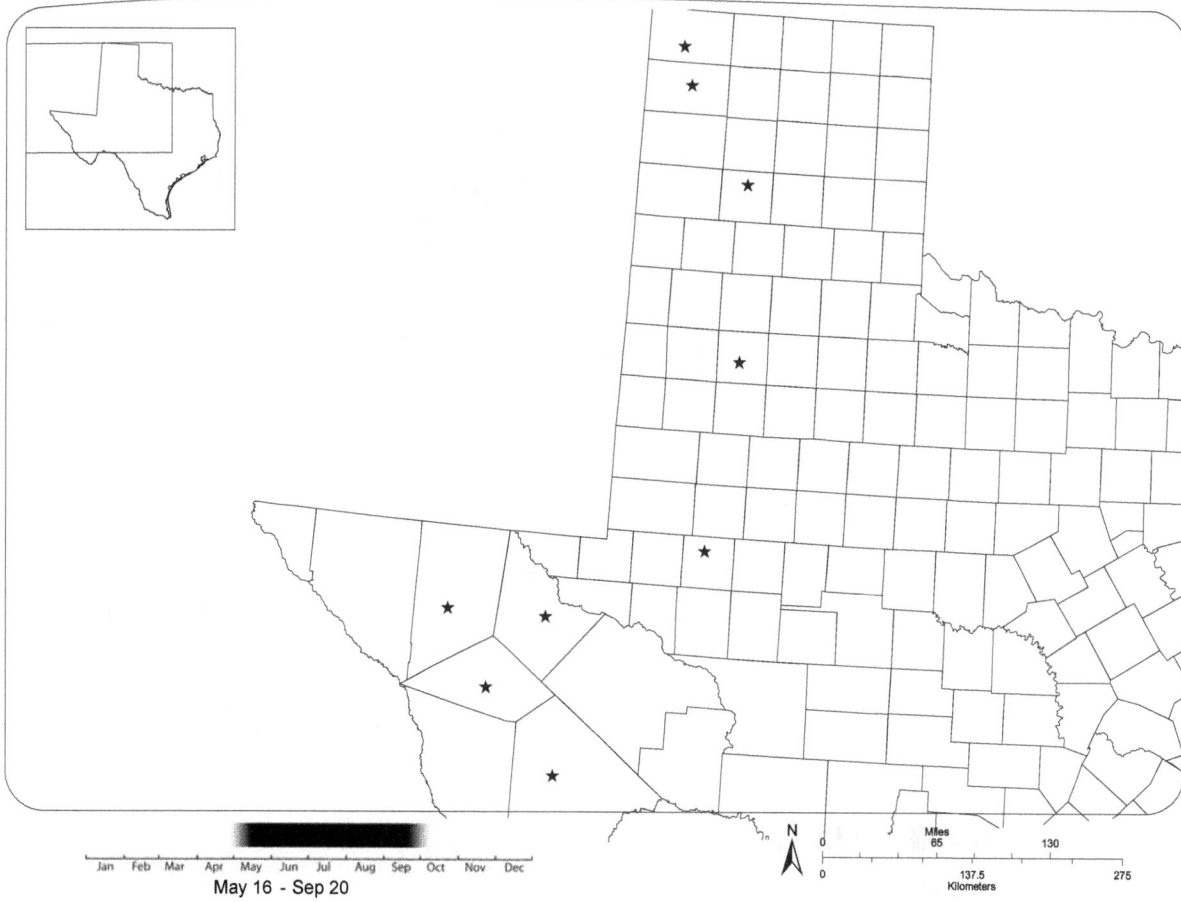

May 16 - Sep 20

HABITAT: Ponds, springs and slow moving streams with heavy marginal vegetation.

Brewster
Culberson
Dallam
Hartley
Jeff Davis
Lubbock
Midland
Randall
Reeves

Ischnura demorsa (Hagen) / Mexican Forktail

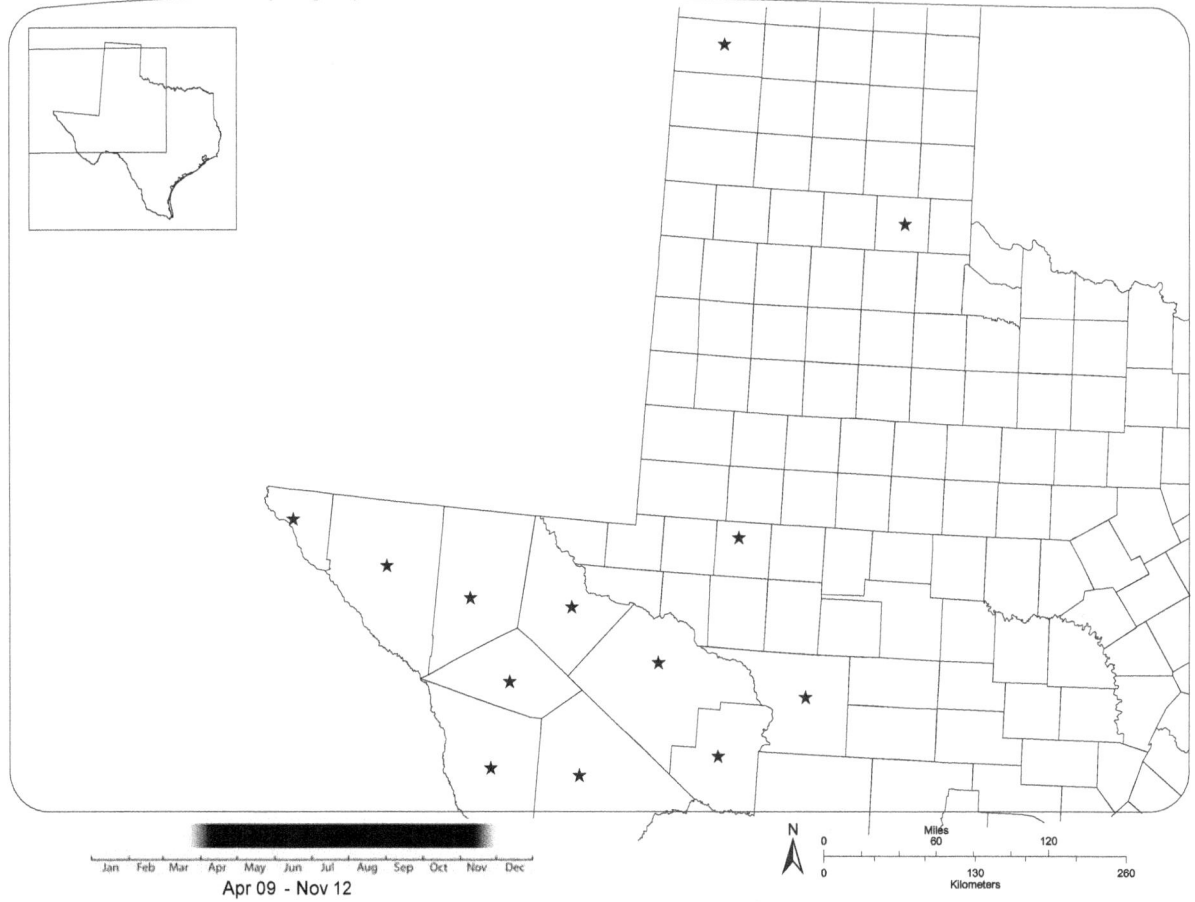

Jan Feb Mar Apr May Jun Jul Aug Sep Oct Nov Dec

Apr 09 - Nov 12

HABITAT: Creeks, streams, springs and slow reaches of rivers with moderate vegetation.

Brewster
Crockett
Culberson
El Paso
Hall
Hartley
Hudspeth
Jeff Davis
Midland
Pecos
Presidio
Reeves
Terrell

Ischnura denticollis (Burmeister) / Black-fronted Forktail

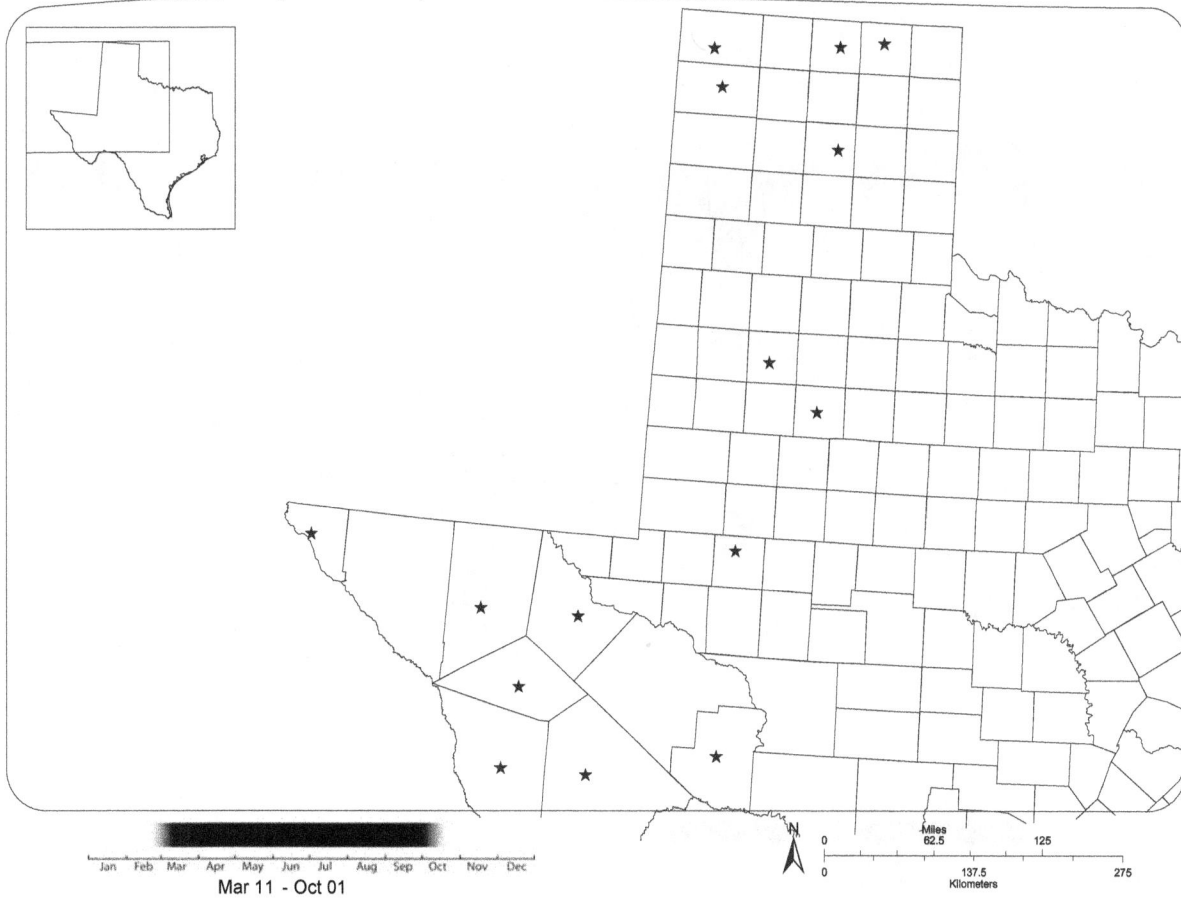

Mar 11 - Oct 01

HABITAT: Vegetated streams or ponds, often associated with springs, especially at northern latitudes.

Brewster
Carson
Culberson
Dallam
El Paso
Garza
Hansford
Hartley
Jeff Davis
Lubbock
Midland
Ochiltree
Presidio
Reeves
Terrell

Ischnura hastata (Say) / Citrine Forktail

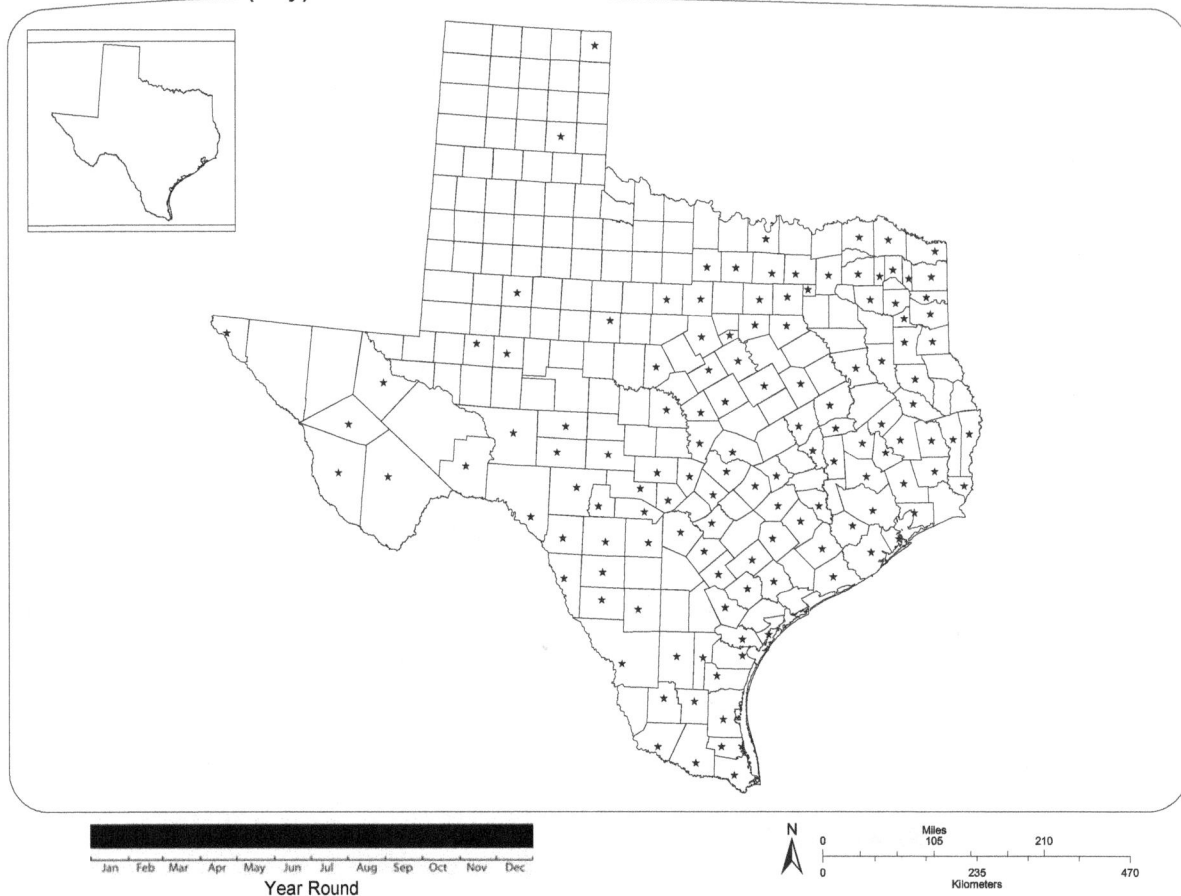

Jan Feb Mar Apr May Jun Jul Aug Sep Oct Nov Dec
Year Round

HABITAT: Heavily vegetated ponds and lakes and other permanent or temporary bodies of water.

Anderson	Crockett	Hidalgo	Marion	Somervell
Angelina	Dallas	Hopkins	Matagorda	Starr
Aransas	Denton	Hunt	Maverick	Stephens
Austin	DeWitt	Jack	McLennan	Sutton
Bandera	Dimmit	Jasper	Medina	Tarrant
Bastrop	Donley	Jeff Davis	Midland	Taylor
Bee	Duval	Jim Hogg	Montgomery	Terrell
Bexar	Edwards	Jim Wells	Morris	Titus
Blanco	El Paso	Johnson	Nacogdoches	Travis
Borden	Ellis	Karnes	Newton	Trinity
Bosque	Erath	Kendall	Nueces	Tyler
Bowie	Fayette	Kenedy	Orange	Upshur
Brazoria	Fort Bend	Kerr	Palo Pinto	Uvalde
Brazos	Franklin	Kimble	Panola	Val Verde
Brewster	Galveston	Kinney	Polk	Victoria
Brooks	Gillespie	Kleberg	Presidio	Walker
Brown	Glasscock	La Salle	Real	Webb
Burnet	Goliad	Lamar	Red River	Wharton
Cameron	Gregg	Lampasas	Reeves	Willacy
Cass	Grimes	Lavaca	Robertson	Williamson
Chambers	Guadalupe	Lee	Rockwall	Wilson
Cherokee	Hamilton	Leon	Rusk	Wise
Collin	Hardin	Liberty	San Jacinto	Wood
Colorado	Harris	Limestone	San Patricio	Zavala
Cooke	Harrison	Lipscomb	San Saba	
Coryell	Hays	Madison	Schleicher	

Ischnura kellicotti Williamson / Lilypad Forktail

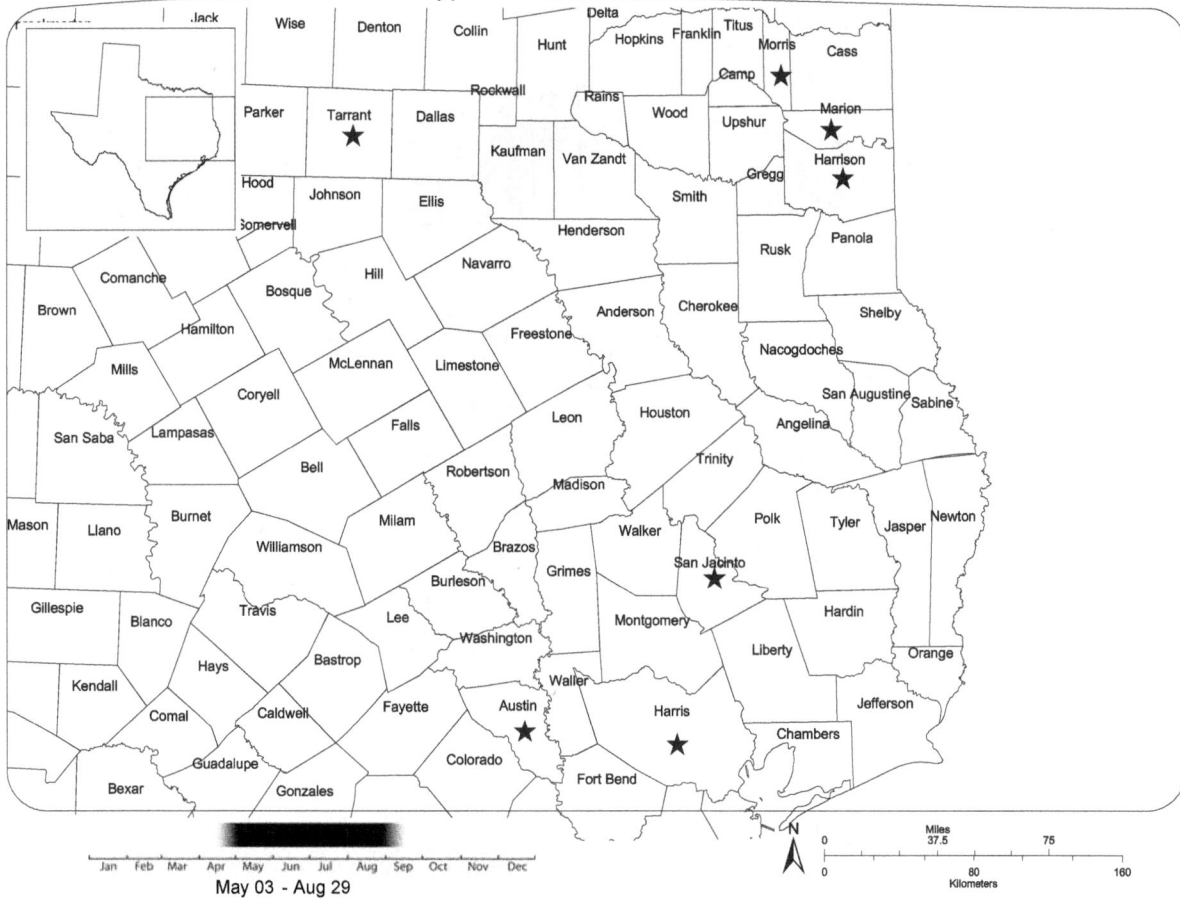

May 03 - Aug 29

HABITAT: Strongly associated with floating lily pads in lakes.

Austin
Harris
Harrison
Marion
Morris
San Jacinto
Tarrant

Ischnura posita (Hagen) / Fragile Forktail

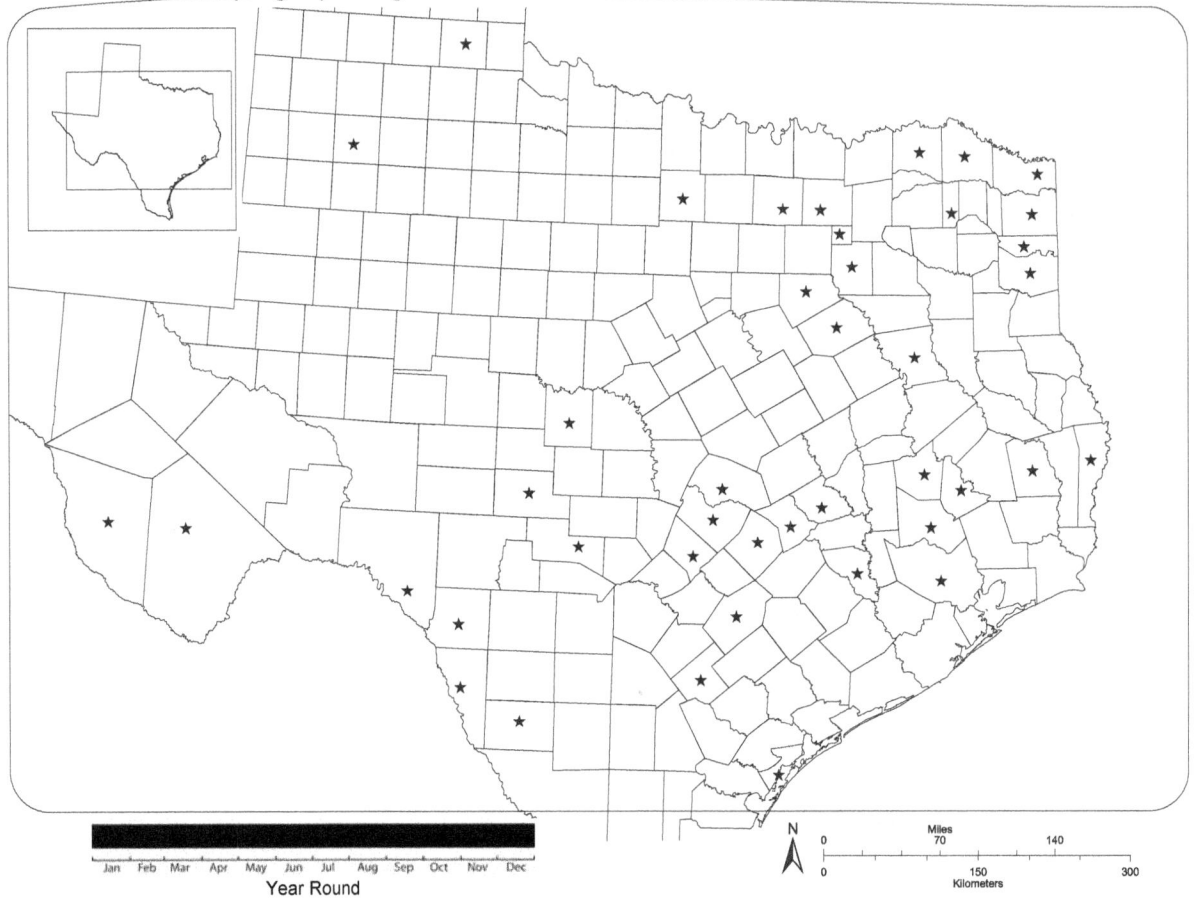

Jan Feb Mar Apr May Jun Jul Aug Sep Oct Nov Dec
Year Round

HABITAT: Heavily vegetated ponds, marshes and slow moving waters.

Anderson	Lubbock
Aransas	Marion
Austin	Maverick
Bastrop	McCulloch
Bowie	Montgomery
Brewster	Navarro
Burleson	Newton
Cass	Presidio
Collin	Red River
Denton	Rockwall
Dimmit	San Jacinto
Ellis	Travis
Franklin	Tyler
Gonzales	Val Verde
Hall	Walker
Harris	Williamson
Harrison	
Hays	
Jack	
Karnes	
Kaufman	
Kerr	
Kimble	
Kinney	
Lamar	
Lee	

Ischnura prognata (Hagen) / Furtive Forktail

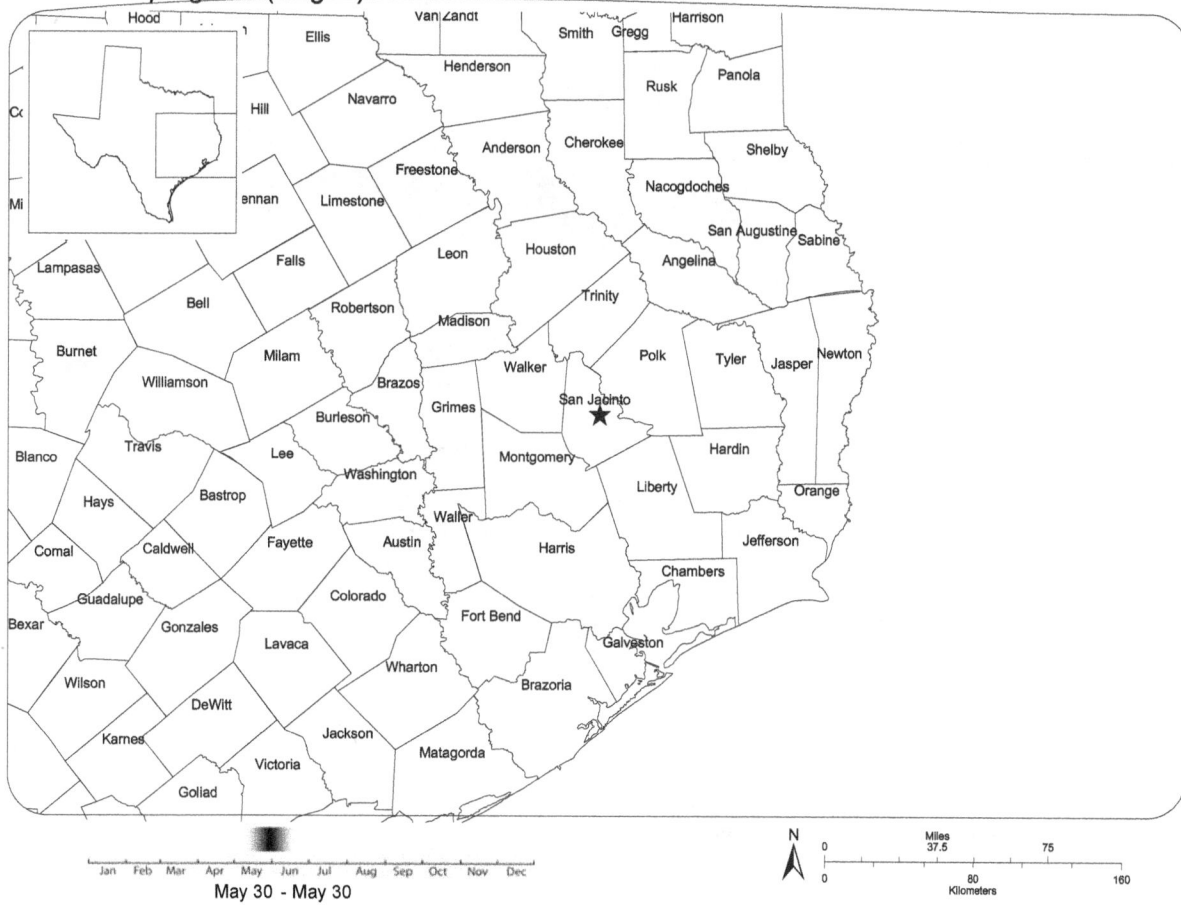

May 30 - May 30

HABITAT: Heavily shaded ponds, swamps and sloughs.

San Jacinto

Ischnura ramburii (Selys) / Rambur's Forktail

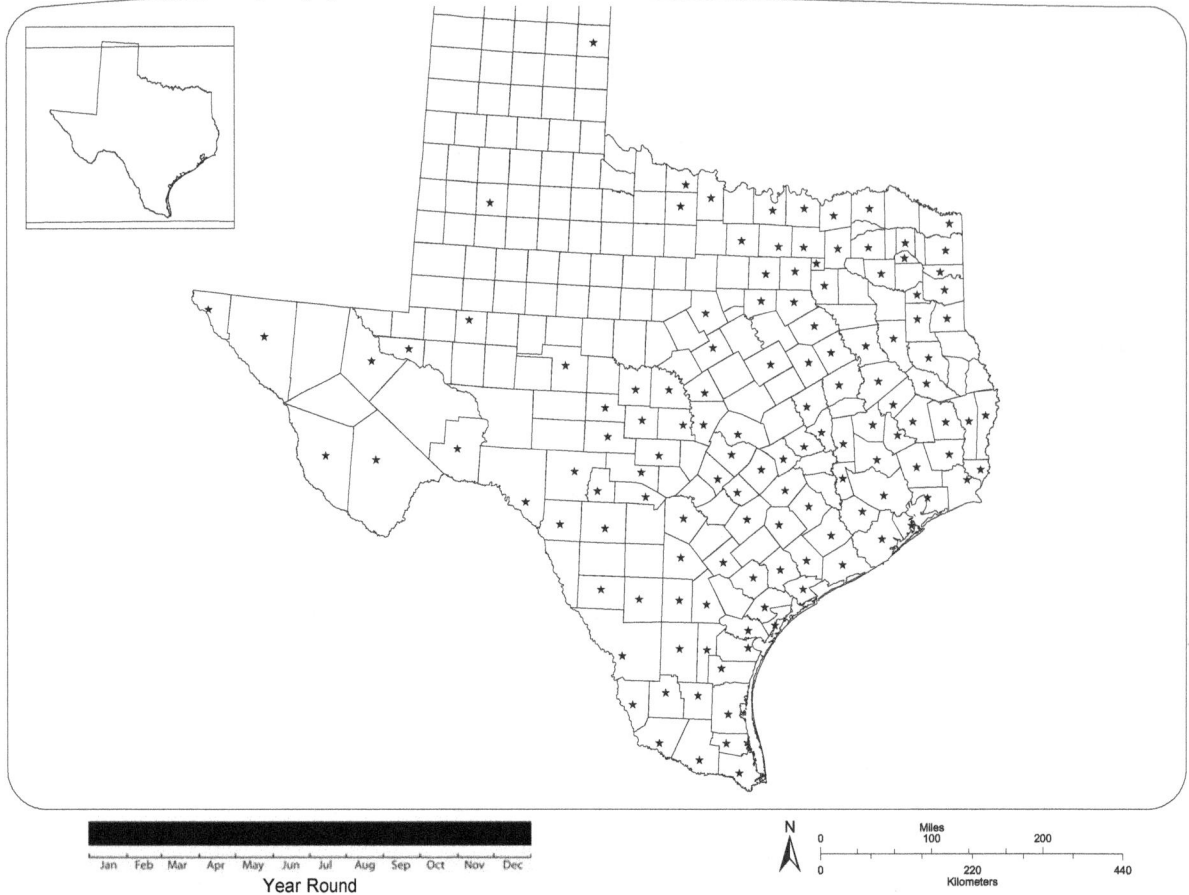

Year Round

HABITAT: Heavily vegetated ponds, lakes, marshes and slow reaches of streams exposed to sunlight including brackish waters.

Anderson	Dallas	Hopkins	Llano	San Patricio
Angelina	Denton	Houston	Lubbock	San Saba
Aransas	Dimmit	Hudspeth	Marion	Starr
Archer	Duval	Hunt	Mason	Tarrant
Atascosa	Edwards	Jackson	Matagorda	Terrell
Bandera	El Paso	Jasper	McCulloch	Titus
Bastrop	Ellis	Jefferson	McLennan	Tom Green
Bexar	Erath	Jim Hogg	McMullen	Travis
Bowie	Fannin	Jim Wells	Menard	Trinity
Brazoria	Fayette	Johnson	Midland	Tyler
Brazos	Fort Bend	Karnes	Montgomery	Uvalde
Brewster	Freestone	Kaufman	Nacogdoches	Val Verde
Brooks	Galveston	Kenedy	Navarro	Victoria
Burleson	Gillespie	Kerr	Newton	Walker
Burnet	Goliad	Kimble	Nueces	Waller
Caldwell	Gonzales	Kinney	Orange	Ward
Calhoun	Grayson	Kleberg	Panola	Webb
Cameron	Gregg	La Salle	Polk	Wharton
Camp	Grimes	Lamar	Presidio	Wichita
Cass	Hamilton	Lampasas	Real	Willacy
Chambers	Hardin	Lavaca	Reeves	Williamson
Cherokee	Harris	Lee	Refugio	Wise
Clay	Harrison	Leon	Robertson	Wood
Collin	Hays	Liberty	Rockwall	Zapata
Colorado	Hemphill	Limestone	Rusk	
Cooke	Hidalgo	Live Oak	San Jacinto	

Ischnura verticalis (Say) / Eastern Forktail

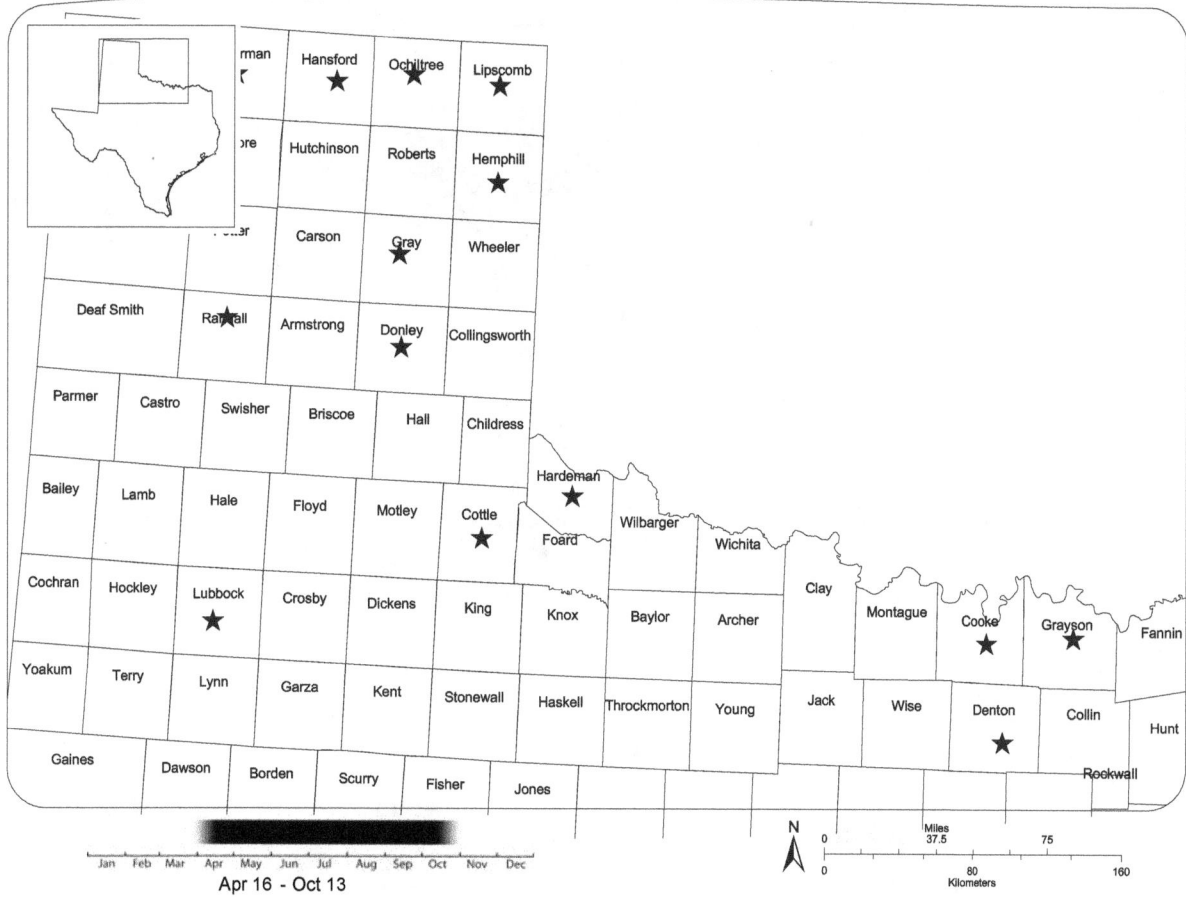

Apr 16 - Oct 13

HABITAT: Ponds, lakes, slow moving streams and marshes.

Cooke
Cottle
Dallam
Denton
Donley
Gray
Grayson
Hansford
Hardeman
Hartley
Hemphill
Lipscomb
Lubbock
Ochiltree
Randall
Sherman

Leptobasis melinogaster Gonzales / Cream-tipped Swampdamsel

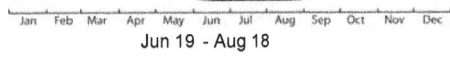

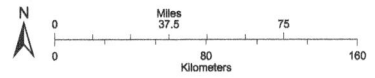

Jun 19 - Aug 18

HABITAT: Slow reaches of streams and bacwater areas.

Cameron
Hidalgo
Kleberg

Nehalennia integricollis Calvert / Southern Sprite

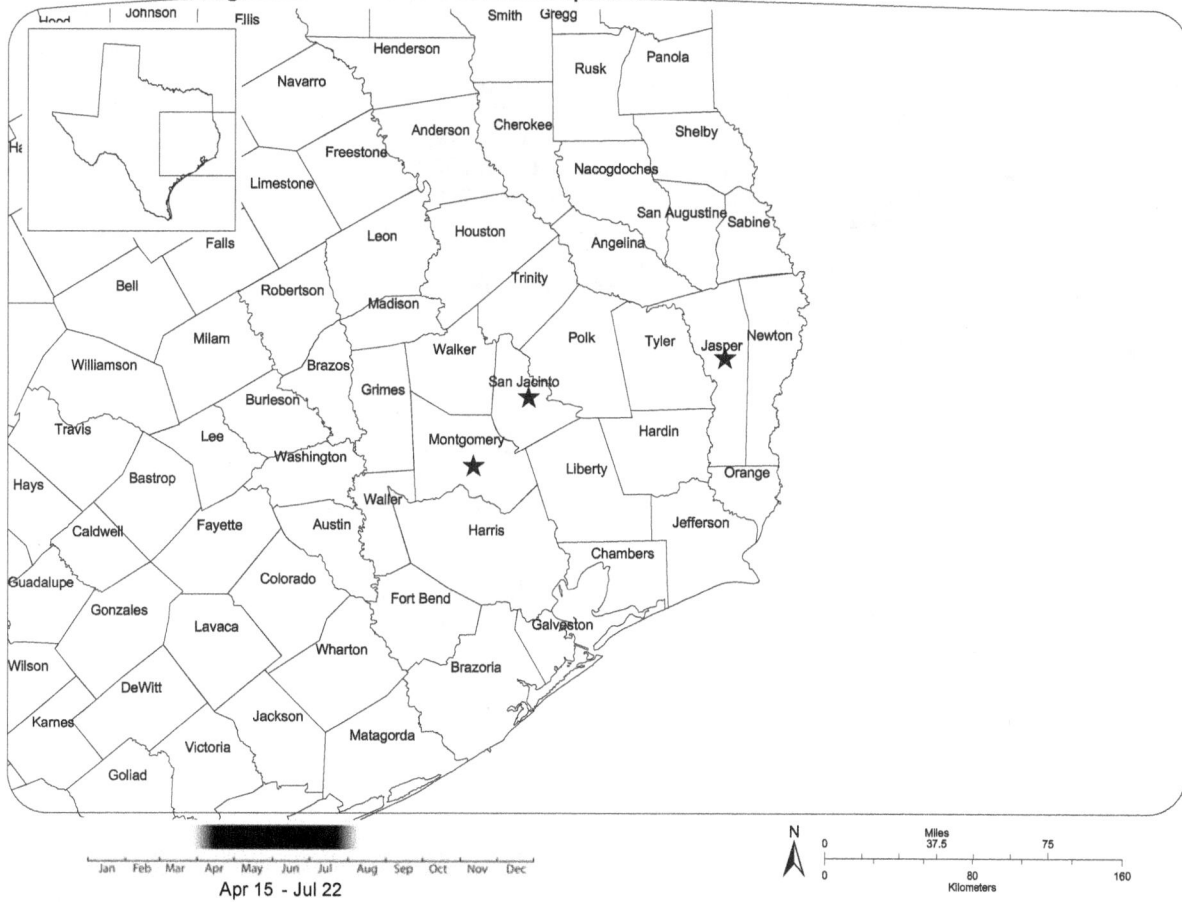

Apr 15 - Jul 22

HABITAT: Ponds, lakes, bogs and slow reaches of streams with moderately dense vegetation.

Jasper
Montgomery
San Jacinto

Nehalennia pallidula Calvert / Everglades Sprite

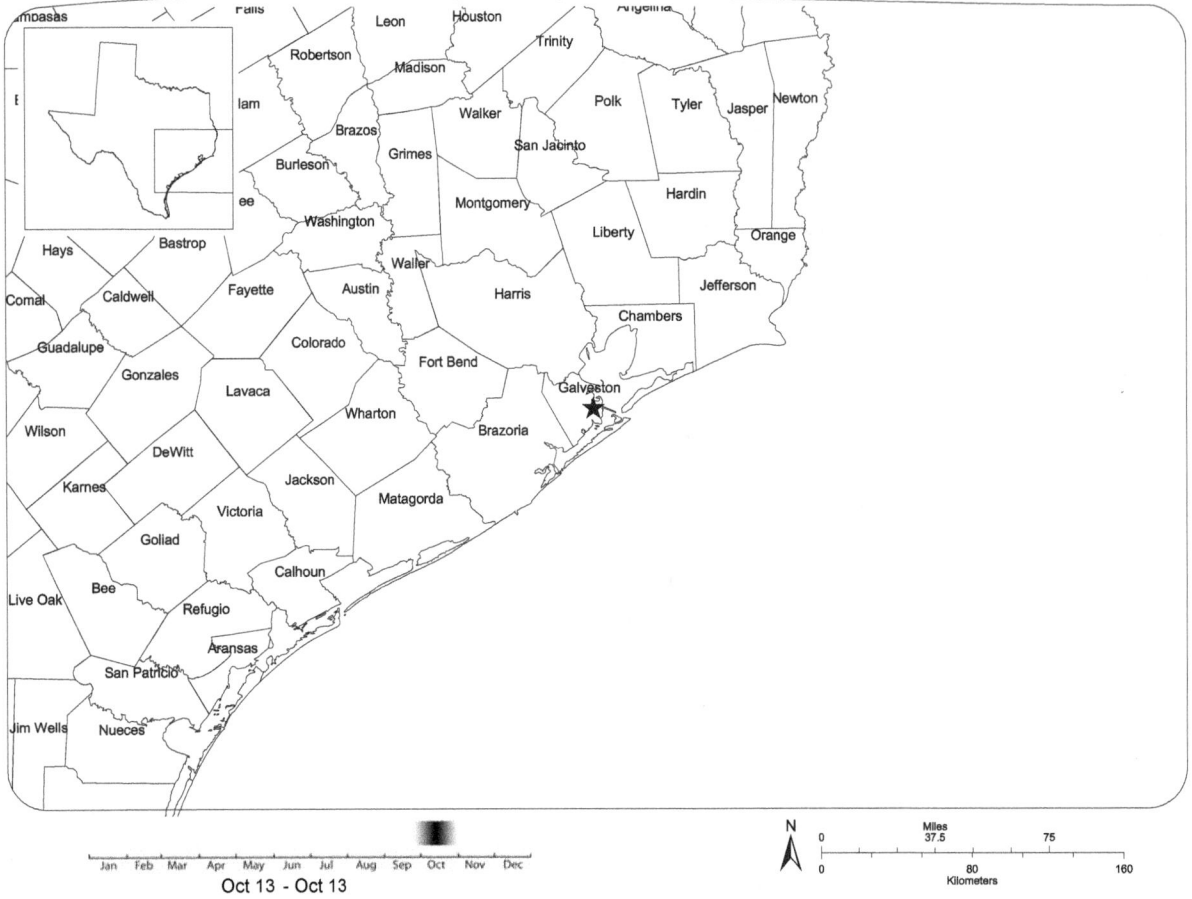

Oct 13 - Oct 13

HABITAT: Primary habitat in Florida is the Everglades, but it is known from ponds and rock pits.

Galveston

Neoerythromma cultellatum (Sèlys) / Caribbean Yellowface

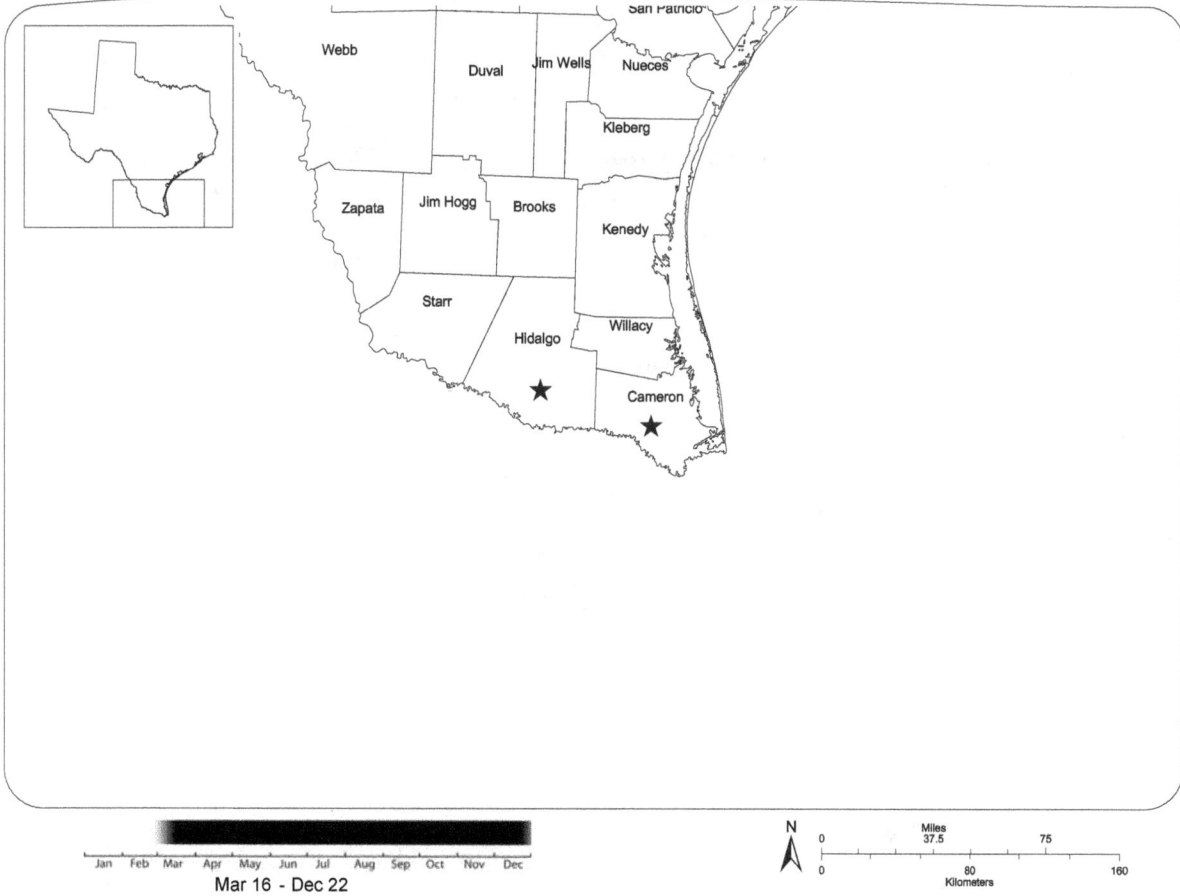

Mar 16 - Dec 22

HABITAT: Ponds and slow reaches of streams or rivers with abundant floating debris.

Cameron
Hidalgo

Telebasis byersi Westfall / Duckweed Firetail

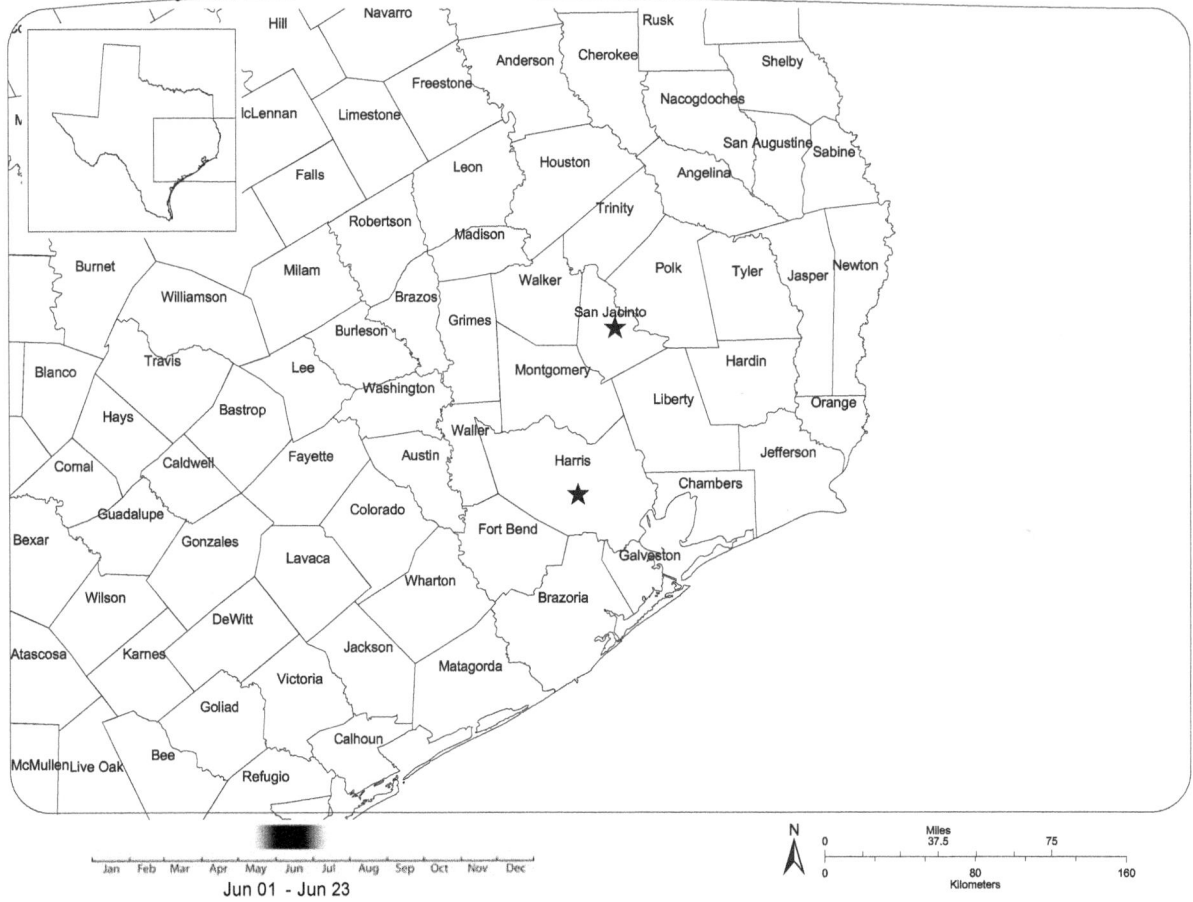

Jun 01 - Jun 23

HABITAT: Swampy, partially shaded areas with abundant floating duckweed.

Harris
San Jacinto

Telebasis salva (Hagen) / Desert Firetail

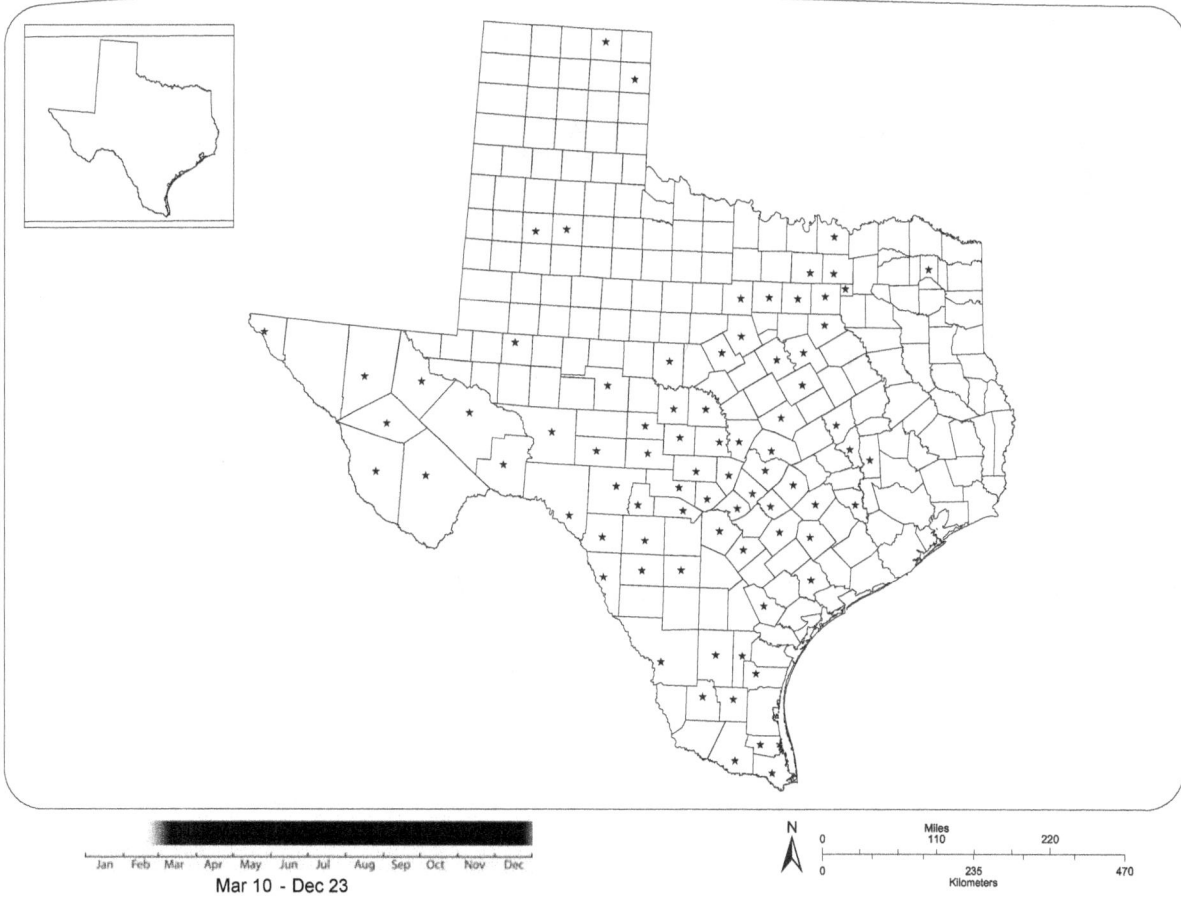

Mar 10 - Dec 23

HABITAT: Ponds, lakes, pools, springs and slow reaches of streams with open sunlight and abundant emergent vegetation.

Austin	Ellis	McLennan	Zavala
Bandera	Erath	Menard	
Bastrop	Fayette	Midland	
Bee	Frio	Ochiltree	
Bell	Gillespie	Palo Pinto	
Bexar	Gonzales	Parker	
Blanco	Grayson	Pecos	
Bosque	Grimes	Presidio	
Brazos	Hays	Real	
Brewster	Hemphill	Reeves	
Brooks	Hidalgo	Robertson	
Burnet	Hill	Rockwall	
Caldwell	Jeff Davis	San Saba	
Cameron	Jim Hogg	Sutton	
Coleman	Jim Wells	Tarrant	
Collin	Kendall	Terrell	
Comal	Kerr	Titus	
Comanche	Kimble	Tom Green	
Crockett	Kinney	Travis	
Crosby	Kleberg	Uvalde	
Culberson	Lavaca	Val Verde	
Dallas	Llano	Victoria	
Denton	Lubbock	Webb	
Duval	Mason	Willacy	
Edwards	Maverick	Williamson	
El Paso	McCulloch	Wilson	

Tachopteryx thoreyi (Hagen in Selys) / Gray Petaltail

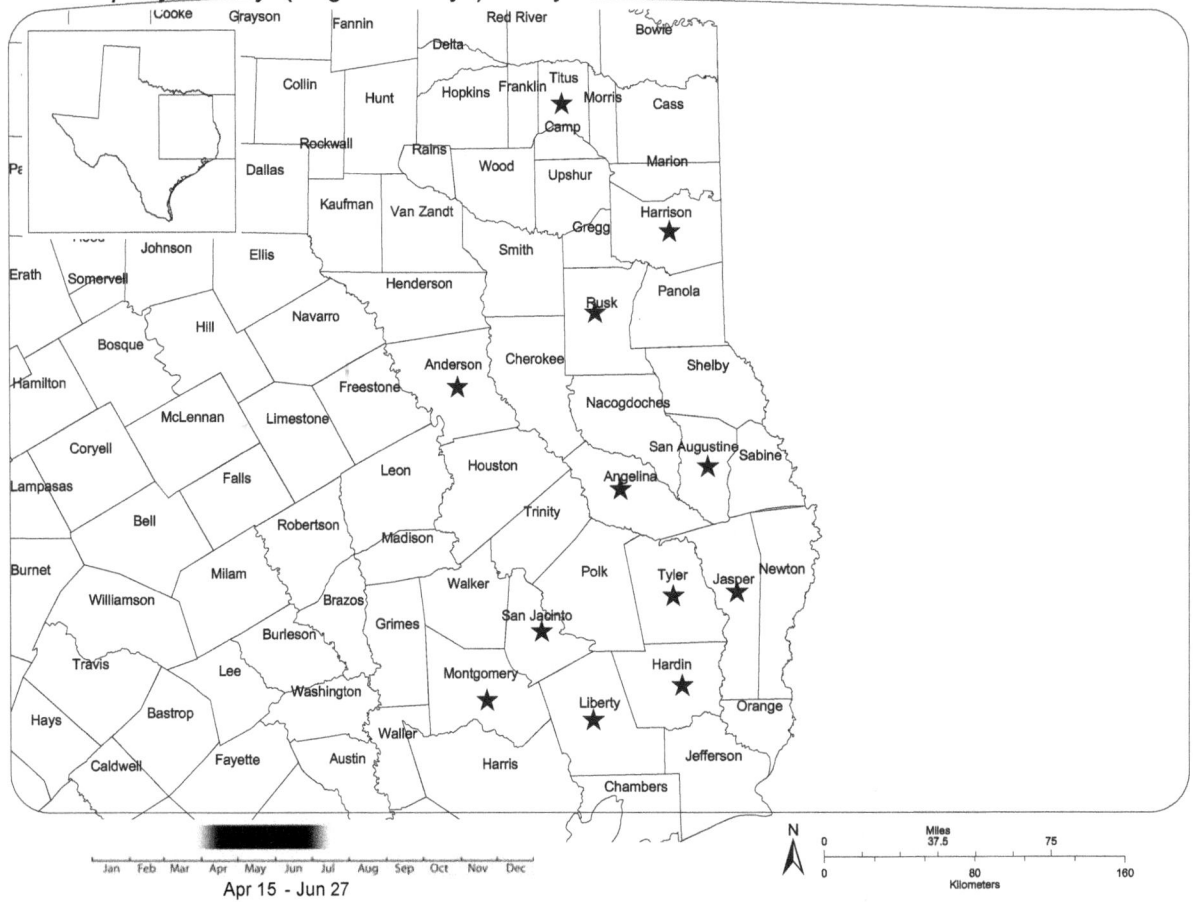

Apr 15 - Jun 27

HABITAT: Permanent springs and seepages of hardwood forests.

Anderson
Angelina
Hardin
Harrison
Jasper
Liberty
Montgomery
Rusk
San Augustine
San Jacinto
Titus
Tyler

Aeshna persephone Donnelly / Persephone's Darner

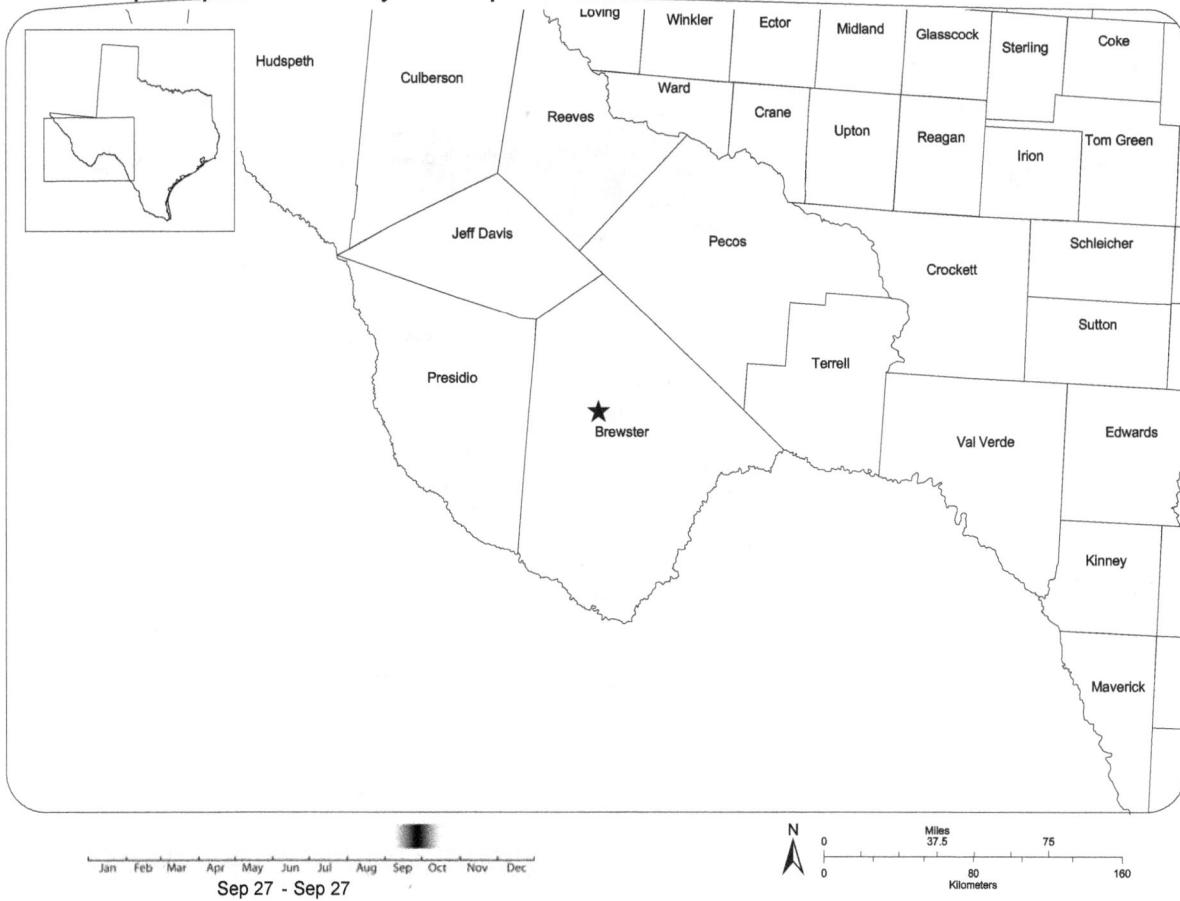

Sep 27 - Sep 27

HABITAT: Partially shaded desert mountain streams.

Brewster

Aeshna umbrosa Walker / Shadow Darner

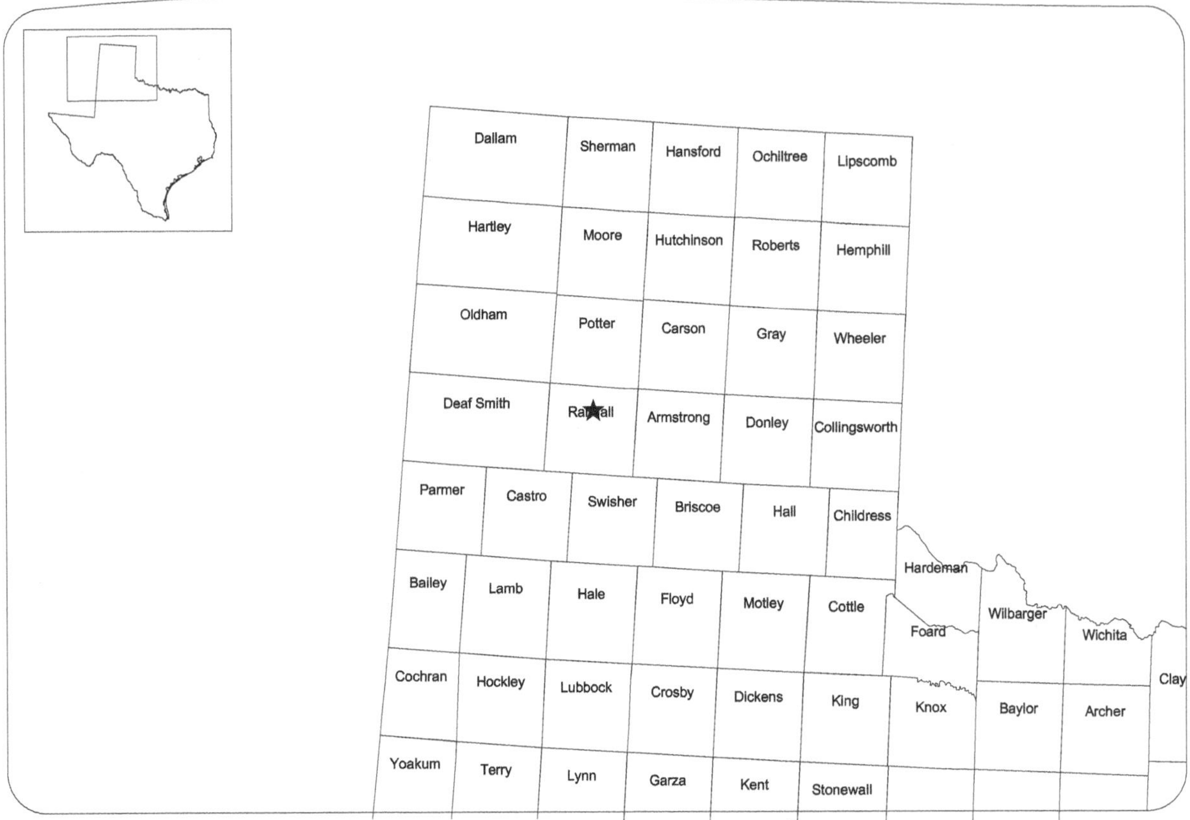

Jul 26 - Jul 26

HABITAT: Partly shaded, slow-flowing forest streams and ditches.

Randall

Anax amazili (Burmeister) / Amazon Darner

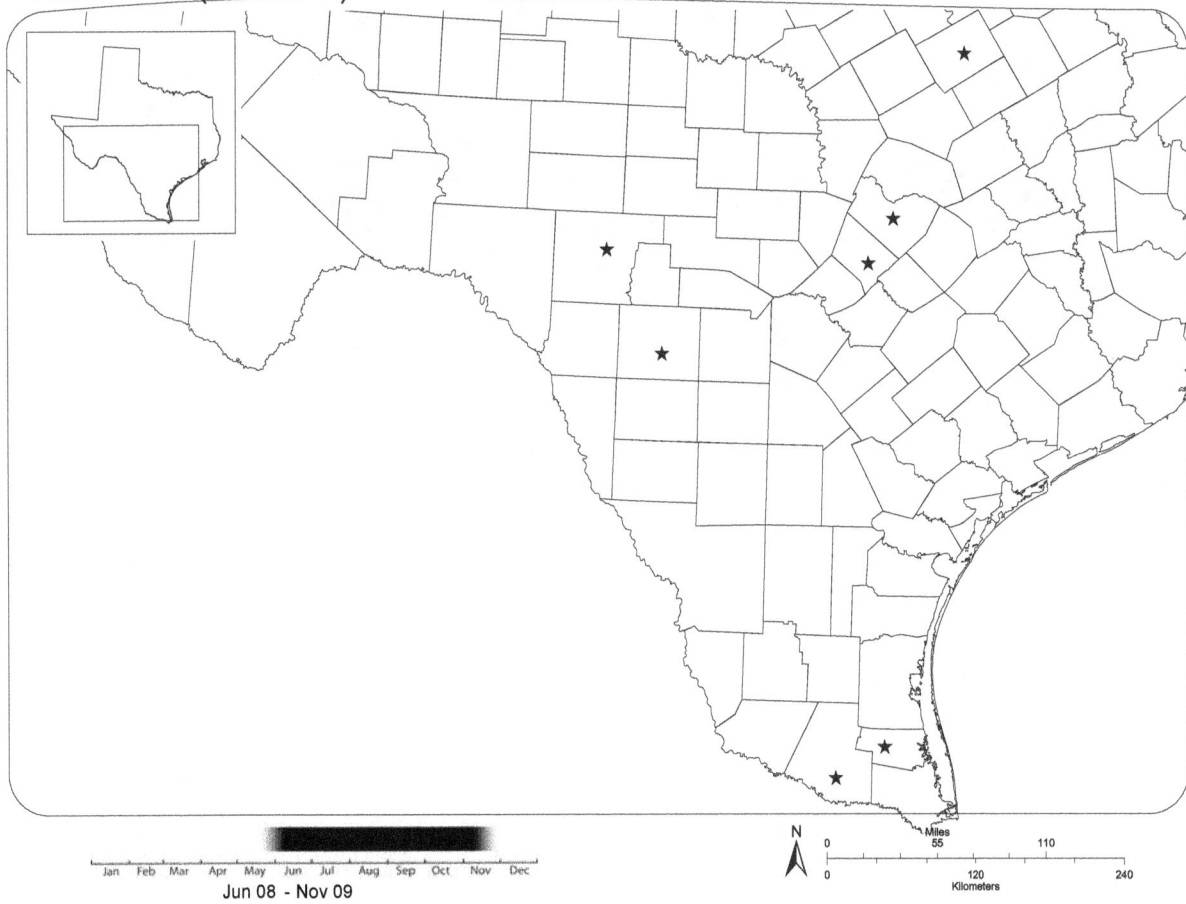

Jun 08 - Nov 09

HABITAT: Tropical ponds and lakes with weeds including brackish pools.

Edwards
Hays
Hidalgo
McLennan
Reeves
Travis
Uvalde
Willacy

Anax concolor Brauer / Blue-spotted Comet Darner

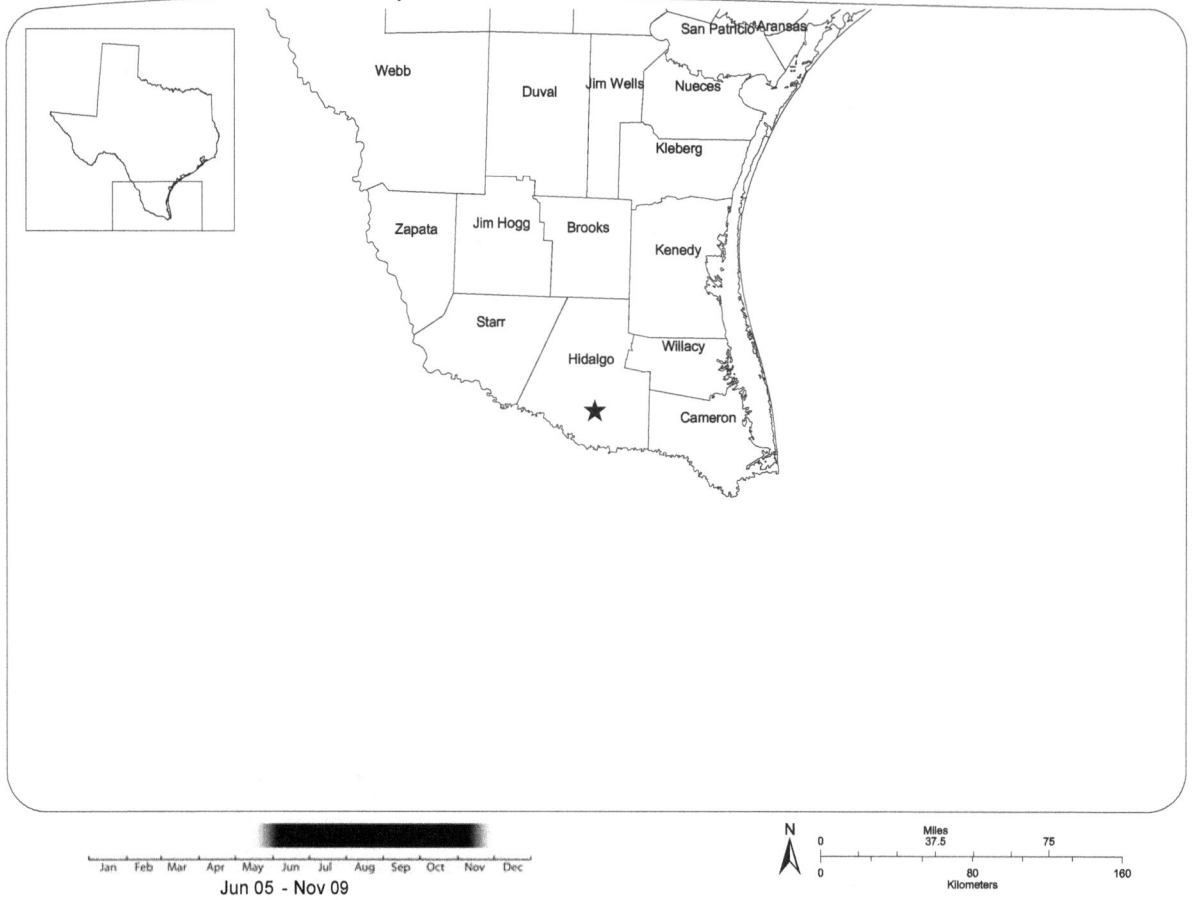

Jun 05 - Nov 09

HABITAT: Primarily fishless temporary and semi-permanent grassy ponds and pools.

Hidalgo

Anax junius (Drury) / Common Green Darner

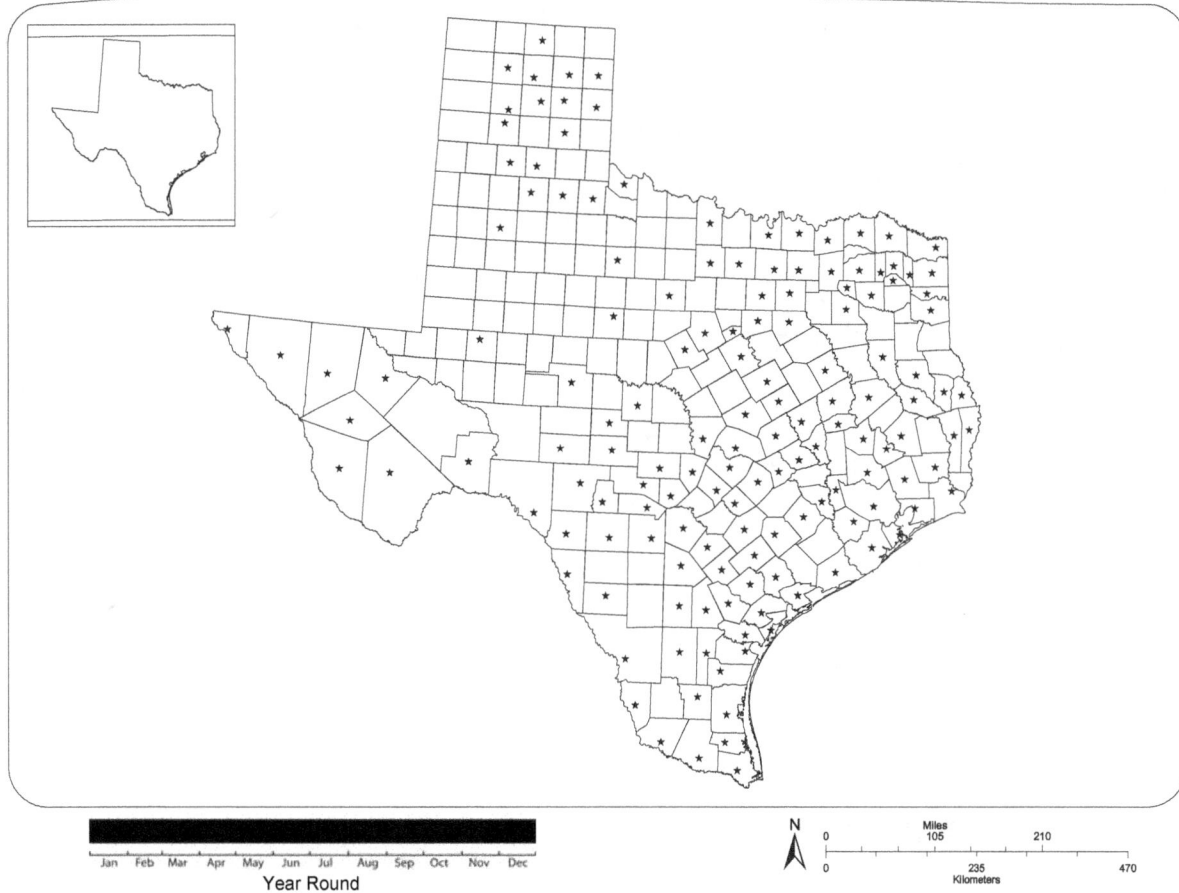

Jan Feb Mar Apr May Jun Jul Aug Sep Oct Nov Dec
Year Round

HABITAT: Permanent and temporary ponds, lakes, bays and slow-flowing streams with emergent vegetation.

Angelina	Cherokee	Goliad	Kenedy	Motley	Terrell
Aransas	Clay	Gonzales	Kerr	Nacogdoches	Titus
Atascosa	Collin	Gray	Kimble	Newton	Tom Green
Austin	Colorado	Grayson	Kinney	Nueces	Travis
Bandera	Comanche	Hansford	Kleberg	Polk	Uvalde
Bastrop	Cooke	Hardeman	Lamar	Potter	Val Verde
Bee	Cottle	Hardin	Lavaca	Presidio	Van Zandt
Bell	Culberson	Harris	Lee	Rains	Victoria
Bexar	Dallas	Harrison	Leon	Randall	Walker
Blanco	Denton	Haskell	Liberty	Real	Waller
Bosque	DeWitt	Hays	Live Oak	Red River	Webb
Bowie	Dimmit	Hemphill	Lubbock	Reeves	Wheeler
Brazoria	Donley	Hidalgo	Madison	Refugio	Willacy
Brazos	Duval	Hopkins	Marion	Roberts	Williamson
Brewster	Edwards	Houston	Matagorda	Robertson	Wilson
Briscoe	El Paso	Hudspeth	Maverick	Sabine	Wise
Brooks	Ellis	Hunt	McCulloch	San Augustine	Wood
Burleson	Erath	Hutchinson	McLennan	San Jacinto	Zapata
Burnet	Falls	Jack	McMullen	San Patricio	
Caldwell	Fannin	Jasper	Medina	Somervell	
Calhoun	Floyd	Jeff Davis	Menard	Starr	
Cameron	Fort Bend	Jefferson	Midland	Stephens	
Camp	Franklin	Jim Wells	Milam	Sutton	
Carson	Freestone	Johnson	Montgomery	Swisher	
Cass	Galveston	Karnes	Moore	Tarrant	
Chambers	Gillespie	Kendall	Morris	Taylor	

Anax longipes (Hagen) / Comet Darner

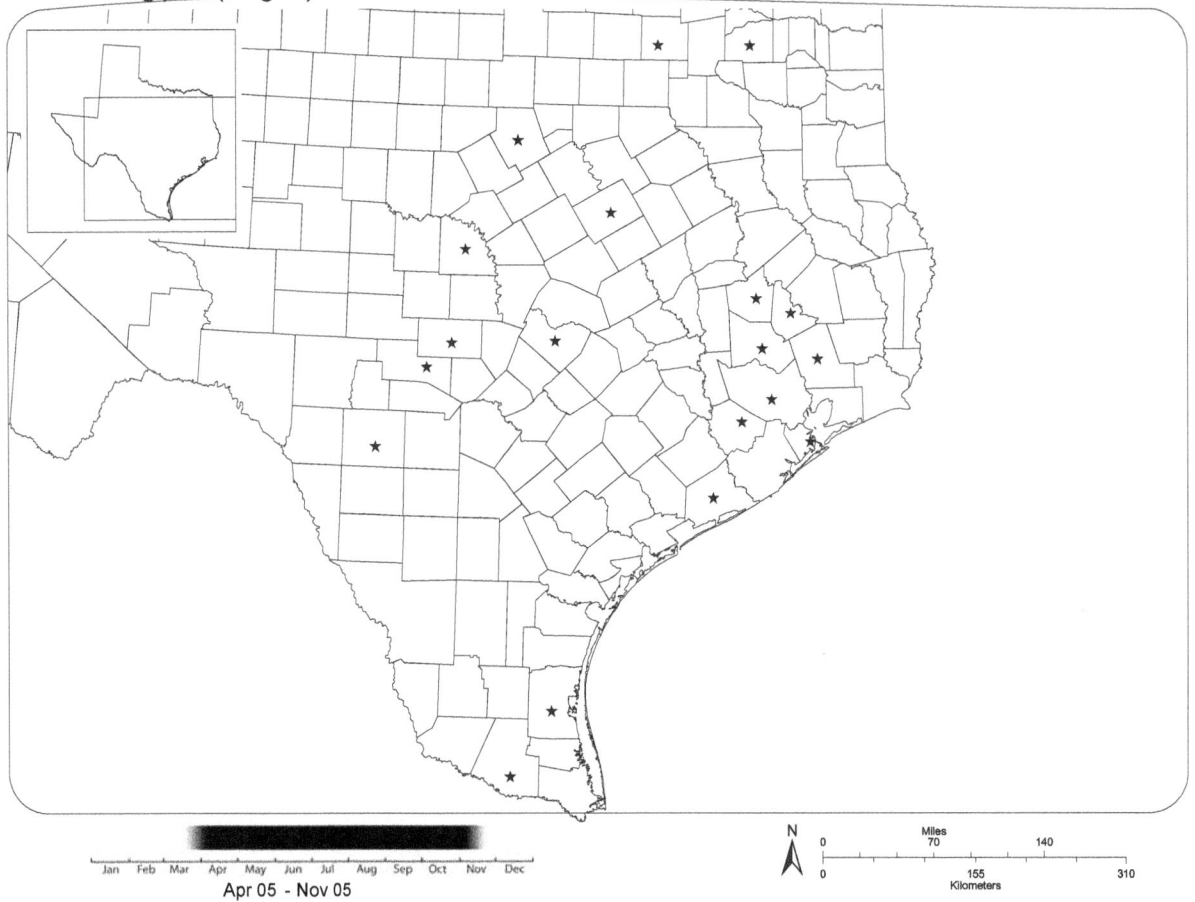

Apr 05 - Nov 05

HABITAT: Primarily fishless temporary and semi-permanent grassy ponds and pools.

Collin
Erath
Fort Bend
Galveston
Gillespie
Harris
Hidalgo
Hopkins
Kenedy
Kerr
Liberty
Matagorda
McLennan
Montgomery
San Jacinto
San Saba
Travis
Uvalde
Walker

Anax walsinghami McLachlan / Giant Darner

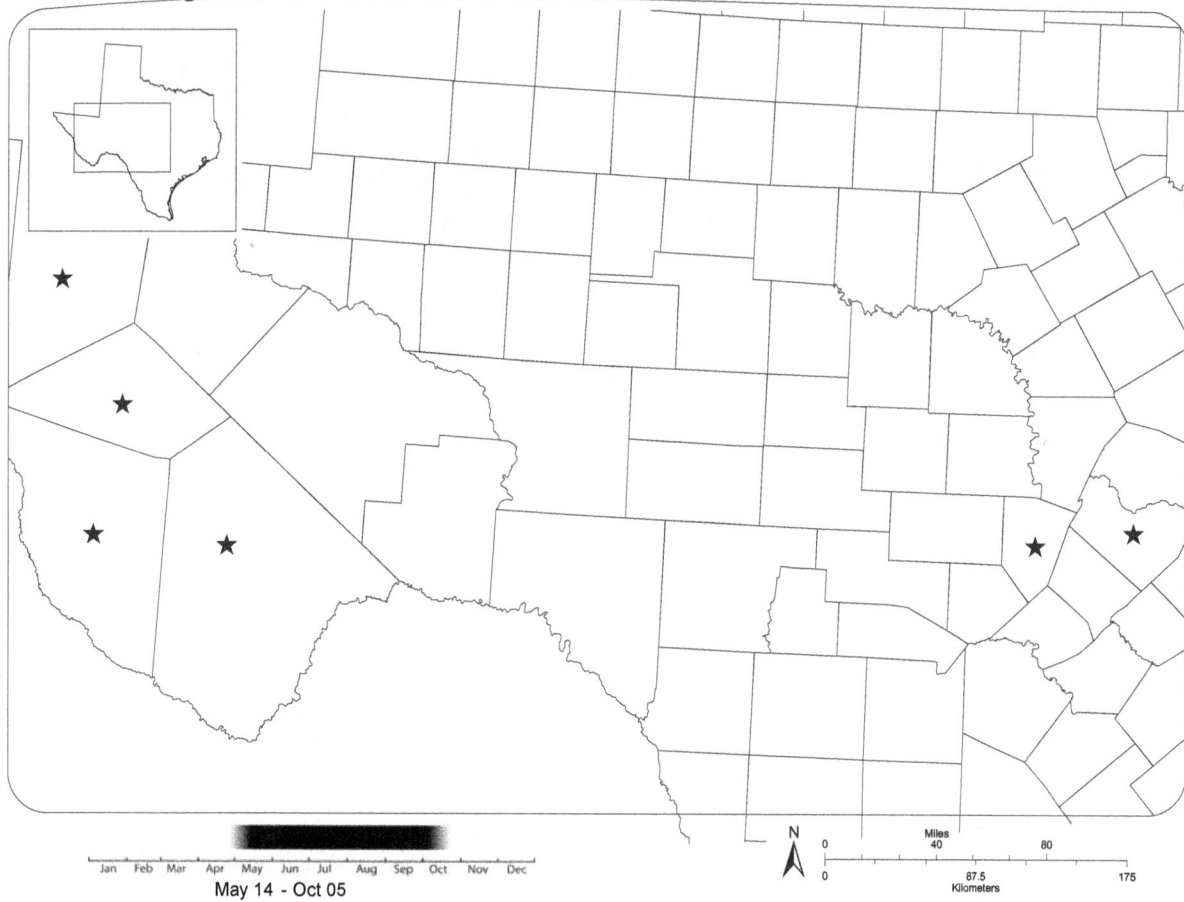

May 14 - Oct 05

HABITAT: Slow-flowing open, spring-fed, streams, ponds and pools

Blanco
Brewster
Culberson
Jeff Davis
Presidio
Travis

Basiaeschna janata (Say) / Springtime Darner

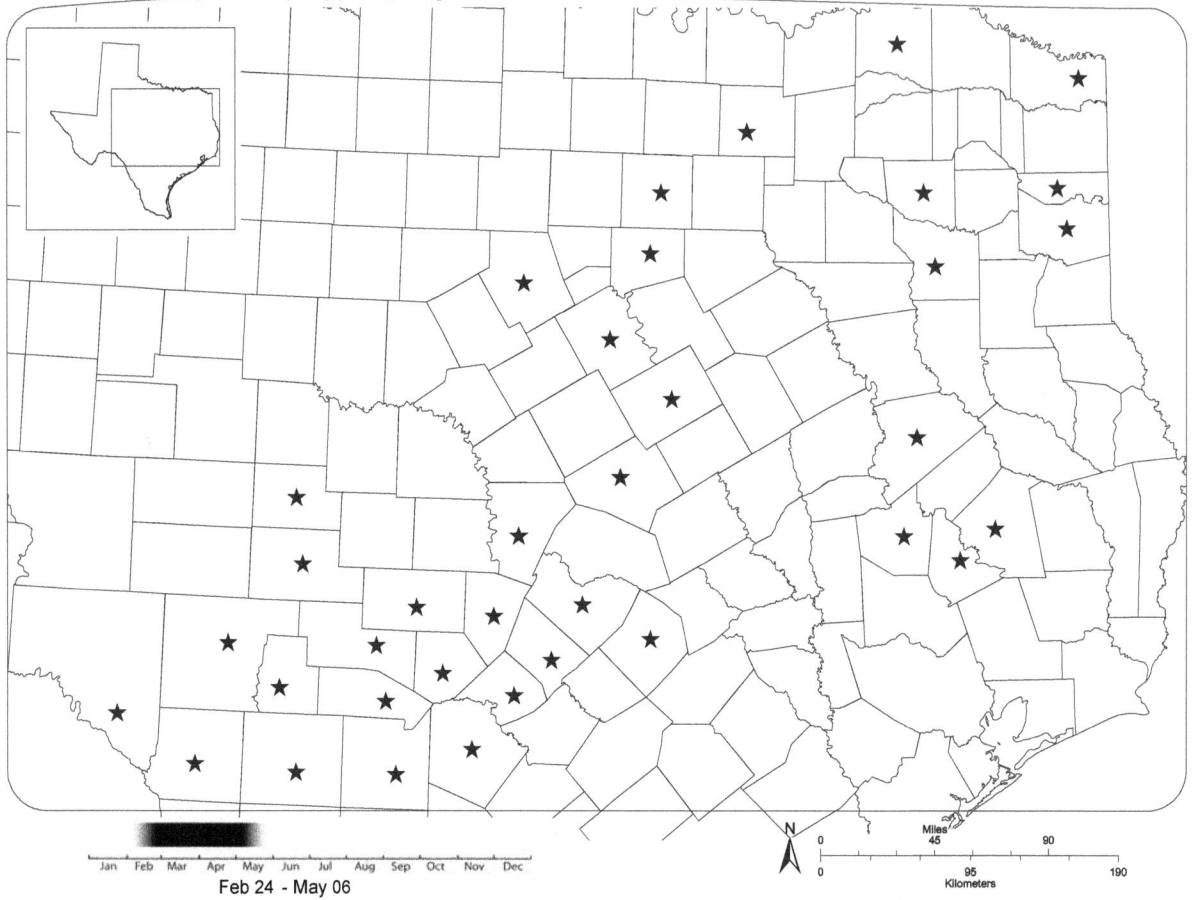

Jan Feb Mar Apr May Jun Jul Aug Sep Oct Nov Dec

Feb 24 - May 06

HABITAT: Small forest lakes and streams and rivers with slow current.

Bandera
Bastrop
Bell
Bexar
Blanco
Bosque
Bowie
Burnet
Collin
Comal
Edwards
Erath
Gillespie
Harrison
Hays
Houston
Johnson
Kendall
Kerr
Kimble
Kinney
Lamar
Marion
McLennan
Medina
Menard

Polk
Real
San Jacinto
Smith
Tarrant
Travis
Uvalde
Val Verde
Walker
Wood

Boyeria vinosa (Say) / Fawn Darner

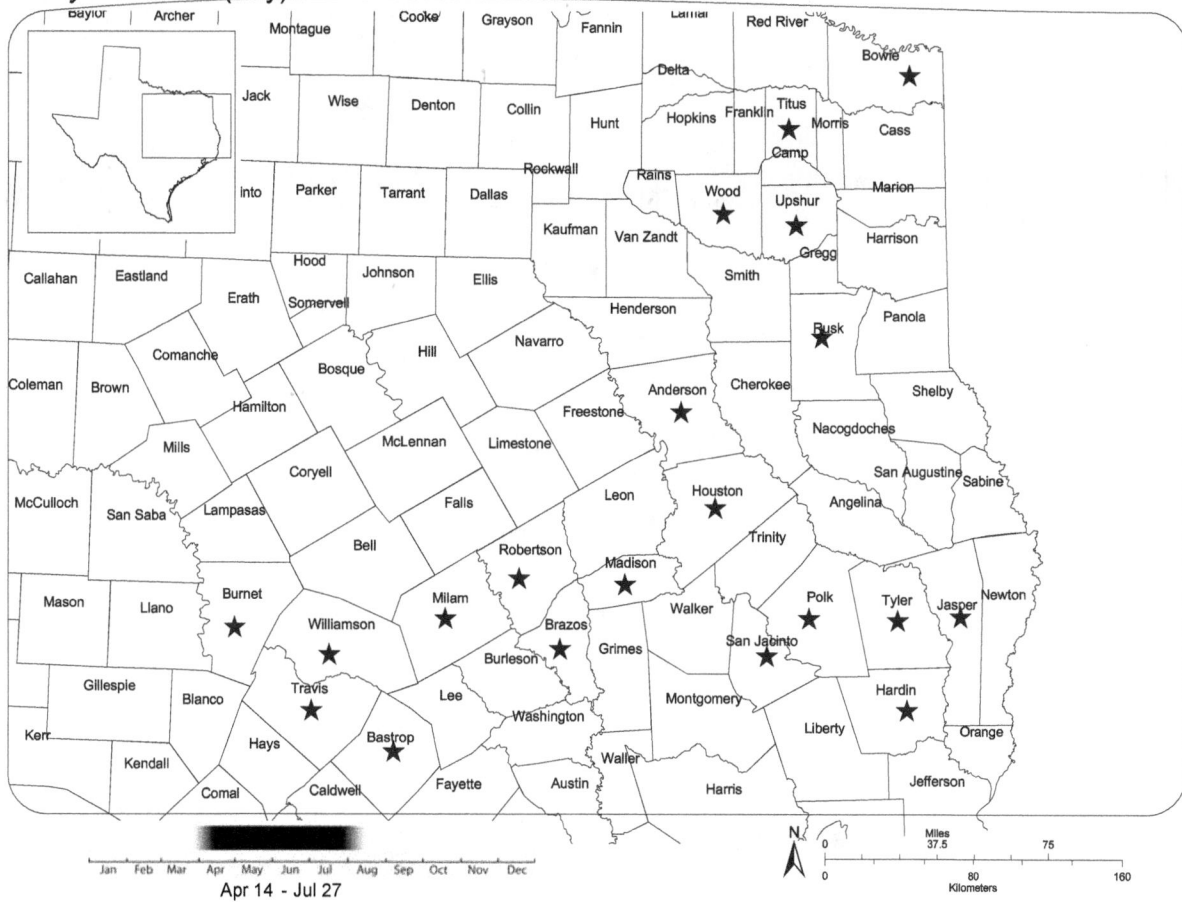

Apr 14 - Jul 27

HABITAT: Forest streams, rivers and lake shores with sufficient shade.

Anderson
Bastrop
Bowie
Brazos
Burnet
Hardin
Houston
Jasper
Madison
Milam
Polk
Robertson
Rusk
San Jacinto
Titus
Travis
Tyler
Upshur
Williamson
Wood

Coryphaeschna adnexa (Hagen) / Blue-faced Darner

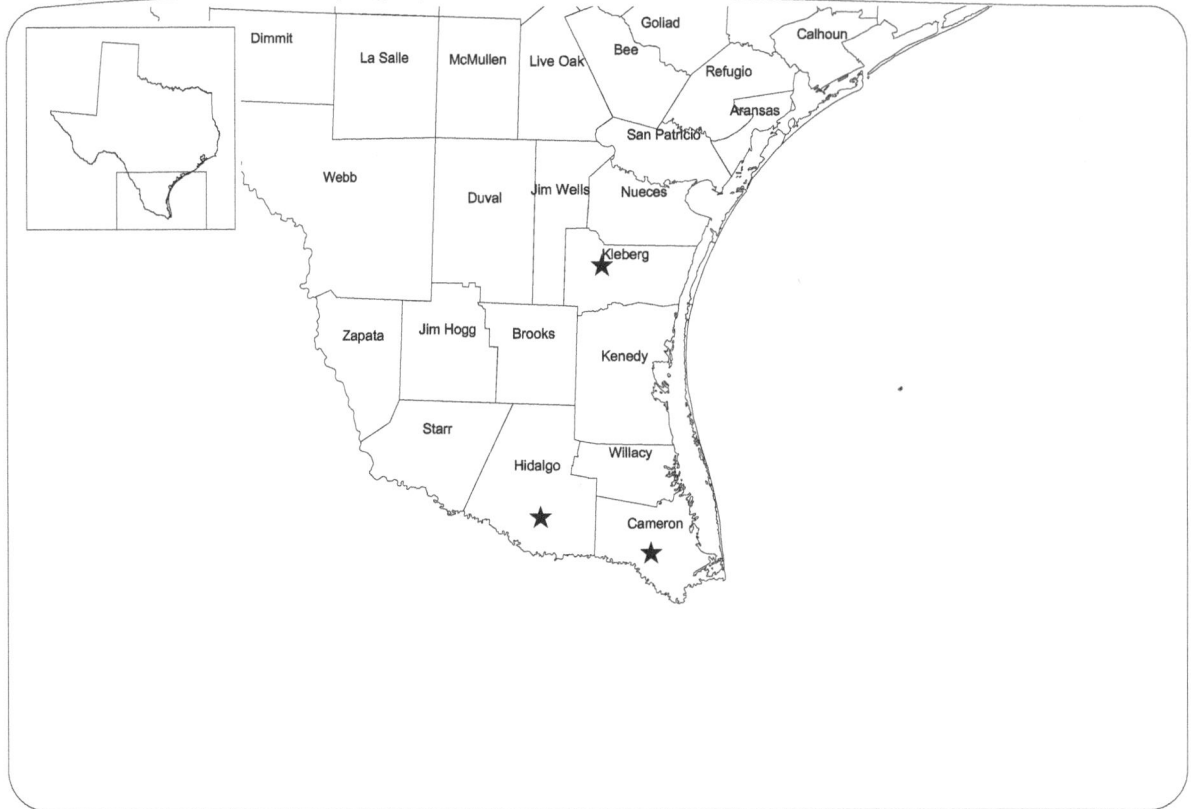

Jun 06 - Oct 27

HABITAT: Heavily vegetated lakes, canals and marshy areas.

Cameron
Hidalgo
Kleberg

Coryphaeschna ingens (Rambur) / Regal Darner

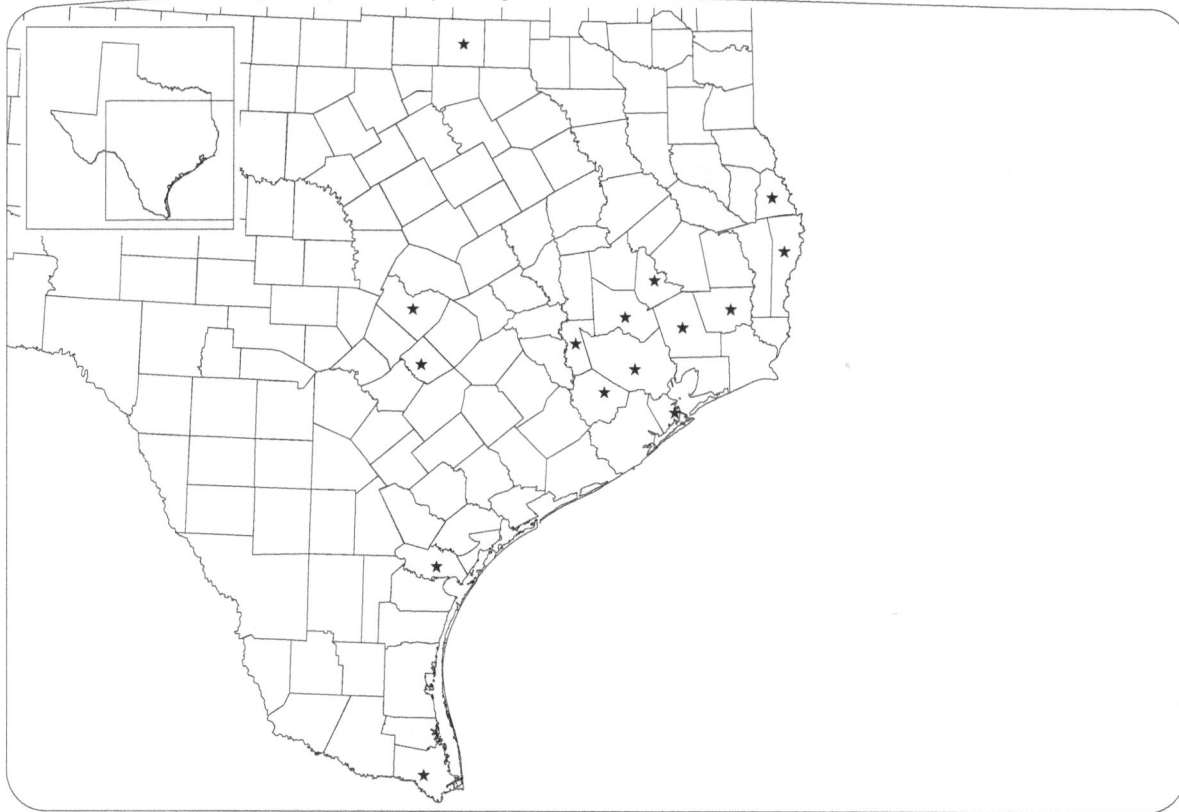

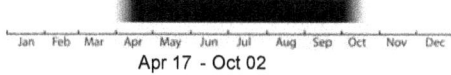

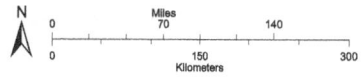

Apr 17 - Oct 02

HABITAT: Lakes and slow flowing streams with heavy vegetation.

Caldwell
Cameron
Fort Bend
Galveston
Hardin
Harris
Liberty
Montgomery
Newton
Sabine
San Jacinto
San Patricio
Tarrant
Travis
Waller

Epiaeschna heros (Fabricius) / Swamp Darner

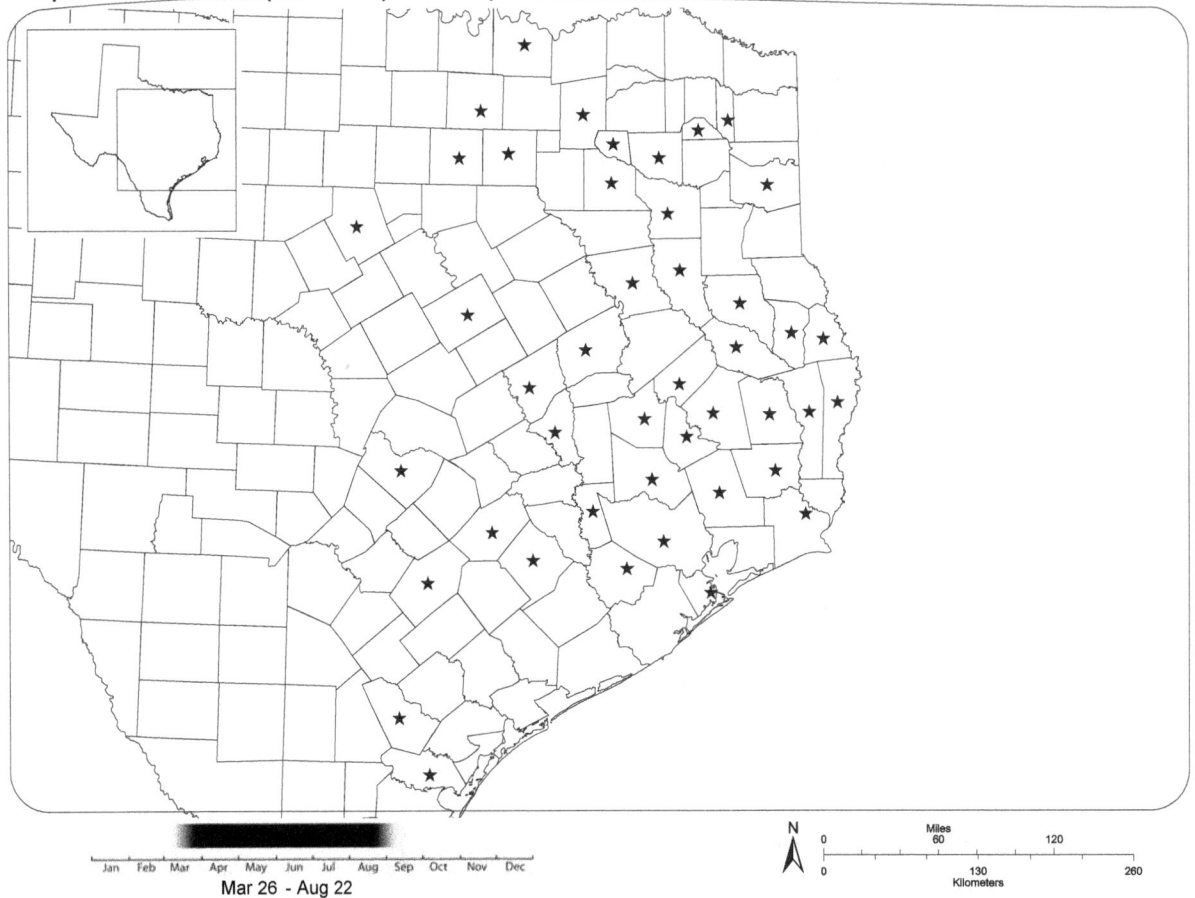

Mar 26 - Aug 22

HABITAT: Heavily wooded ponds, streams and ox-bows including ephemeral pools and ponds

Anderson
Angelina
Bee
Brazos
Camp
Cherokee
Colorado
Dallas
Denton
Erath
Fayette
Fort Bend
Galveston
Gonzales
Grayson
Hardin
Harris
Harrison
Hunt
Jasper
Jefferson
Leon
Liberty
McLennan
Montgomery
Morris

Nacogdoches
Newton
Polk
Rains
Robertson
Sabine
San Augustine
San Jacinto
San Patricio
Smith
Tarrant
Travis
Trinity
Tyler
Van Zandt
Walker
Waller
Wood

Gomphaeschna furcillata (say) / Harlequin Darner

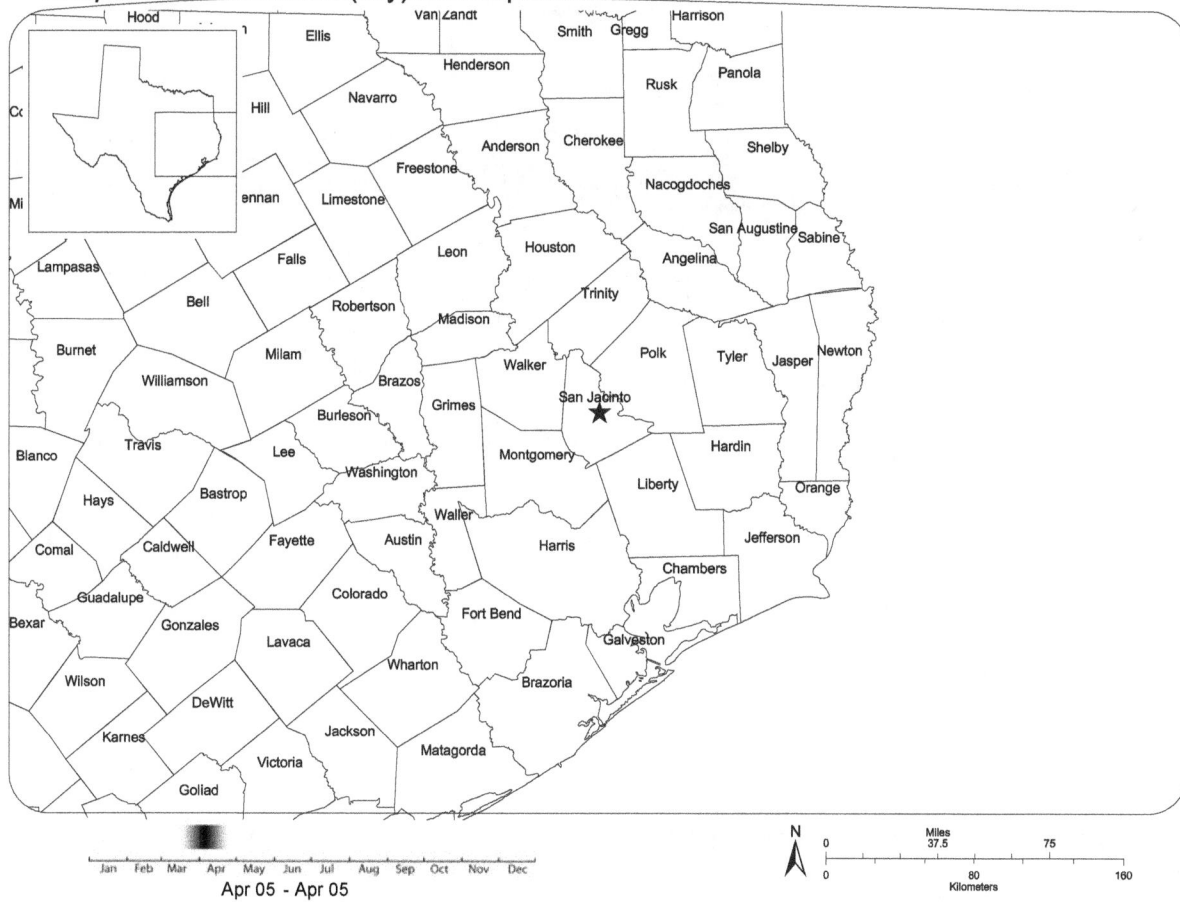

Apr 05 - Apr 05

HABITAT: Shallow sphagnum bogs and swamps.

San Jacinto

Gynacantha mexicana Selys / Bar-sided Darner

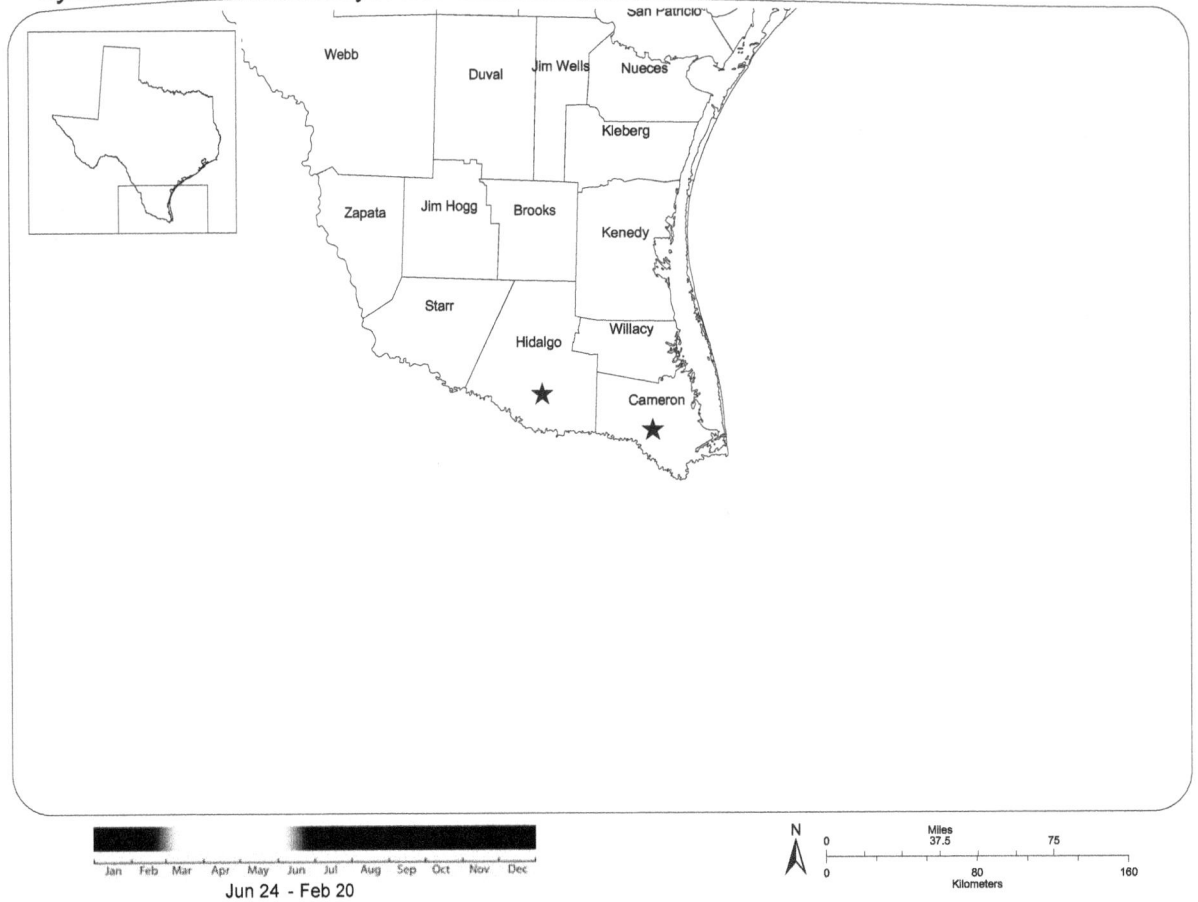

Jun 24 - Feb 20

HABITAT: Ephemeral ponds and pools.

Cameron
Hidalgo

Nasiaeschna pentacantha (Rambur) / Cyrano Darner

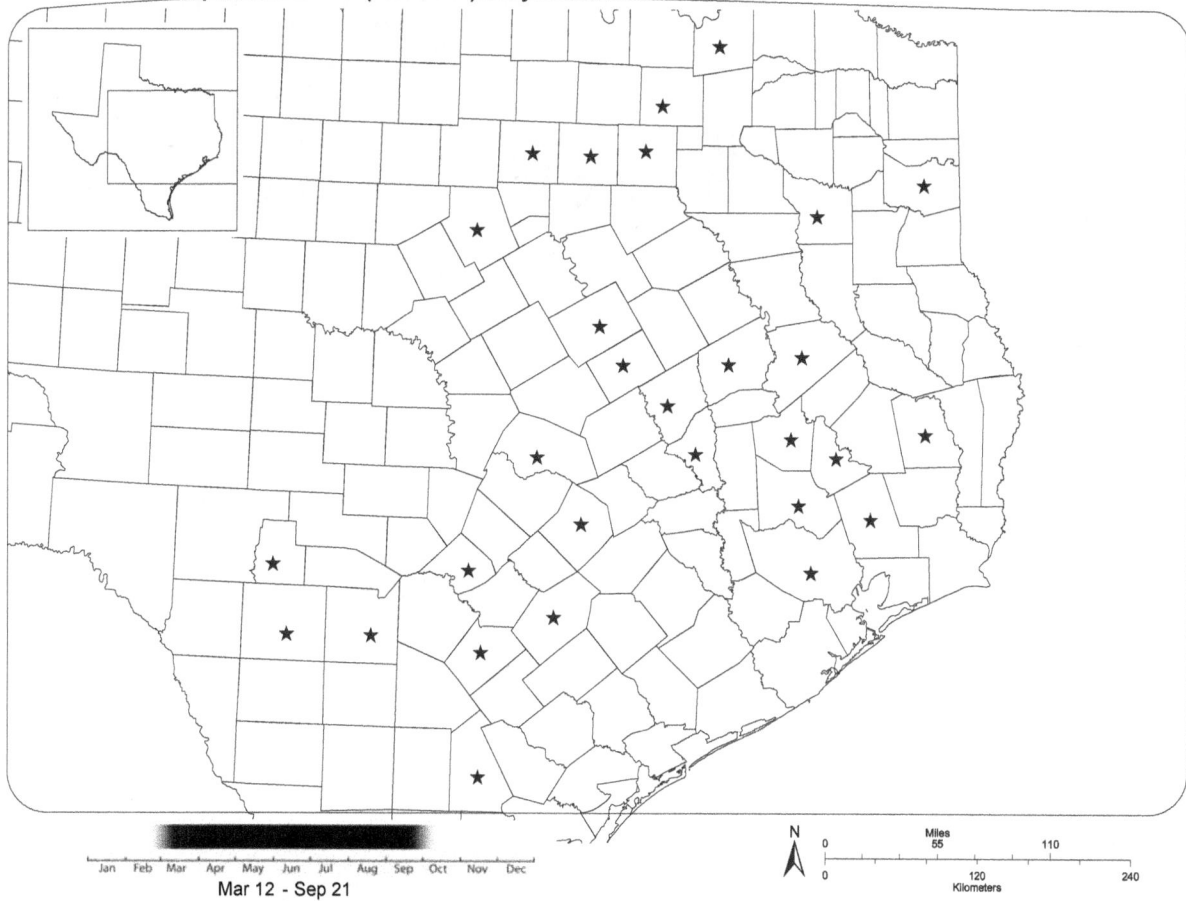

Mar 12 - Sep 21

HABITAT: Sheltered forest ponds, streams and lake coves.

Bastrop
Brazos
Collin
Comal
Dallas
Erath
Falls
Fannin
Gonzales
Harris
Harrison
Houston
Leon
Liberty
Live Oak
McLennan
Medina
Montgomery
Parker
Real
Robertson
San Jacinto
Smith
Tarrant
Tyler
Uvalde

Walker
Williamson
Wilson

Rhionaeschna dugesi Calvert / Arroyo Darner

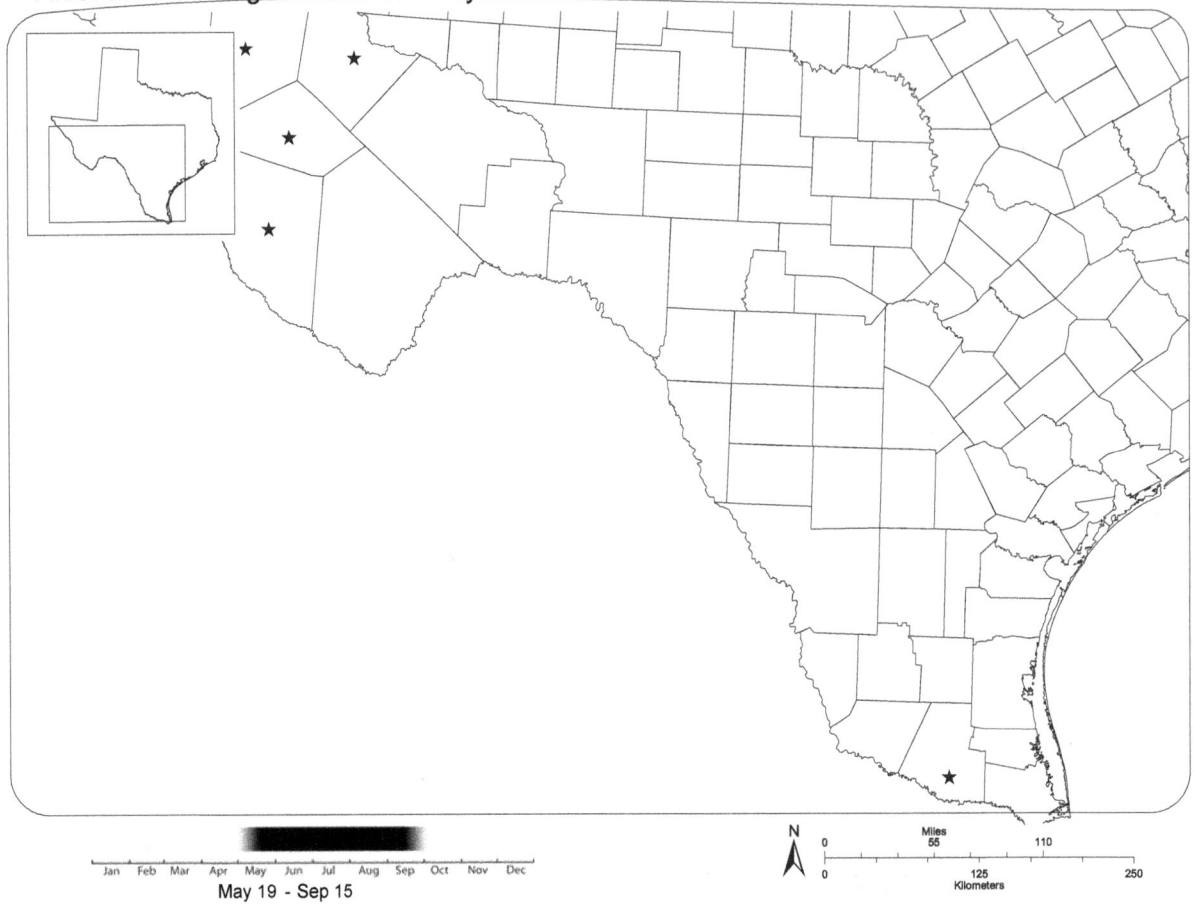

May 19 - Sep 15

HABITAT: Pools of slow flowing permanent mountain streams, rivulets, and arroyos. Often found higher up the watershed towards the headwaters than Blue-eyed Darner.

Culberson
Hidalgo
Jeff Davis
Presidio
Reeves

Rhionaeschna multicolor Hagen / Blue-eyed Darner

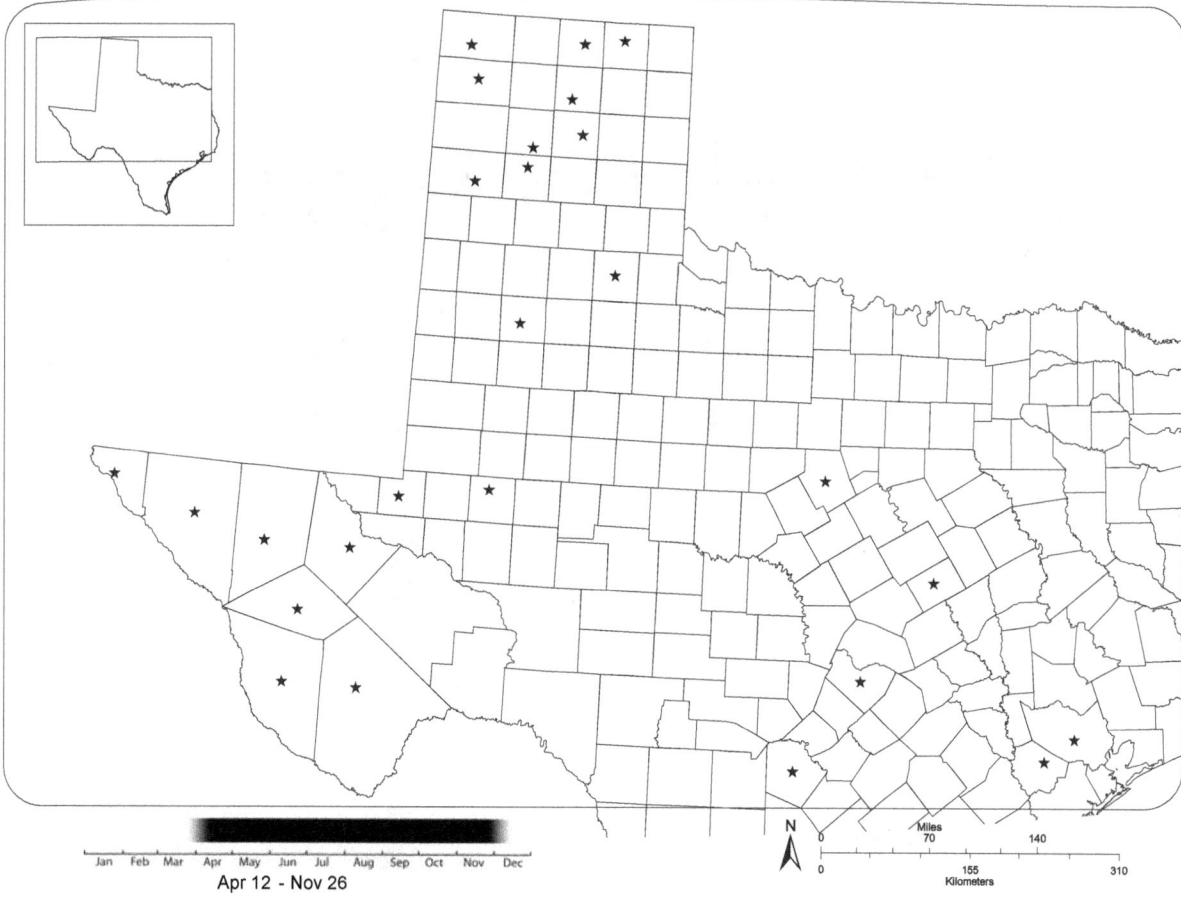

Apr 12 - Nov 26

HABITAT: Open sunlit areas of slow-flowing streams, sloughs, lakes and ponds, including alkaline ones, with moderate vegetation

Bexar
Brewster
Carson
Culberson
Dallam
Deaf Smith
El Paso
Erath
Falls
Fort Bend
Hansford
Harris
Hartley
Hudspeth
Hutchinson
Jeff Davis
Lubbock
Midland
Motley
Ochiltree
Potter
Presidio
Randall
Reeves
Travis
Winkler

Rhionaeschna psilus Calvert / Turquoise-tipped Darner

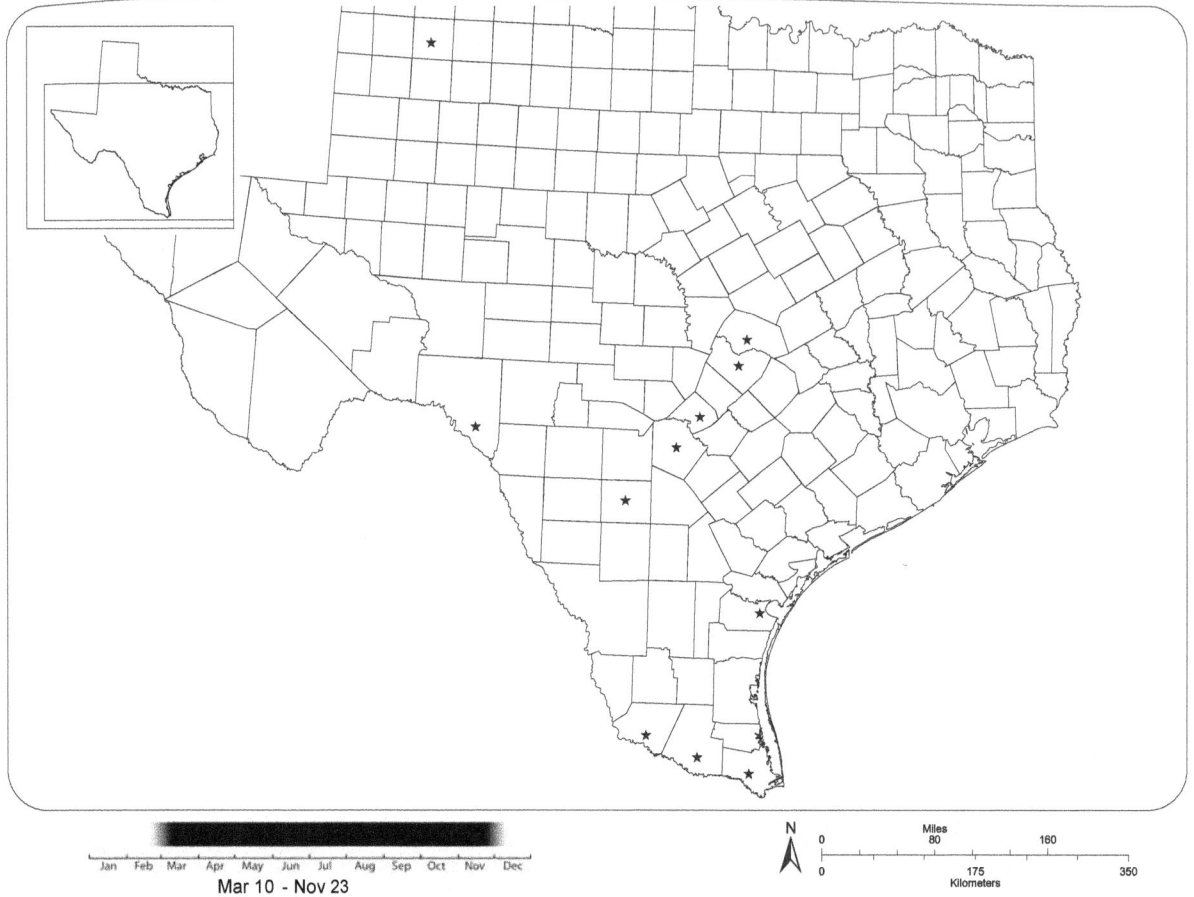

Mar 10 - Nov 23

HABITAT: Slow-flowing, open sunlit permanent and temporary streams and ponds.

Bexar
Cameron
Comal
Frio
Hidalgo
Lubbock
Nueces
Starr
Travis
Val Verde
Williamson

Triacanthagyna septima (Selys) / Pale-green Darner

Jun 22 - Oct 22

HABITAT: Fishless, ephemeral ponds with sufficient emergent vegetation and shade.

Hidalgo

Aphylla angustifolia Garrison / Broad-striped Forceptail

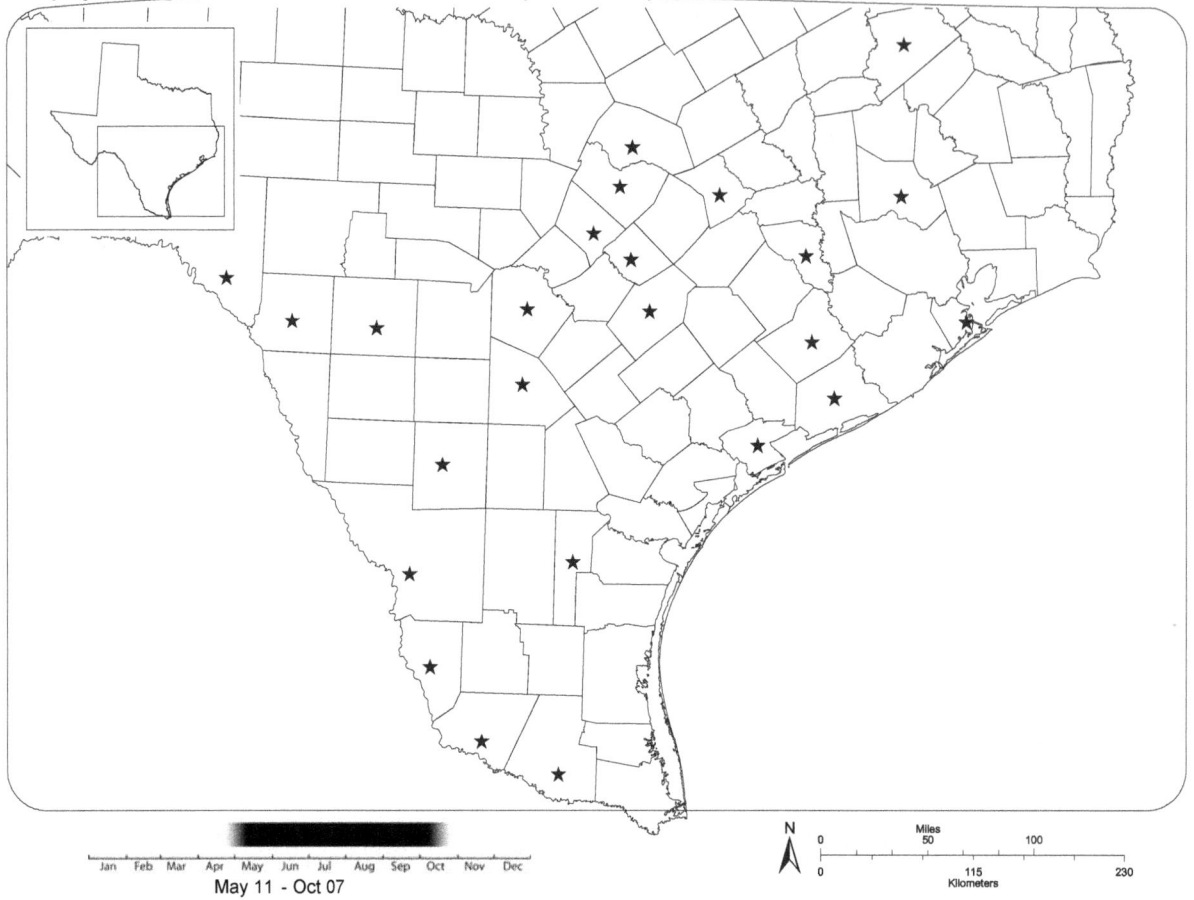

Jan Feb Mar Apr May Jun Jul Aug Sep Oct Nov Dec

May 11 - Oct 07

HABITAT: Lakes, ponds and pools of intermittent streams with muddy bottoms.

Atascosa
Austin
Bexar
Caldwell
Calhoun
Galveston
Gonzales
Hays
Hidalgo
Houston
Jim Wells
Kinney
La Salle
Lee
Matagorda
Montgomery
Starr
Travis
Uvalde
Val Verde
Webb
Wharton
Williamson
Zapata

Aphylla protracta (Hagen in Selys) / Narrow-striped Forceptail

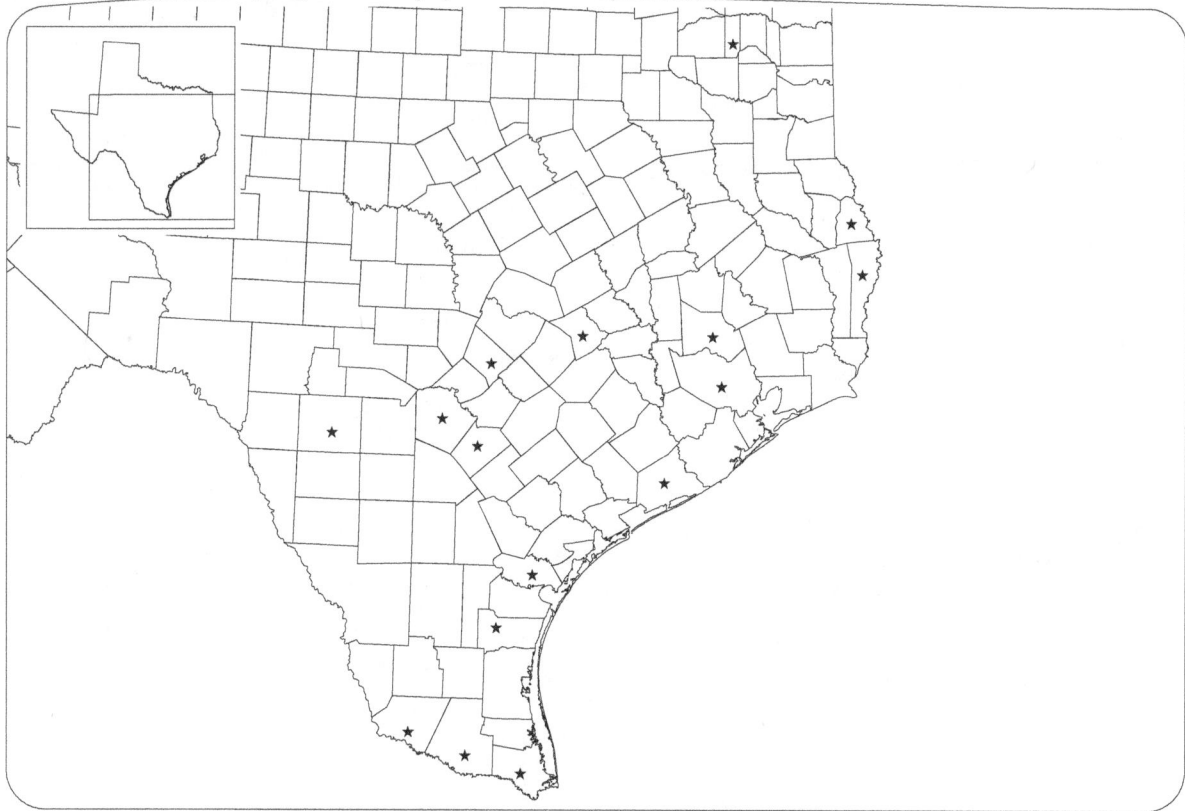

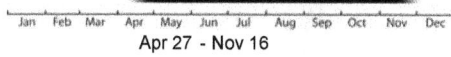

Apr 27 - Nov 16

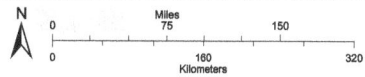

HABITAT: Lakes, ponds and pools of intermittent streams with muddy bottoms.

Bexar
Cameron
Franklin
Harris
Hays
Hidalgo
Kleberg
Lee
Matagorda
Montgomery
Newton
Sabine
San Patricio
Starr
Uvalde
Wilson

Aphylla williamsoni (Gloyd) / Two-striped Forceptail

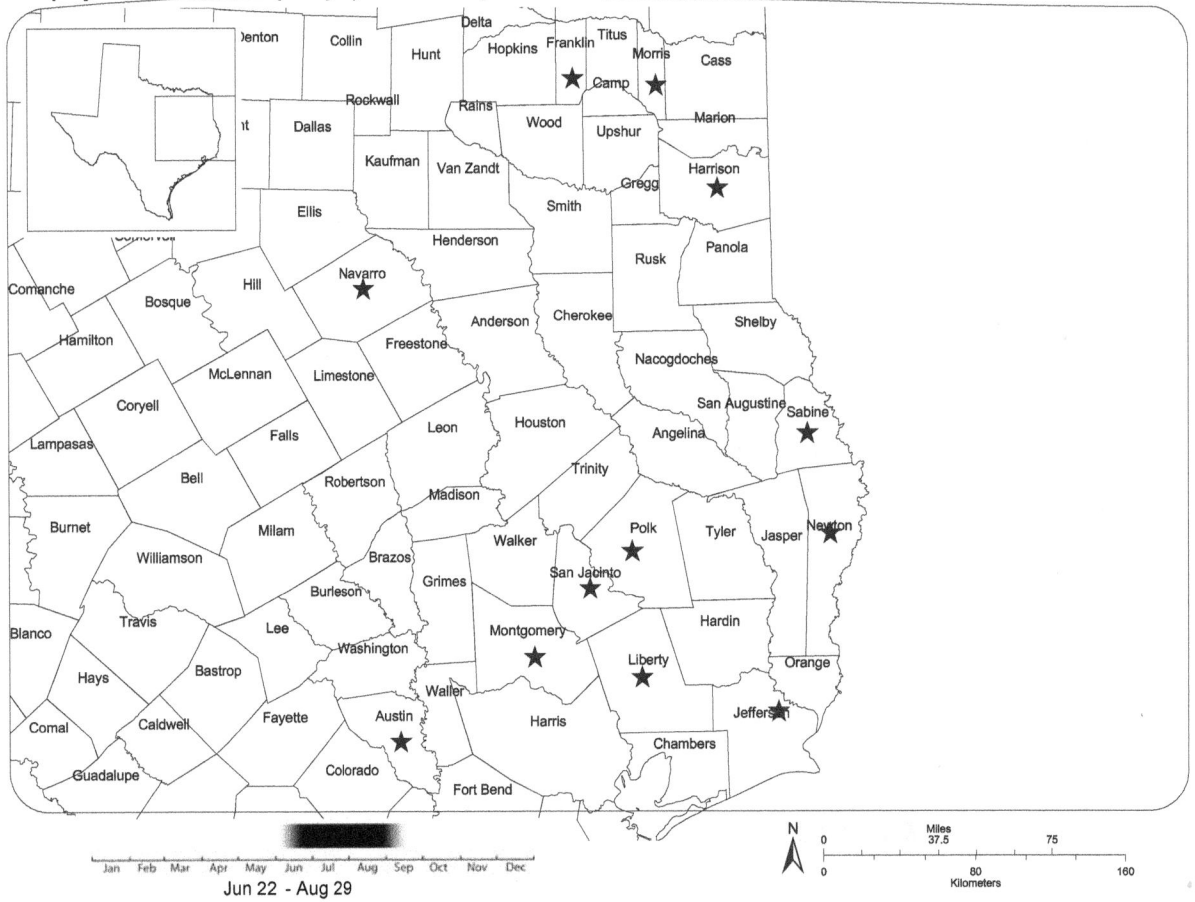

Jun 22 - Aug 29

HABITAT: Ponds, lakes, borrow pits and sluggish streams.

Austin
Franklin
Harrison
Jefferson
Liberty
Montgomery
Morris
Navarro
Newton
Polk
Sabine
San Jacinto

Arigomphus lentulus (Needham) / Stillwater Clubtail

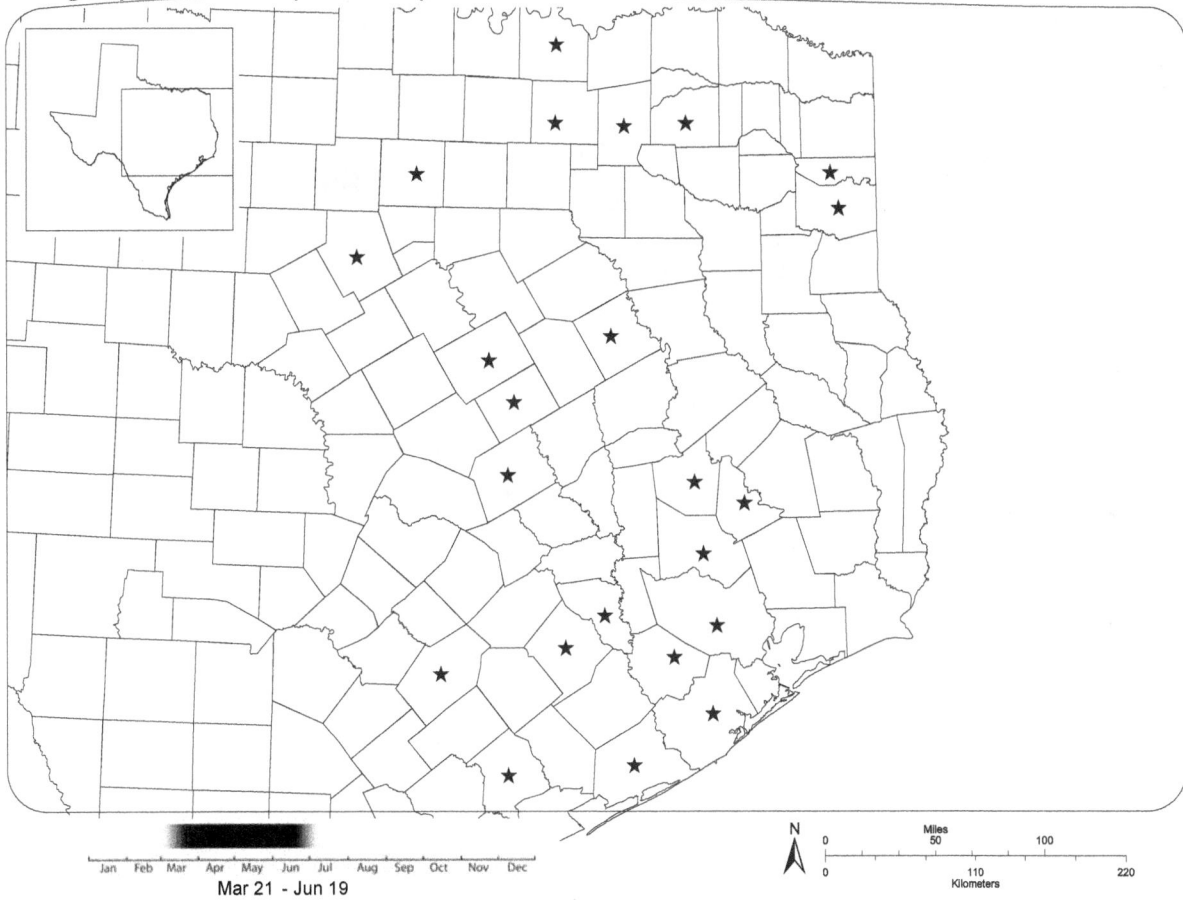

Mar 21 - Jun 19

HABITAT: Semi-permanent and artificial ponds, lakes and slow areas of streams with muddy bottoms.

Austin
Brazoria
Collin
Colorado
Erath
Falls
Fort Bend
Freestone
Gonzales
Grayson
Harris
Harrison
Hopkins
Hunt
Marion
Matagorda
McLennan
Milam
Montgomery
Parker
San Jacinto
Victoria
Walker

Arigomphus maxwelli (Ferguson) / Bayou Clubtail

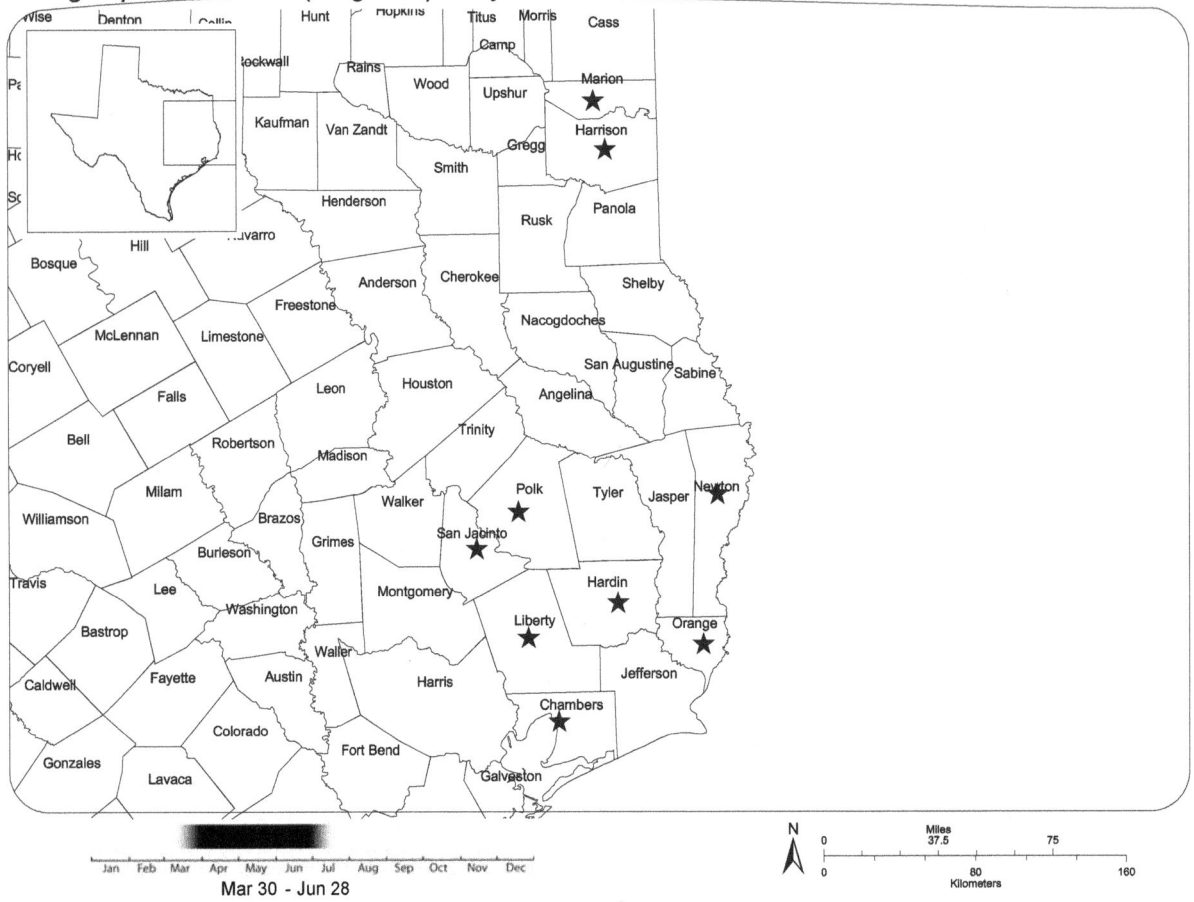

Mar 30 - Jun 28

HABITAT: Ditches, bayous and semi-permanent lakes and ponds with muddy bottoms.

Chambers
Hardin
Harrison
Liberty
Marion
Newton
Orange
Polk
San Jacinto

Arigomphus submedianus (Williamson) / Jade Clubtail

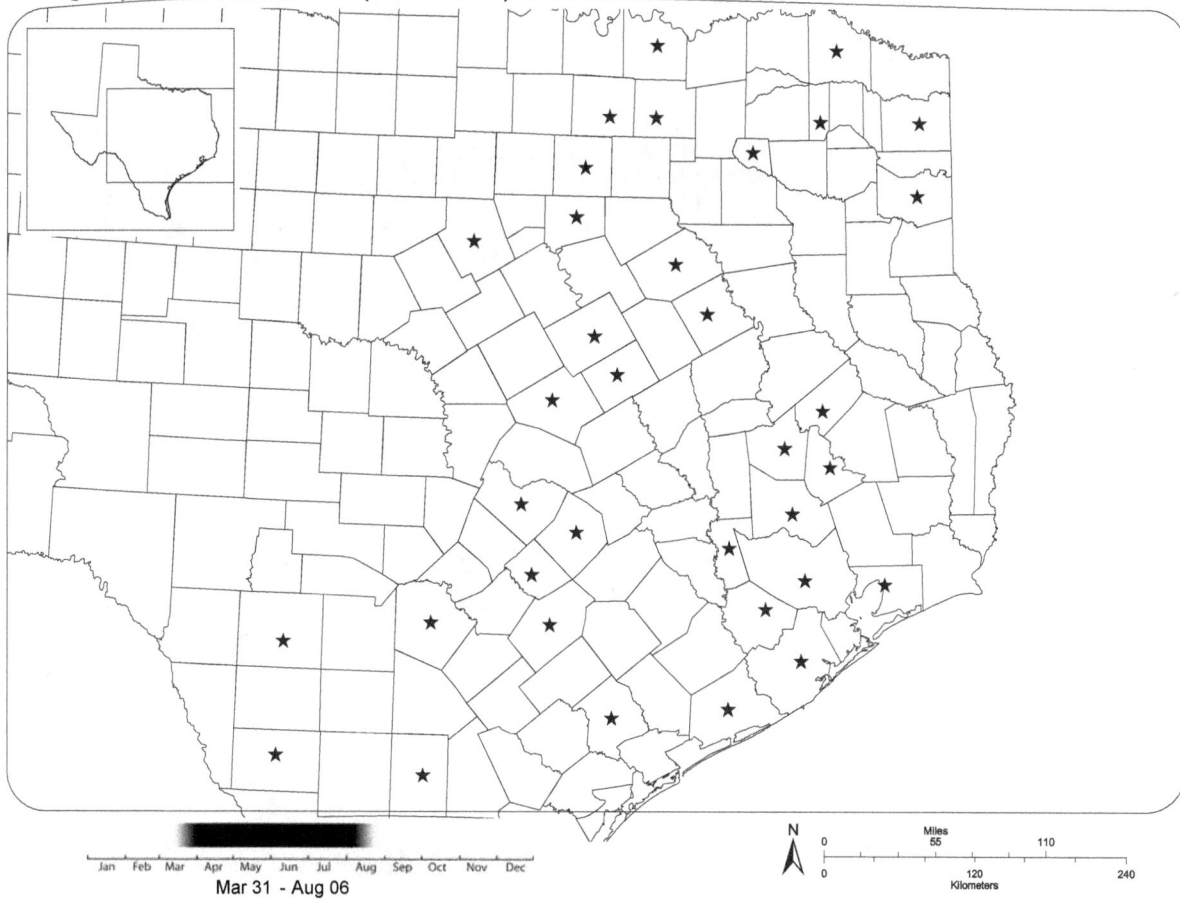

Jan Feb Mar Apr May Jun Jul Aug Sep Oct Nov Dec

Mar 31 - Aug 06

HABITAT: Semi-permanent and artificial ponds, lakes and slow-areas of streams with muddy bottoms.

Bastrop	Red River
Bell	San Jacinto
Bexar	Tarrant
Brazoria	Travis
Caldwell	Trinity
Cass	Uvalde
Chambers	Victoria
Collin	Walker
Denton	Waller
Dimmit	
Erath	
Falls	
Fort Bend	
Franklin	
Freestone	
Gonzales	
Grayson	
Harris	
Harrison	
Johnson	
Matagorda	
McLennan	
McMullen	
Montgomery	
Navarro	
Rains	

Dromogomphus spinosus Selys / Black-shouldered Spinyleg

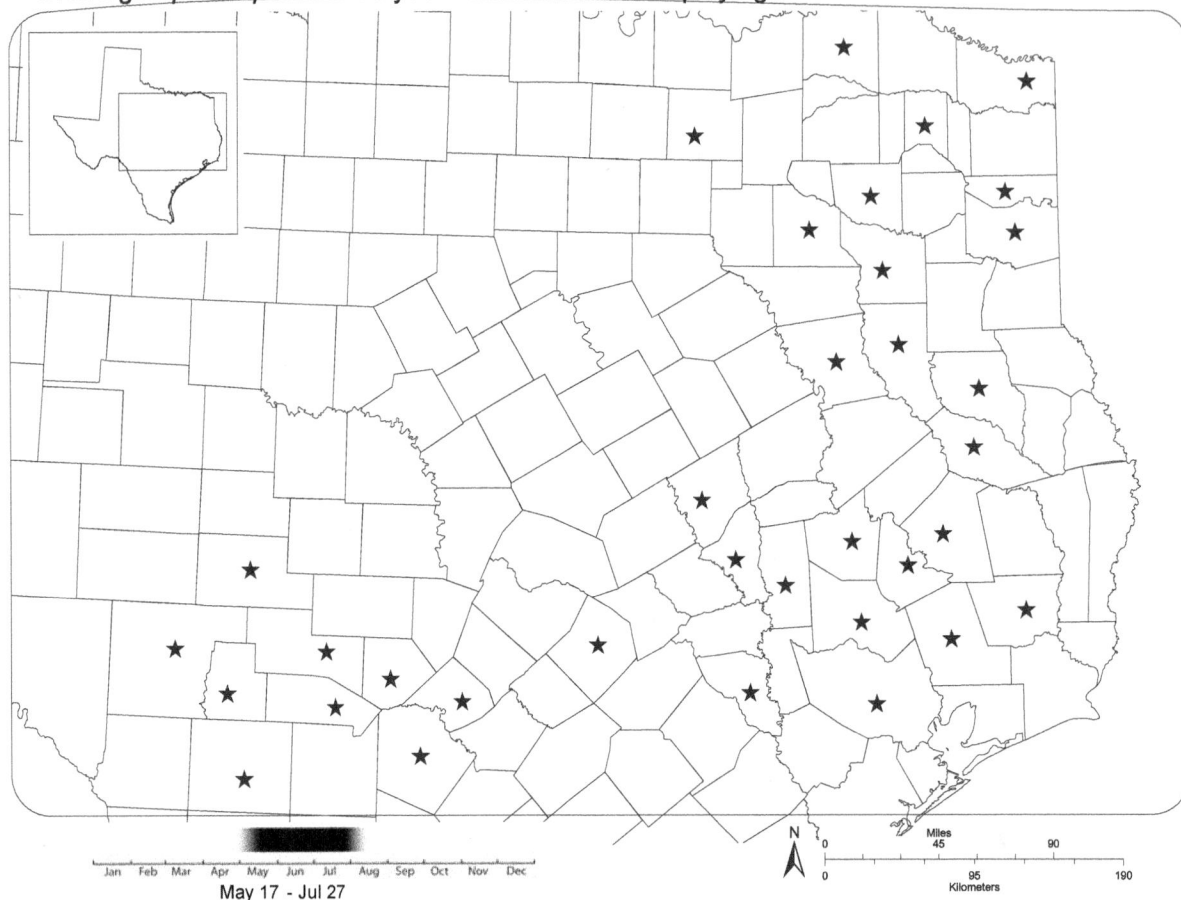

May 17 - Jul 27

HABITAT: Small to large streams and oxbows with slow to rapid flow and sandy or muddy bottoms.

Anderson
Angelina
Austin
Bandera
Bastrop
Bexar
Bowie
Brazos
Cherokee
Collin
Comal
Edwards
Grimes
Hardin
Harris
Harrison
Kendall
Kerr
Kimble
Lamar
Liberty
Marion
Montgomery
Nacogdoches
Polk
Real

Robertson
San Jacinto
Smith
Titus
Uvalde
Van Zandt
Walker
Wood

Dromogomphus spoliatus (Hagen in Selys) / Flag-tailed Spinyleg

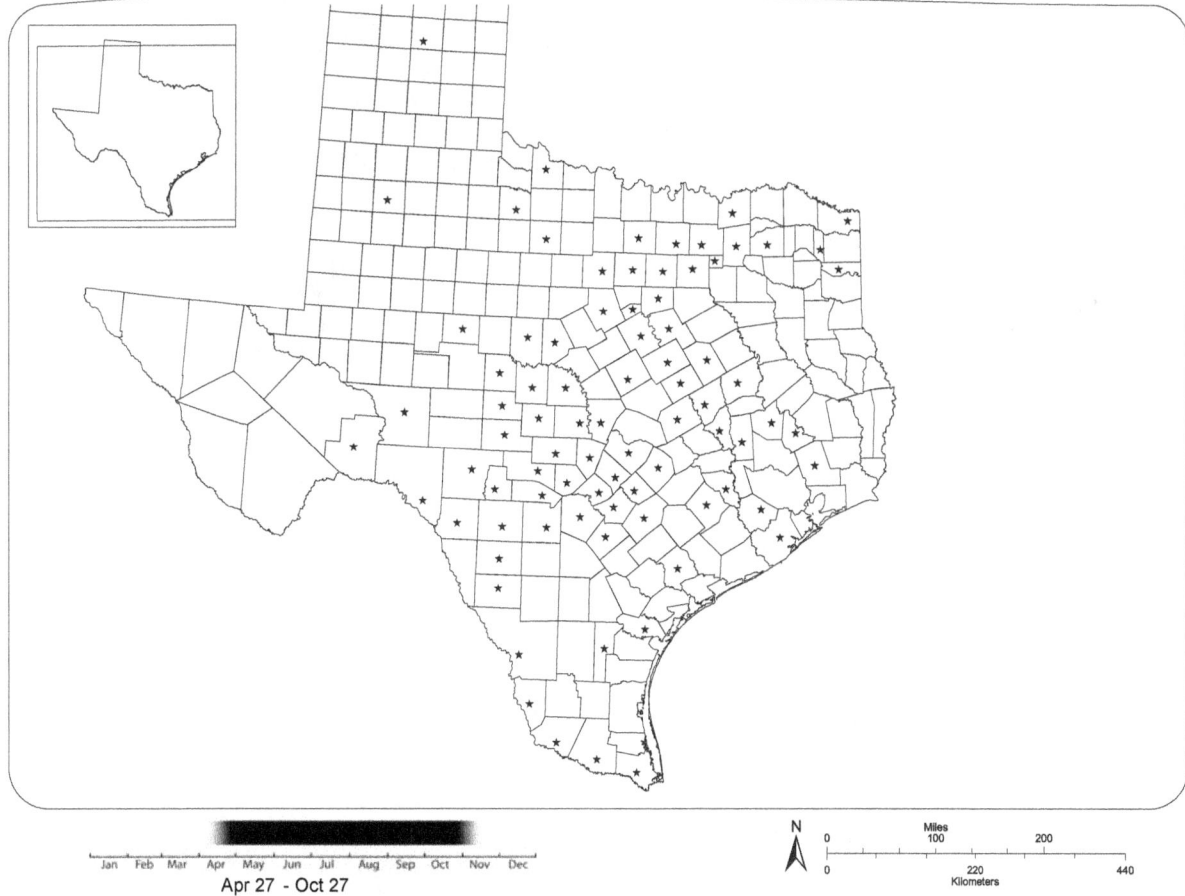

Apr 27 - Oct 27

HABITAT: Small, clear sandy or mud bottomed streams with a regular current.

Austin	Falls	Mason	Wilbarger
Bandera	Fannin	McCulloch	Wilson
Bastrop	Fort Bend	McLennan	Wise
Bexar	Gillespie	Medina	Zapata
Blanco	Gonzales	Menard	Zavala
Bosque	Grimes	Milam	
Bowie	Guadalupe	Morris	
Brazoria	Hays	Palo Pinto	
Brazos	Hidalgo	Parker	
Brown	Hill	Real	
Burnet	Hopkins	Robertson	
Caldwell	Hunt	Rockwall	
Cameron	Hutchinson	San Jacinto	
Coke	Jim Wells	San Patricio	
Coleman	Johnson	San Saba	
Collin	Kendall	Somervell	
Colorado	Kerr	Starr	
Comal	Kimble	Tarrant	
Concho	Kinney	Terrell	
Coryell	Knox	Throckmorton	
Crockett	Leon	Travis	
Dallas	Liberty	Uvalde	
Denton	Limestone	Val Verde	
Dimmit	Llano	Victoria	
Edwards	Lubbock	Walker	
Erath	Marion	Webb	

Erpetogomphus compositus Hagen in Selys / White-belted Ringtail

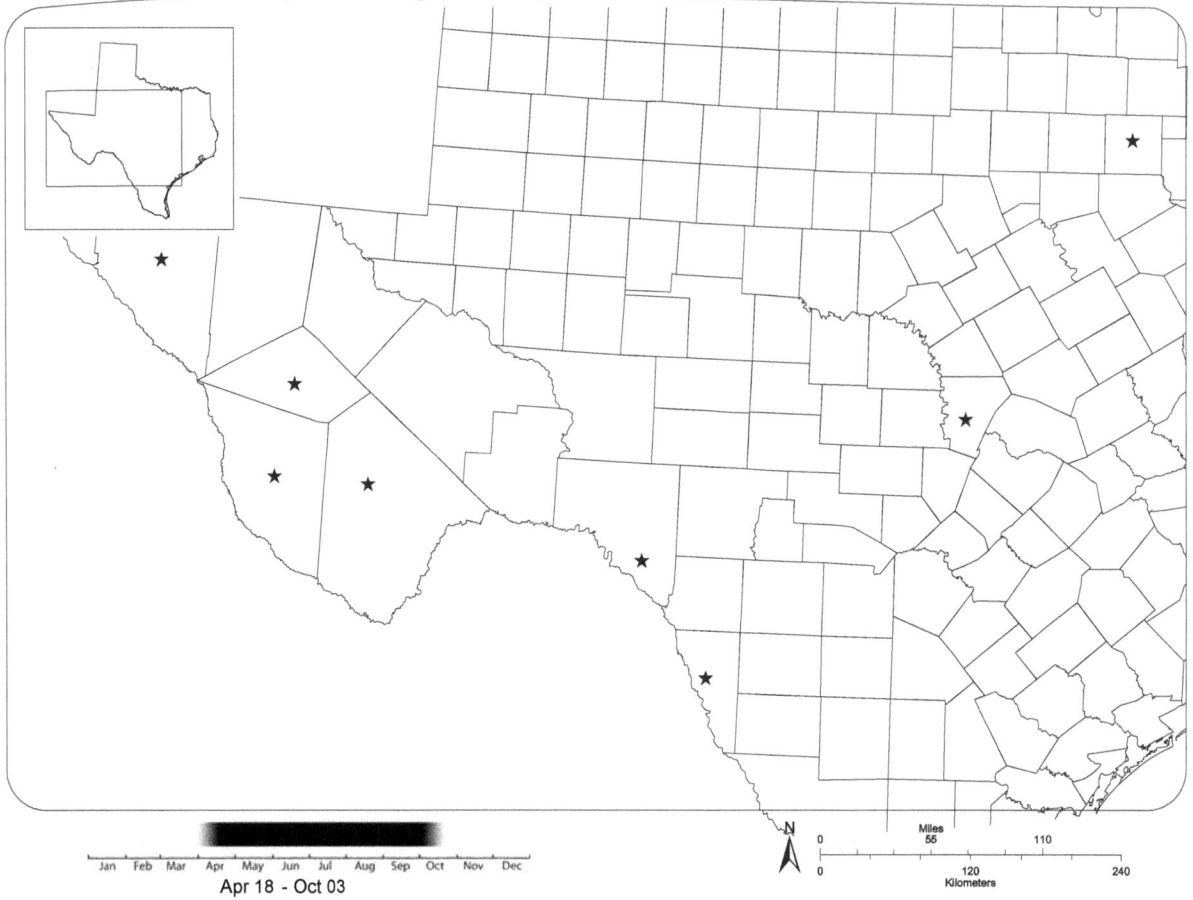

Jan Feb Mar Apr May Jun Jul Aug Sep Oct Nov Dec

Apr 18 - Oct 03

HABITAT: Desert streams, creeks and irrigation ditches with wide sandy or rocky margins.

Brewster
Burnet
Dallas
El Paso
Hudspeth
Jeff Davis
Maverick
Presidio
Val Verde

Erpetogomphus designatus Hagen in Selys / Eastern Ringtail

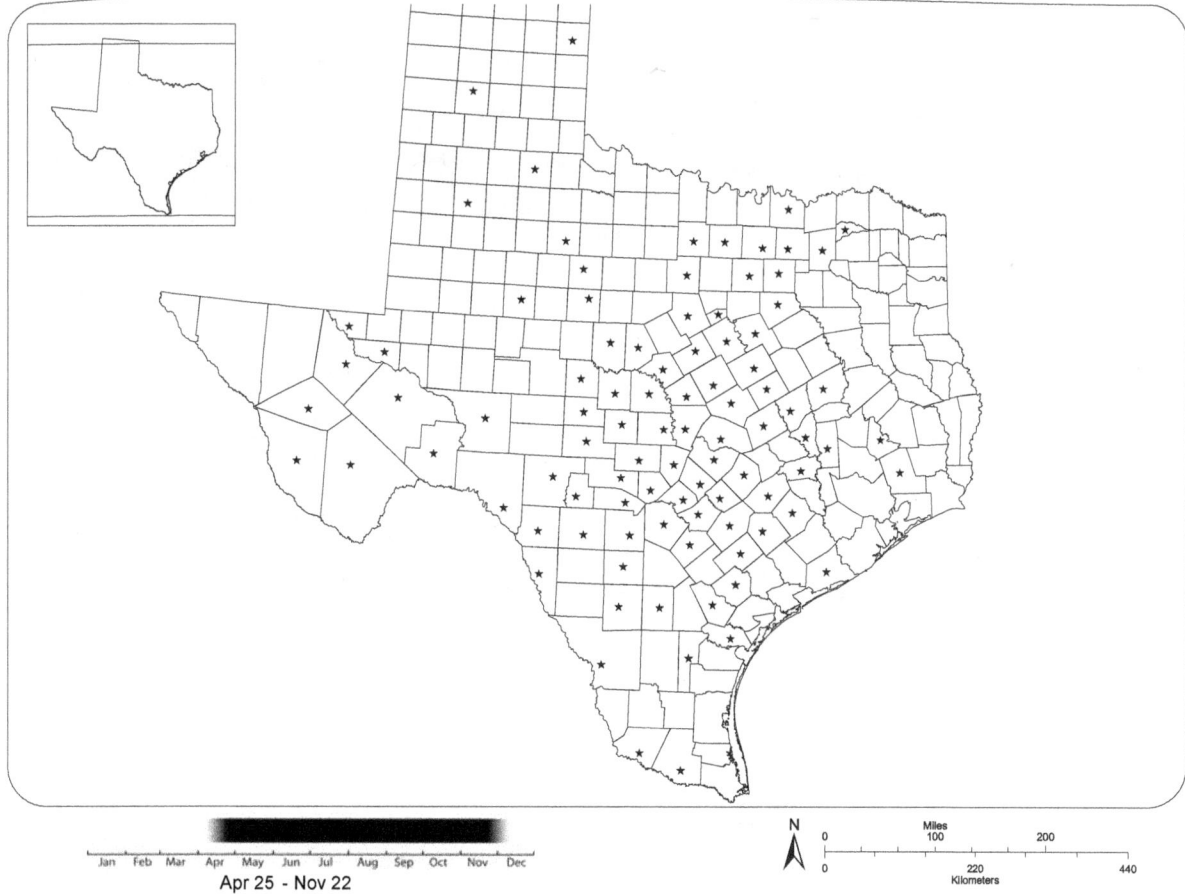

Jan Feb Mar Apr May Jun Jul Aug Sep Oct Nov Dec

Apr 25 - Nov 22

HABITAT: Clear streams and rivers of deciduous forests with moderate current.

Bandera	Falls	Leon	San Saba
Bastrop	Fayette	Liberty	Somervell
Bee	Frio	Llano	Starr
Bell	Gillespie	Loving	Stonewall
Bexar	Goliad	Lubbock	Tarrant
Blanco	Gonzales	Mason	Taylor
Bosque	Grayson	Matagorda	Terrell
Brazos	Grimes	Maverick	Travis
Brewster	Guadalupe	McCulloch	Uvalde
Brown	Hamilton	McLennan	Val Verde
Burnet	Hays	McMullen	Ward
Caldwell	Hemphill	Medina	Washington
Coleman	Hidalgo	Menard	Webb
Collin	Hill	Milam	Williamson
Colorado	Hunt	Mills	Wilson
Comal	Jack	Mitchell	Wise
Concho	Jeff Davis	Motley	
Coryell	Jim Wells	Palo Pinto	
Crockett	Jones	Pecos	
Dallas	Kendall	Presidio	
Delta	Kerr	Randall	
Denton	Kimble	Real	
DeWitt	Kinney	Reeves	
Edwards	La Salle	Robertson	
Ellis	Lampasas	San Jacinto	
Erath	Lavaca	San Patricio	

Erpetogomphus eutainia Calvert / Blue-faced Ringtail

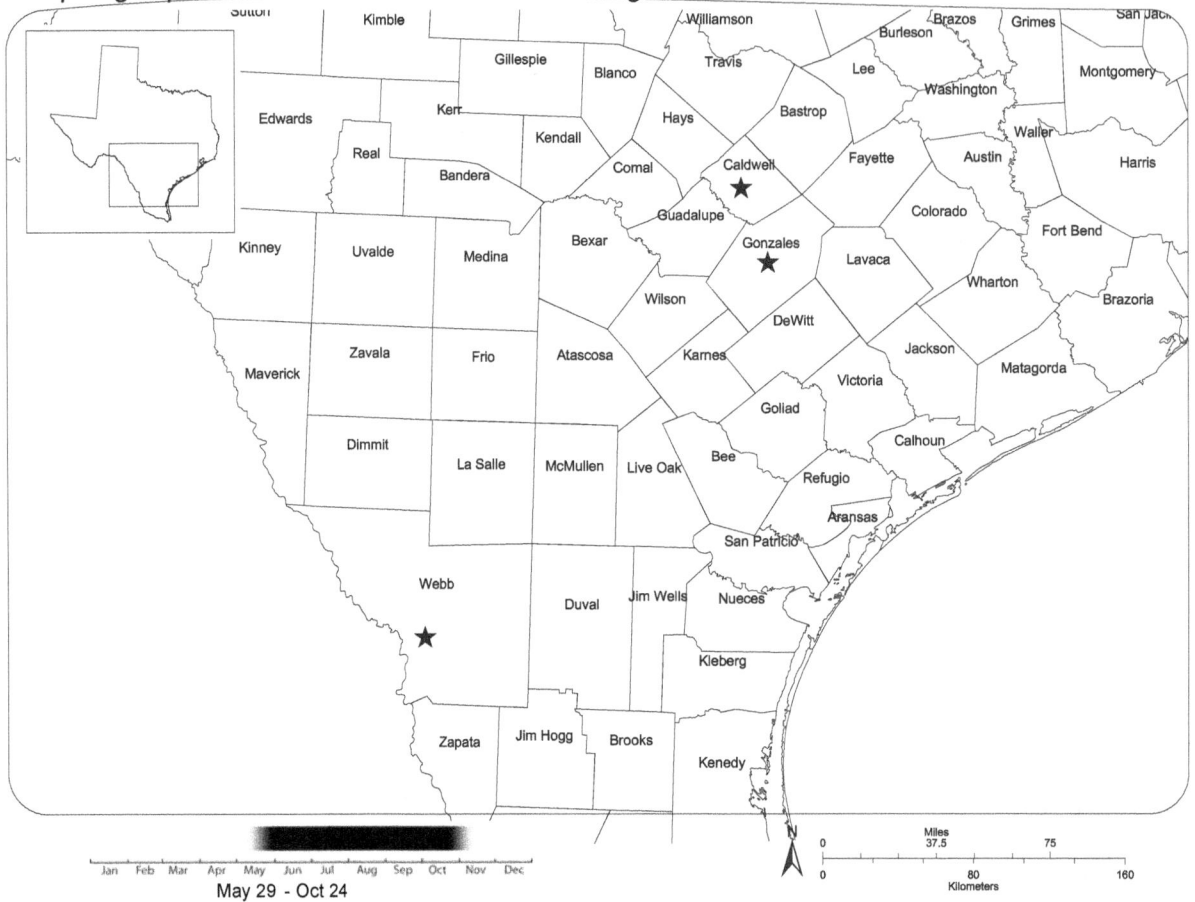

May 29 - Oct 24

HABITAT: Small rivulets and streams of central Texas, with swift current and cobble bottoms. Presently restricted to the San Marcos and Guadalupe Rivers of central Texas and a single record in south Texas on the Rio Grande.

Caldwell
Gonzales
Webb

Erpetogomphus heterodon Garrison / Dashed Ringtail

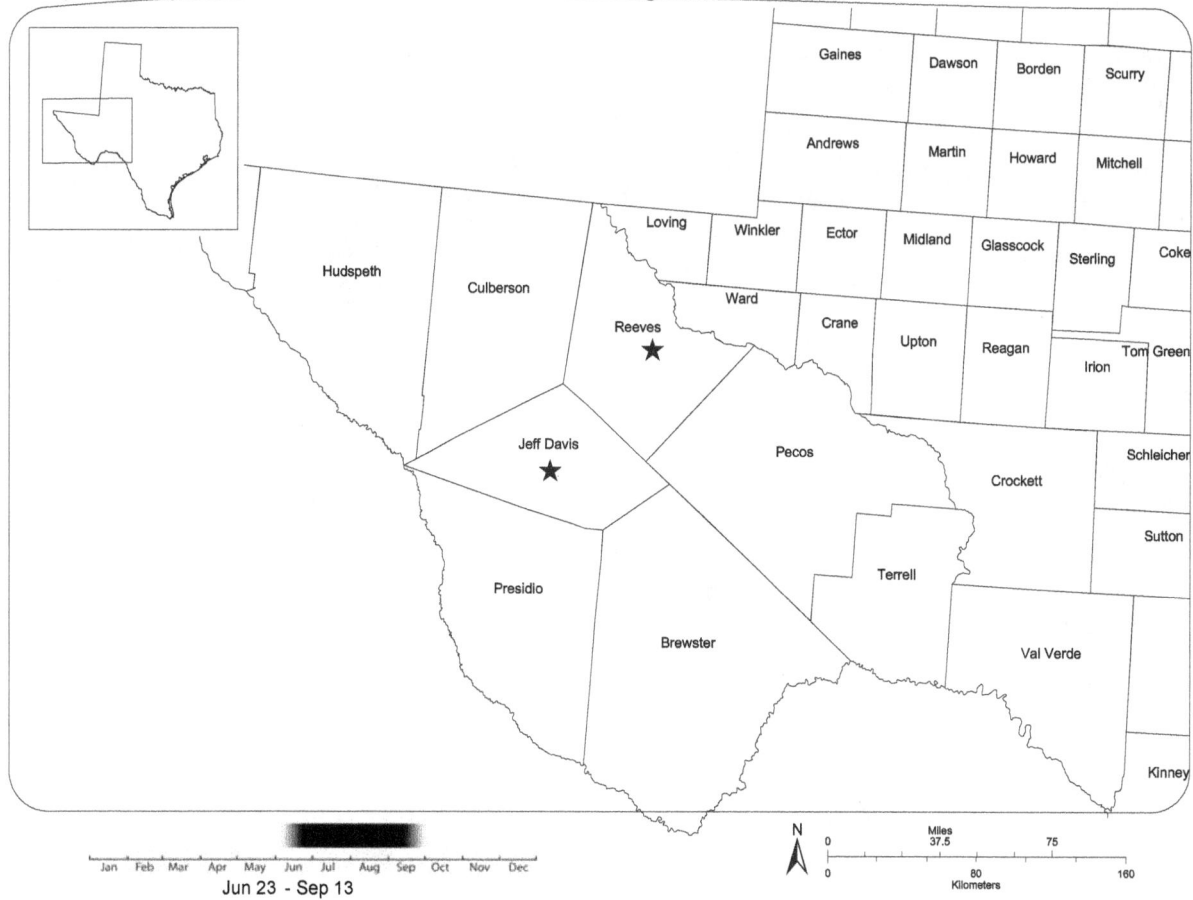

Jun 23 - Sep 13

HABITAT: Higher altitude rivers and streams with swift current and rocky or cobble bottoms.

Jeff Davis
Reeves

Erpetogomphus lampropeltis natrix Williamson and Williamson / Serpent Ringtail

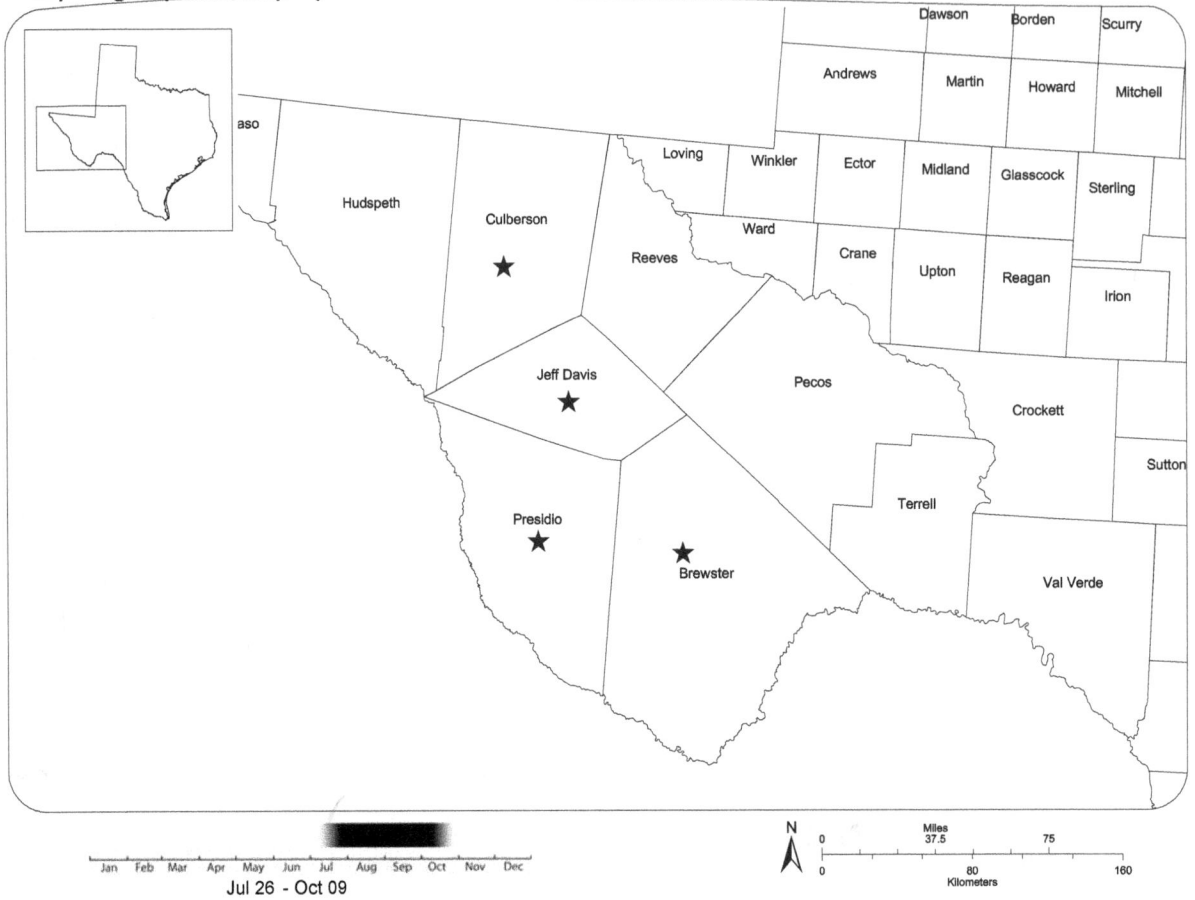

Jul 26 - Oct 09

HABITAT: Rivers and streams with swift current and rocky or cobble bottoms.

Brewster
Culberson
Jeff Davis
Presidio

Gomphus apomyius Donnelly / Banner Clubtail

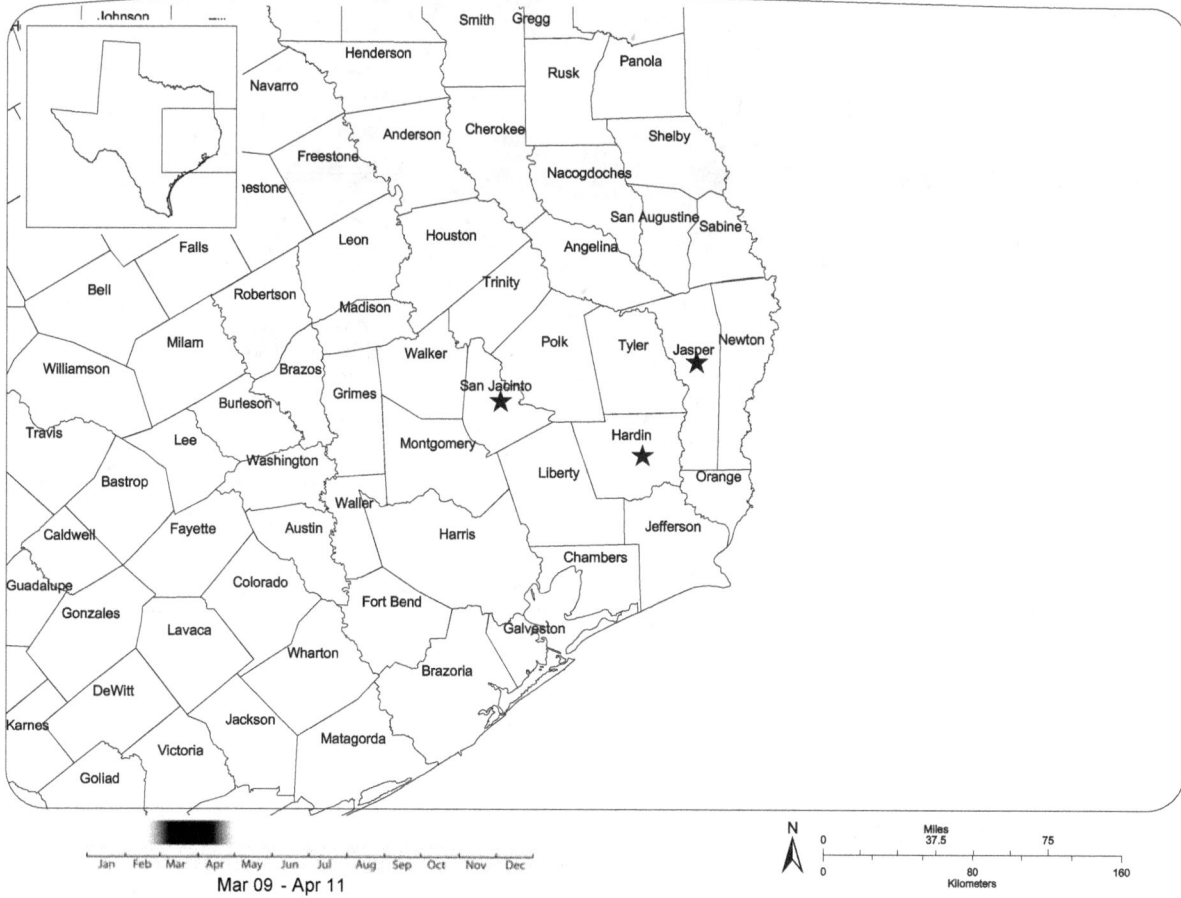

Mar 09 - Apr 11

HABITAT: Small, shaded streams with loose flowing sand.

Hardin
Jasper
San Jacinto

Gomphus externus Hagen in Selys / Plains Clubtail

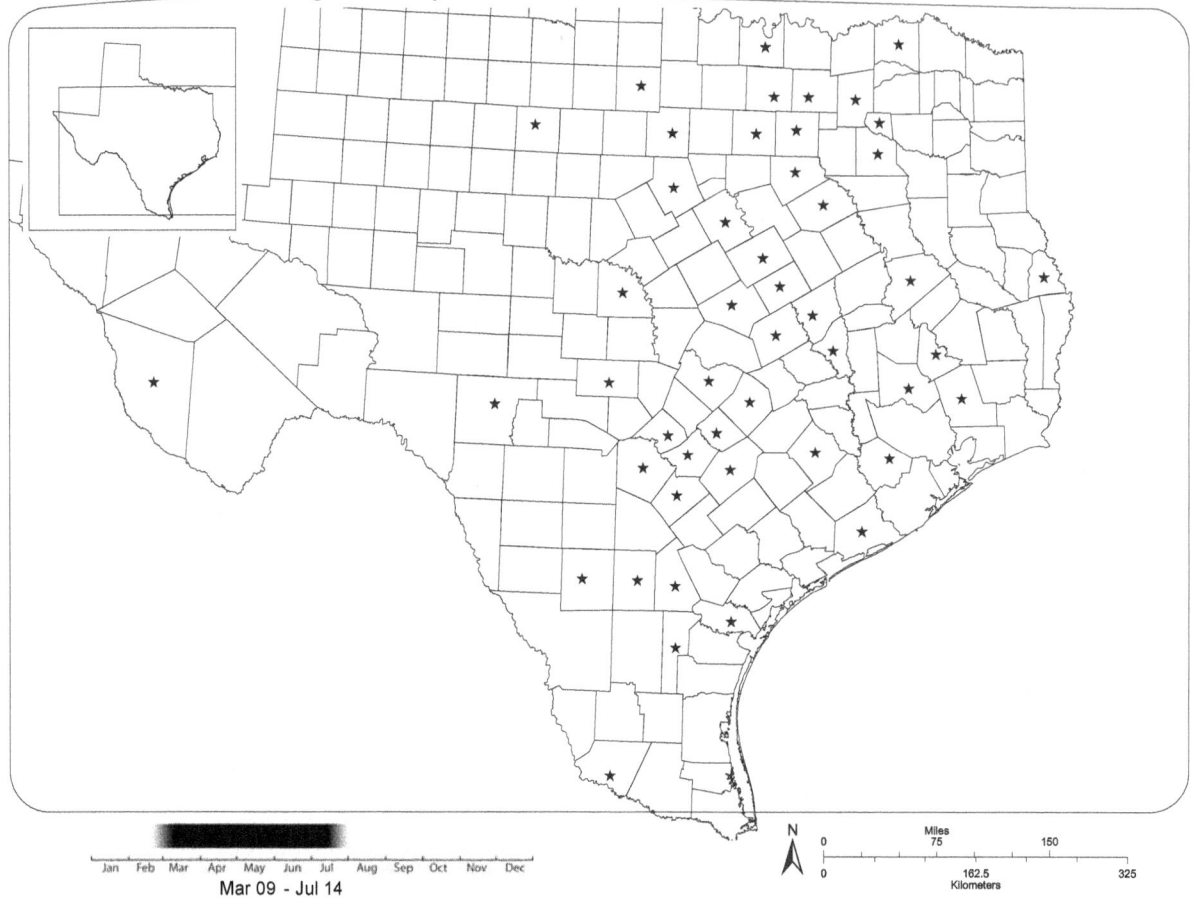

Mar 09 - Jul 14

HABITAT: Large muddy bottomed rivers and streams with moderate flow.

Bastrop	Liberty
Bell	Live Oak
Bexar	Matagorda
Bosque	McLennan
Brazos	McMullen
Caldwell	Milam
Collin	Montgomery
Colorado	Navarro
Comal	Palo Pinto
Cooke	Presidio
Dallas	Rains
Denton	Robertson
Edwards	Sabine
Ellis	San Jacinto
Erath	San Patricio
Falls	San Saba
Fort Bend	Starr
Gillespie	Tarrant
Gonzales	Travis
Guadalupe	Van Zandt
Houston	Wilson
Hunt	Young
Jim Wells	
Jones	
La Salle	
Lamar	

Gomphus gonzalezi Dunkle / Tamaulipan Clubtail

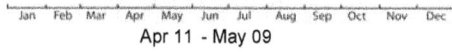

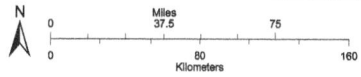

Apr 11 - May 09

HABITAT: Muddy canal-like channels and clear, spring fed deep rivers.

Cameron
Starr

Gomphus graslinellus Walsh / Pronghorn Clubtail

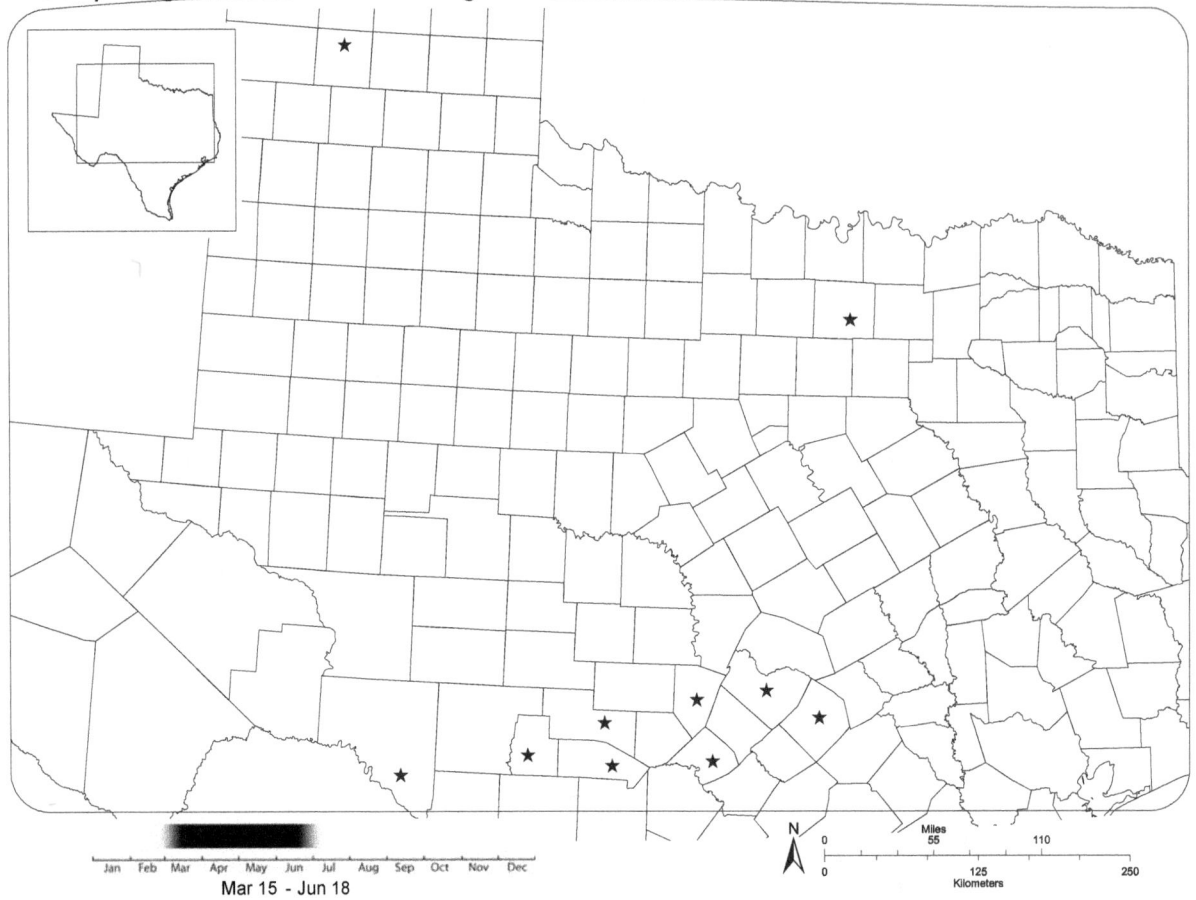

Mar 15 - Jun 18

HABITAT: Ponds, lakes and slow-reaches of small and large streams.

Bandera
Bastrop
Blanco
Comal
Denton
Kerr
Randall
Real
Travis
Val Verde

Gomphus hybridus Williamson / Cocoa Clubtail

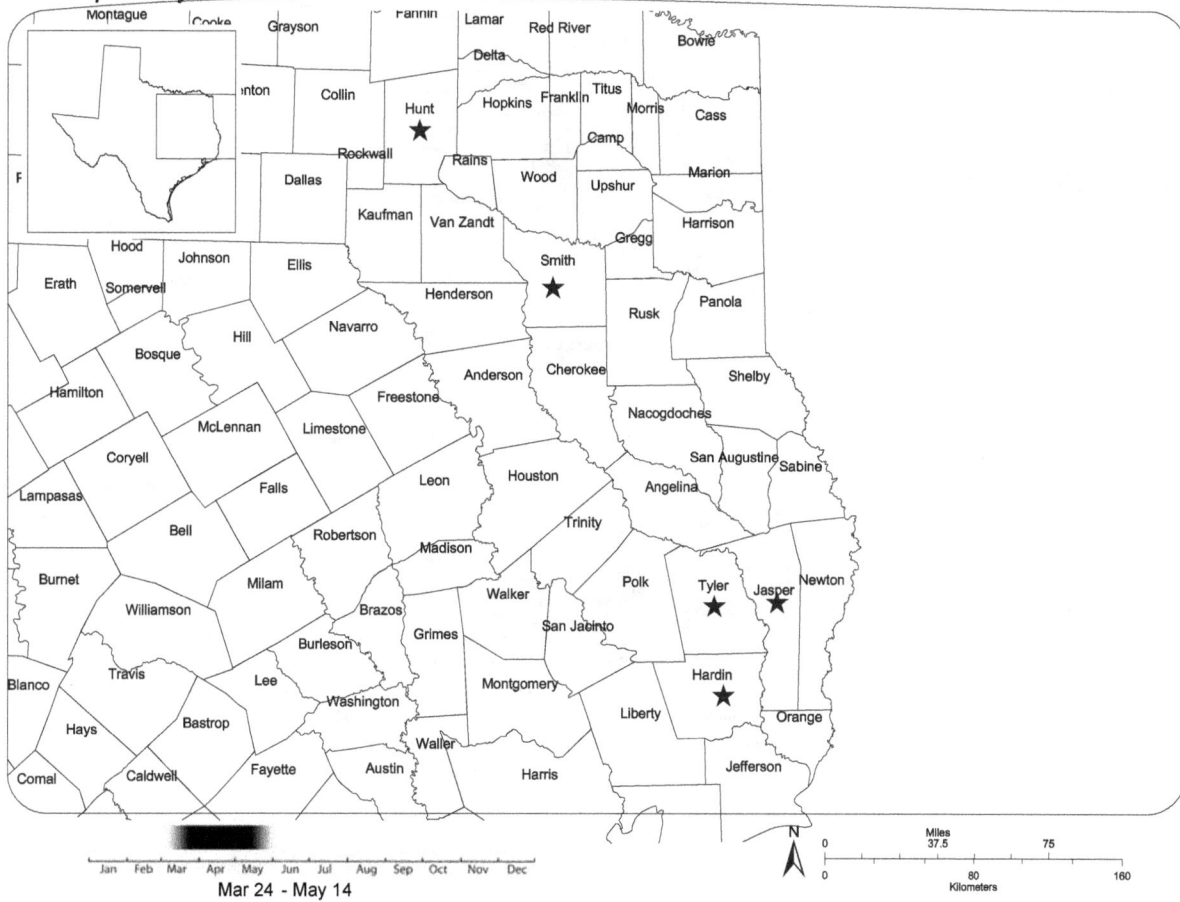

Mar 24 - May 14

HABITAT: Large turbid rivers with moderate current and sandy bottoms.

Hardin
Hunt
Jasper
Smith
Tyler

Gomphus lividus Selys / Ashy Clubtail

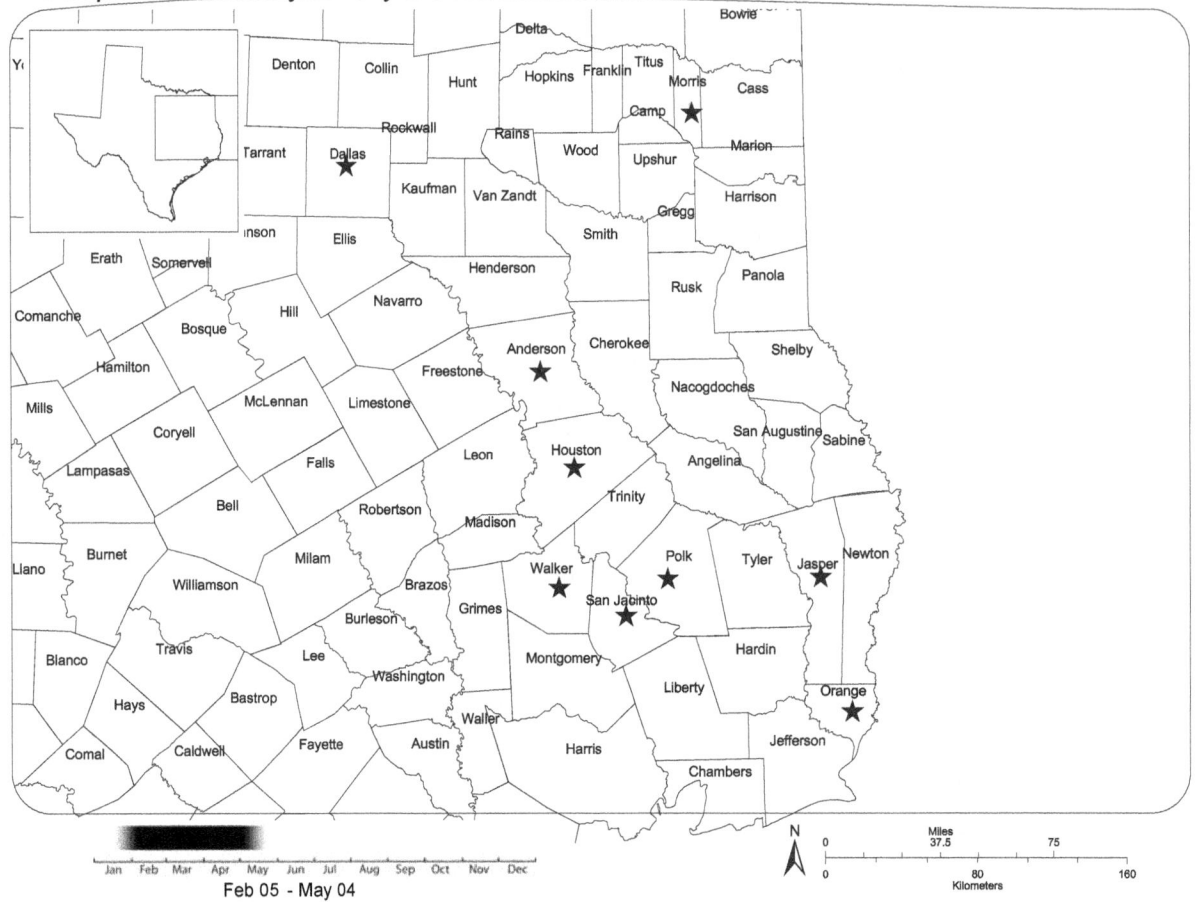

Feb 05 - May 04

HABITAT: Sand or mud-bottomed streams and rivers with moderate current; sheltered inlets and bays of lakes.

Anderson
Dallas
Houston
Jasper
Morris
Orange
Polk
San Jacinto
Walker

Gomphus militaris Hagen in Selys / Sulphur-tipped Clubtail

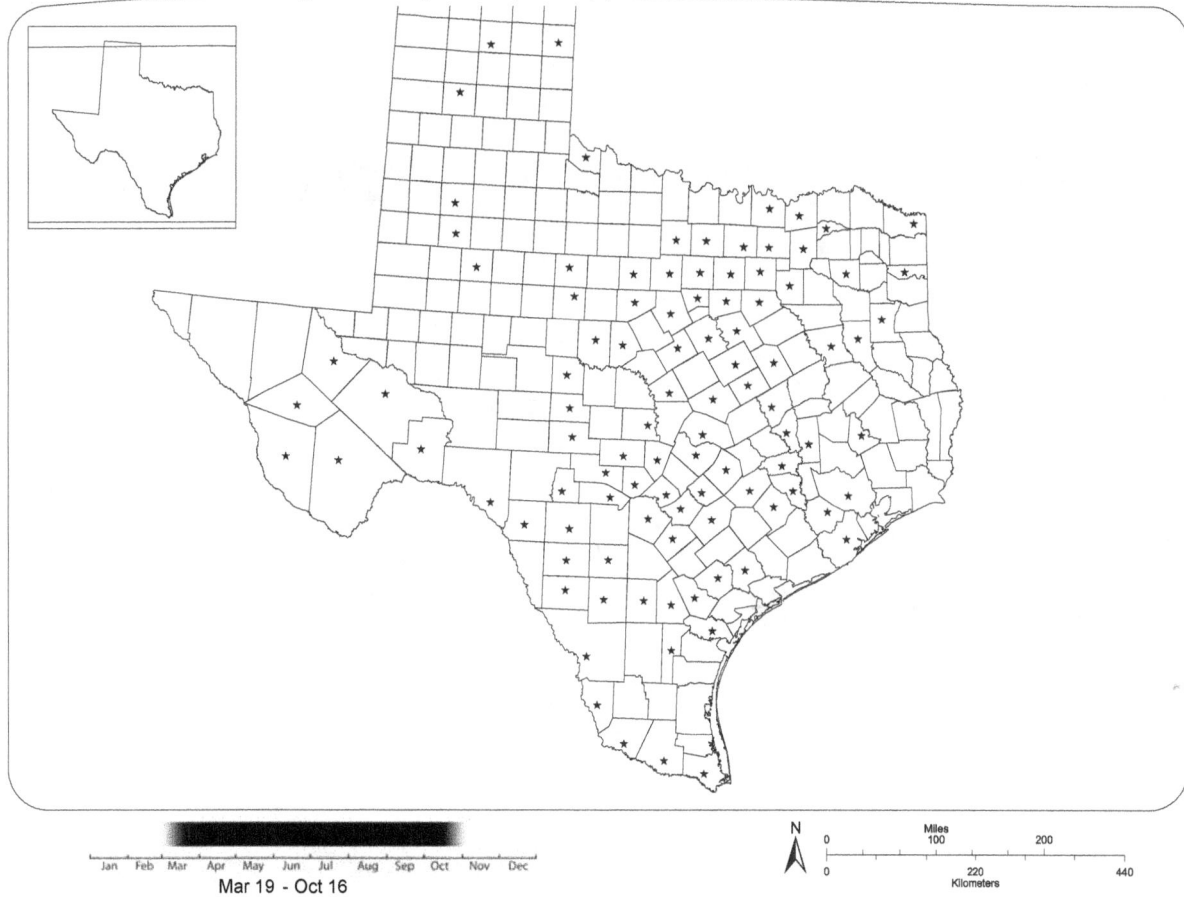

Mar 19 - Oct 16

HABITAT: Ponds, lakes, streams and creeks with muddy bottoms.

Anderson	Dimmit	Jim Wells	Robertson
Austin	Eastland	Johnson	Rusk
Bandera	Ellis	Jones	San Jacinto
Bastrop	Erath	Kaufman	San Patricio
Bee	Falls	Kendall	Starr
Bell	Fannin	Kerr	Stephens
Bexar	Fayette	Kimble	Tarrant
Blanco	Fort Bend	Kinney	Taylor
Borden	Frio	La Salle	Terrell
Bosque	Gillespie	Lampasas	Travis
Bowie	Goliad	Limestone	Uvalde
Brazoria	Gonzales	Live Oak	Val Verde
Brazos	Grayson	Llano	Victoria
Brewster	Grimes	Lubbock	Washington
Brown	Guadalupe	Lynn	Webb
Caldwell	Hamilton	Marion	Williamson
Cameron	Hardeman	McLennan	Wilson
Cherokee	Harris	McMullen	Wise
Coleman	Hemphill	Menard	Wood
Collin	Hidalgo	Palo Pinto	Zapata
Colorado	Hill	Parker	Zavala
Comal	Hood	Pecos	
Concho	Hunt	Presidio	
Dallas	Hutchinson	Randall	
Delta	Jack	Real	
Denton	Jeff Davis	Reeves	

189

Gomphus modestus Needham / Gulf Coast Clubtail

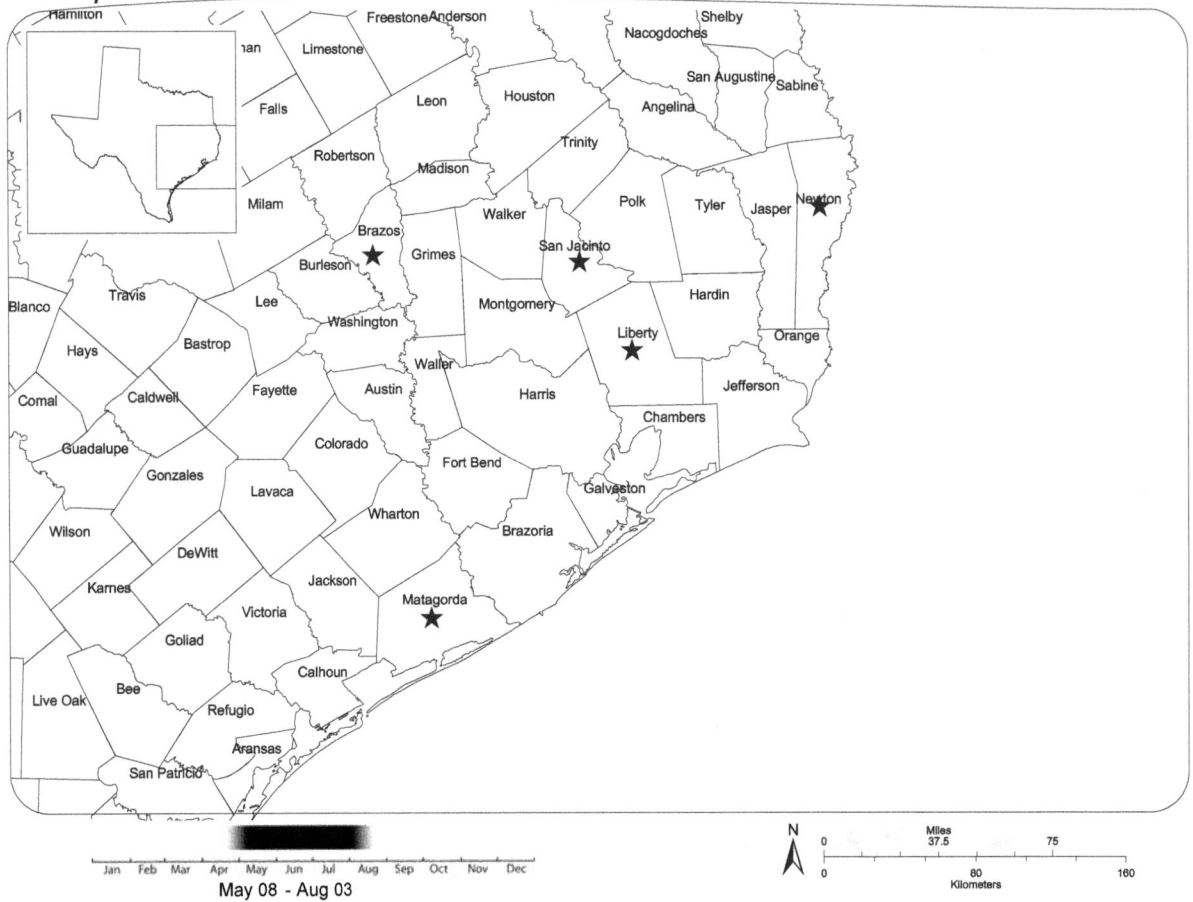

May 08 - Aug 03

HABITAT: Medium-sized coastal streams and rivers with mud or sand bottoms.

Brazos
Liberty
Matagorda
Newton
San Jacinto

Gomphus oklahomensis Pritchard / Oklahoma Clubtail

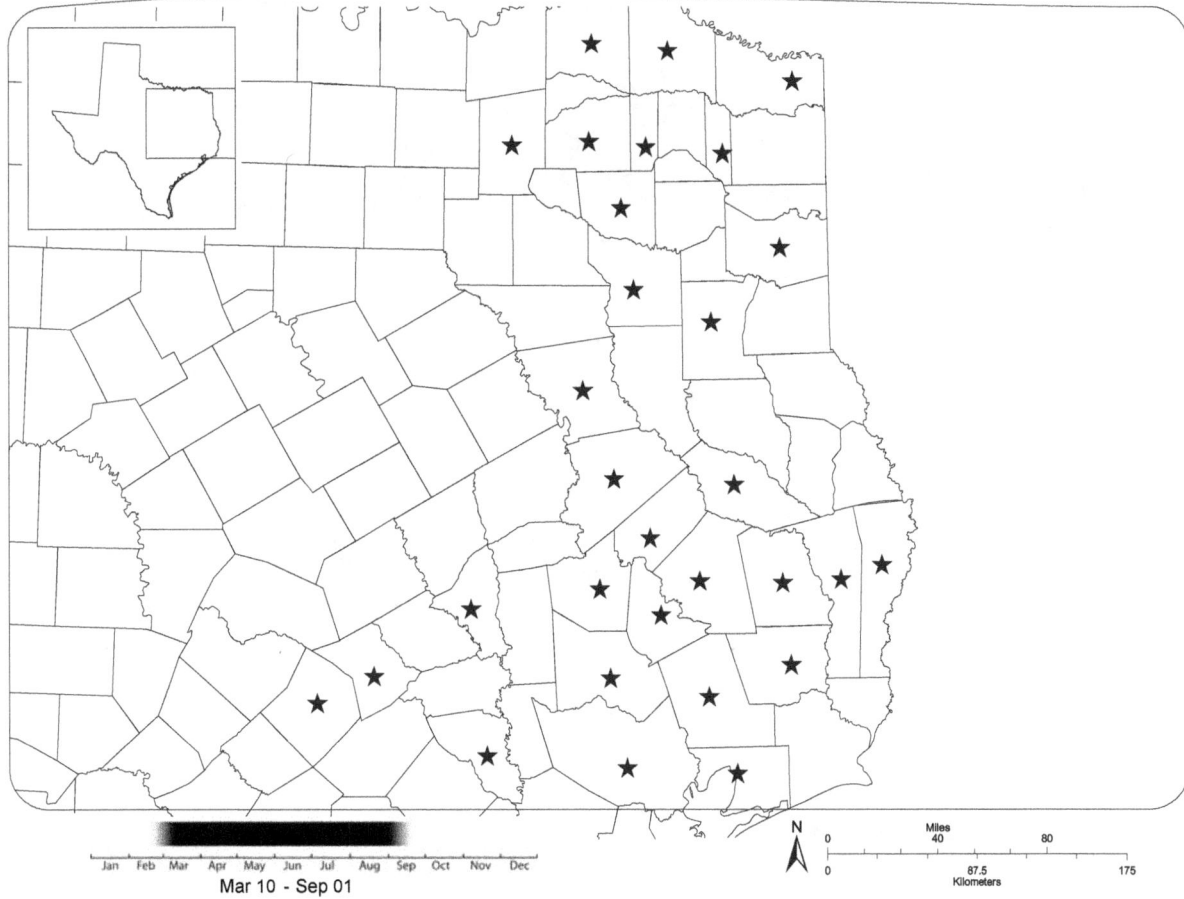

Mar 10 - Sep 01

HABITAT: Small creeks and streams with moderate current and sand or mud bottoms.

Anderson	Trinity
Angelina	Tyler
Austin	Walker
Bastrop	Wood
Bowie	
Brazos	
Chambers	
Franklin	
Hardin	
Harris	
Harrison	
Hopkins	
Houston	
Hunt	
Jasper	
Lamar	
Lee	
Liberty	
Montgomery	
Morris	
Newton	
Polk	
Red River	
Rusk	
San Jacinto	
Smith	

Gomphus vastus Walsh / Cobra Clubtail

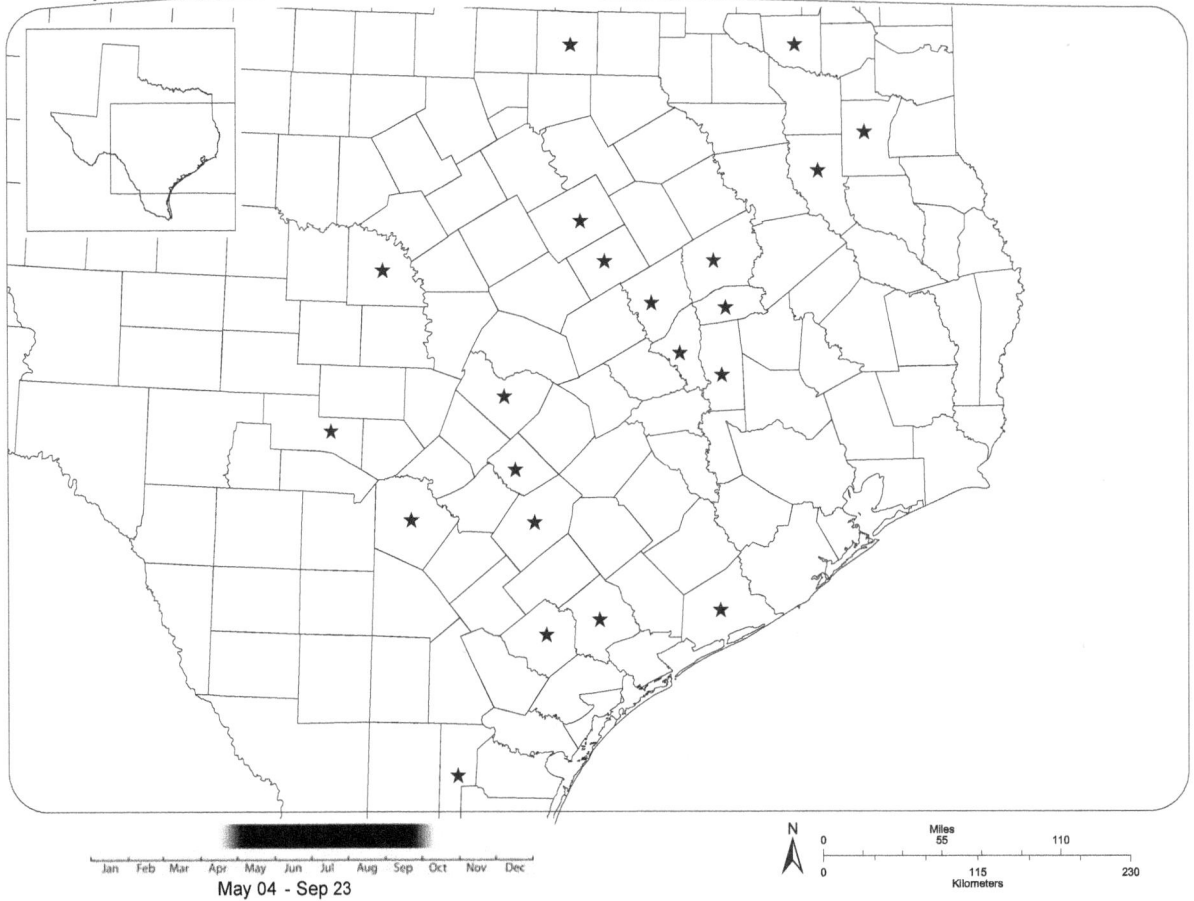

May 04 - Sep 23

HABITAT: Medium-sized rivers or lakes with areas of alternating sand and gravel.

Bexar
Brazos
Caldwell
Cherokee
Falls
Goliad
Gonzales
Grimes
Jim Wells
Kerr
Leon
Madison
Matagorda
McLennan
Robertson
Rusk
San Saba
Tarrant
Travis
Victoria
Wood

Hagenius brevistylus Selys / Dragonhunter

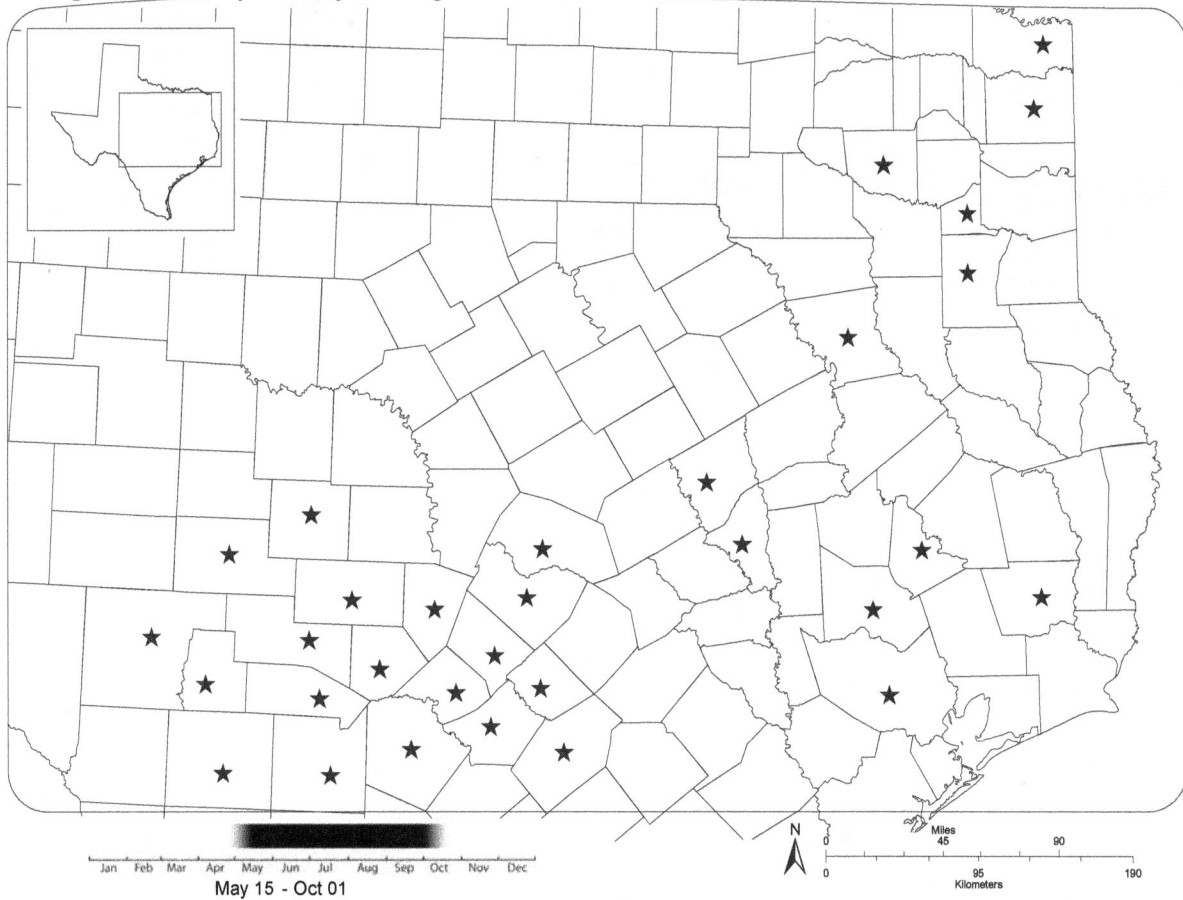

May 15 - Oct 01

HABITAT: Streams, rivers and creeks with moderate to fast current and undercut banks.

Anderson
Bandera
Bexar
Blanco
Bowie
Brazos
Caldwell
Cass
Comal
Edwards
Gillespie
Gonzales
Gregg
Guadalupe
Hardin
Harris
Hays
Kendall
Kerr
Kimble
Mason
Medina
Montgomery
Real
Robertson
Rusk

San Jacinto
Travis
Uvalde
Williamson
Wood

Phyllocycla breviphylla Belle / Ringed Forceptail

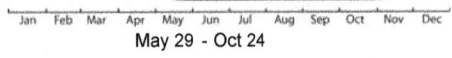

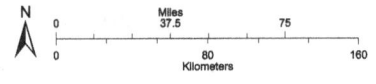

May 29 - Oct 24

HABITAT: Slow moving streams and backwaters.

Cameron
Hidalgo
Starr

Phyllogomphoides albrighti (Needham) / Five-striped Leaftail

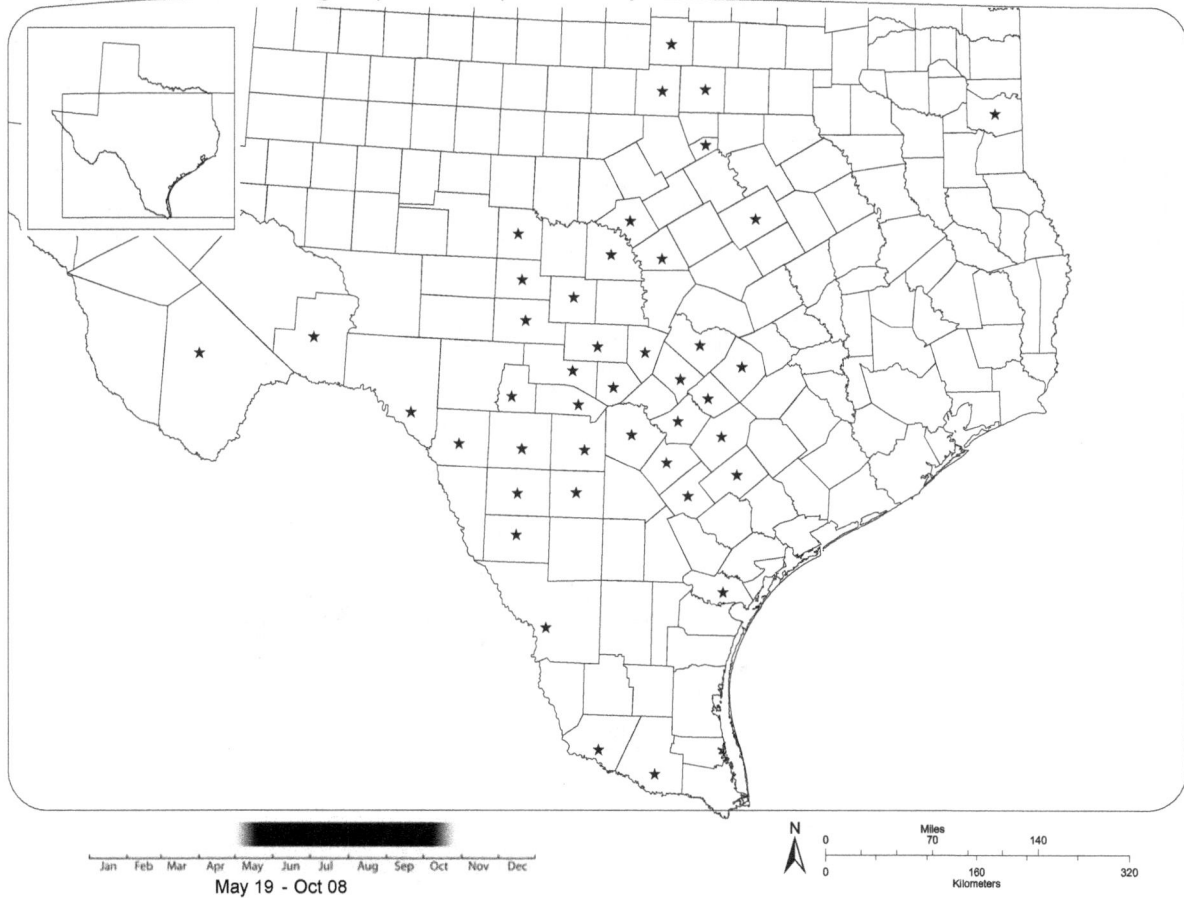

May 19 - Oct 08

HABITAT: Streams and rivers with swift current and cobble or muddy bottoms, emarginated by vegetation.

Bandera	Menard
Bastrop	Mills
Bexar	Palo Pinto
Blanco	Parker
Brewster	Real
Caldwell	San Patricio
Concho	San Saba
DeWitt	Somervell
Dimmit	Starr
Frio	Terrell
Gillespie	Travis
Gonzales	Uvalde
Guadalupe	Val Verde
Harrison	Webb
Hays	Wilson
Hidalgo	Zavala
Jack	
Karnes	
Kendall	
Kerr	
Kimble	
Kinney	
Lampasas	
Mason	
McLennan	
Medina	

Phyllogomphoides stigmatus (Say) / Four-striped Leaftail

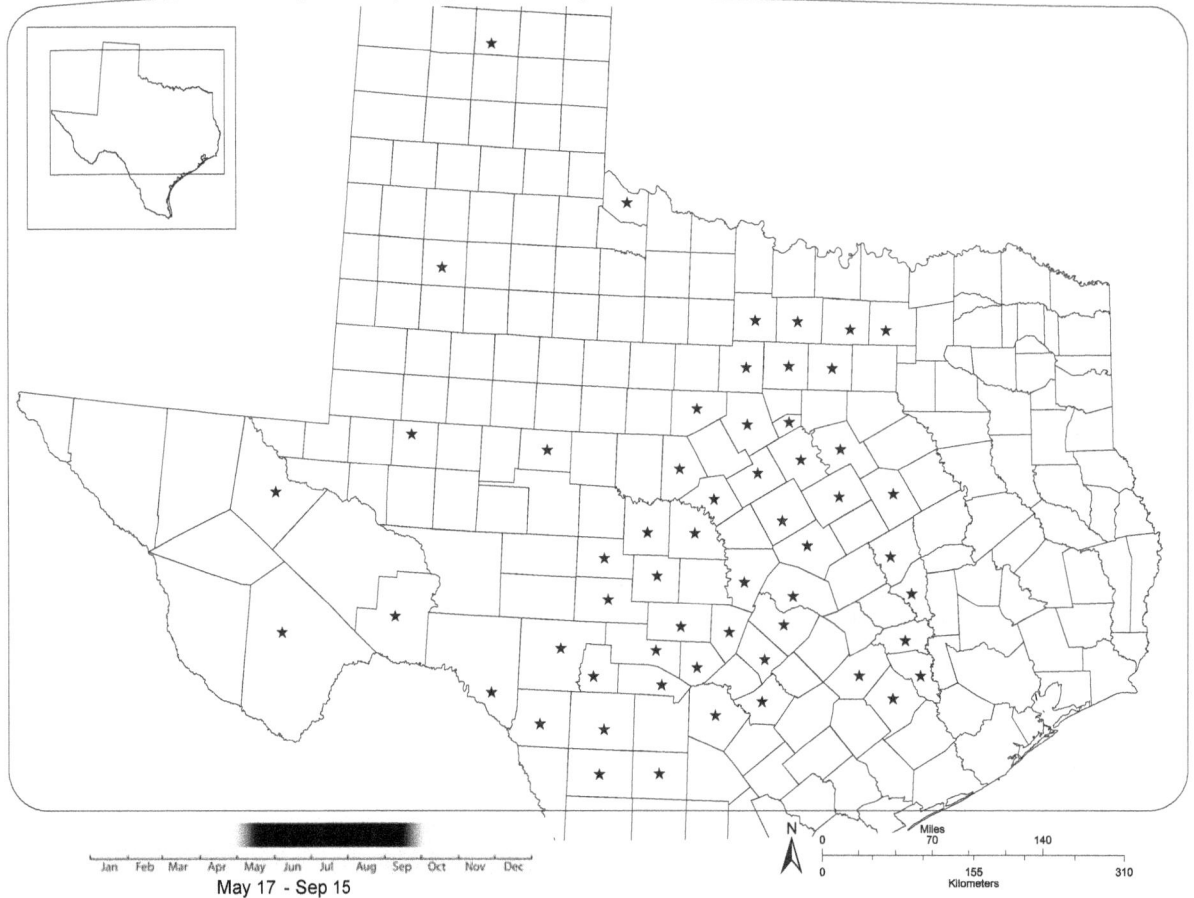

May 17 - Sep 15

HABITAT: Ponds and slow reaches of streams with muddy bottom and heavy vegetation.

Austin	Hutchinson	Washington
Bandera	Jack	Williamson
Bell	Kendall	Wise
Bexar	Kerr	Zavala
Blanco	Kimble	
Bosque	Kinney	
Brazos	Limestone	
Brewster	Lubbock	
Brown	Mason	
Burnet	McCulloch	
Coke	McLennan	
Collin	Menard	
Colorado	Midland	
Coryell	Mills	
Denton	Palo Pinto	
Eastland	Parker	
Edwards	Real	
Erath	Reeves	
Fayette	Robertson	
Frio	San Saba	
Gillespie	Somervell	
Guadalupe	Tarrant	
Hamilton	Terrell	
Hardeman	Travis	
Hays	Uvalde	
Hill	Val Verde	

Progomphus borealis McLachlan in Selys / Gray Sanddragon

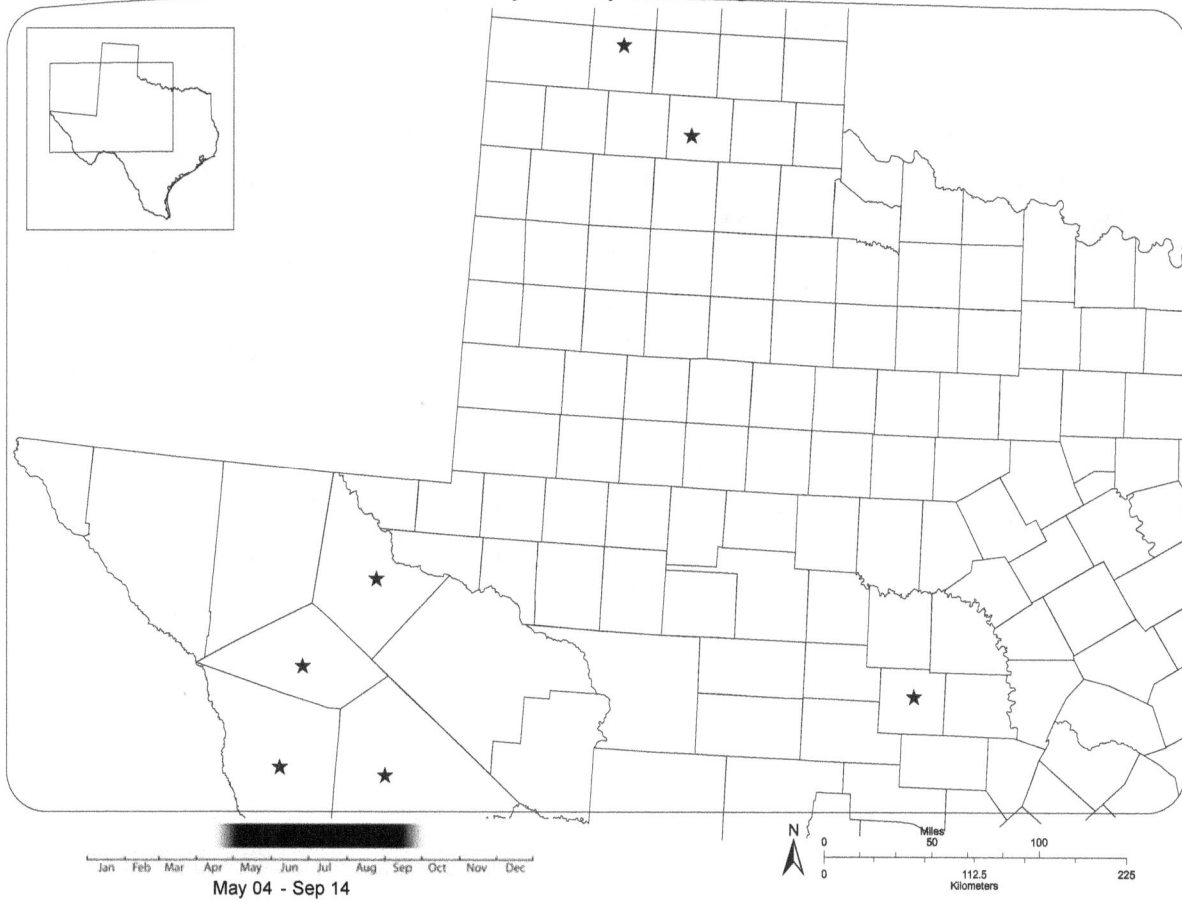

May 04 - Sep 14

HABITAT: Shallow desert, sandy-bottomed streams.

Brewster
Briscoe
Jeff Davis
Mason
Presidio
Randall
Reeves

Progomphus obscurus (Rambur) / Common Sanddragon

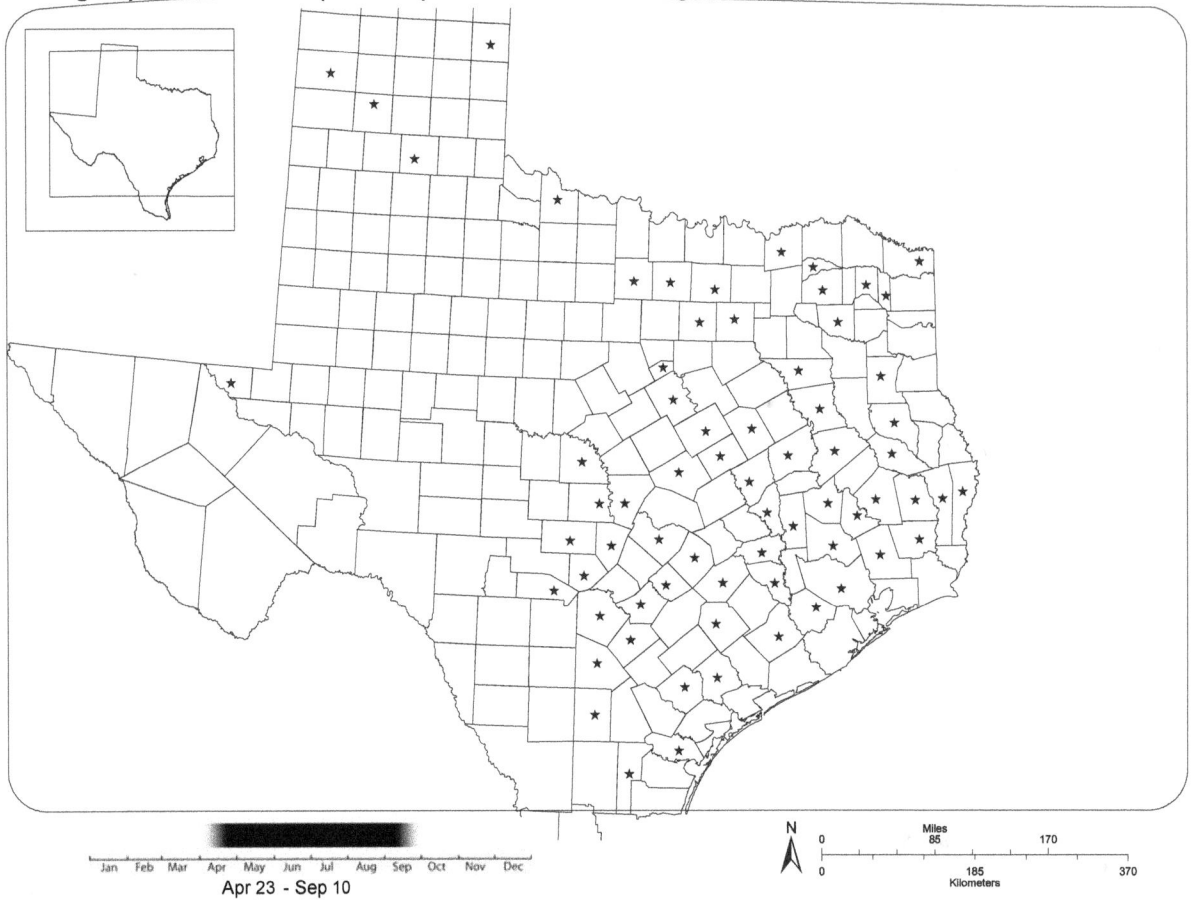

Apr 23 - Sep 10

HABITAT: Shallow streams and lakes with a sandy bottoms.

Anderson	Hardin	Rusk
Angelina	Harris	San Jacinto
Atascosa	Hemphill	San Patricio
Austin	Henderson	San Saba
Bandera	Hopkins	Somervell
Bastrop	Houston	Tarrant
Bell	Jack	Titus
Bexar	Jasper	Travis
Blanco	Jim Wells	Tyler
Bosque	Kendall	Victoria
Bowie	Lavaca	Walker
Brazos	Leon	Washington
Briscoe	Liberty	Wharton
Burnet	Limestone	Wilbarger
Caldwell	Llano	Wilson
Dallas	Loving	Wise
Delta	McLennan	Wood
Denton	McMullen	
Falls	Montgomery	
Fannin	Morris	
Fayette	Nacogdoches	
Fort Bend	Newton	
Gillespie	Oldham	
Goliad	Polk	
Grimes	Randall	
Guadalupe	Robertson	

Stylurus intricatus (Hagen in Selys) / Brimstone Clubtail

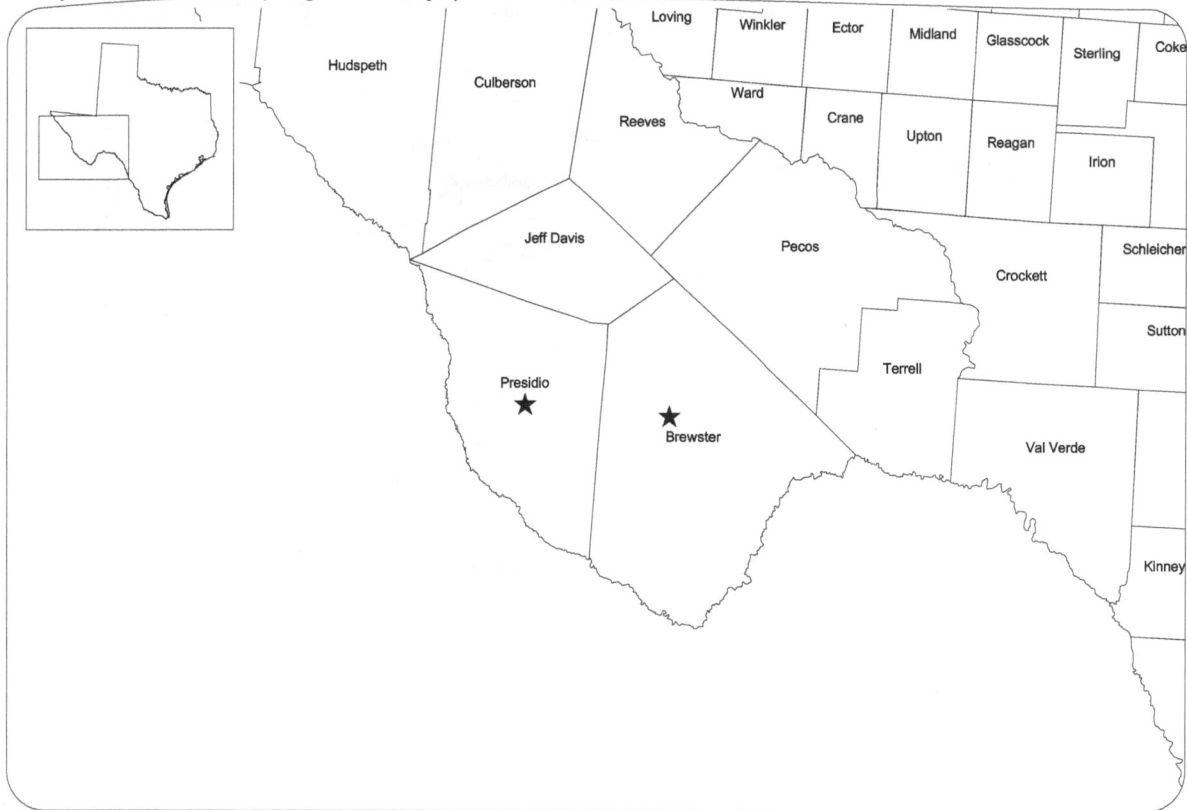

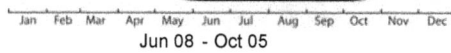

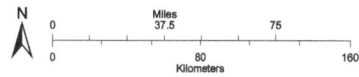

Jun 08 - Oct 05

HABITAT: Slow flowing, open, desert streams and rivers.

Brewster
Presidio

Stylurus laurae Williamson / Laura's Clubtail

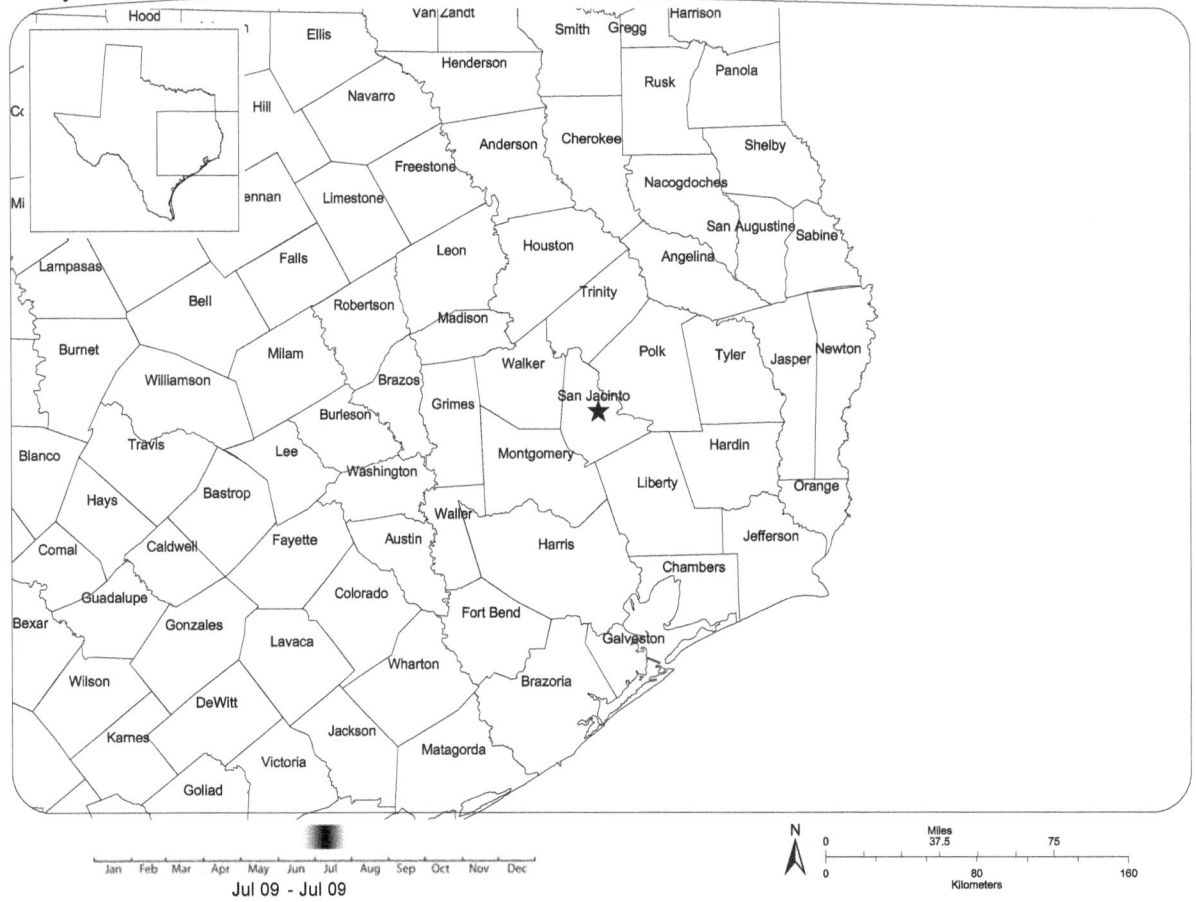

Jul 09 - Jul 09

HABITAT: Shallow, well shaded, rivers and streams with cobble, sand or mud substrate.

San Jacinto

Stylurus plagiatus (Selys) / Russet-tipped Clubtail

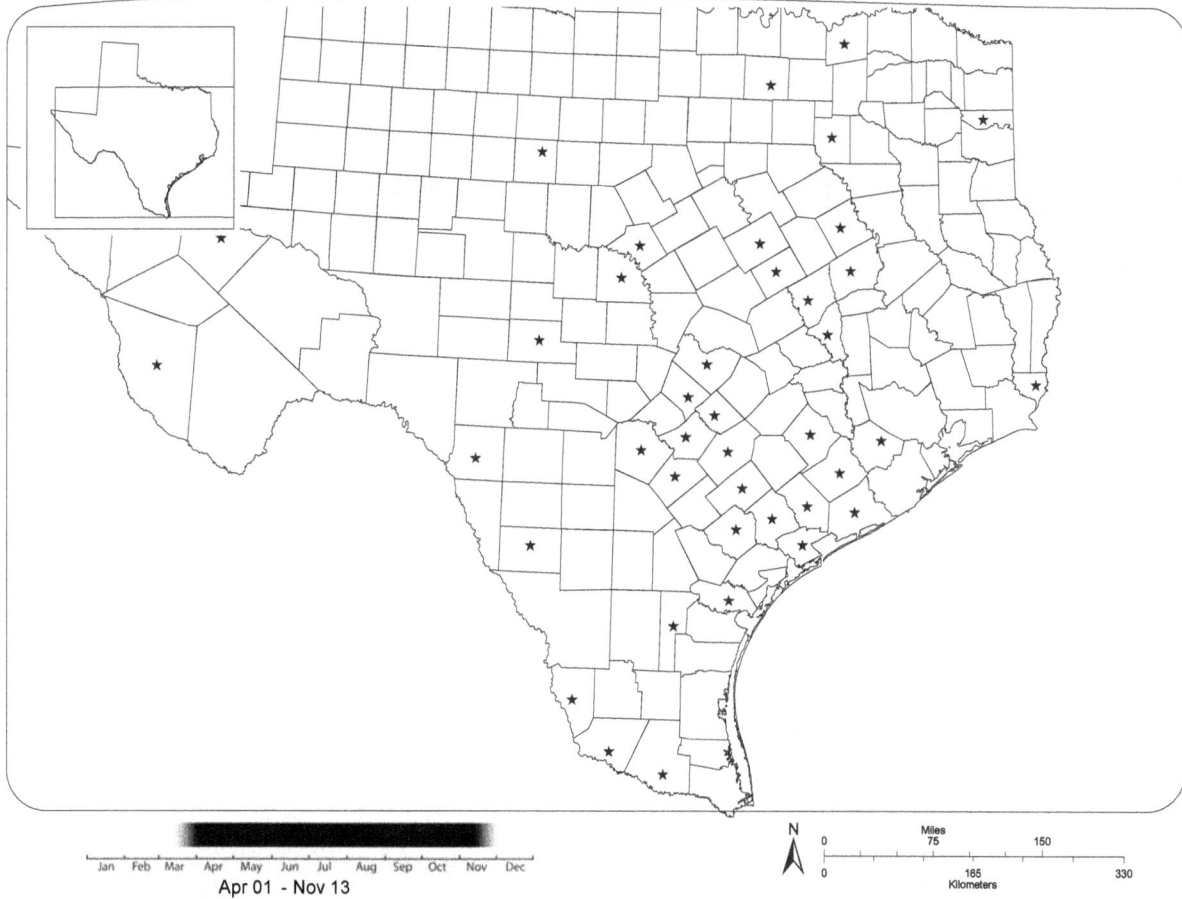

Apr 01 - Nov 13

HABITAT: Weedy rivers, streams and lakes with moderate to little current.

Bexar	Mills
Brazos	Orange
Caldwell	Presidio
Calhoun	Reeves
Colorado	Robertson
Denton	San Patricio
DeWitt	San Saba
Dimmit	Starr
Falls	Taylor
Fannin	Travis
Fort Bend	Victoria
Freestone	Wharton
Goliad	Wilson
Gonzales	Zapata
Guadalupe	
Hays	
Hidalgo	
Jackson	
Jim Wells	
Kaufman	
Kimble	
Kinney	
Leon	
Marion	
Matagorda	
McLennan	

Cordulegaster maculata Selys / Twin-spotted Spiketail

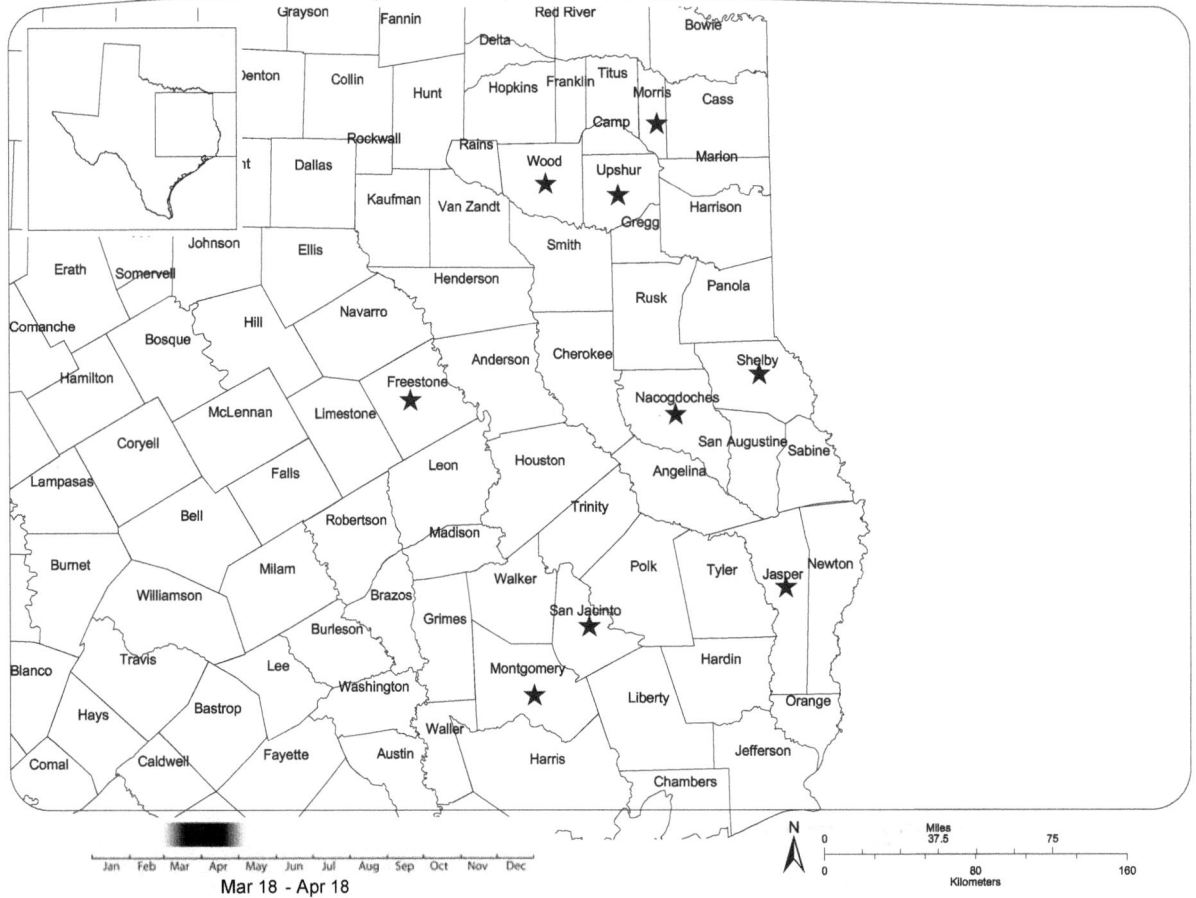

Mar 18 - Apr 18

HABITAT: Small, rapidly flowing spring-fed forest streams and seepages with sandy or muck bottoms.

Freestone
Jasper
Montgomery
Morris
Nacogdoches
San Jacinto
Shelby
Upshur
Wood

Cordulegaster obliqua (Say) / Arrowhead Spiketail

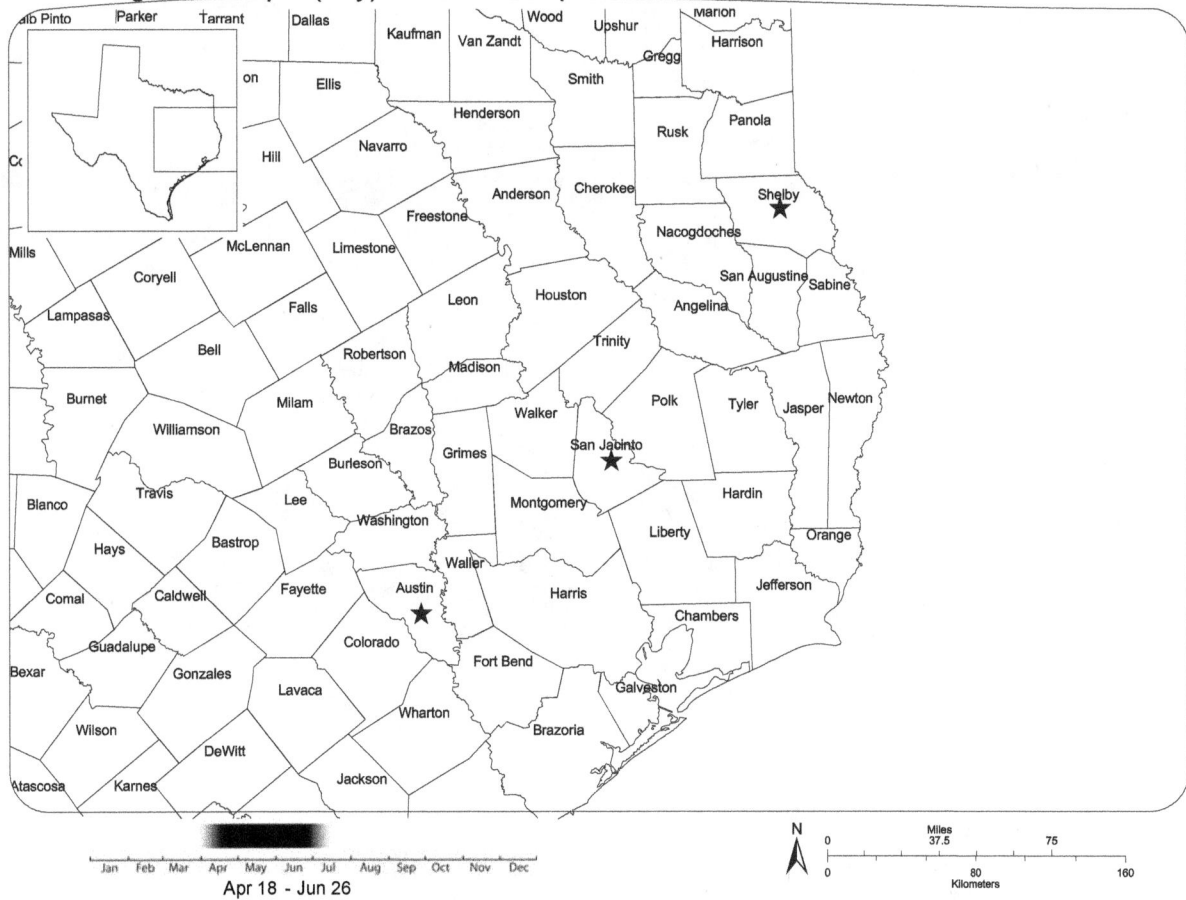

Apr 18 - Jun 26

HABITAT: Small, rapidly flowing spring-fed forest streams and seepages with sandy or muck bottoms.

Austin
San Jacinto
Shelby

Didymops transversa (Say) / Stream Cruiser

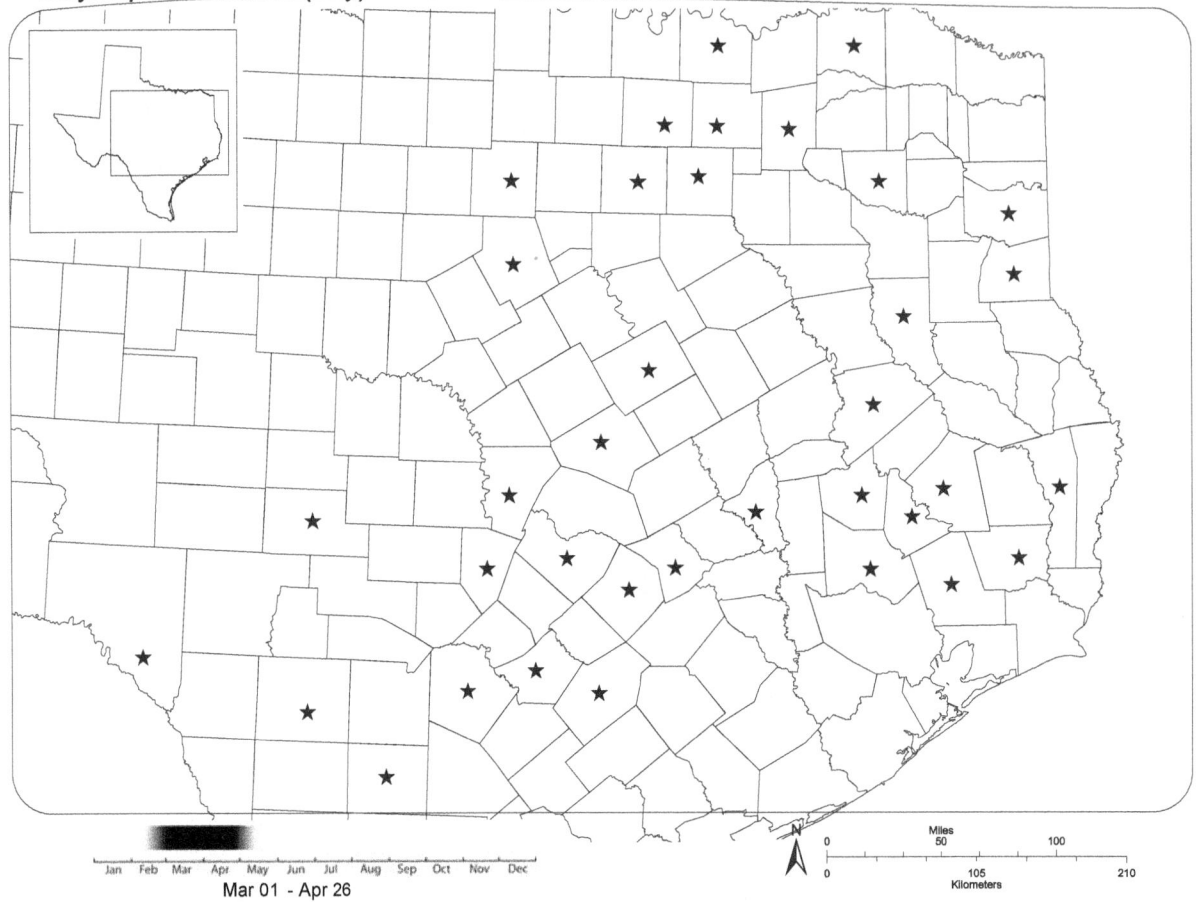

Mar 01 - Apr 26

HABITAT: Medium to large streams and rivers.

Bastrop
Bell
Bexar
Blanco
Brazos
Burnet
Cherokee
Collin
Dallas
Denton
Erath
Frio
Gonzales
Grayson
Guadalupe
Hardin
Harrison
Houston
Hunt
Jasper
Kimble
Lamar
Lee
Liberty
McLennan
Montgomery

Palo Pinto
Panola
Polk
San Jacinto
Tarrant
Travis
Uvalde
Val Verde
Walker
Wood

Macromia annulata Hagen / Bronzed River Cruiser

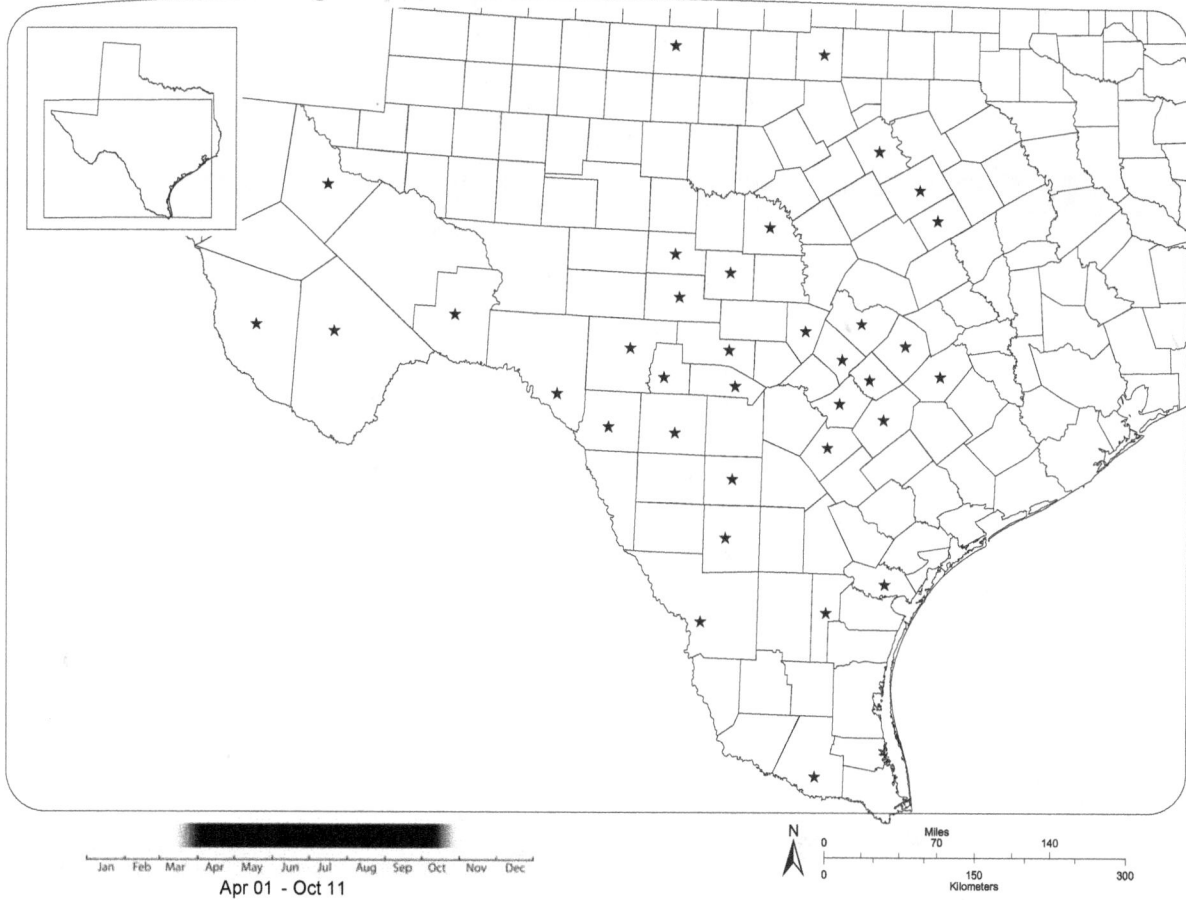

Apr 01 - Oct 11

HABITAT: Large rivers and streams.

Bandera
Bastrop
Blanco
Bosque
Brewster
Caldwell
Edwards
Falls
Fayette
Frio
Gonzales
Guadalupe
Hays
Hidalgo
Jim Wells
Jones
Kerr
Kimble
Kinney
La Salle
Mason
McLennan
Menard
Palo Pinto
Presidio
Real

Reeves
San Patricio
San Saba
Terrell
Travis
Uvalde
Val Verde
Webb
Wilson

Macromia illinoiensis georgina (Selys) / Georgia River Cruiser

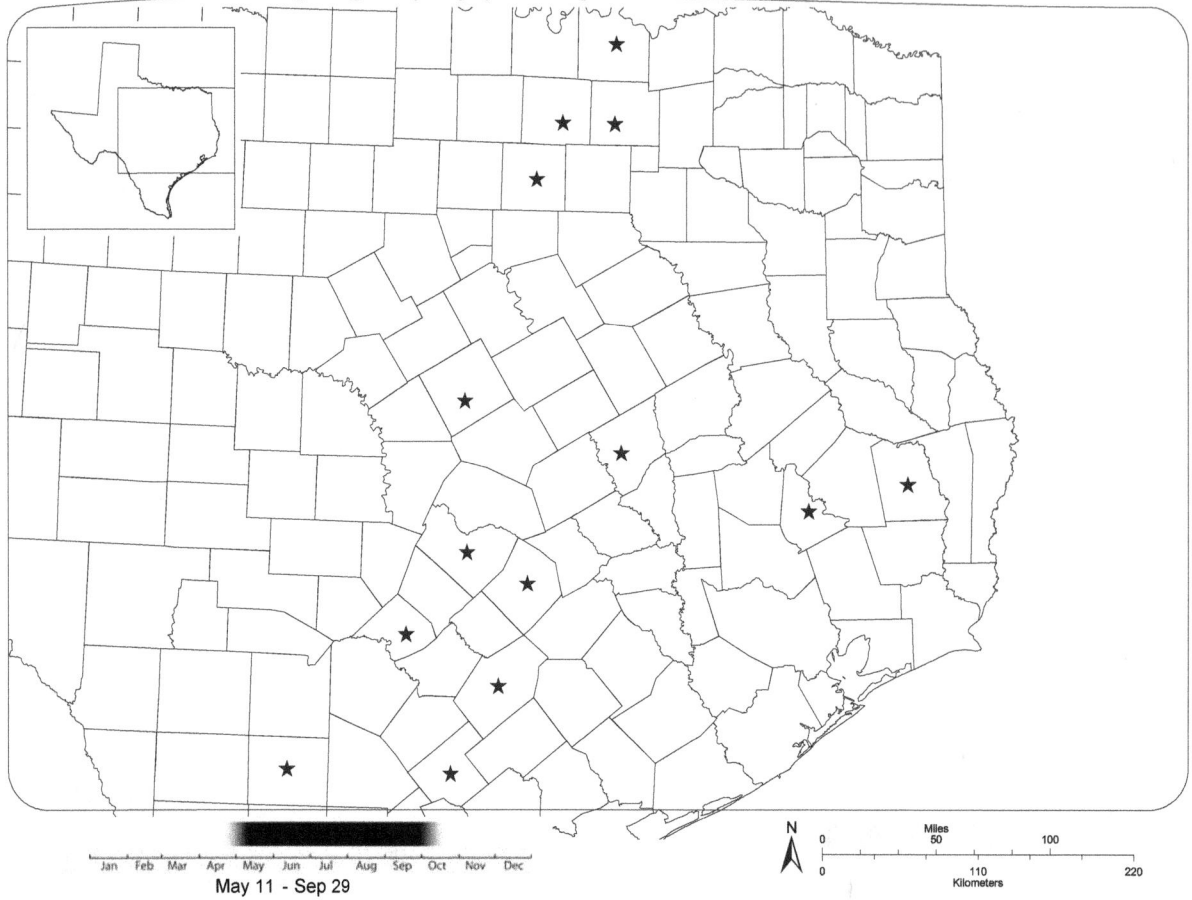

May 11 - Sep 29

HABITAT:

Bastrop
Collin
Comal
Coryell
Denton
Frio
Gonzales
Grayson
Karnes
Robertson
San Jacinto
Tarrant
Travis
Tyler

Macromia pacifica Hagen / Gilded River Cruiser

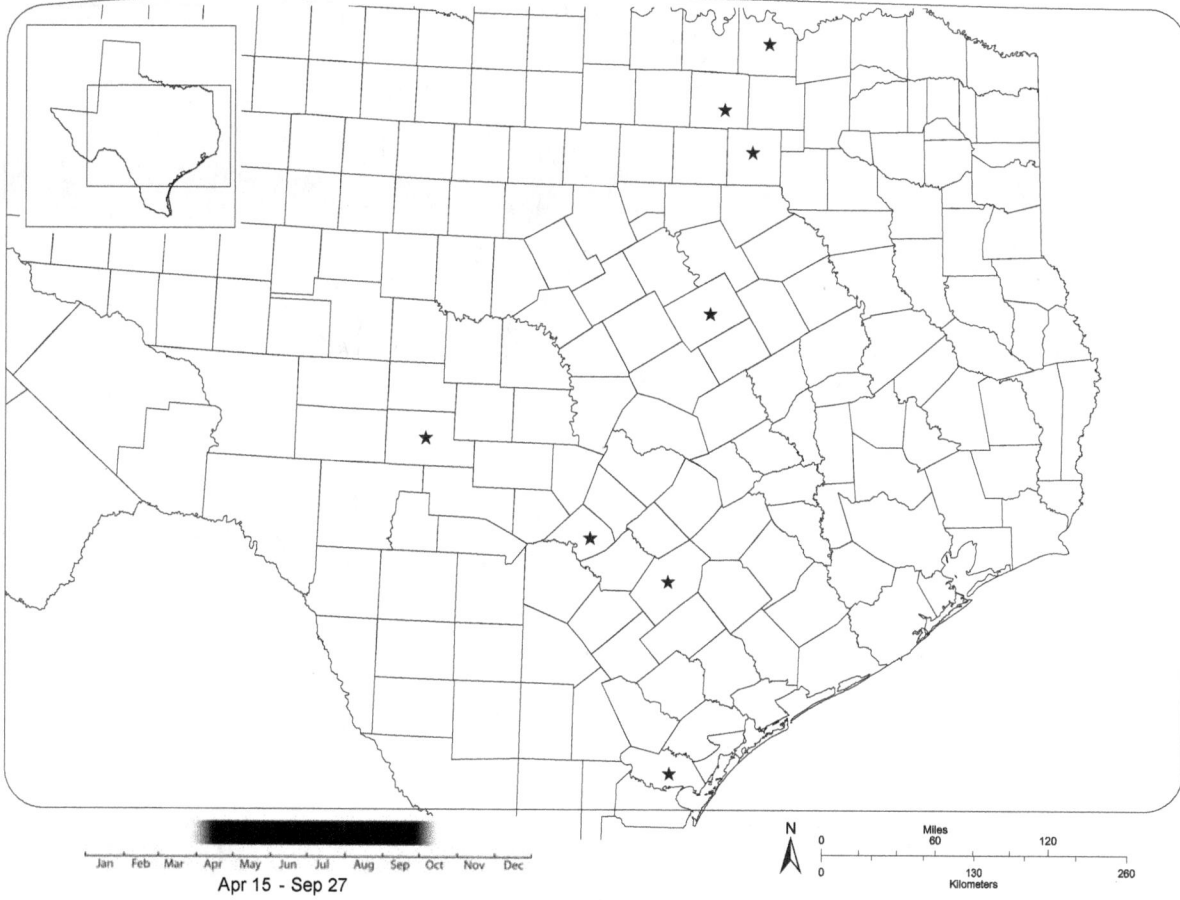

Jan Feb Mar Apr May Jun Jul Aug Sep Oct Nov Dec

Apr 15 - Sep 27

HABITAT: Moderate-sized streams and rivers with pools and areas of slow flow.

Comal
Dallas
Denton
Gonzales
Grayson
Kimble
McLennan
San Patricio

Macromia taeniolata Rambur / Royal River Cruiser

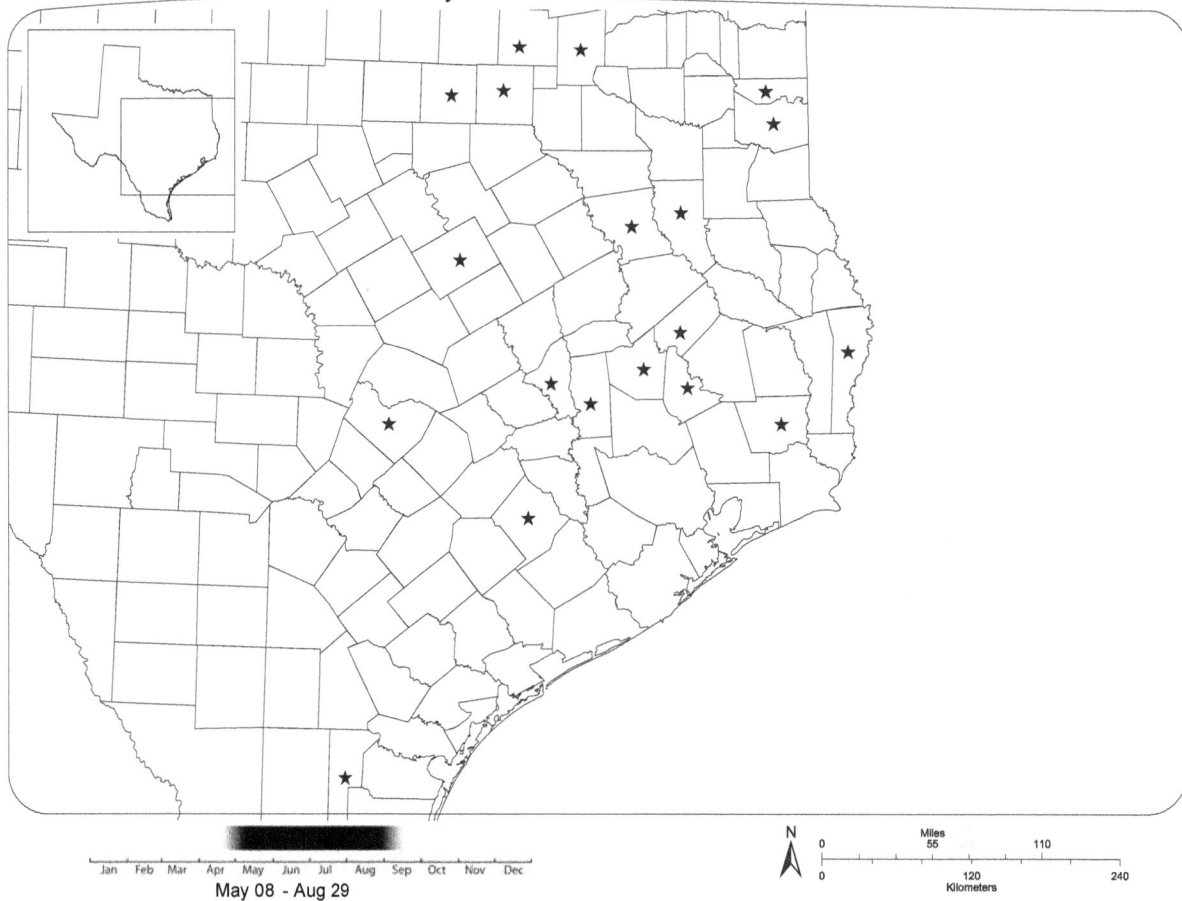

May 08 - Aug 29

HABITAT: Rivers, streams and lakes.

Anderson
Brazos
Cherokee
Collin
Colorado
Dallas
Grimes
Hardin
Harrison
Hunt
Jim Wells
Marion
McLennan
Newton
San Jacinto
Tarrant
Travis
Trinity
Walker

Epitheca costalis (Selys) / Stripe-winged Baskettail

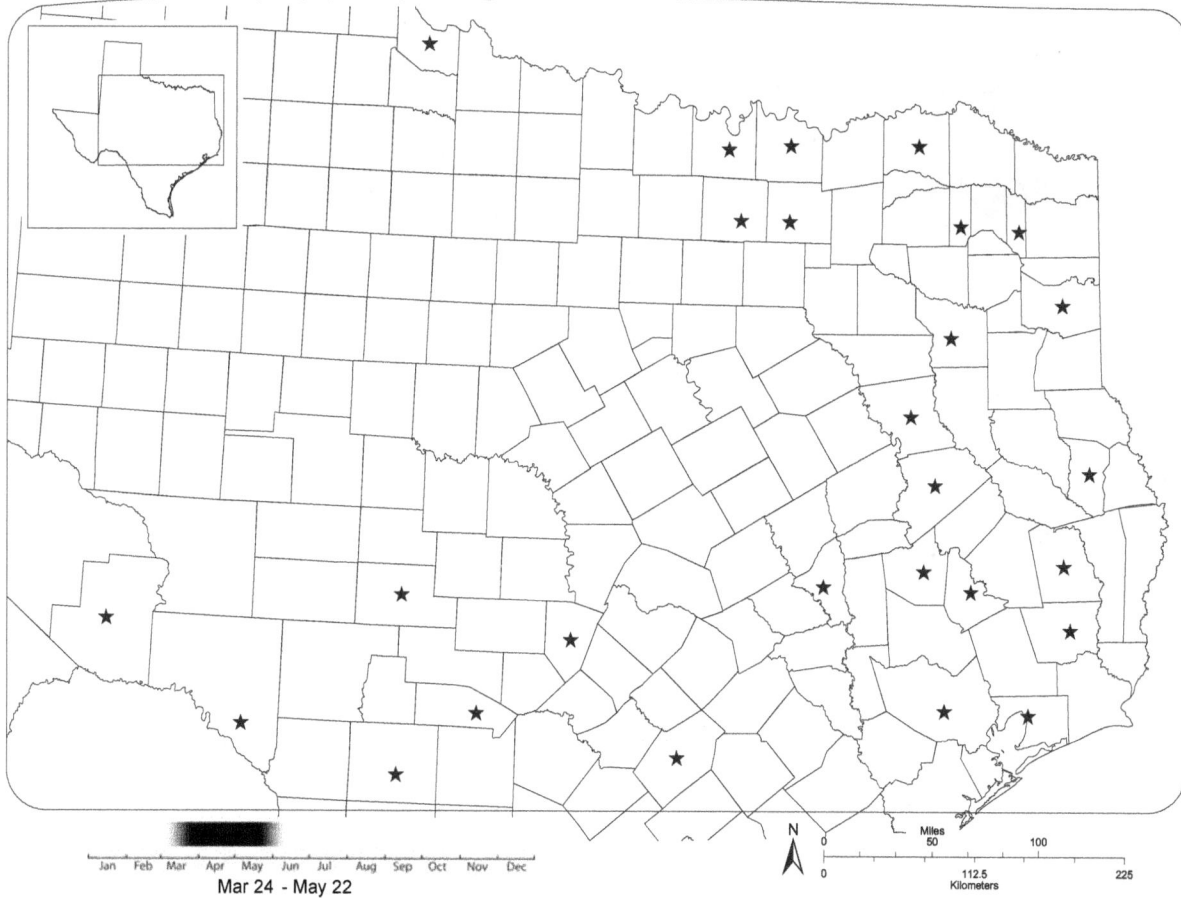

Jan Feb Mar Apr May Jun Jul Aug Sep Oct Nov Dec

Mar 24 - May 22

HABITAT: Lakes, ponds and slow reaches of streams and rivers.

Anderson   Walker
Bandera
Blanco
Brazos
Chambers
Collin
Cooke
Denton
Franklin
Gonzales
Grayson
Hardeman
Hardin
Harris
Harrison
Houston
Kimble
Lamar
Morris
San Augustine
San Jacinto
Smith
Terrell
Tyler
Uvalde
Val Verde

Epitheca cynosura (Say) / Common Baskettail

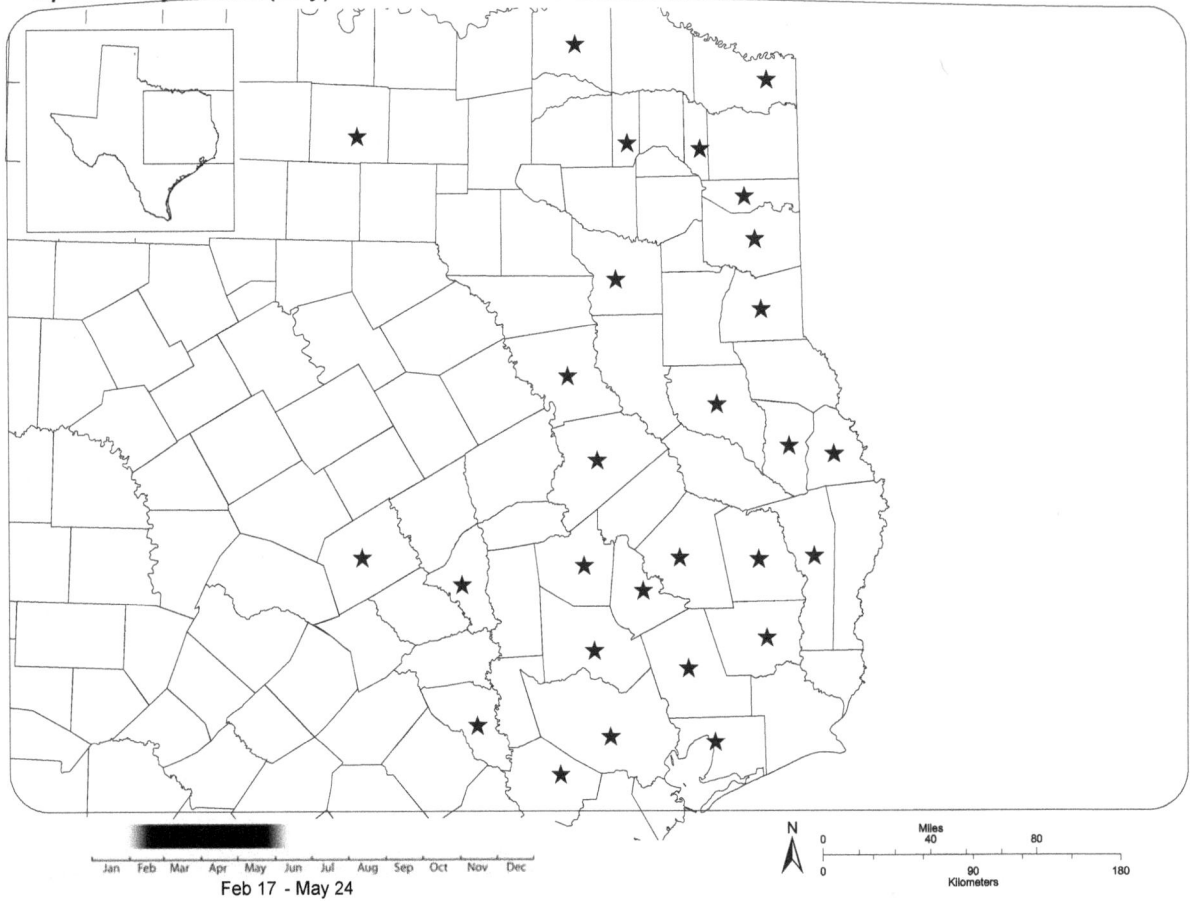

Feb 17 - May 24

HABITAT: Almost any permanent or temporary, quiet water, including ponds, lakes, marshes, streams and rivers, with submerged and emergent vegetation.

Anderson

Austin

Bowie

Brazos

Chambers

Denton

Fort Bend

Franklin

Hardin

Harris

Harrison

Houston

Jasper

Lamar

Liberty

Marion

Milam

Montgomery

Morris

Nacogdoches

Panola

Polk

Sabine

San Augustine

San Jacinto

Smith

Tyler

Walker

Epitheca petechialis (Muttkowski) / Dot-winged Baskettail

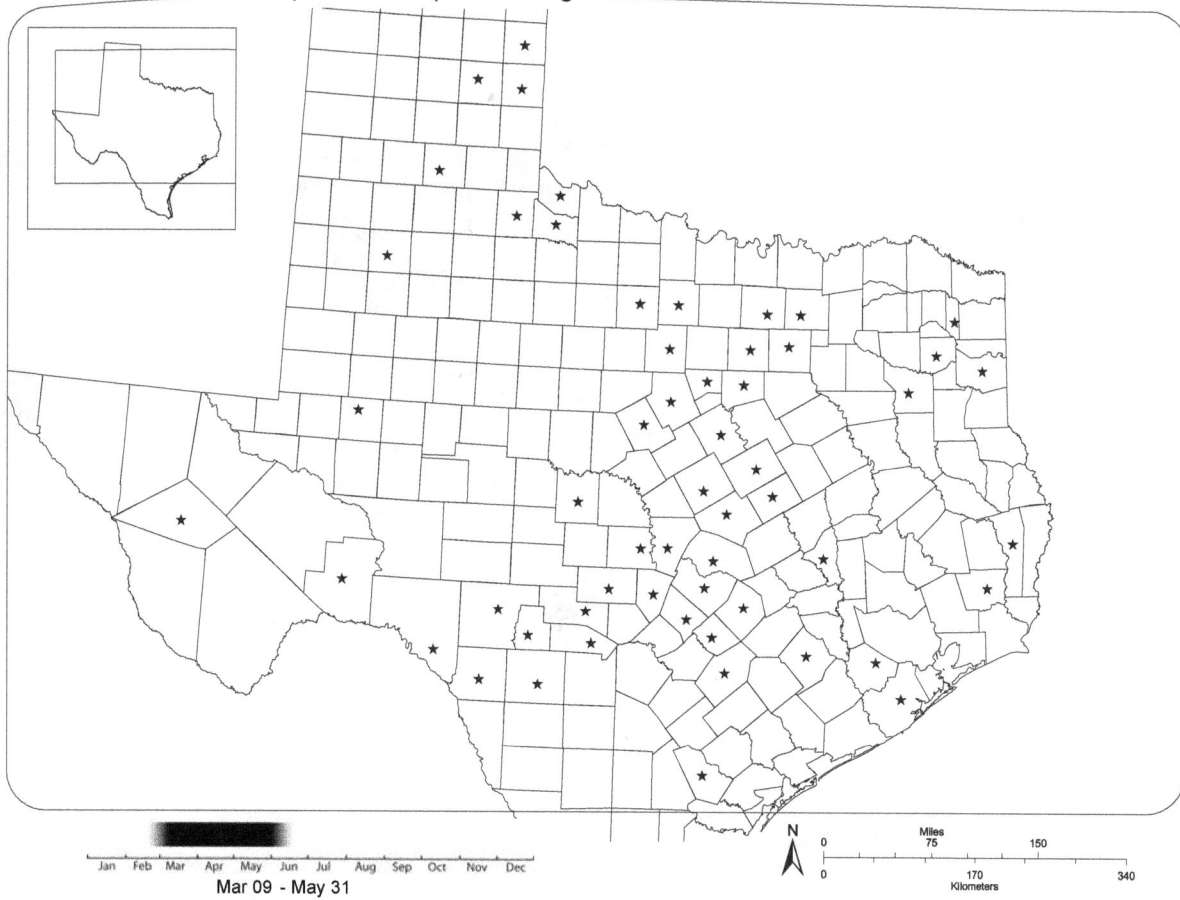

Jan Feb Mar Apr May Jun Jul Aug Sep Oct Nov Dec

Mar 09 - May 31

HABITAT: Lakes, ponds and slow reaches of streams and rivers.

Bandera	Hardeman	Val Verde
Bastrop	Hardin	Wheeler
Bee	Harrison	Williamson
Bell	Hays	Young
Blanco	Hemphill	
Bosque	Hood	
Brazoria	Jack	
Brazos	Jasper	
Briscoe	Jeff Davis	
Burnet	Johnson	
Caldwell	Kerr	
Collin	Kinney	
Colorado	Llano	
Comanche	Lubbock	
Coryell	McCulloch	
Cottle	McLennan	
Dallas	Midland	
Denton	Morris	
Edwards	Palo Pinto	
Erath	Real	
Falls	Smith	
Foard	Tarrant	
Fort Bend	Terrell	
Gillespie	Travis	
Gonzales	Upshur	
Gray	Uvalde	

Epitheca princeps Hagen / Prince Baskettail

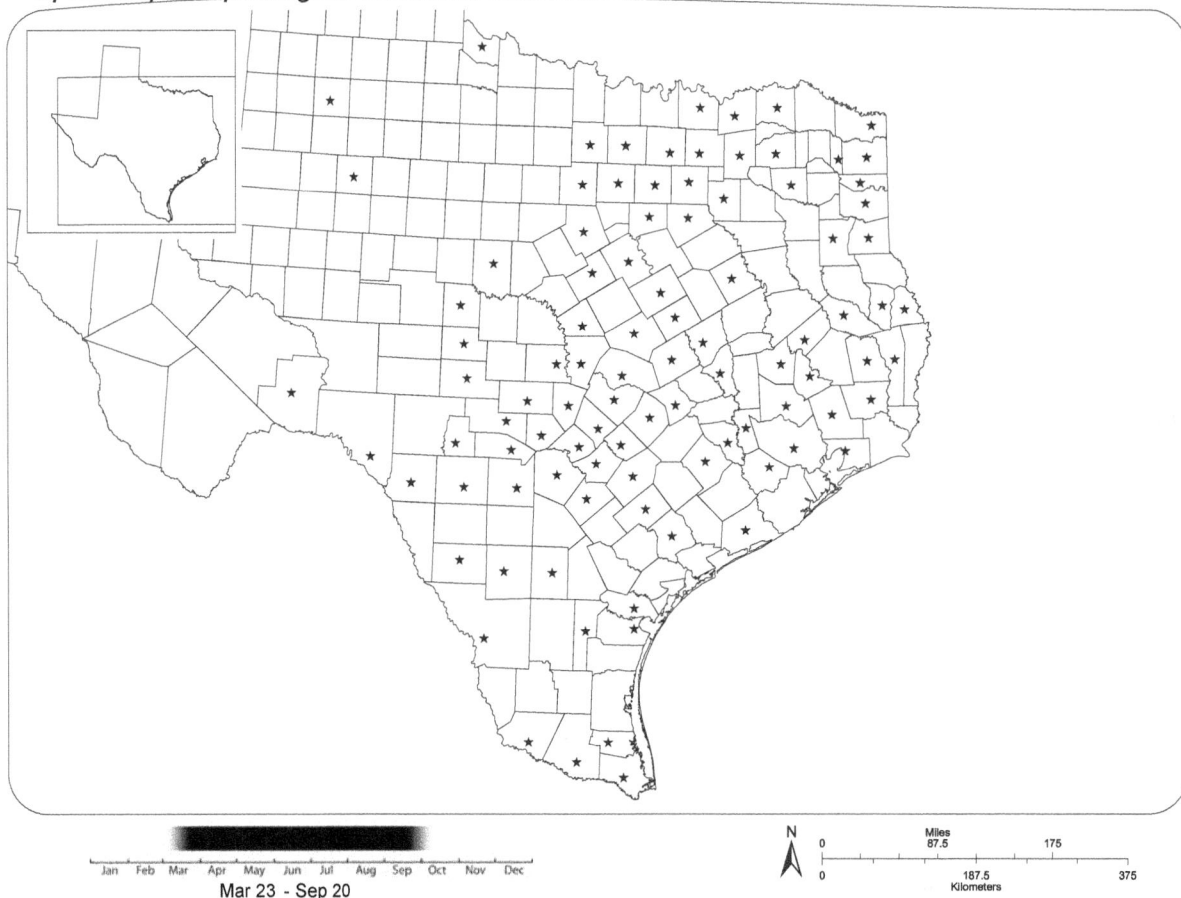

Jan Feb Mar Apr May Jun Jul Aug Sep Oct Nov Dec

Mar 23 - Sep 20

HABITAT: Quiet reaches of streams, rivers, ponds and lakes.

Angelina	Erath	Kinney	San Jacinto
Austin	Falls	La Salle	San Patricio
Bandera	Fannin	Lamar	Starr
Bastrop	Fort Bend	Lampasas	Tarrant
Bell	Freestone	Lee	Terrell
Bexar	Gillespie	Liberty	Travis
Blanco	Gonzales	Llano	Trinity
Borden	Grayson	Lubbock	Tyler
Bosque	Guadalupe	Marion	Uvalde
Bowie	Hamilton	Matagorda	Val Verde
Brazos	Hardeman	McLennan	Victoria
Burnet	Hardin	McMullen	Walker
Caldwell	Harris	Medina	Waller
Cameron	Harrison	Menard	Webb
Cass	Hays	Milam	Willacy
Chambers	Hidalgo	Montgomery	Williamson
Coleman	Hopkins	Morris	Wilson
Collin	Hunt	Nueces	Wise
Colorado	Jack	Palo Pinto	Wood
Comal	Jasper	Panola	
Concho	Jim Wells	Parker	
Dallas	Johnson	Real	
Denton	Kaufman	Robertson	
DeWitt	Kendall	Rusk	
Dimmit	Kerr	Sabine	
Ellis	Kimble	San Augustine	

Epitheca semiaquea (Burmeister) / Mantled Baskettail

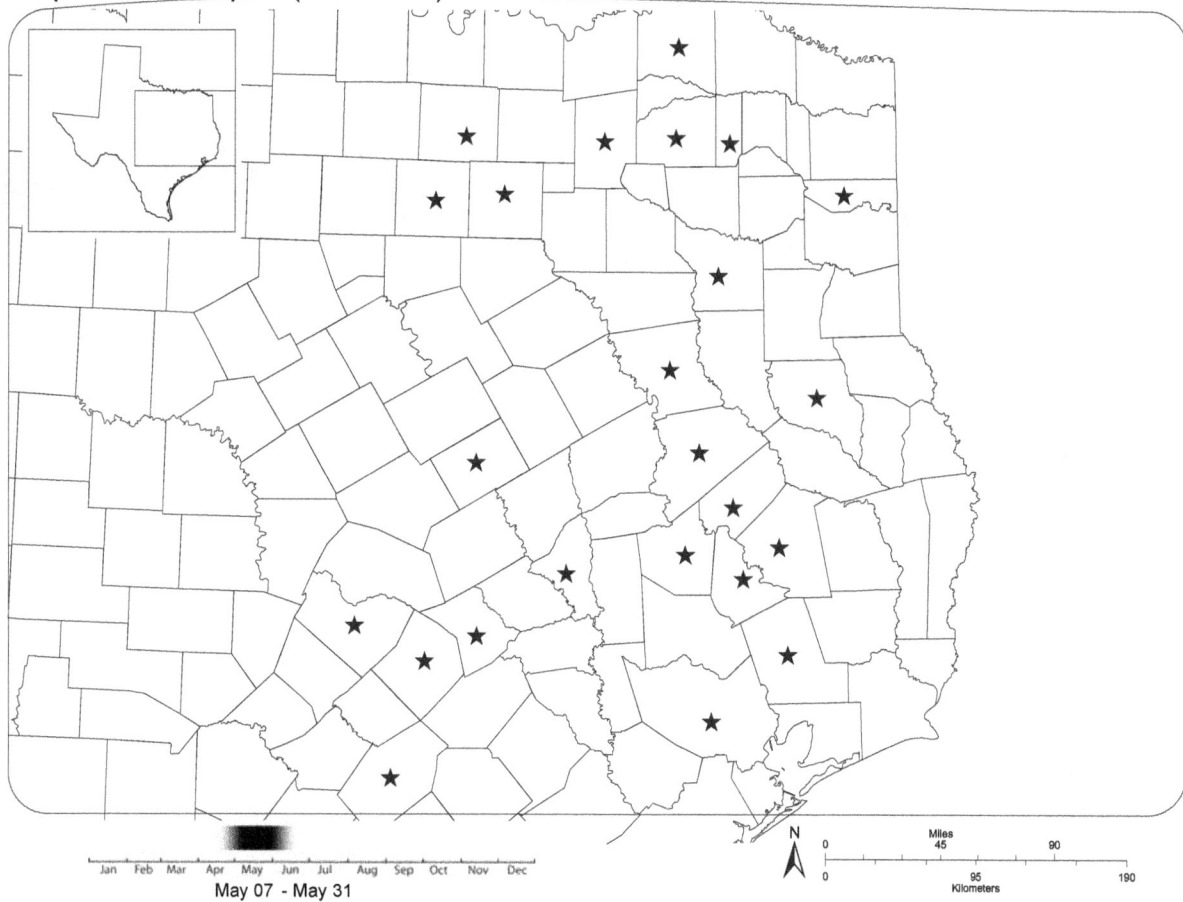

Jan Feb Mar Apr May Jun Jul Aug Sep Oct Nov Dec

May 07 - May 31

HABITAT: Lakes and ponds with submerged and emergent vegetation.

Anderson
Bastrop
Brazos
Dallas
Denton
Falls
Franklin
Gonzales
Harris
Hopkins
Houston
Hunt
Lamar
Lee
Liberty
Marion
Nacogdoches
Polk
San Jacinto
Smith
Tarrant
Travis
Trinity
Walker

Epitheca spinosa (Hagen in Selys) / Robust Baskettail

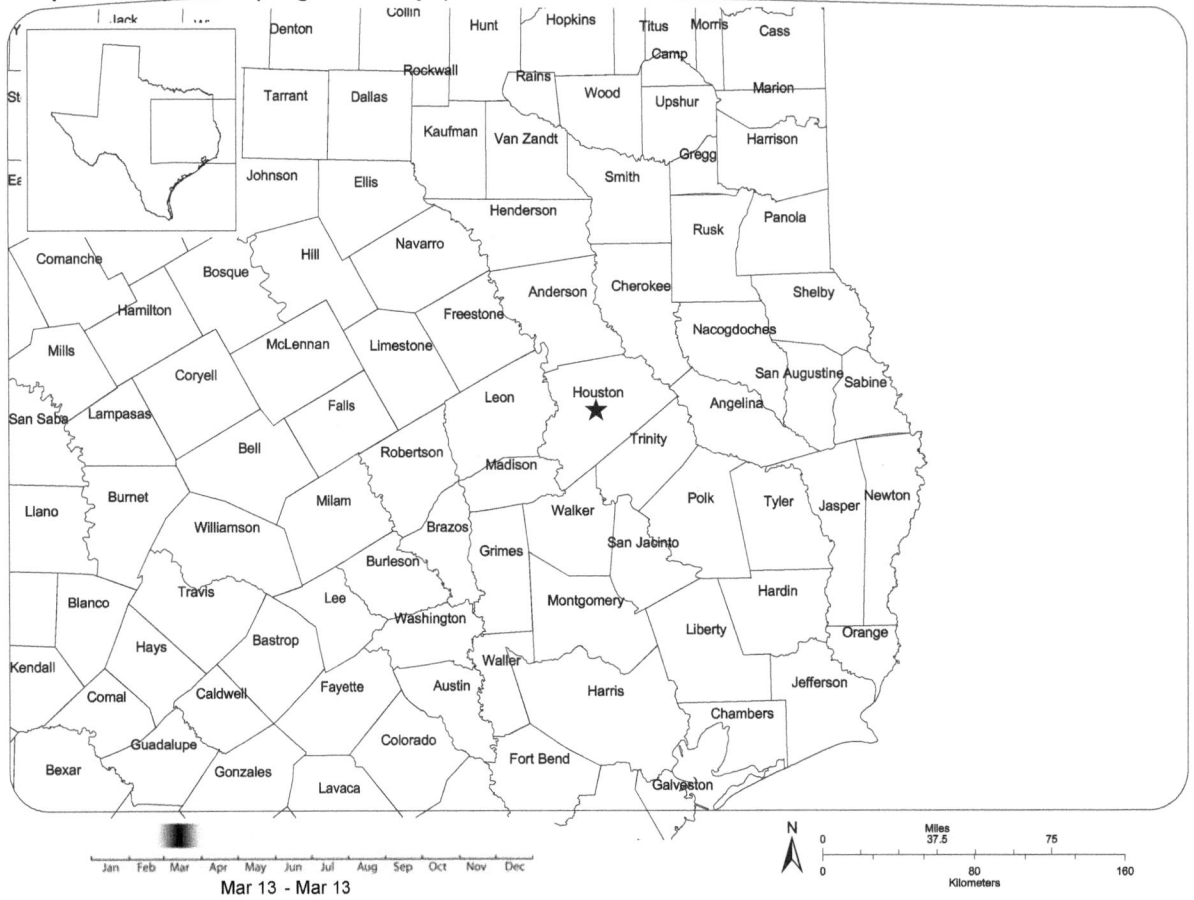

Mar 13 - Mar 13

HABITAT: Lakes, ponds and wooded swamps with little flow.

Houston

Helocordulia selysii (Hagen in Selys) / Selys' Sundragon

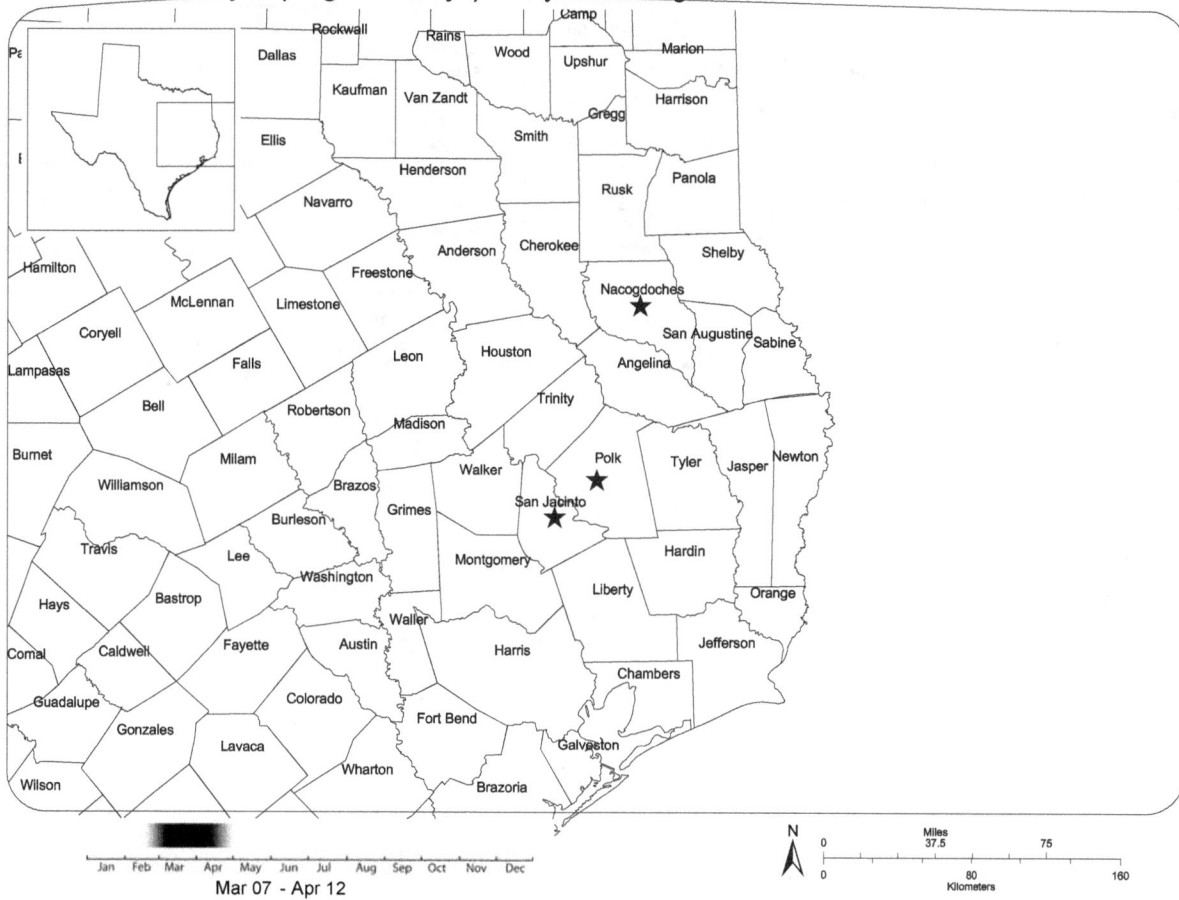

Mar 07 - Apr 12

HABITAT: Small, cool forest streams with sandy bottoms.

Nacogdoches
Polk
San Jacinto

Neurocordulia alabamensis Hodges in Needham and Westfall / Alabama Shadowdragon

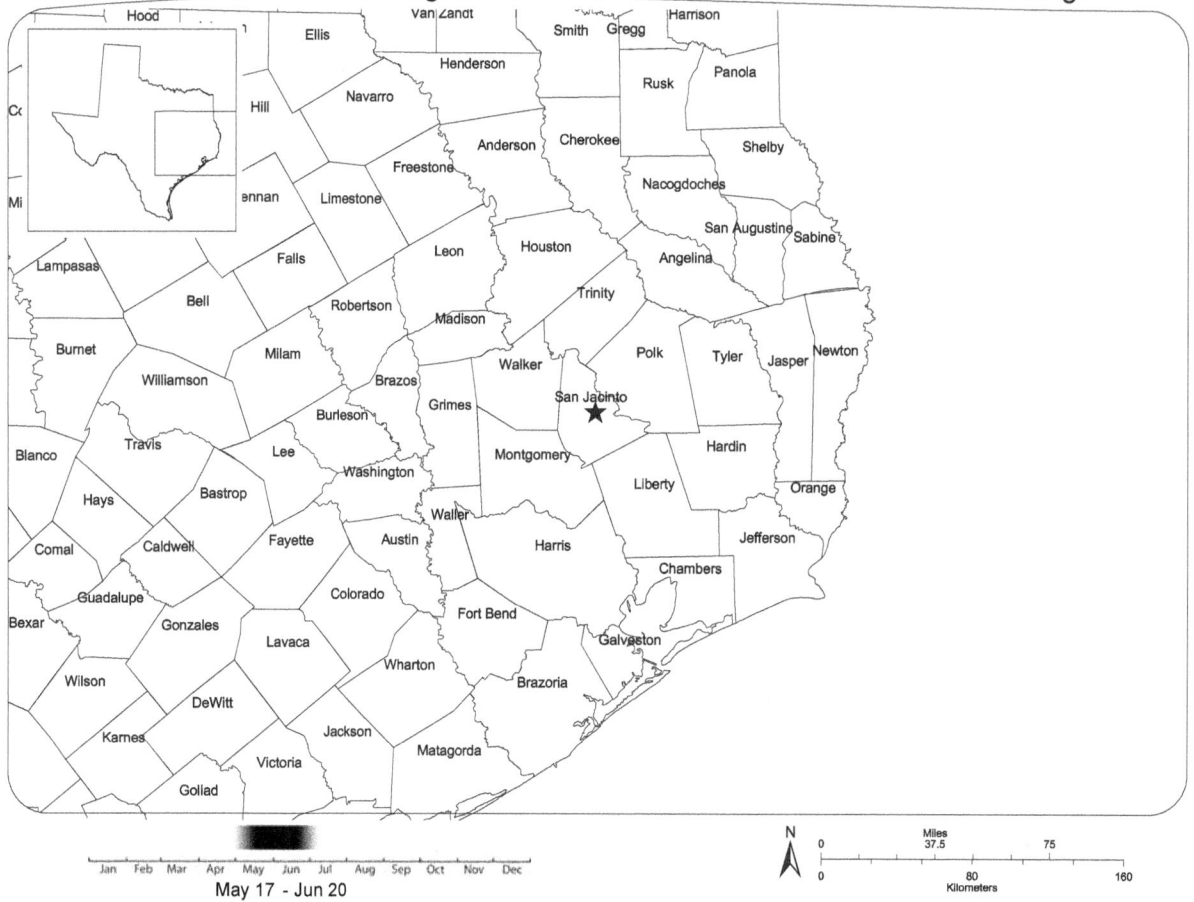

May 17 - Jun 20

HABITAT: Small to medium-sized slow flowing or spring-fed forest streams, frequently tannin stained.

San Jacinto

Neurocordulia molesta (Walsh) / Smoky Shadowdragon

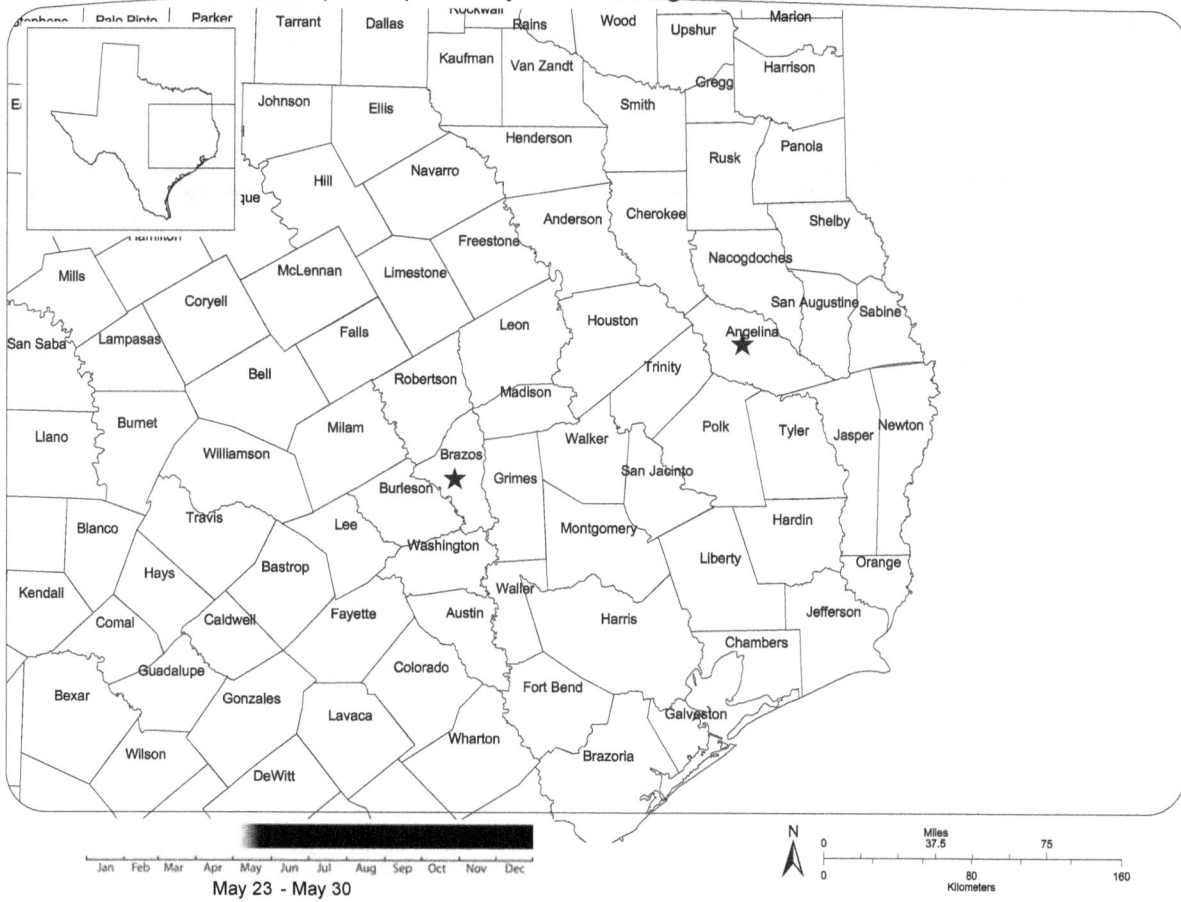

May 23 - May 30

HABITAT: Rivers and medium-sized streams with strong current.

Angelina
Brazos

Neurocordulia xanthosoma (Williamson) / Orange Shadowdragon

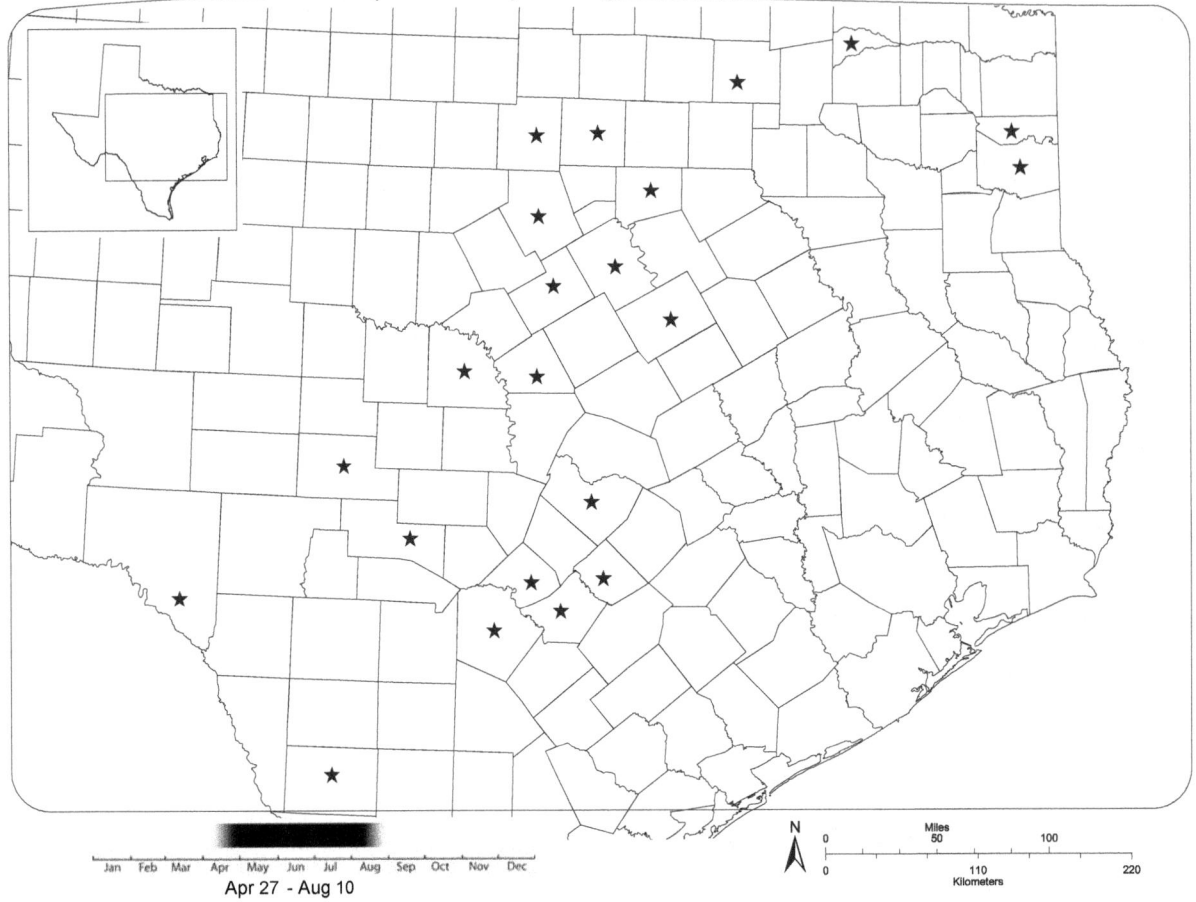

Apr 27 - Aug 10

HABITAT: Medium-sized turbid rivers and streams with strong current.

Bexar
Bosque
Caldwell
Collin
Comal
Delta
Dimmit
Erath
Guadalupe
Hamilton
Harrison
Johnson
Kerr
Kimble
Lampasas
Marion
McLennan
Palo Pinto
Parker
San Saba
Travis
Val Verde

Somatochlora filosa (Hagen) / Fine-lined Emerald

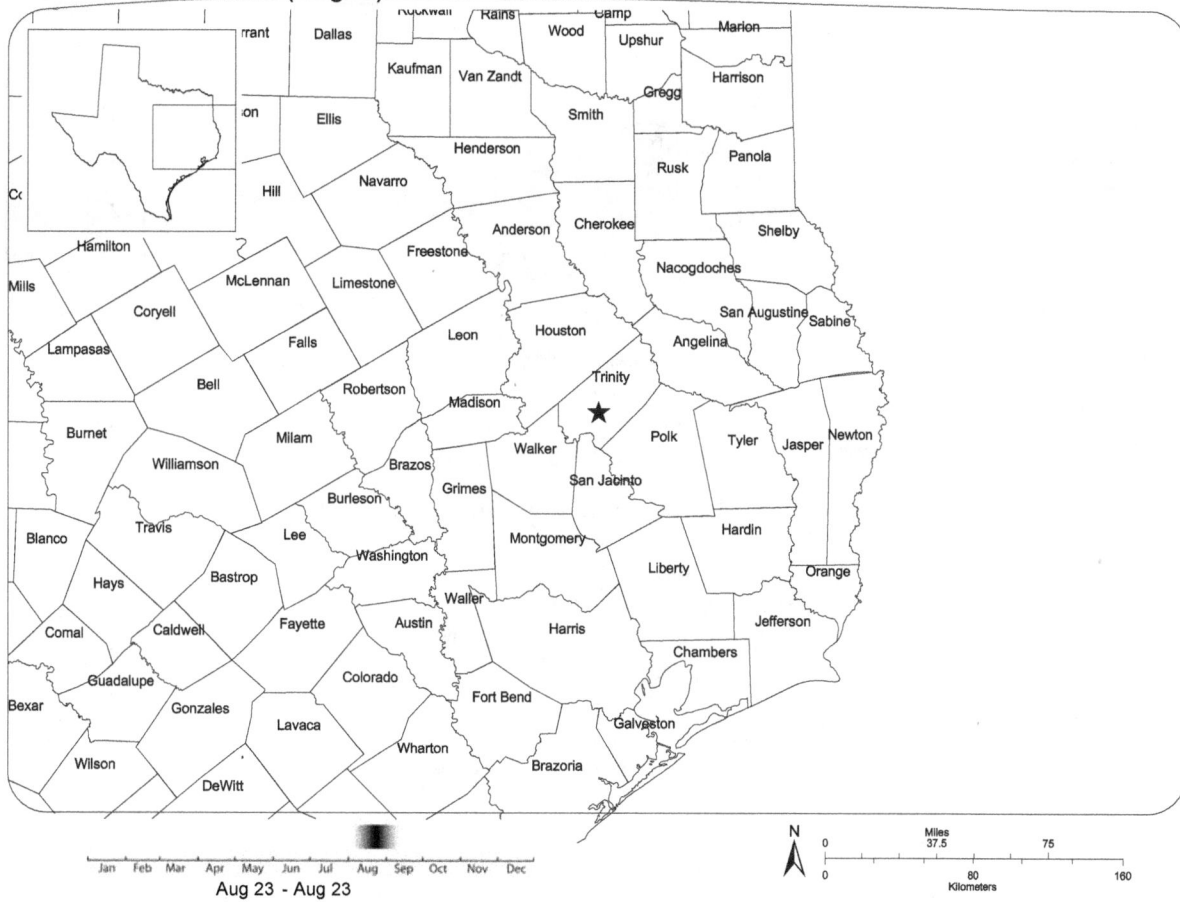

Aug 23 - Aug 23

HABITAT: Probably spring-fed seeps and forest streams.

Trinity

Somatochlora georgiana Walker / Coppery Emerald

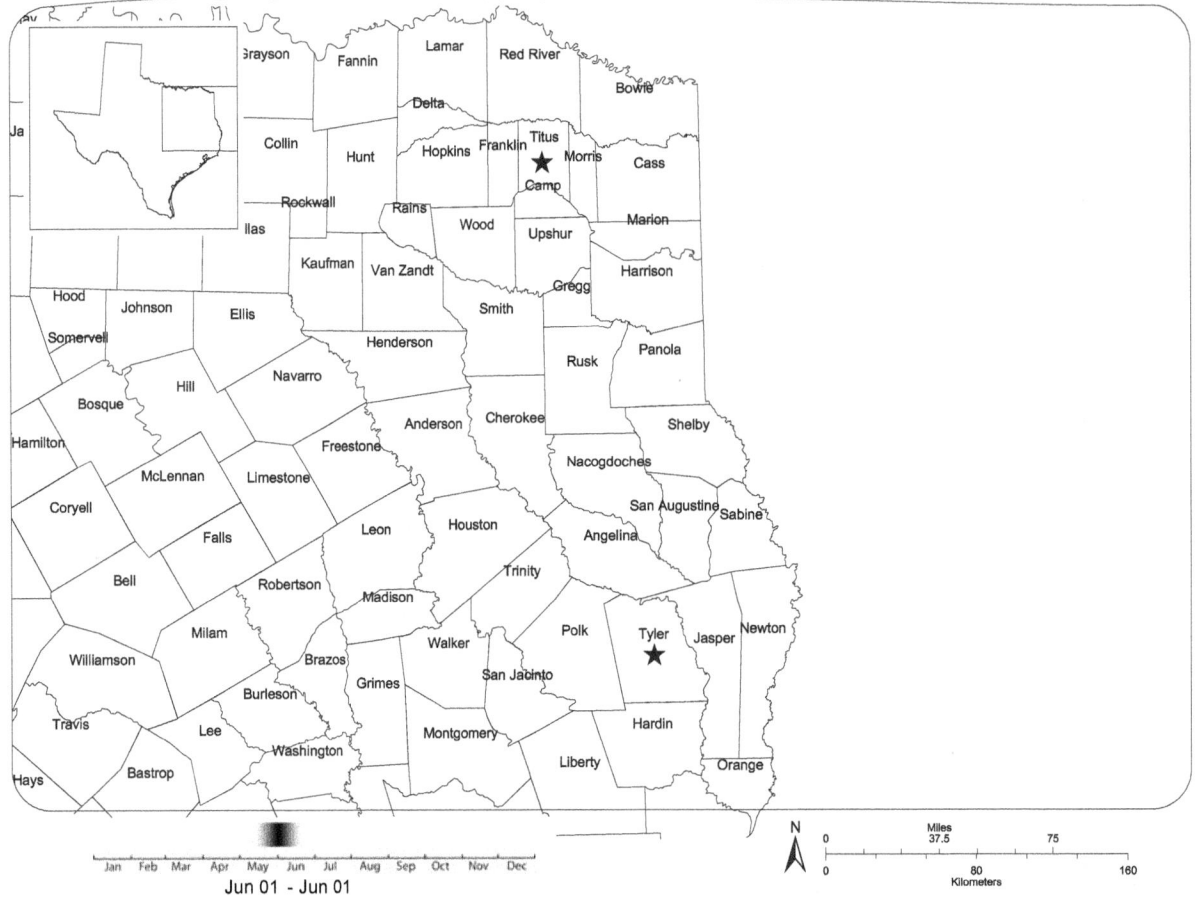

Jun 01 - Jun 01

HABITAT: Pools and slow flowing tannin-stained forest streams.

Titus
Tyler

Somatochlora linearis (Hagen) / Mocha Emerald

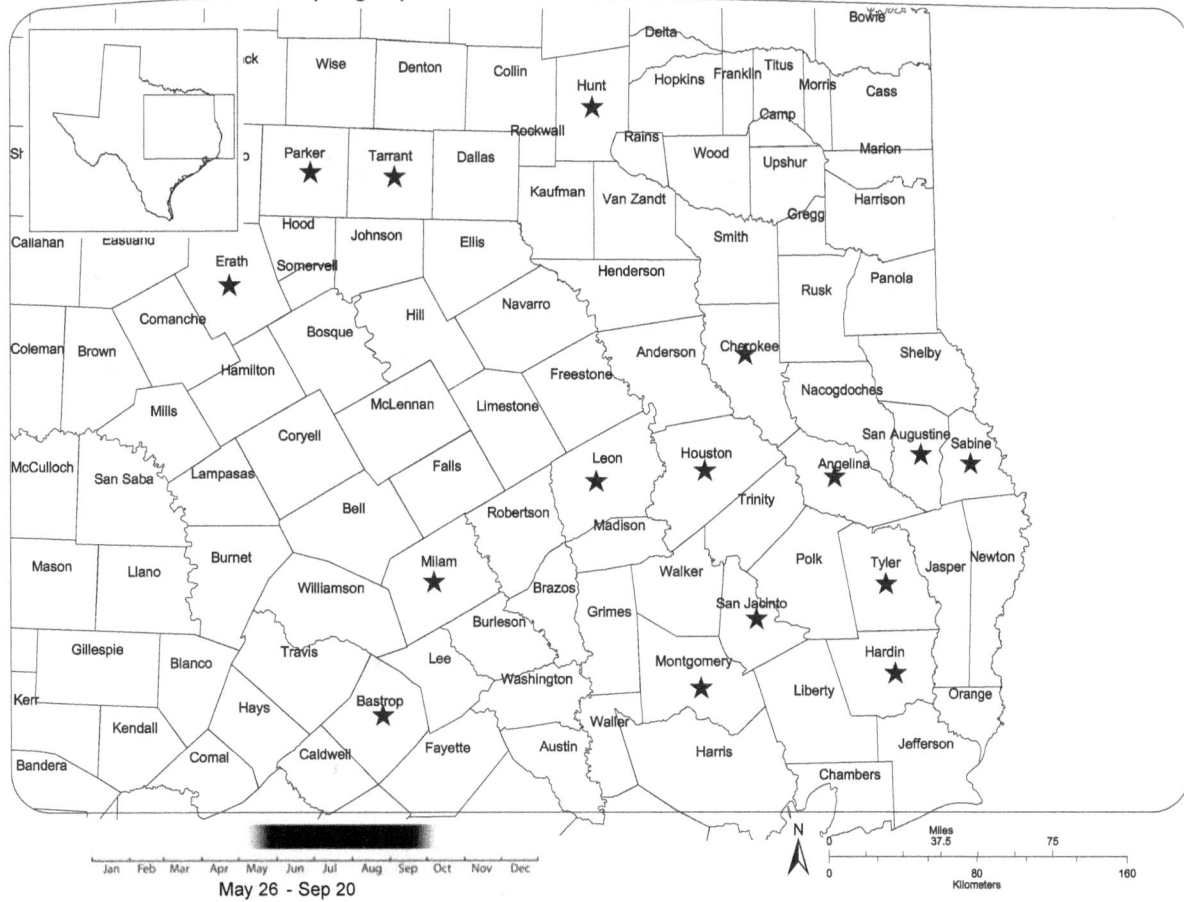

May 26 - Sep 20

HABITAT: Permanent and temporary forest streams.

Angelina
Bastrop
Cherokee
Erath
Hardin
Houston
Hunt
Leon
Milam
Montgomery
Parker
Sabine
San Augustine
San Jacinto
Tarrant
Tyler

Somatochlora margarita Donnelly / Texas Emerald

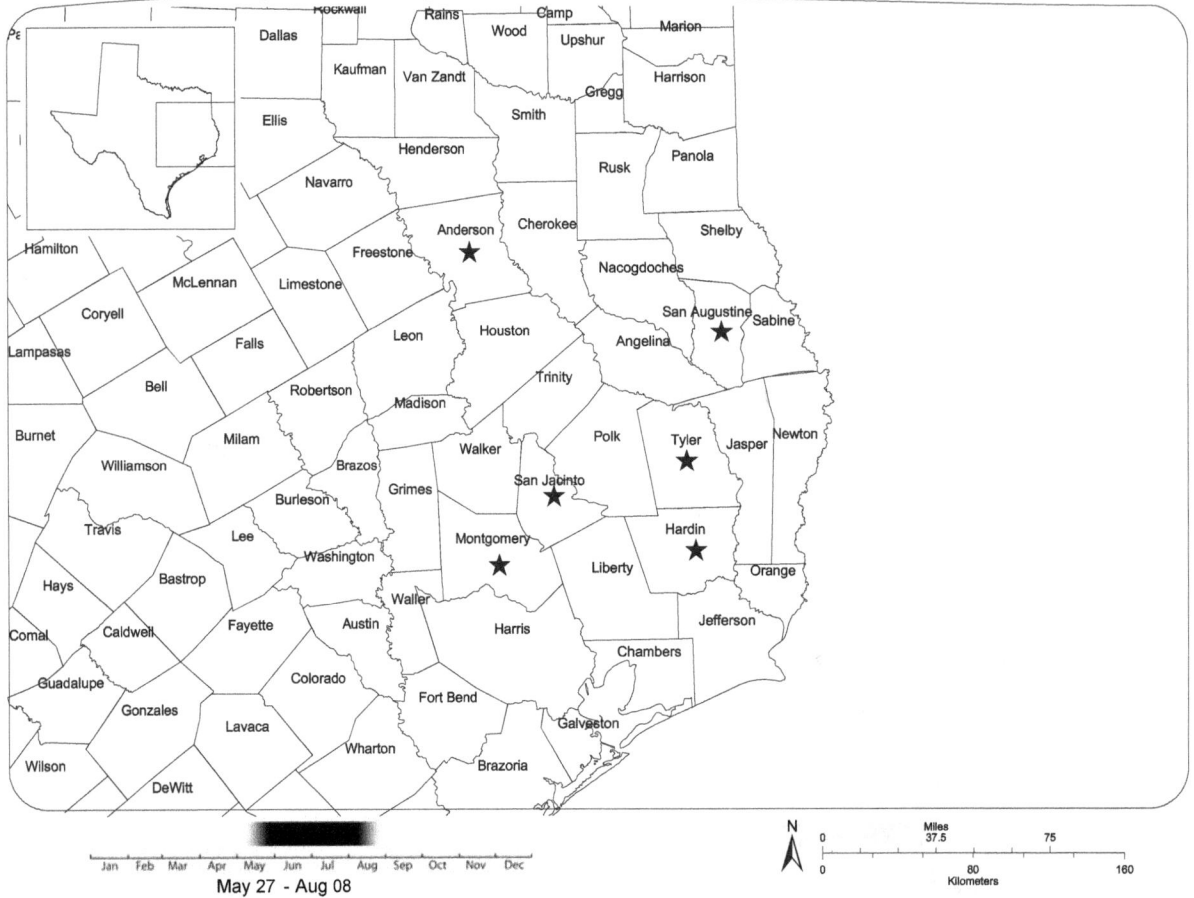

May 27 - Aug 08

HABITAT: Small, sandy forest streams, with moderate current.

Anderson
Hardin
Montgomery
San Augustine
San Jacinto
Tyler

Somatochlora tenebrosa (Say) / Clamp-tipped Emerald

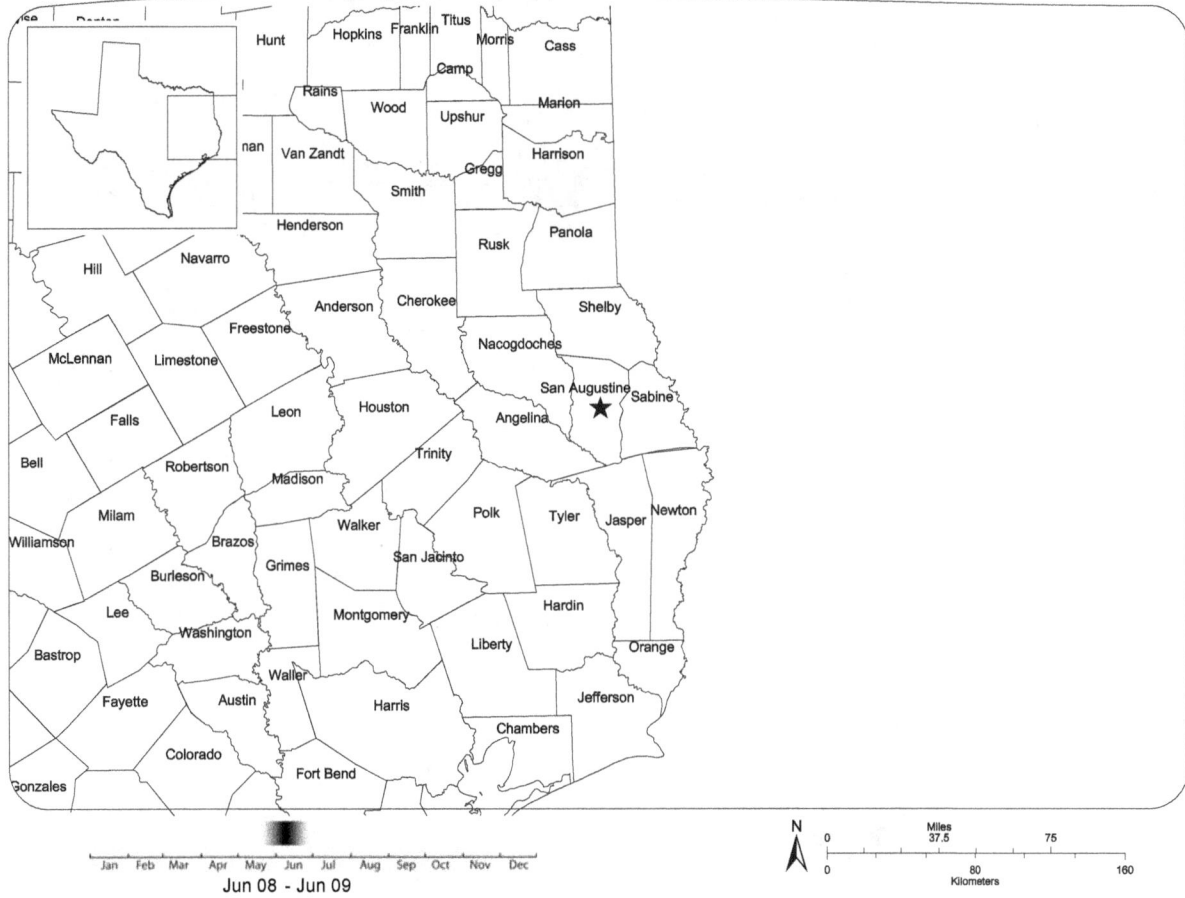

Jun 08 - Jun 09

HABITAT: Small forested streams with intermittent riffles and pools.

San Augustine

Brachymesia furcata (Hagen) / Red-tailed Pennant

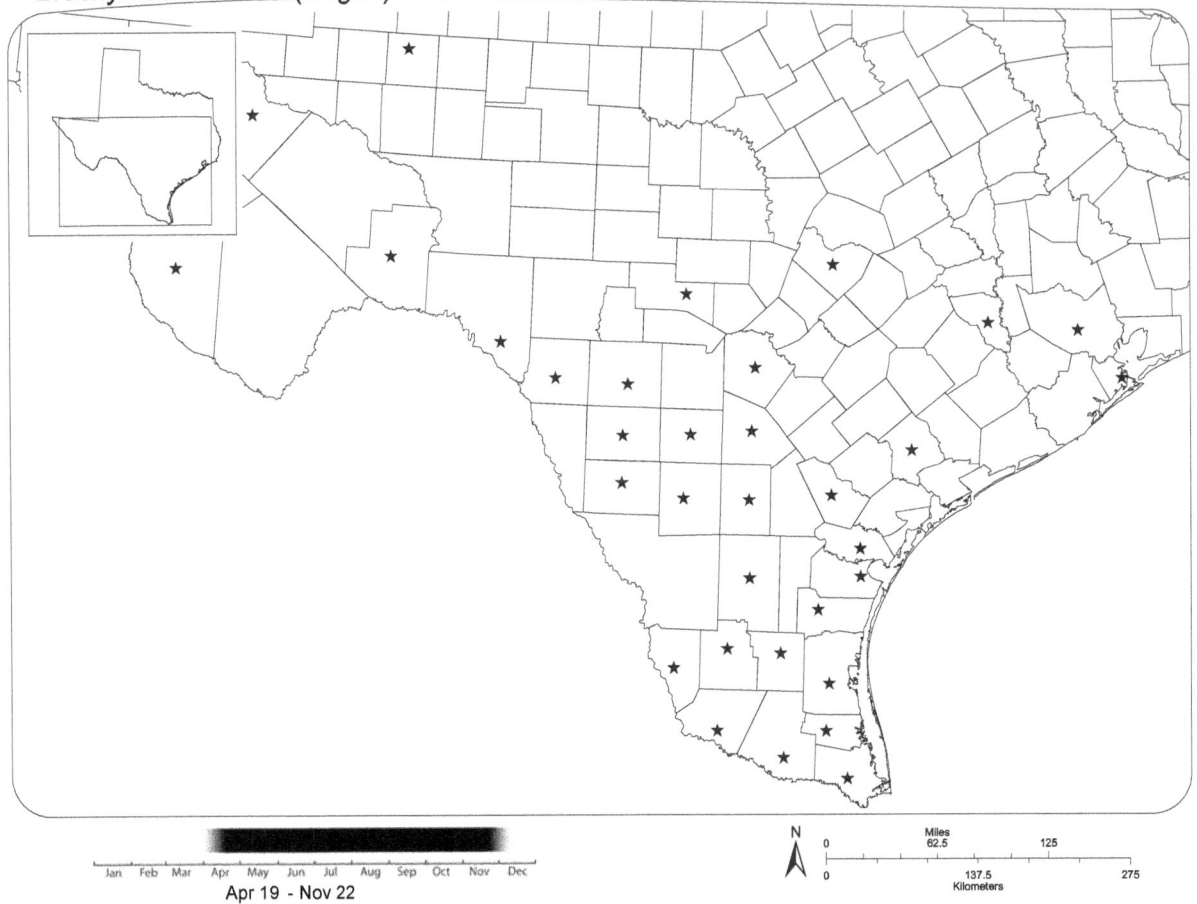

Apr 19 - Nov 22

HABITAT: Ponds, lakes and ditches with permanent or semipermanent water including brackish waters.

Atascosa
Austin
Bee
Bexar
Brooks
Cameron
Dimmit
Duval
Frio
Galveston
Harris
Hidalgo
Hudspeth
Jim Hogg
Kenedy
Kerr
Kinney
Kleberg
La Salle
McMullen
Midland
Nueces
Presidio
Reeves
San Patricio
Starr

Terrell
Travis
Uvalde
Val Verde
Victoria
Willacy
Zapata
Zavala

Brachymesia gravida (Calvert) / Four-spotted Pennant

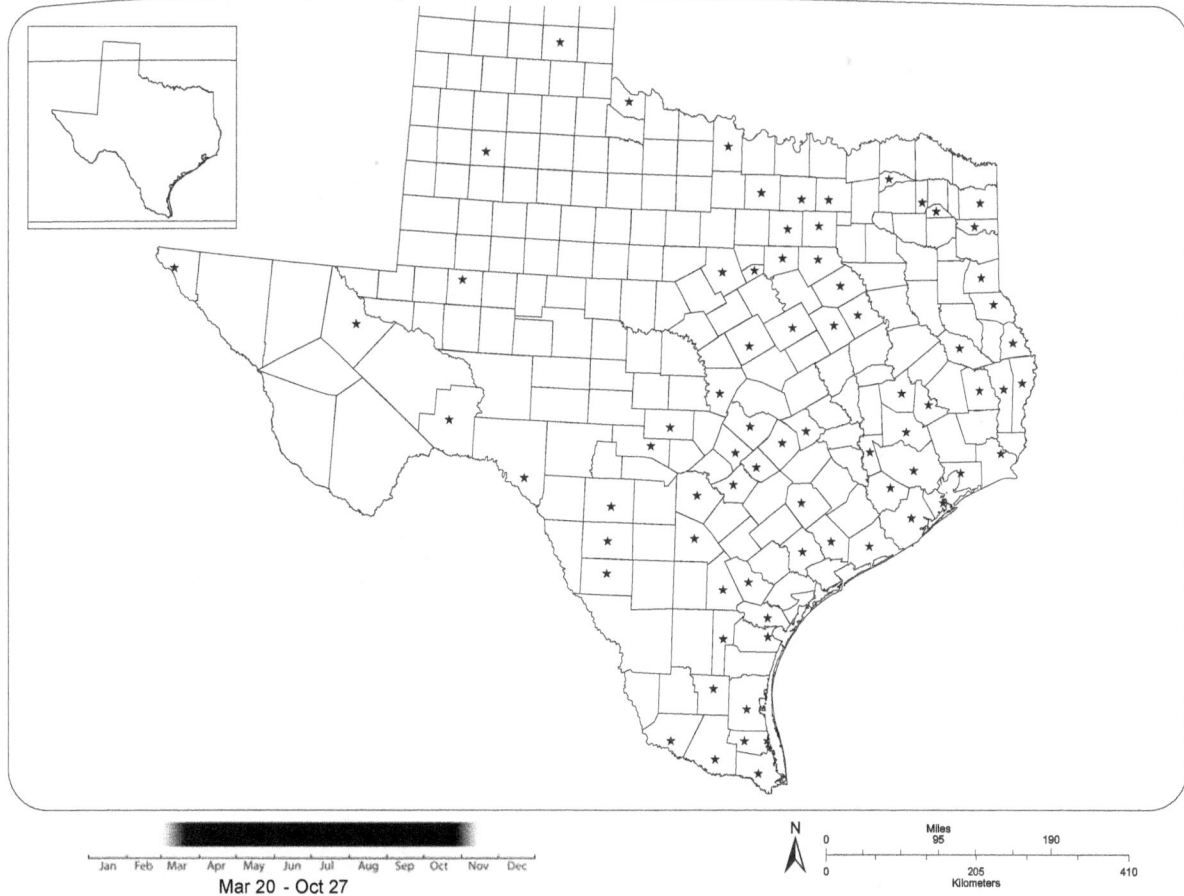

Jan Feb Mar Apr May Jun Jul Aug Sep Oct Nov Dec

Mar 20 - Oct 27

HABITAT: Ponds, lakes and roadside ditches, including brackish waters.

Angelina	Freestone	Newton
Atascosa	Galveston	Nueces
Bastrop	Gillespie	Panola
Bee	Guadalupe	Reeves
Bexar	Hardeman	Sabine
Brazoria	Harris	San Jacinto
Brooks	Hays	San Patricio
Burnet	Hidalgo	Shelby
Caldwell	Jackson	Somervell
Cameron	Jasper	Starr
Camp	Jefferson	Tarrant
Cass	Jim Wells	Terrell
Chambers	Johnson	Travis
Clay	Kenedy	Tyler
Collin	Kerr	Uvalde
Coryell	Lavaca	Val Verde
Dallas	Lee	Victoria
Delta	Limestone	Walker
Denton	Live Oak	Waller
Dimmit	Lubbock	Willacy
Donley	Marion	Wise
El Paso	Matagorda	Zavala
Ellis	McLennan	
Erath	Midland	
Fort Bend	Montgomery	
Franklin	Navarro	

Brachymesia herbida (Gundlach) / Tawny Pennant

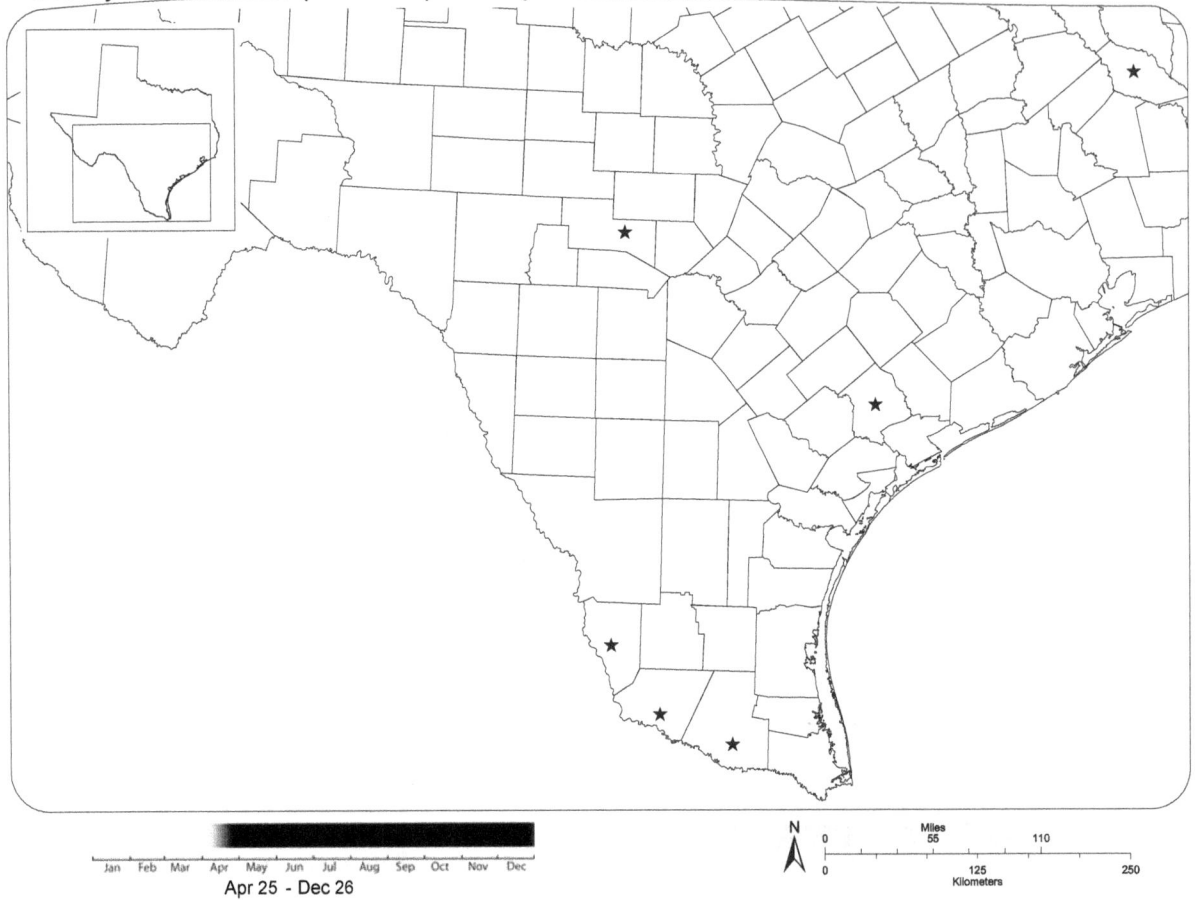

Apr 25 - Dec 26

HABITAT: Ponds, lakes, marshes and roadside ditches including brackish waters.

Angelina
Hidalgo
Kerr
Presidio
Starr
Victoria
Zapata

Brechmorhoga mendax (Hagen) / Pale-faced Clubskimmer

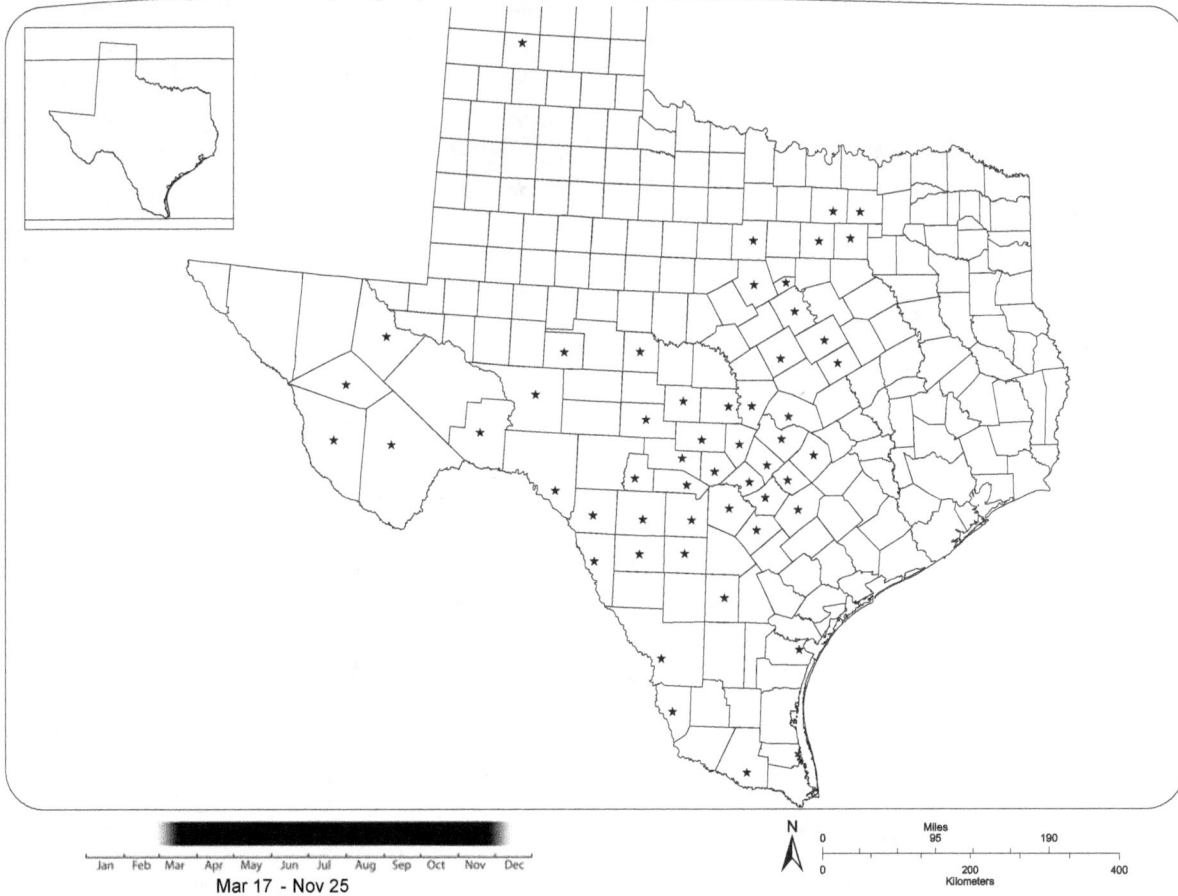

Jan Feb Mar Apr May Jun Jul Aug Sep Oct Nov Dec

Mar 17 - Nov 25

HABITAT: Sand and cobble streams and rivers.

Bandera	Kerr
Bastrop	Kimble
Bexar	Kinney
Blanco	Llano
Bosque	Mason
Brewster	Maverick
Burnet	McLennan
Caldwell	McMullen
Collin	Medina
Comal	Nueces
Concho	Palo Pinto
Coryell	Presidio
Crockett	Randall
Dallas	Real
Denton	Reeves
Erath	Somervell
Falls	Tarrant
Frio	Terrell
Gillespie	Travis
Gonzales	Uvalde
Guadalupe	Val Verde
Hays	Webb
Hidalgo	Williamson
Irion	Wilson
Jeff Davis	Zapata
Kendall	Zavala

Cannaphila insularis funerea (Carpenter) / Gray-waisted Skimmer

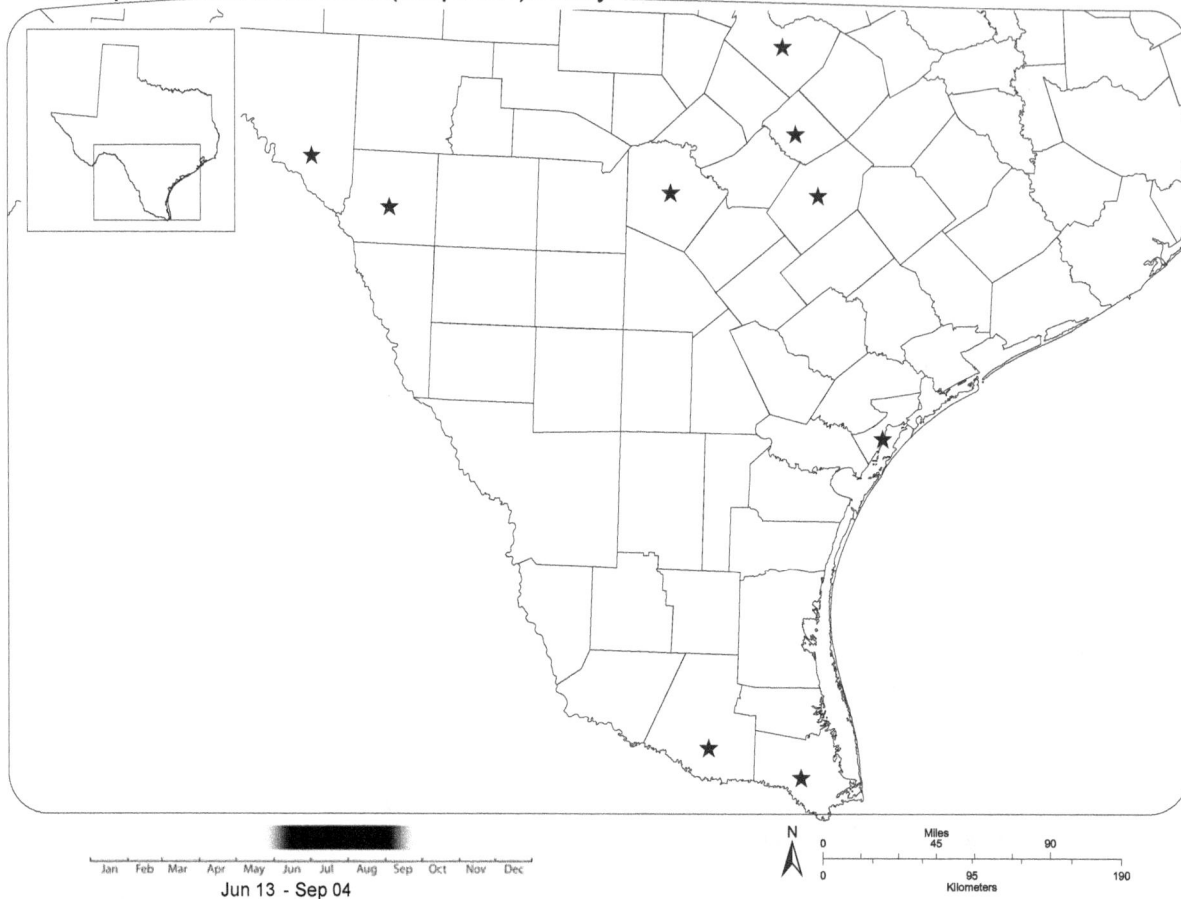

Jan Feb Mar Apr May Jun Jul Aug Sep Oct Nov Dec

Jun 13 - Sep 04

HABITAT: Marshy ponds, lakes and streams.

Aransas
Bexar
Caldwell
Cameron
Gonzales
Hidalgo
Kinney
Travis
Val Verde

Celithemis amanda (Hagen) / Amanda's Pennant

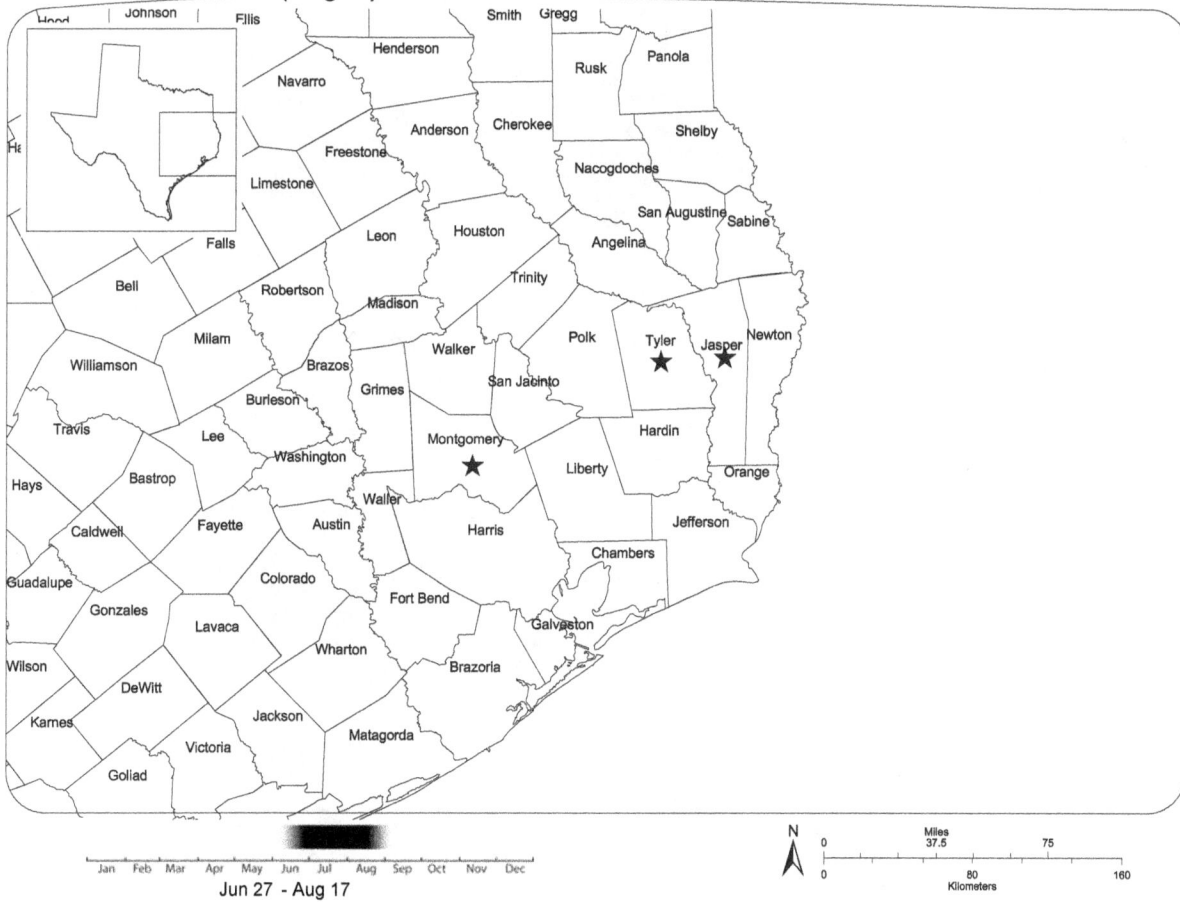

Jun 27 - Aug 17

HABITAT: Calm lakes, ponds, and marshes with emergent vegetation.

Jasper
Montgomery
Tyler

Celithemis elisa (Hagen) / Calico Pennant

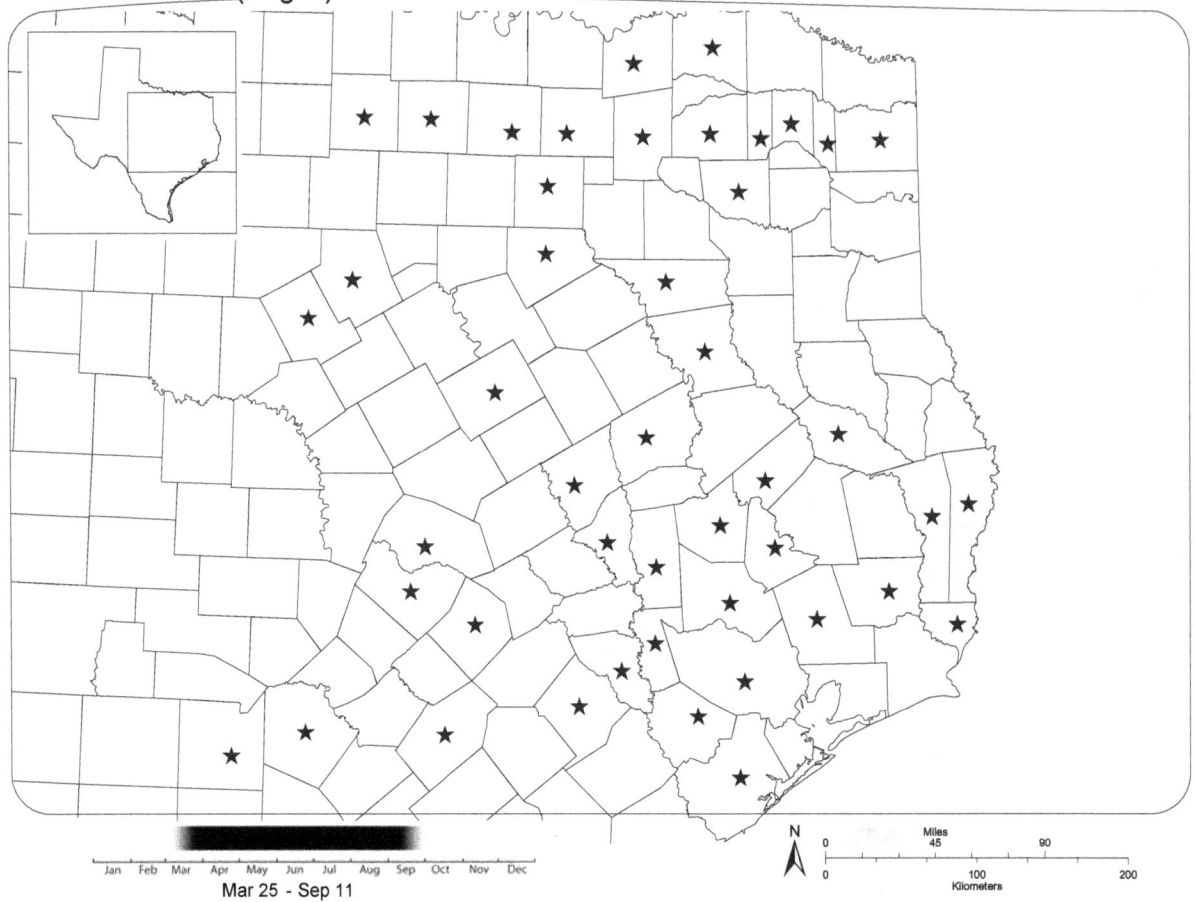

Mar 25 - Sep 11

HABITAT: Lakes, Ponds and borrow pits with emergent vegetation and calm, clear waters.

Anderson	Jasper
Angelina	Lamar
Austin	Leon
Bastrop	Liberty
Bexar	McLennan
Brazoria	Medina
Brazos	Montgomery
Cass	Morris
Collin	Newton
Colorado	Orange
Comanche	Robertson
Dallas	San Jacinto
Denton	Titus
Ellis	Travis
Erath	Trinity
Fannin	Walker
Fort Bend	Waller
Franklin	Williamson
Gonzales	Wise
Grimes	Wood
Hardin	
Harris	
Henderson	
Hopkins	
Hunt	
Jack	

Celithemis eponina (Drury) / Halloween Pennant

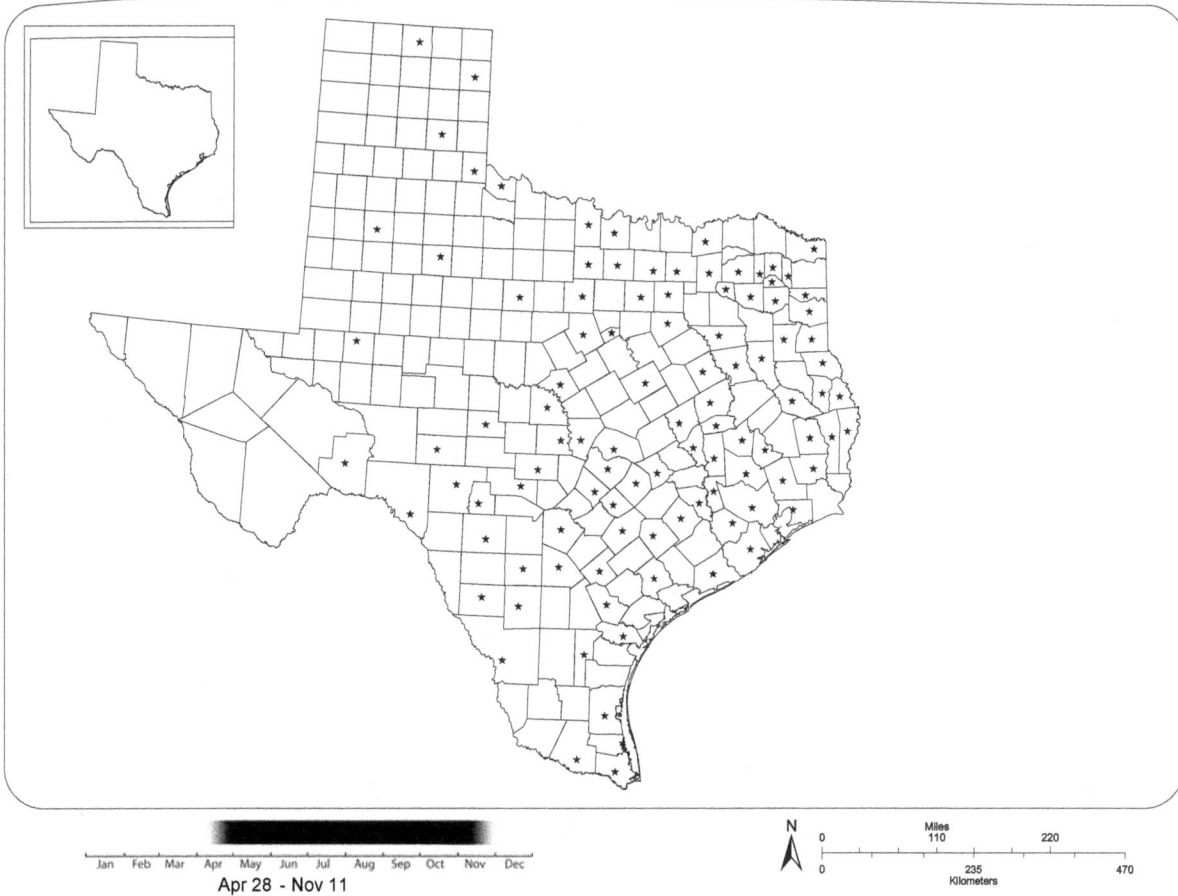

Apr 28 - Nov 11

HABITAT: Lakes, ponds, borrow pits and marshes with emergent vegetation

Anderson	Erath	Kerr	San Augustine
Angelina	Fannin	La Salle	San Jacinto
Atascosa	Fort Bend	Lavaca	San Patricio
Austin	Franklin	Lee	San Saba
Bastrop	Freestone	Leon	Shackelford
Bee	Frio	Liberty	Shelby
Bexar	Gillespie	Llano	Somervell
Bowie	Gonzales	Lubbock	Sutton
Brazoria	Grimes	Madison	Tarrant
Brazos	Hansford	Marion	Terrell
Burnet	Hardeman	Matagorda	Titus
Caldwell	Hardin	McLennan	Travis
Cameron	Harris	Menard	Tyler
Camp	Harrison	Midland	Upshur
Chambers	Hays	Mills	Uvalde
Cherokee	Hemphill	Montague	Val Verde
Childress	Henderson	Montgomery	Victoria
Clay	Hidalgo	Morris	Walker
Collin	Hopkins	Newton	Waller
Colorado	Hunt	Palo Pinto	Webb
Dallas	Jack	Panola	Williamson
Denton	Jasper	Rains	Wise
Dimmit	Jim Wells	Real	Wood
Donley	Karnes	Robertson	
Edwards	Kenedy	Rusk	
Ellis	Kent	Sabine	

Celithemis fasciata Kirby / Banded Pennant

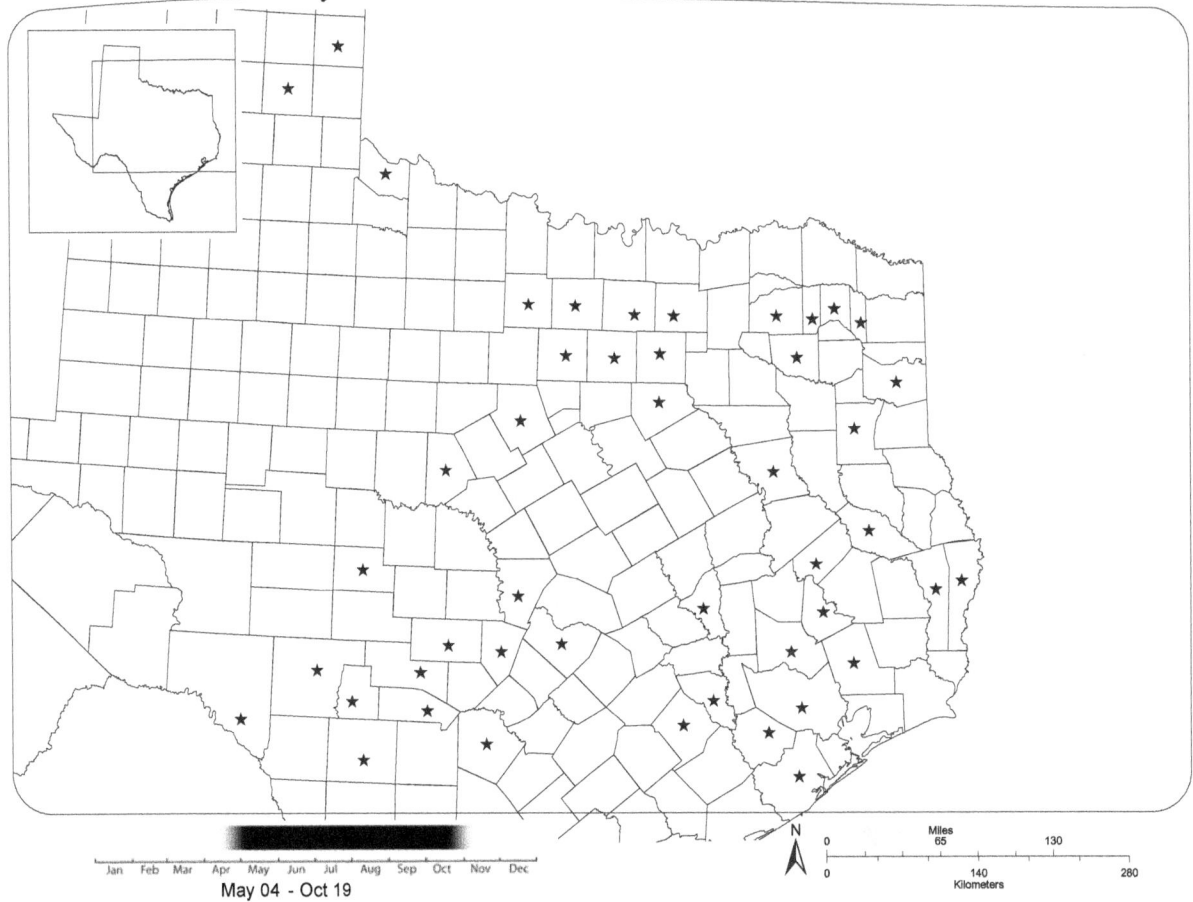

May 04 - Oct 19

HABITAT: Permanent lakes, ponds and borrow pits with emergent vegetation.

Anderson	Jasper
Angelina	Kerr
Austin	Liberty
Bandera	Menard
Bexar	Montgomery
Blanco	Morris
Brazoria	Newton
Brazos	Parker
Brown	Real
Burnet	Rusk
Collin	San Jacinto
Colorado	Tarrant
Dallas	Titus
Denton	Travis
Donley	Trinity
Edwards	Uvalde
Ellis	Val Verde
Erath	Wheeler
Fort Bend	Wise
Franklin	Wood
Gillespie	
Hardeman	
Harris	
Harrison	
Hopkins	
Jack	

Celithemis ornata (Rambur) / Ornate Pennant

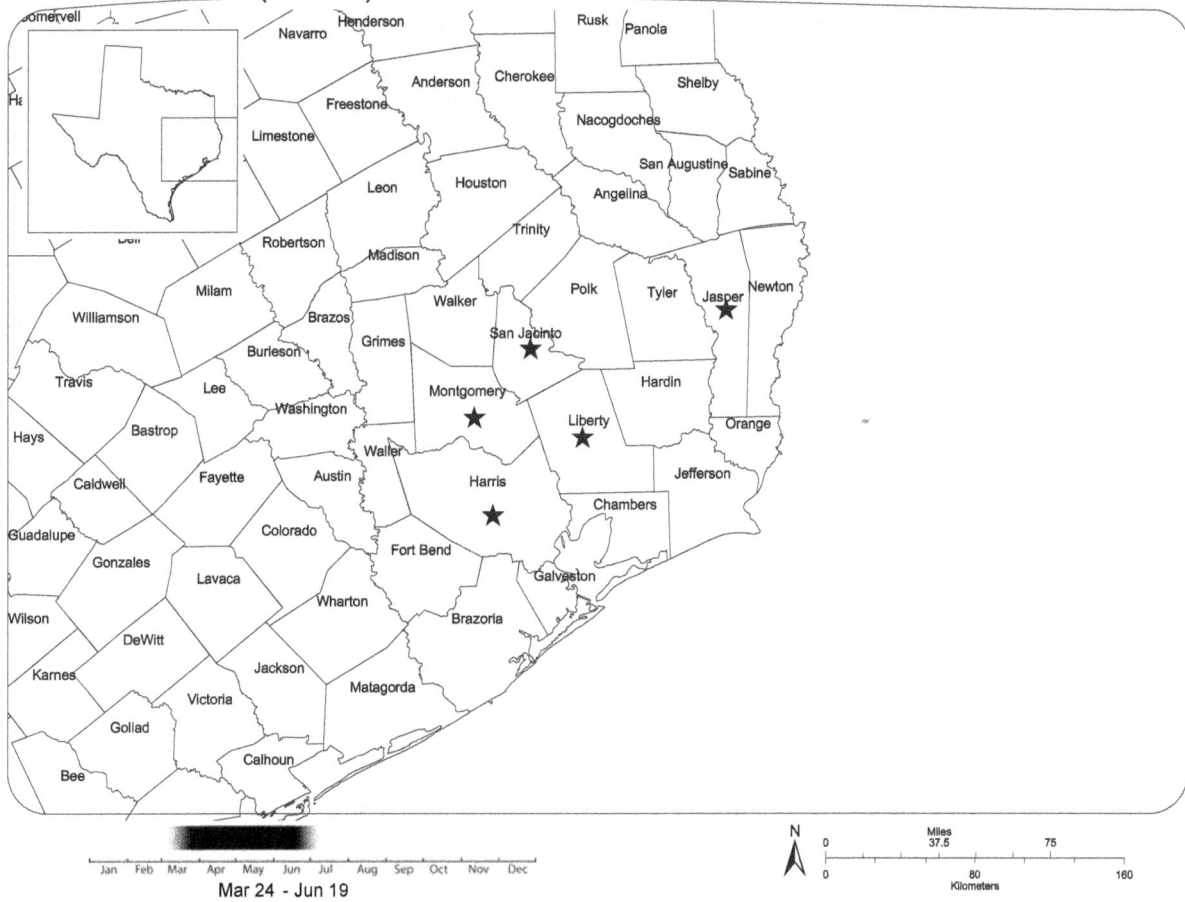

Mar 24 - Jun 19

HABITAT: Lakes, ponds, pools, with calm waters and slow reaches of streams, all with emergent vegetation.

Harris
Jasper
Liberty
Montgomery
San Jacinto

Celithemis verna Pritchard / Double-ringed Pennant

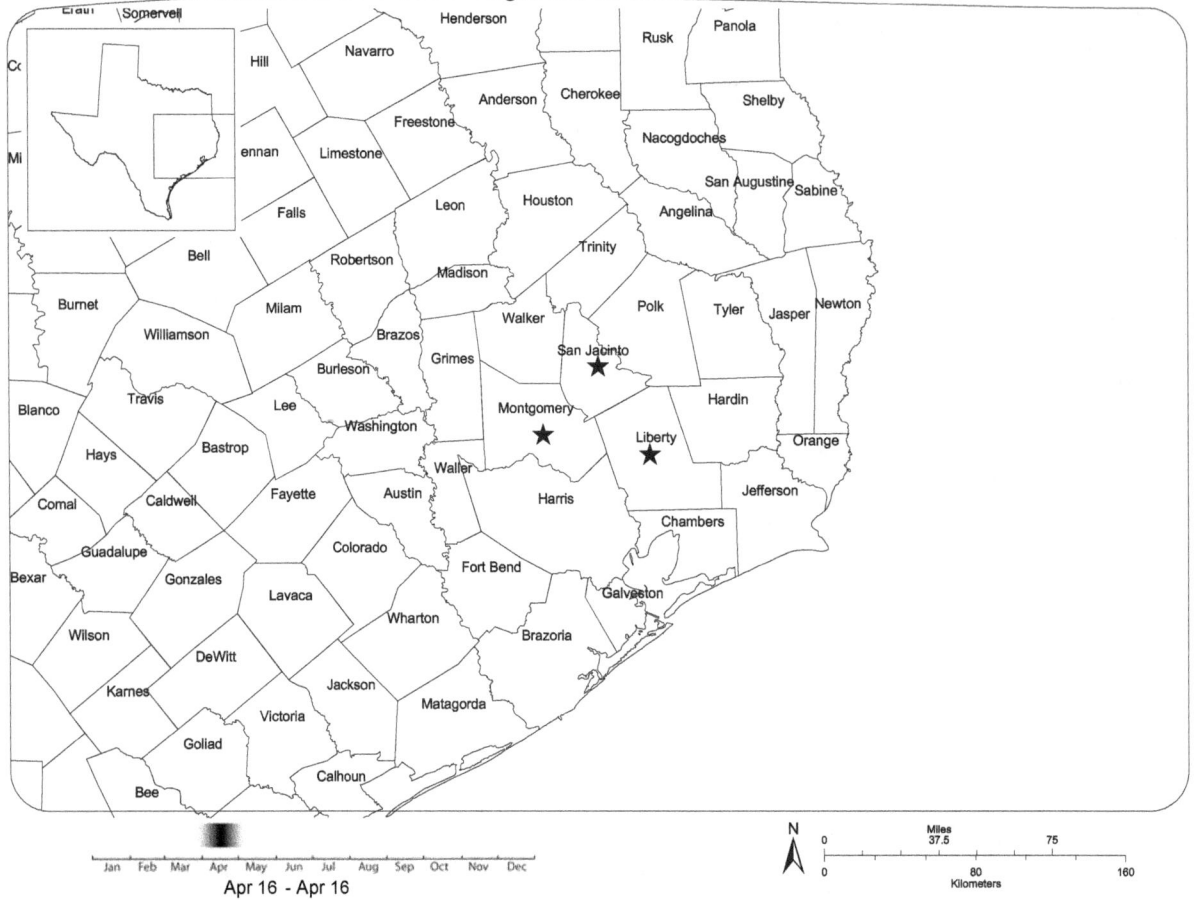

Apr 16 - Apr 16

HABITAT: Newly formed lakes and ponds with emergent vegetation.

Liberty
Montgomery
San Jacinto

Dythemis fugax Hagen / Checkered Setwing

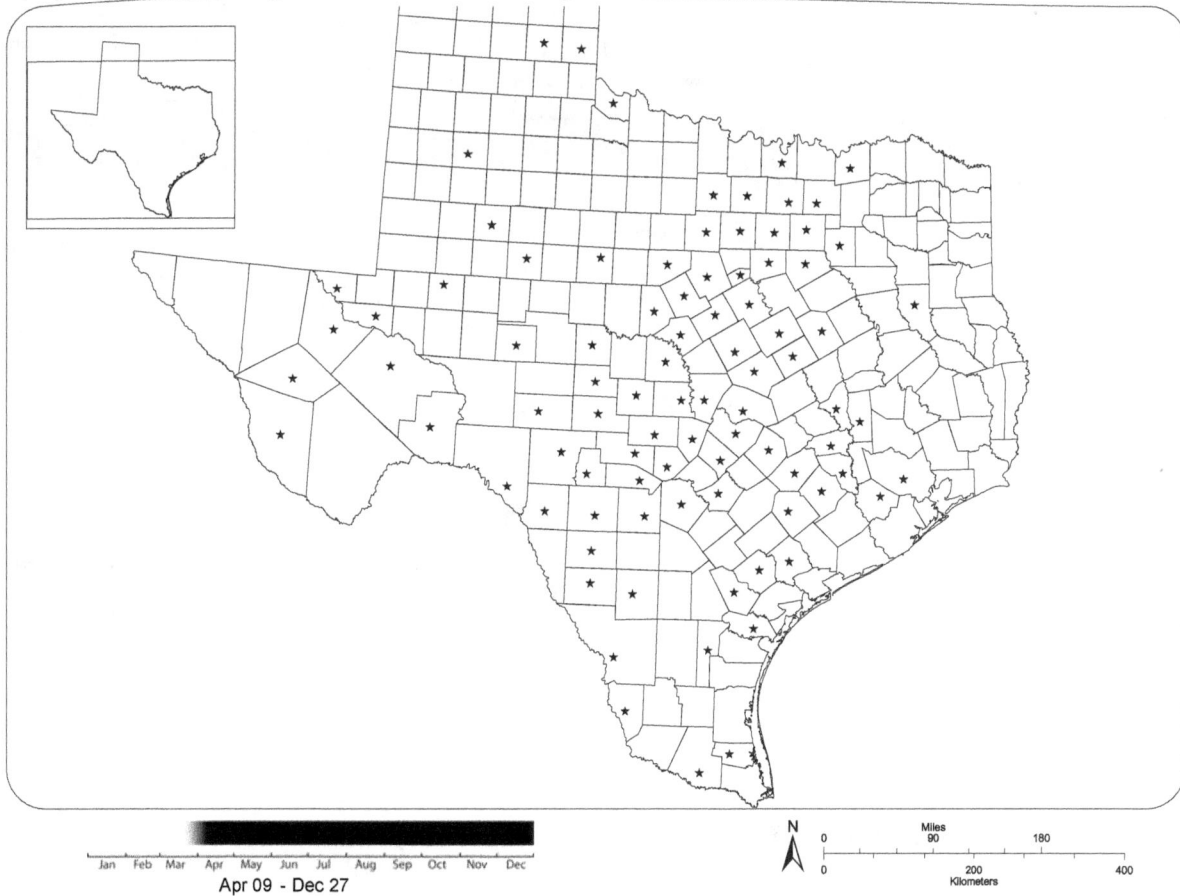

Apr 09 - Dec 27

HABITAT: Ponds and lakes with emergent vegetation.

Austin	Ellis	Lavaca	Uvalde
Bandera	Erath	Limestone	Val Verde
Bastrop	Falls	Llano	Victoria
Bee	Fannin	Loving	Ward
Bell	Fayette	Lubbock	Washington
Bexar	Fort Bend	Mason	Webb
Blanco	Gillespie	McLennan	Willacy
Borden	Goliad	Medina	Williamson
Bosque	Grimes	Menard	Wise
Brazos	Guadalupe	Midland	Zapata
Brown	Hamilton	Mills	Zavala
Burnet	Hardeman	Mitchell	
Cherokee	Harris	Palo Pinto	
Collin	Hays	Parker	
Collingsworth	Hidalgo	Pecos	
Colorado	Irion	Presidio	
Comanche	Jack	Real	
Concho	Jeff Davis	Reeves	
Cooke	Jim Wells	San Patricio	
Coryell	Johnson	San Saba	
Dallas	Kaufman	Somervell	
Denton	Kendall	Sutton	
Dimmit	Kerr	Tarrant	
Donley	Kimble	Taylor	
Eastland	Kinney	Terrell	
Edwards	La Salle	Travis	

Dythemis maya Calvert / Mayan Setwing

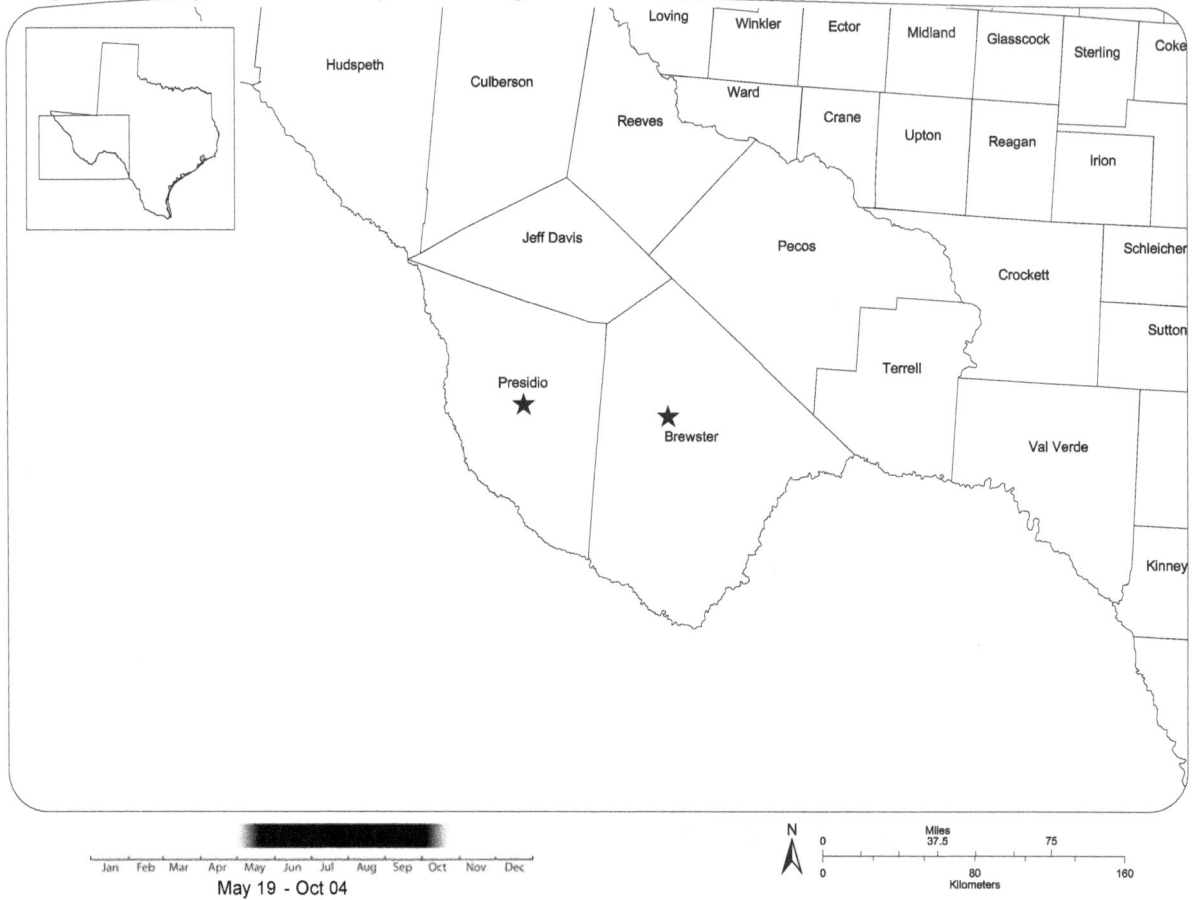

May 19 - Oct 04

HABITAT: Small arid streams with moderate to swift current.

Brewster
Presidio

Dythemis nigrescens Calvert / Black Setwing

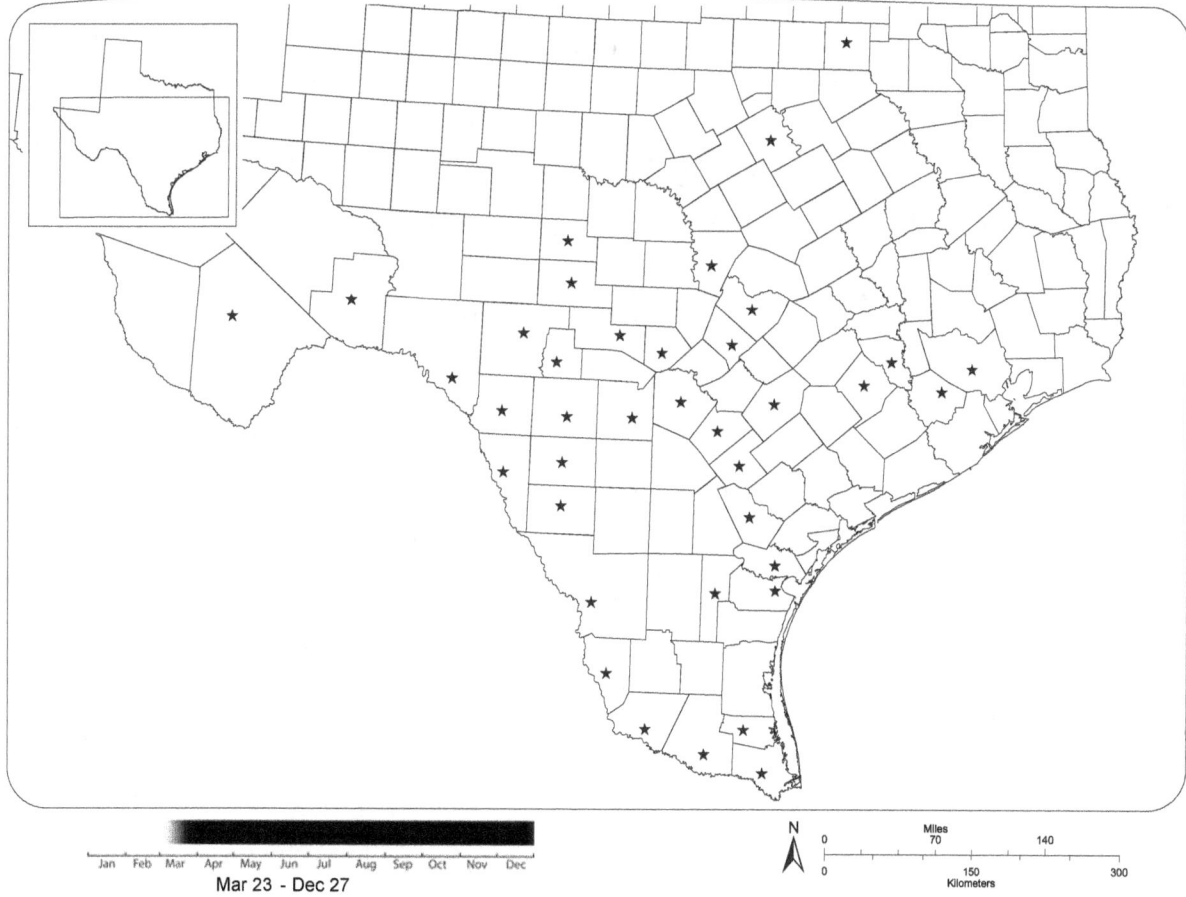

Mar 23 - Dec 27

HABITAT: Creeks, streams and rivers with moderate current.

Austin
Bee
Bexar
Bosque
Brewster
Burnet
Cameron
Colorado
Dallas
Dimmit
Edwards
Fort Bend
Gonzales
Harris
Hays
Hidalgo
Jim Wells
Karnes
Kendall
Kerr
Kimble
Kinney
Maverick
Medina
Menard
Nueces

Real
Reeves
San Patricio
Starr
Terrell
Travis
Uvalde
Val Verde
Webb
Willacy
Wilson
Zapata
Zavala

Dythemis velox Hagen / Swift Setwing

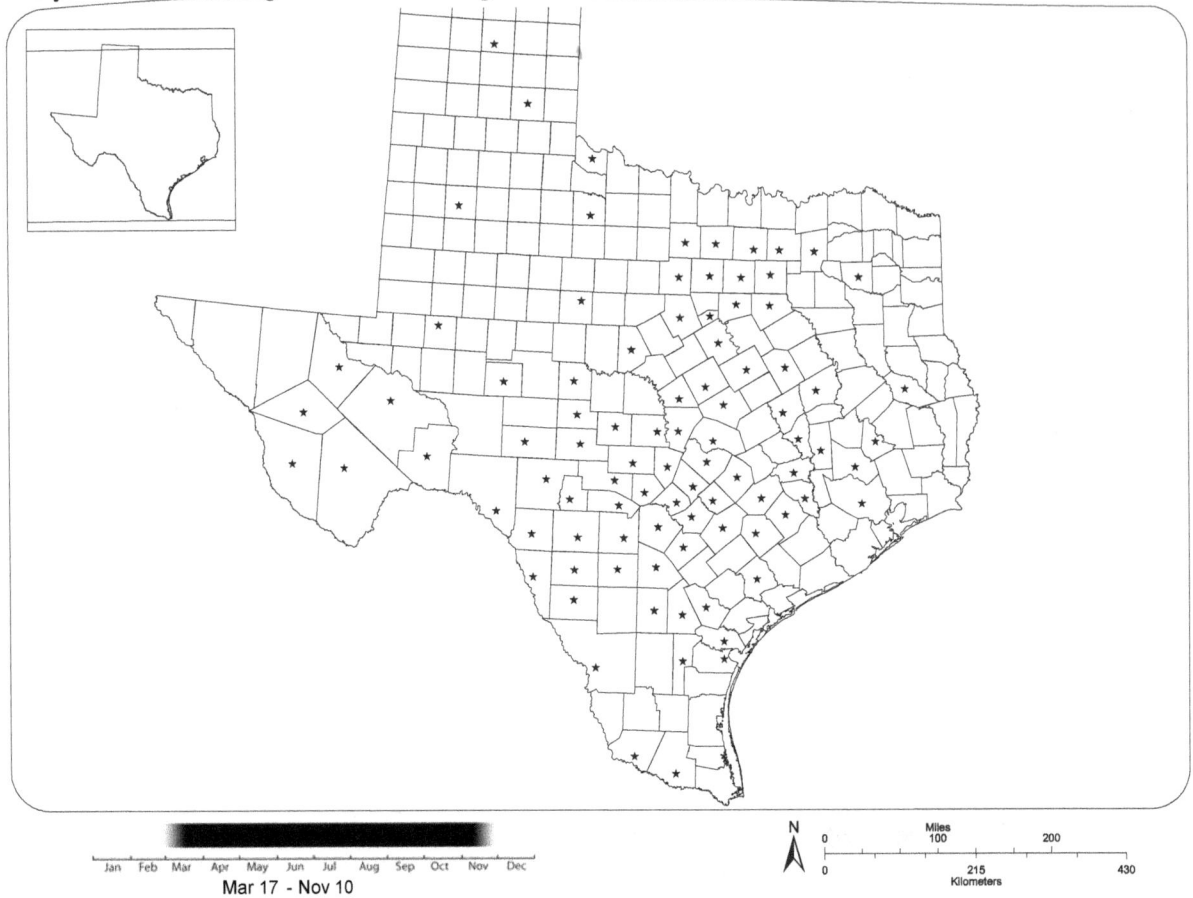

Mar 17 - Nov 10

HABITAT: Lakes, ponds and borrow pits as well as creeks, streams and rivers with moderate current.

Angelina	Erath	Limestone	Taylor
Atascosa	Fayette	Live Oak	Terrell
Austin	Frio	Llano	Travis
Bandera	Gillespie	Lubbock	Uvalde
Bastrop	Gonzales	Mason	Val Verde
Bee	Grimes	Maverick	Victoria
Bell	Guadalupe	McLennan	Washington
Bexar	Hardeman	McMullen	Webb
Blanco	Harris	Medina	Williamson
Bosque	Hays	Menard	Wilson
Brazos	Hidalgo	Midland	Wise
Brewster	Hunt	Montgomery	Wood
Brown	Hutchinson	Nueces	Zavala
Burnet	Irion	Palo Pinto	
Caldwell	Jack	Parker	
Collin	Jeff Davis	Pecos	
Colorado	Jim Wells	Presidio	
Comal	Johnson	Real	
Concho	Kendall	Reeves	
Coryell	Kerr	Robertson	
Dallas	Kimble	San Jacinto	
Denton	Kinney	San Patricio	
Dimmit	Knox	Somervell	
Donley	Lampasas	Starr	
Edwards	Lavaca	Sutton	
Ellis	Leon	Tarrant	

Erythemis attala (Selys) / Black Pondhawk

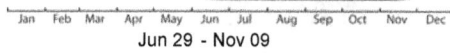

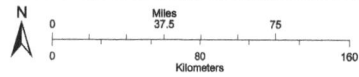

Jun 29 - Nov 09

HABITAT: Ponds, lakes, ditches and slow reaches of rivers and streams.

Cameron
Hidalgo
Kleberg

Erythemis collocata (Hagen) / Western Pondhawk

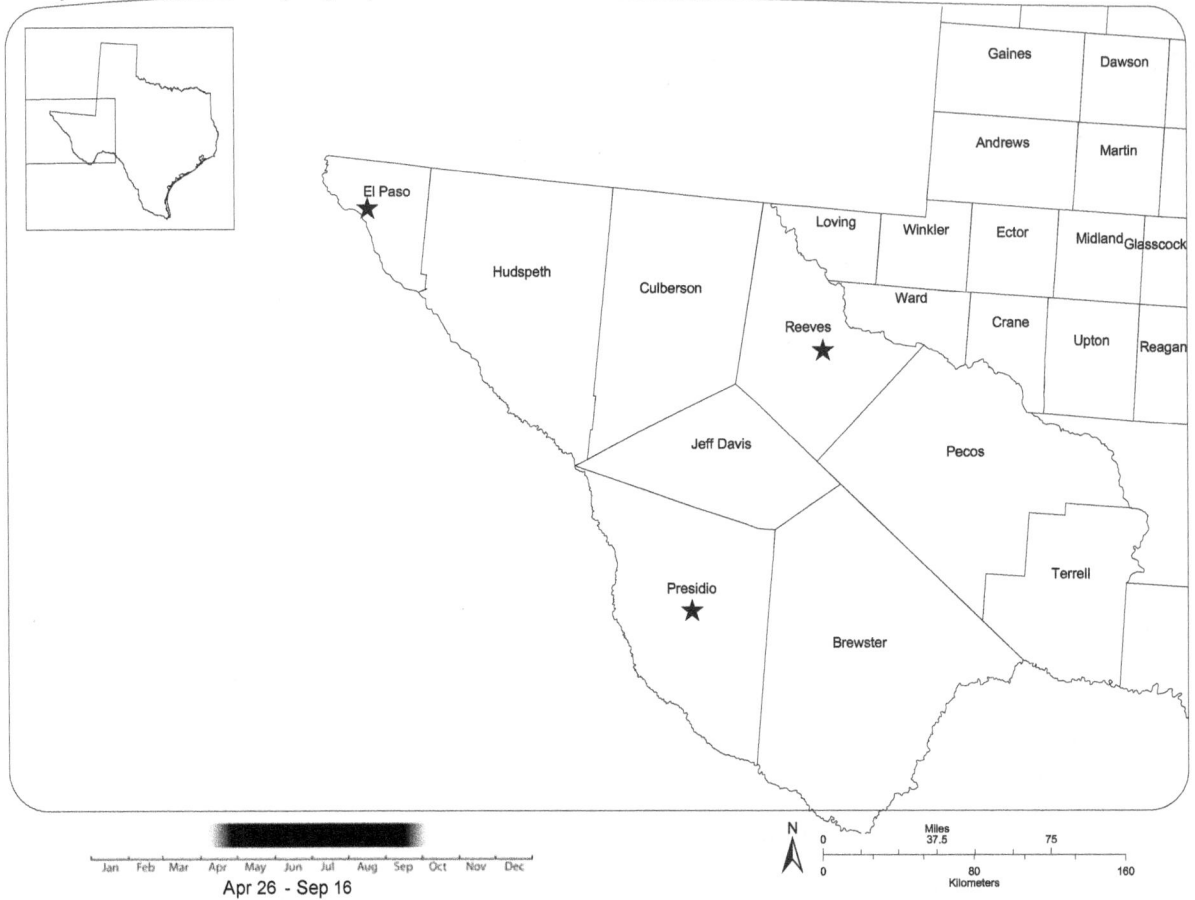

Apr 26 - Sep 16

HABITAT: Ponds, lakes and slow flowing waters of streams and creeks.

El Paso
Presidio
Reeves

Erythemis mithroides (Brauer) / Claret Pondhawk

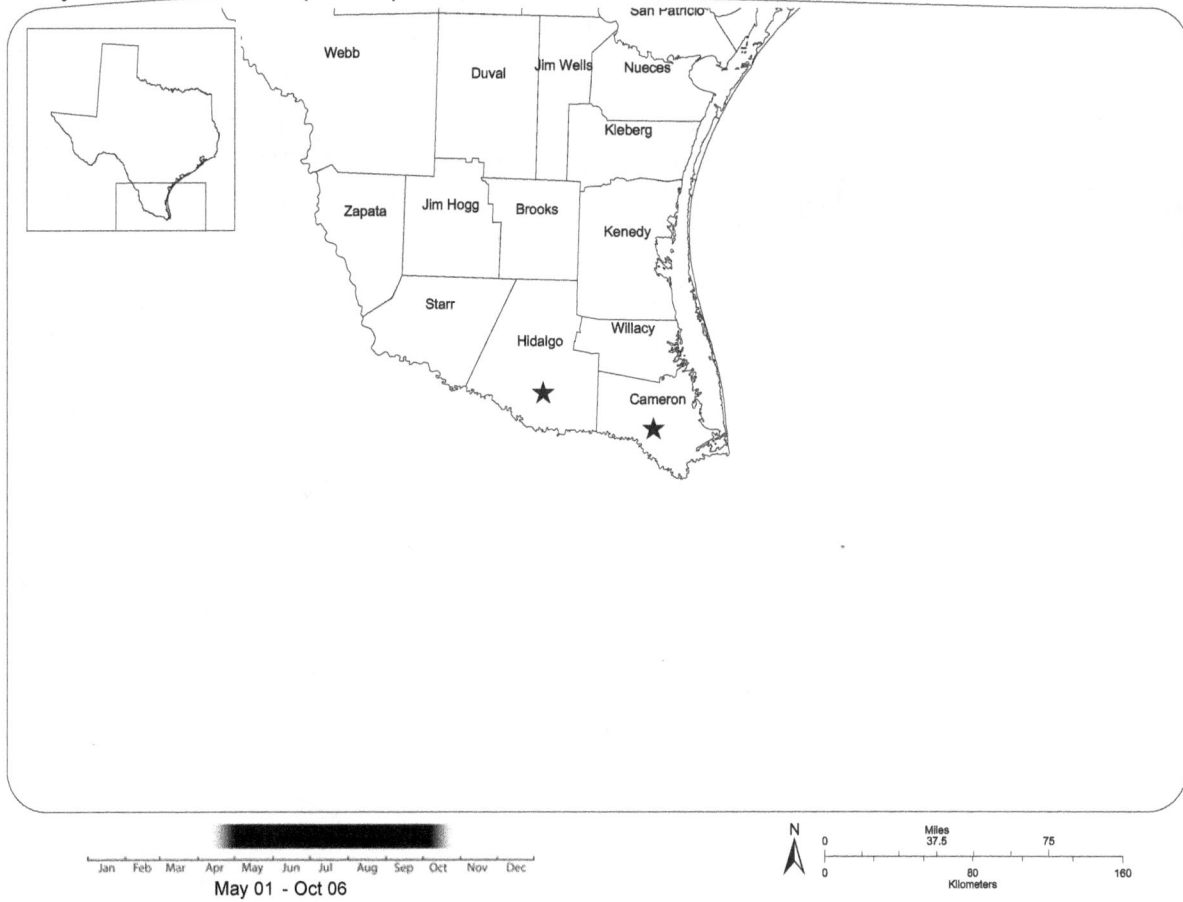

May 01 - Oct 06

HABITAT: Ponds, lakes, ditches and slow reaches of rivers and streams.

Cameron
Hidalgo

Erythemis peruviana (Rambur) / Flame-tailed Pondhawk

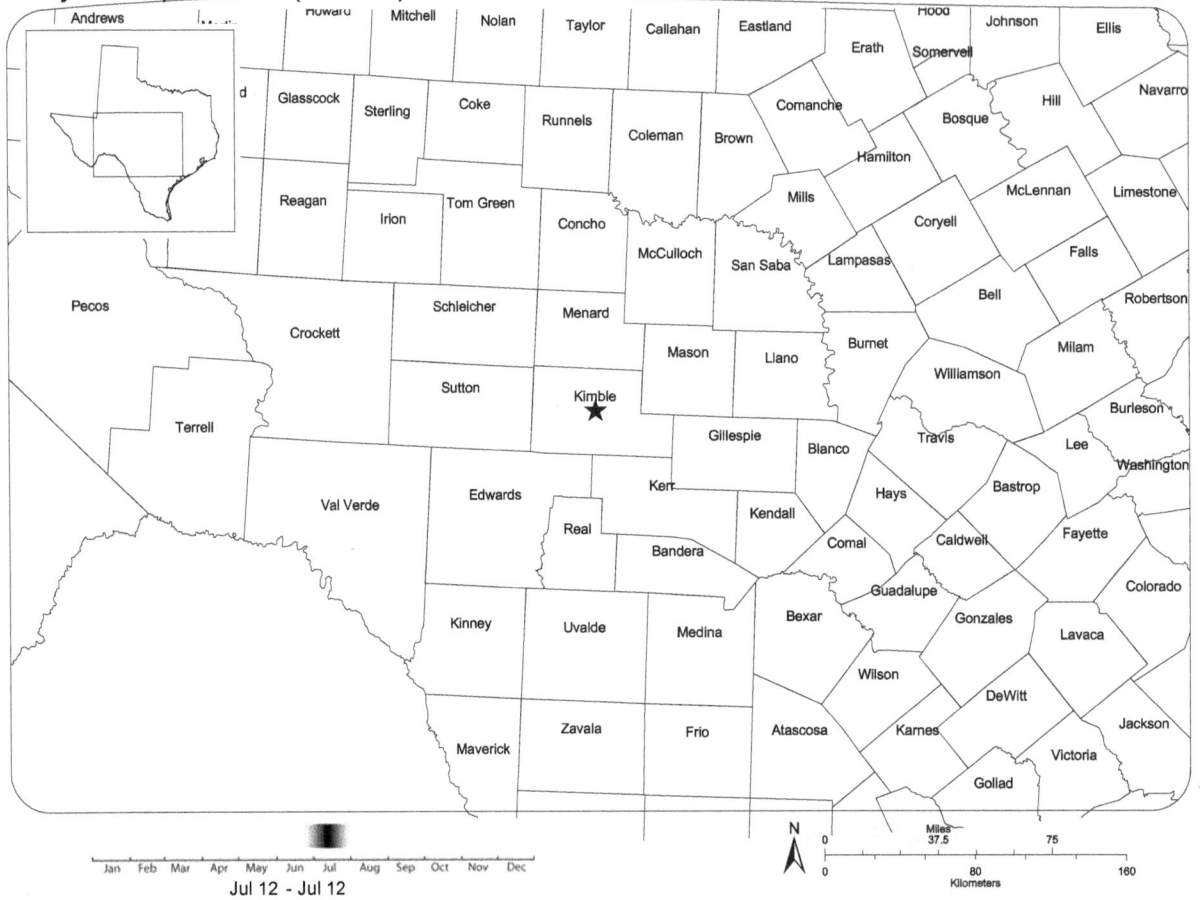

Jul 12 - Jul 12

HABITAT: Ponds, lakes, ditches and slow reaches of rivers and streams.

Kimble

Erythemis plebeja (Burmeister) / Pin-tailed Pondhawk

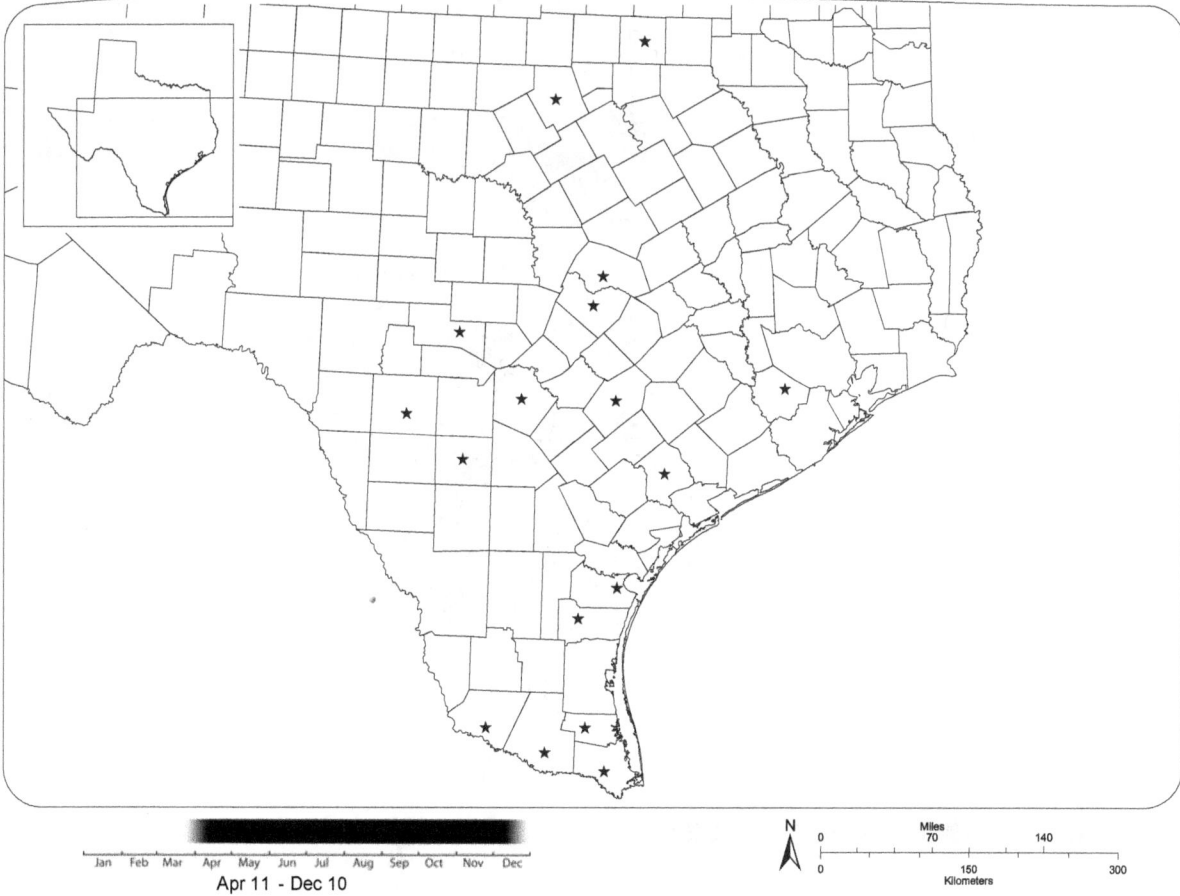

Jan Feb Mar Apr May Jun Jul Aug Sep Oct Nov Dec

Apr 11 - Dec 10

HABITAT: Ponds, lakes, ditches and slow reaches of rivers and streams

Bexar
Cameron
Erath
Fort Bend
Frio
Gonzales
Hidalgo
Kerr
Kleberg
Nueces
Starr
Tarrant
Travis
Uvalde
Victoria
Willacy
Williamson

Erythemis simplicicollis (Say) / Eastern Pondhawk

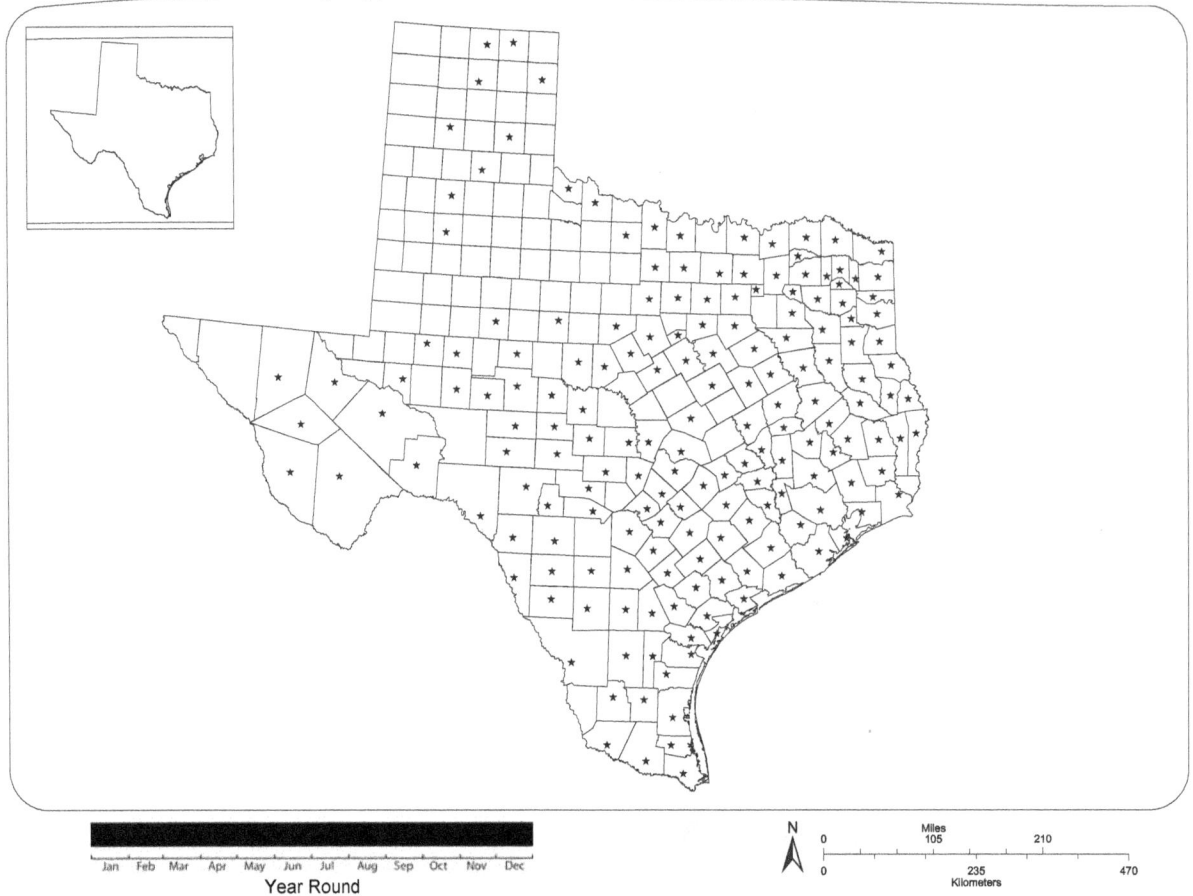

Year Round

HABITAT: Ponds, lakes, ditches, and slow moving creeks, streams and rivers.

Anderson	Cass	Fort Bend	Houston	Llano	Presidio	Travis
Angelina	Chambers	Franklin	Hunt	Lubbock	Rains	Trinity
Aransas	Cherokee	Freestone	Hutchinson	Madison	Randall	Tyler
Archer	Clay	Frio	Irion	Marion	Reagan	Upshur
Atascosa	Coke	Galveston	Jack	Mason	Real	Uvalde
Austin	Coleman	Gillespie	Jackson	Matagorda	Red River	Val Verde
Bandera	Collin	Glasscock	Jasper	Maverick	Reeves	Van Zandt
Bastrop	Colorado	Goliad	Jeff Davis	McCulloch	Refugio	Victoria
Bee	Comal	Gonzales	Jefferson	McLennan	Robertson	Walker
Bell	Comanche	Grayson	Jim Hogg	McMullen	Rockwall	Waller
Bexar	Concho	Gregg	Jim Wells	Menard	Rusk	Washington
Blanco	Crane	Grimes	Johnson	Midland	Sabine	Webb
Bosque	Culberson	Guadalupe	Karnes	Mitchell	San Augustine	Wharton
Bowie	Dallas	Hale	Kenedy	Montague	San Jacinto	Wilbarger
Brazoria	Delta	Hamilton	Kerr	Montgomery	San Patricio	Willacy
Brazos	Denton	Hansford	Kimble	Morris	Schleicher	Williamson
Brewster	DeWitt	Hardeman	Kinney	Nacogdoches	Shelby	Wilson
Briscoe	Dimmit	Hardin	Kleberg	Navarro	Smith	Wise
Brooks	Donley	Harris	La Salle	Newton	Somervell	Wood
Brown	Duval	Harrison	Lamar	Nueces	Starr	Zavala
Burleson	Eastland	Hays	Lavaca	Ochiltree	Sutton	
Burnet	Edwards	Hemphill	Lee	Palo Pinto	Tarrant	
Caldwell	Ellis	Henderson	Leon	Panola	Taylor	
Calhoun	Erath	Hidalgo	Liberty	Parker	Terrell	
Cameron	Fannin	Hill	Limestone	Pecos	Titus	
Camp	Fayette	Hopkins	Live Oak	Polk	Tom Green	

Erythemis vesiculosa (Fabricius) / Great Pondhawk

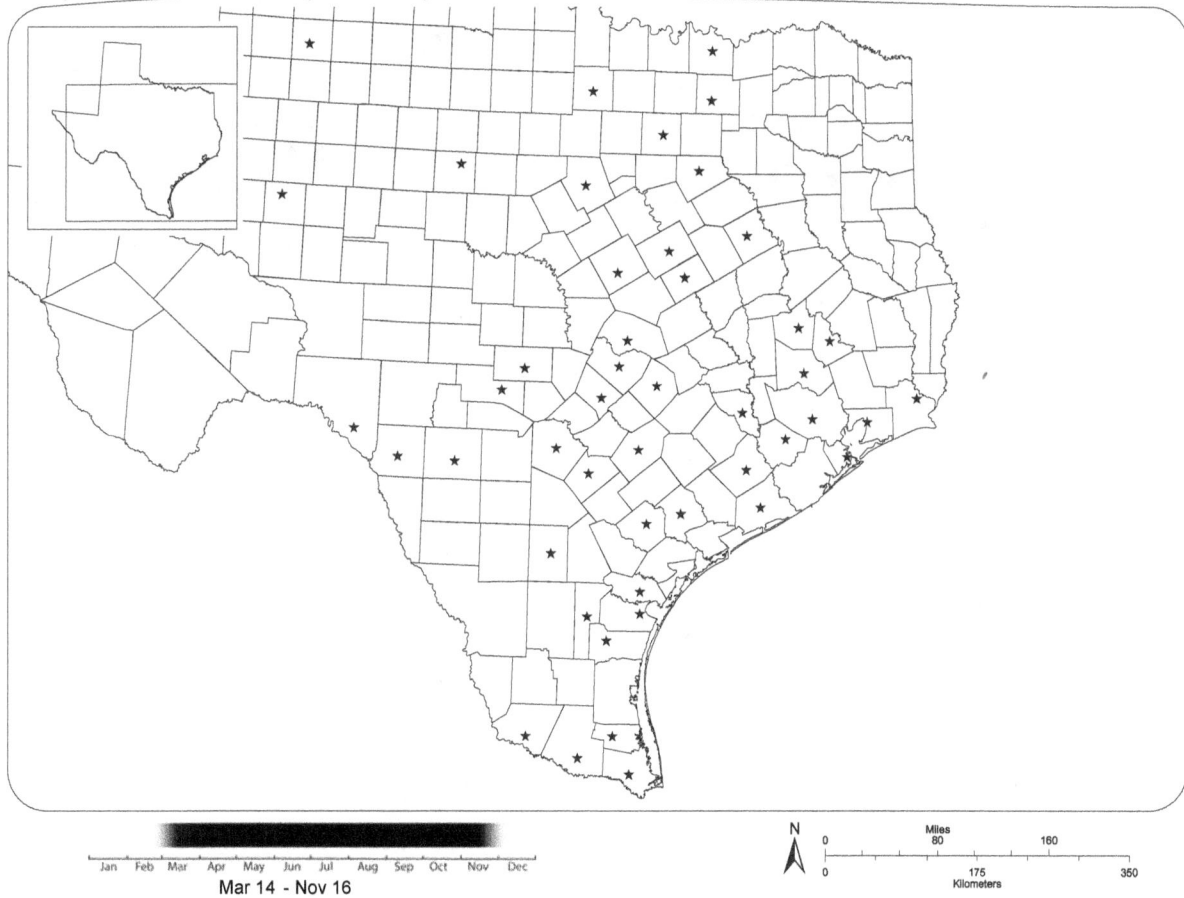

Mar 14 - Nov 16

HABITAT: Ponds, lakes, ditches, and slow moving creeks, streams and rivers.

Austin	Lubbock
Bastrop	Matagorda
Bexar	McLennan
Cameron	McMullen
Chambers	Midland
Collin	Montgomery
Coryell	Nueces
Ellis	San Jacinto
Erath	San Patricio
Falls	Starr
Fort Bend	Tarrant
Freestone	Taylor
Galveston	Travis
Gillespie	Uvalde
Goliad	Val Verde
Gonzales	Victoria
Grayson	Walker
Harris	Wharton
Hays	Willacy
Hidalgo	Williamson
Jack	Wilson
Jefferson	
Jim Wells	
Kerr	
Kinney	
Kleberg	

Erythrodiplax basifusca (Calvert) / Plateau Dragonlet

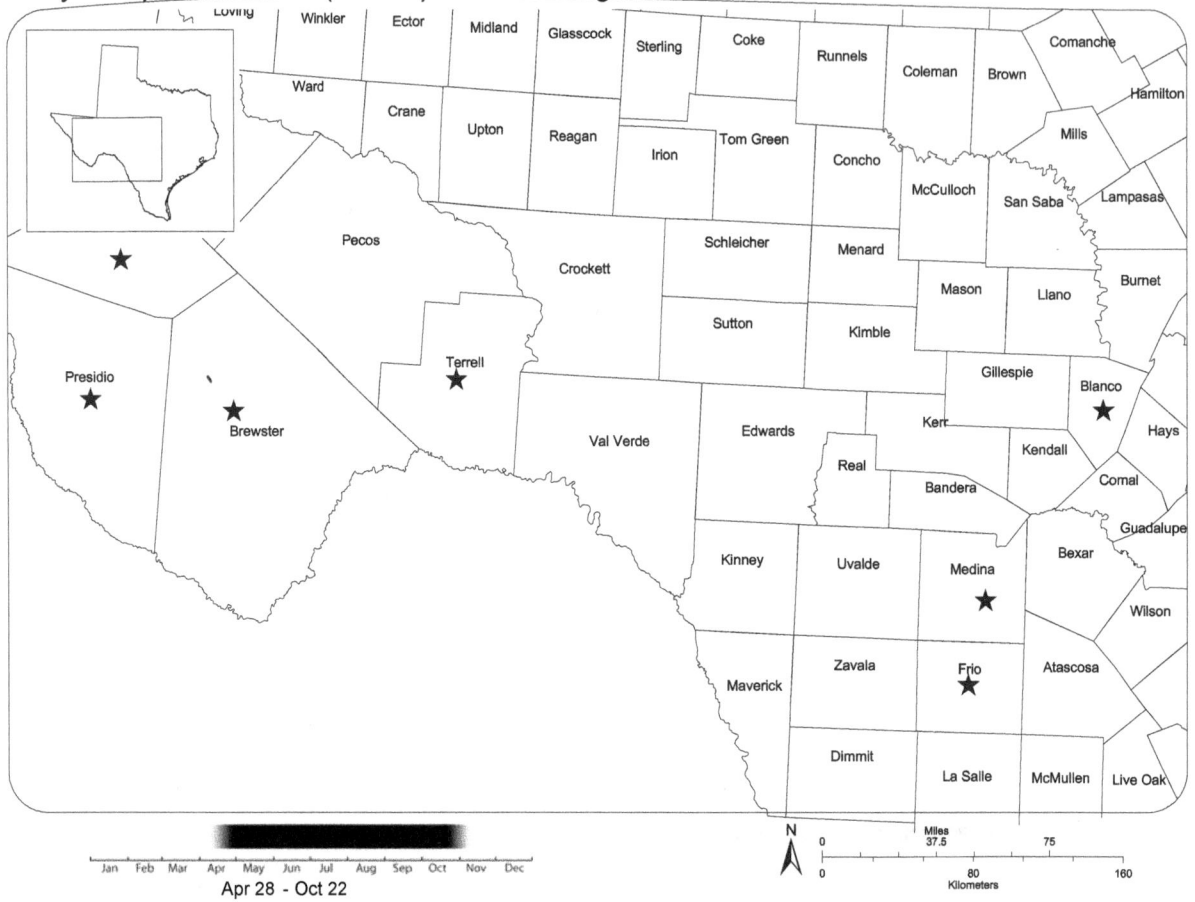

Apr 28 - Oct 22

HABITAT: Marshy creeks, streams and ponds.

Blanco
Brewster
Frio
Jeff Davis
Medina
Presidio
Reeves
Terrell

Erythrodiplax berenice (Hagen) / Seaside Dragonlet

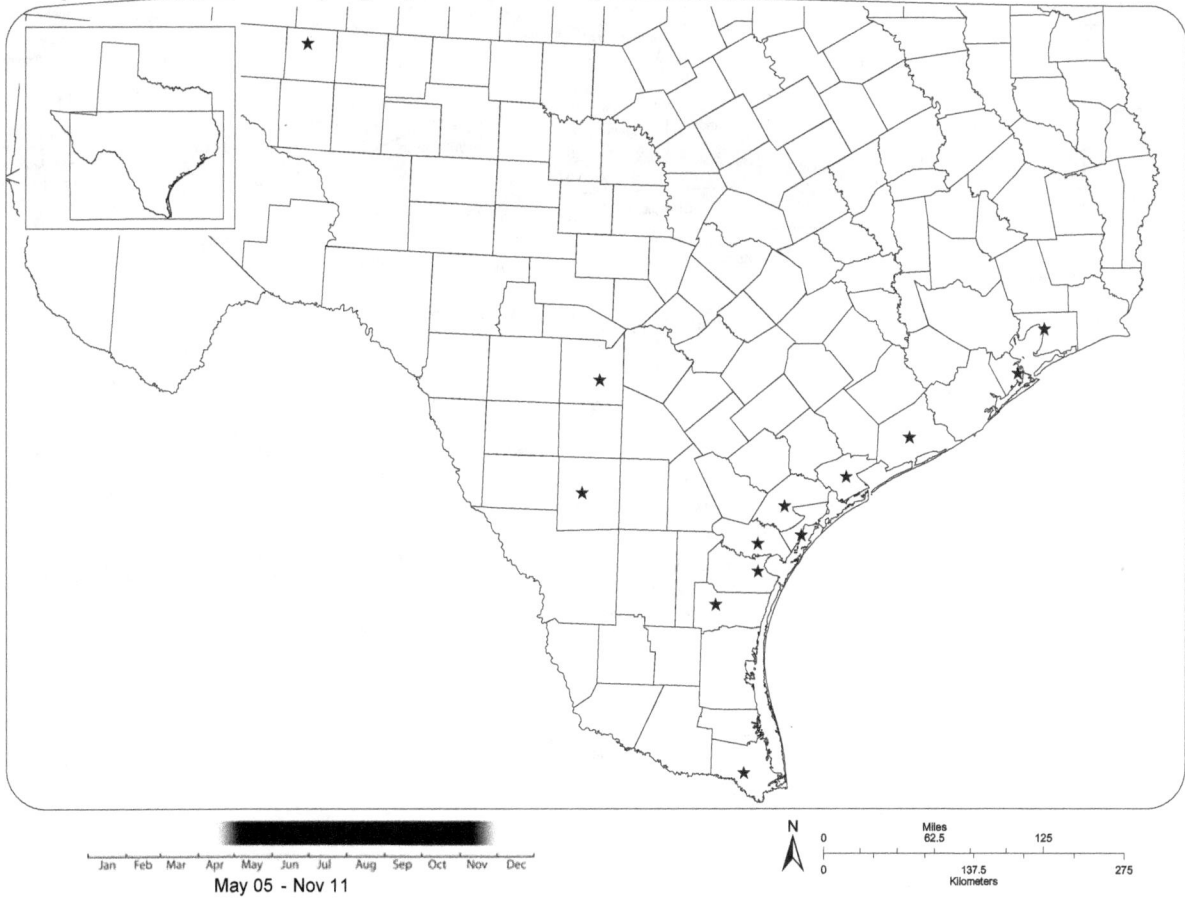

May 05 - Nov 11

HABITAT: Salt marshes, estuaries, bays and occasionally inland lakes high in salinity.

Aransas
Calhoun
Cameron
Chambers
Galveston
Kleberg
La Salle
Matagorda
Medina
Midland
Nueces
Reeves
Refugio
San Patricio
Ward

Erythrodiplax funerea (Hagen) / Black-winged Dragonlet

Jun 15 - Jun 15

HABITAT: Open temporary pools and ponds

Bexar

Erythrodiplax fusca (Rambur) / Red-faced Dragonlet

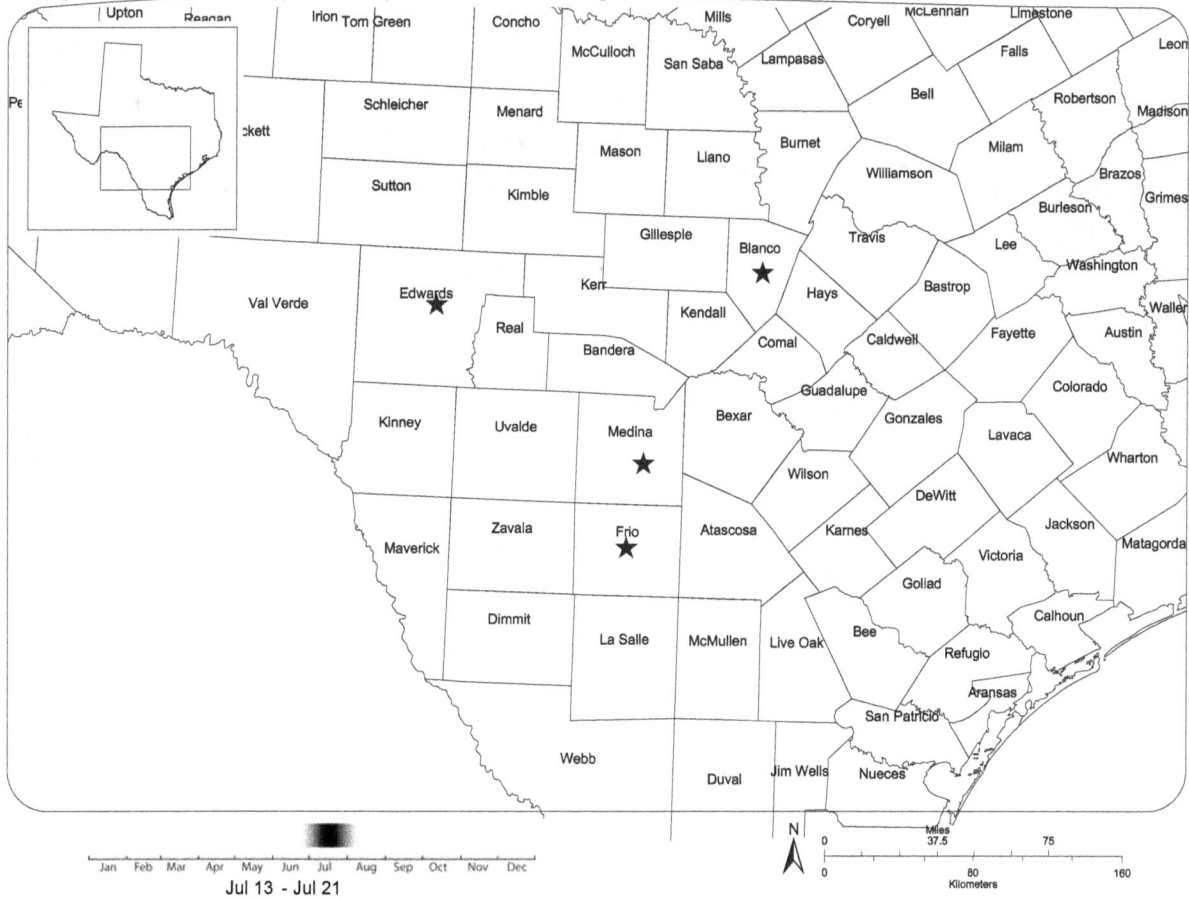

Jul 13 - Jul 21

HABITAT: Marshy swamps, pools, lakes and streams with moderate current and periodic pools.

Blanco
Edwards
Frio
Medina

Erythrodiplax minuscula (Rambur) / Little Blue Dragonlet

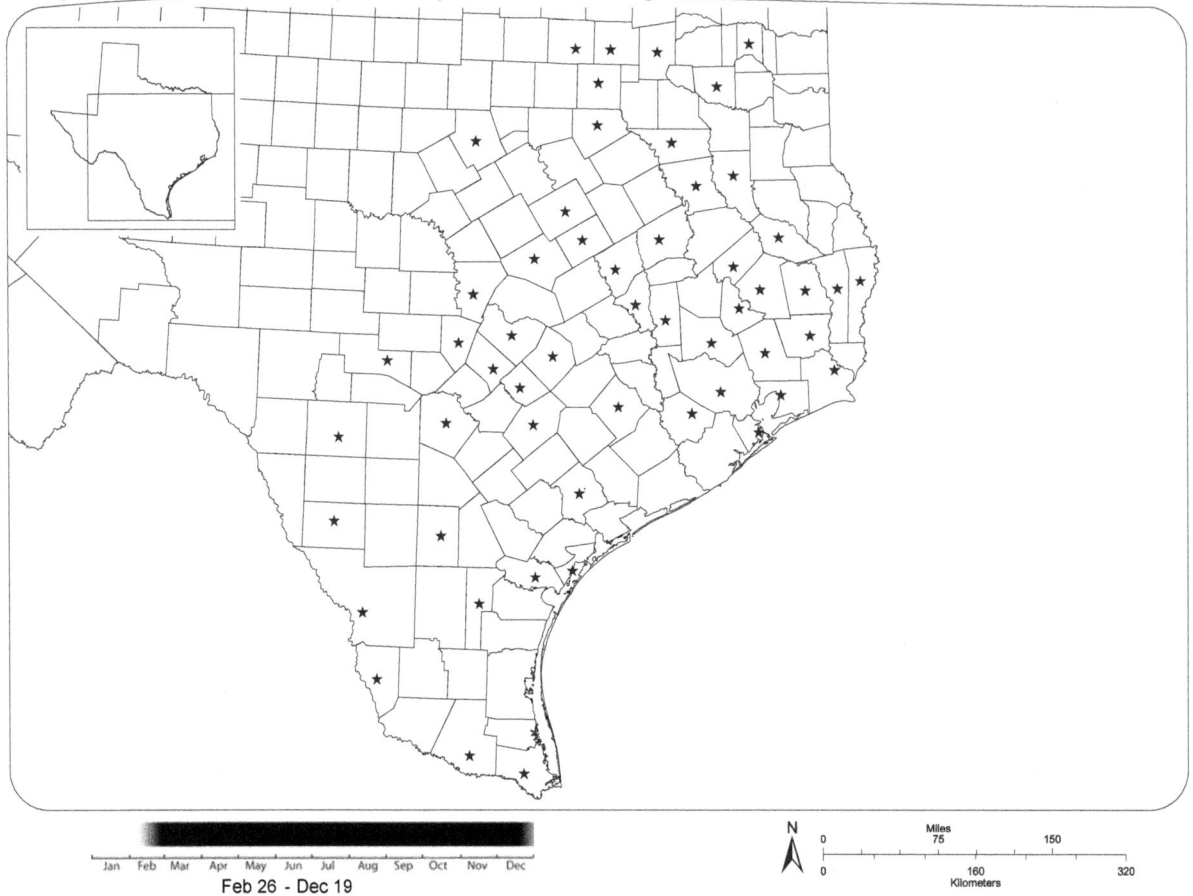

Feb 26 - Dec 19

HABITAT: Marshy ponds, pools, lakes and slow moving streams.

Anderson	Harris	Wood
Angelina	Hays	Zapata
Aransas	Henderson	
Bastrop	Hidalgo	
Bell	Hunt	
Bexar	Jasper	
Blanco	Jefferson	
Brazos	Jim Wells	
Burnet	Kerr	
Caldwell	Leon	
Cameron	Liberty	
Chambers	McLennan	
Cherokee	McMullen	
Collin	Montgomery	
Colorado	Newton	
Dallas	Polk	
Denton	Robertson	
Dimmit	San Jacinto	
Ellis	San Patricio	
Erath	Titus	
Falls	Travis	
Fort Bend	Trinity	
Galveston	Tyler	
Gonzales	Uvalde	
Grimes	Victoria	
Hardin	Webb	

Erythrodiplax umbrata (Linnaeus) / Band-winged Dragonlet

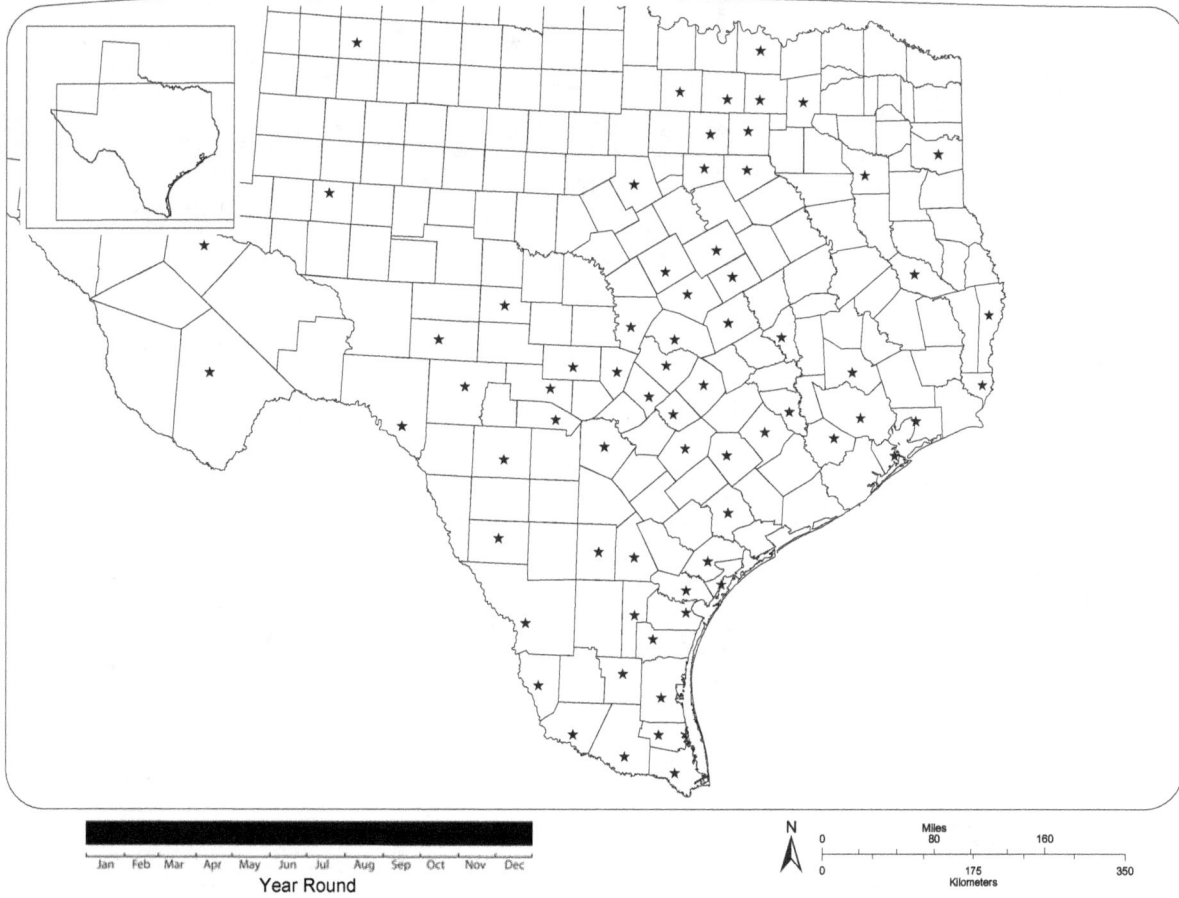

| Jan | Feb | Mar | Apr | May | Jun | Jul | Aug | Sep | Oct | Nov | Dec |

Year Round

HABITAT: Permanent and temporary marshy ponds, pools and lakes.

Angelina	Galveston	Reeves
Aransas	Gillespie	Refugio
Austin	Gonzales	San Patricio
Bandera	Grayson	Smith
Bastrop	Harris	Starr
Bell	Harrison	Sutton
Bexar	Hays	Tarrant
Blanco	Hidalgo	Travis
Brazos	Hunt	Uvalde
Brewster	Jim Wells	Val Verde
Brooks	Johnson	Victoria
Burnet	Kenedy	Webb
Caldwell	Kerr	Willacy
Cameron	Kleberg	Williamson
Chambers	Lavaca	Wise
Collin	Live Oak	Zapata
Colorado	Lubbock	
Coryell	McLennan	
Dallas	McMullen	
Denton	Menard	
Dimmit	Midland	
Edwards	Milam	
Ellis	Montgomery	
Erath	Newton	
Falls	Nueces	
Fort Bend	Orange	

Ladona deplanata (Rambur) / Blue Corporal

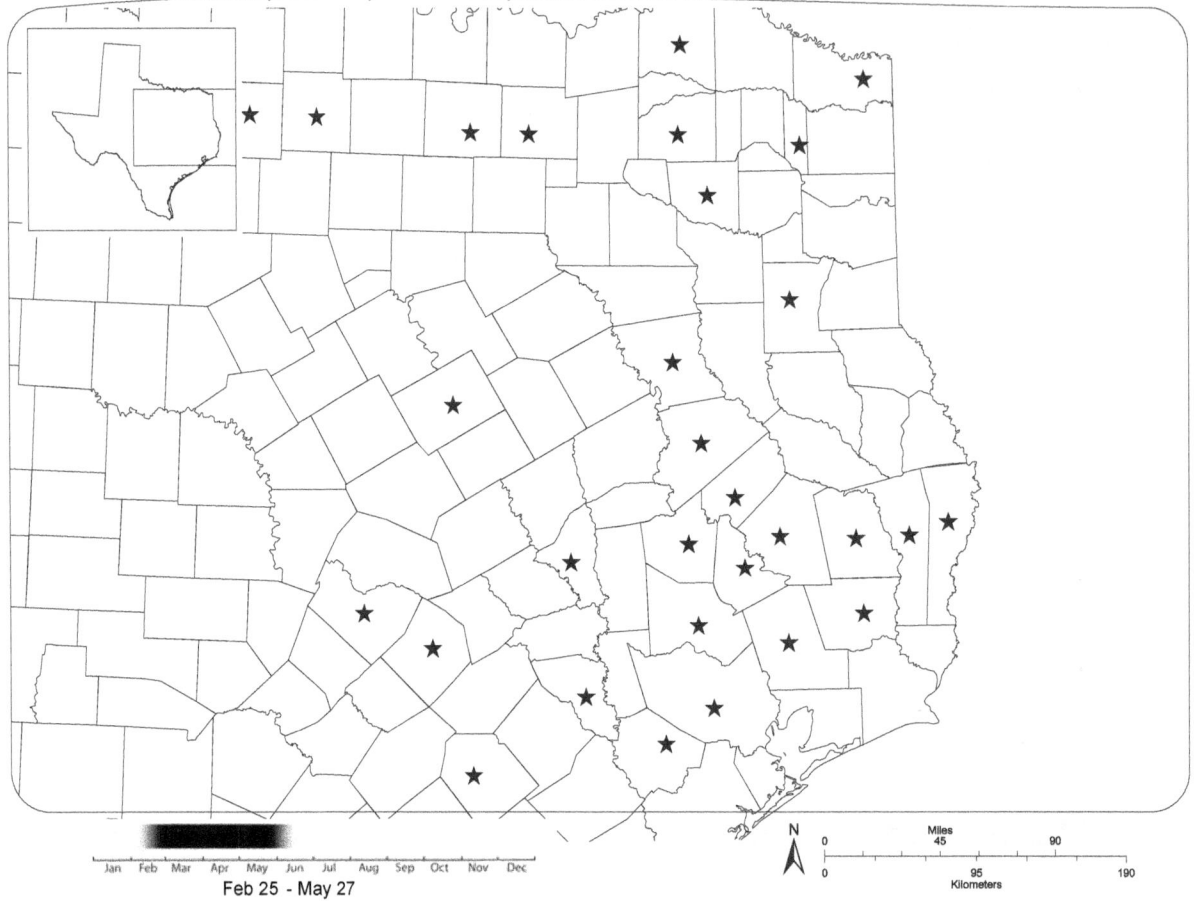

Feb 25 - May 27

HABITAT: Sloughs, ponds, lakes, borrow pits and open areas of slow streams often with sandy bottoms.

Anderson
Austin
Bastrop
Bowie
Brazos
Collin
Denton
Fort Bend
Hardin
Harris
Hopkins
Houston
Jack
Jasper
Lamar
Lavaca
Liberty
McLennan
Montgomery
Morris
Newton
Polk
Rusk
San Jacinto
Travis
Trinity
Tyler
Walker
Wood
Young

Libellula auripennis Burmeister / Golden-winged Skimmer

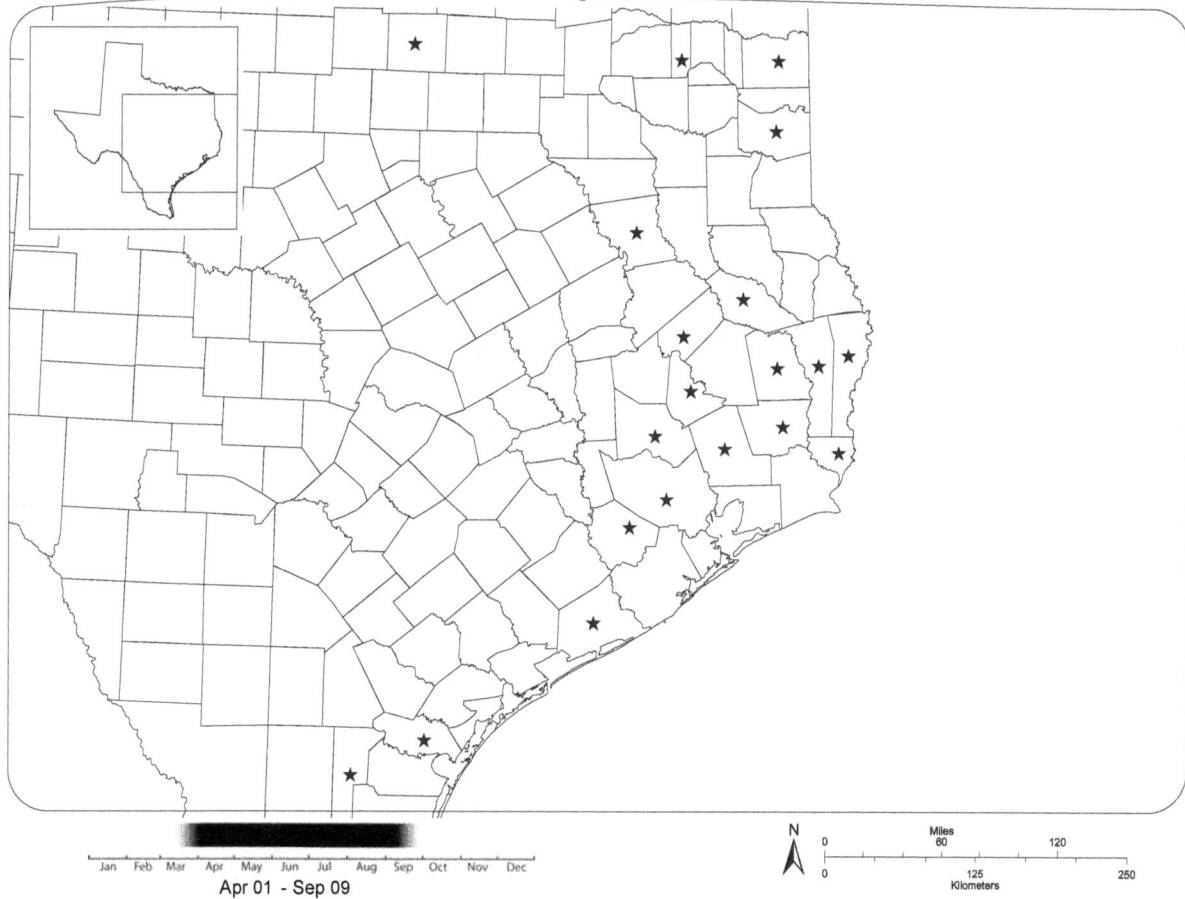

Apr 01 - Sep 09

HABITAT: Ponds, pools, ditches, lakes and occasionally slow flowing streams.

Anderson
Angelina
Cass
Fort Bend
Franklin
Hardin
Harris
Harrison
Jasper
Jim Wells
Liberty
Matagorda
Montgomery
Newton
Orange
San Jacinto
San Patricio
Trinity
Tyler
Wise

Libellula axilena Westwood / Bar-winged Skimmer

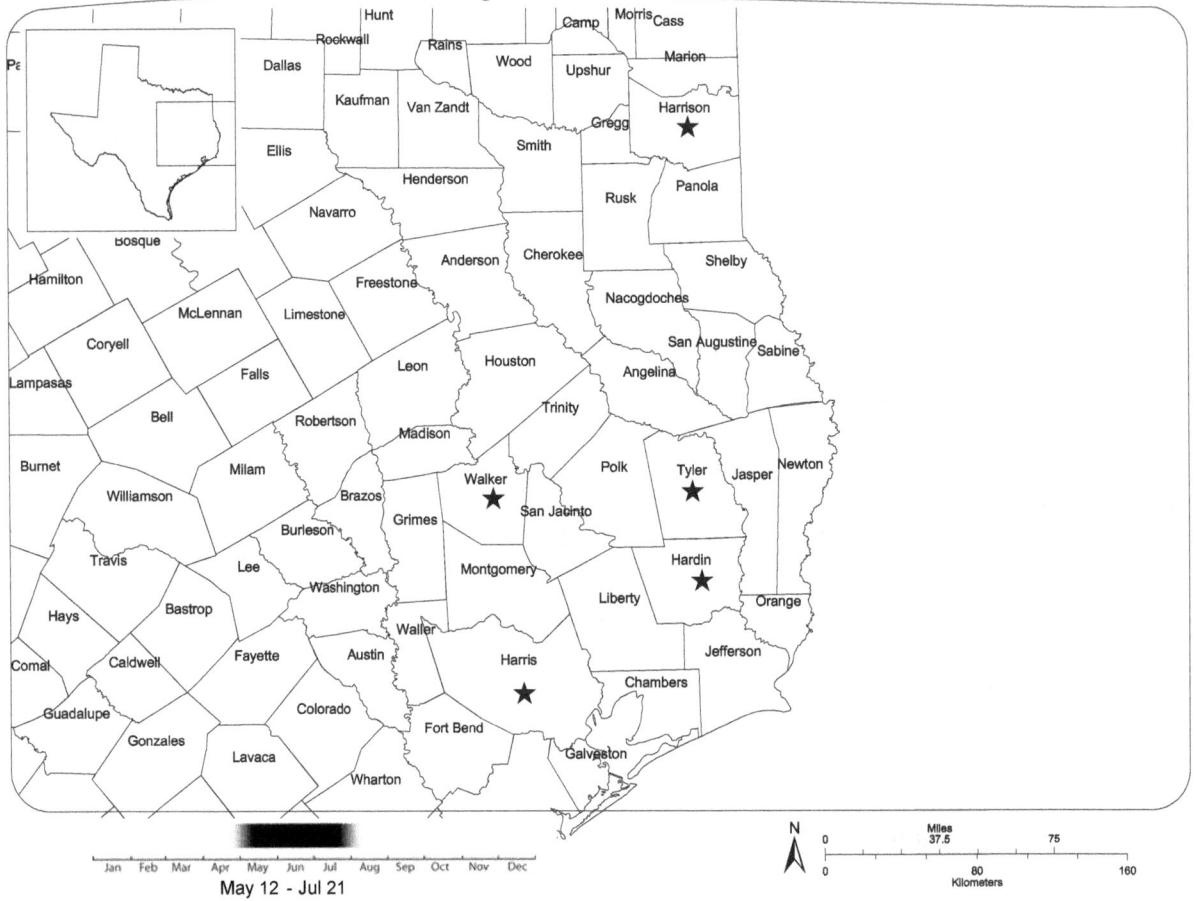

May 12 - Jul 21

HABITAT: Forest ponds, pools and ditches.

Hardin
Harris
Harrison
Tyler
Walker

Libellula comanche Calvert / Comanche Skimmer

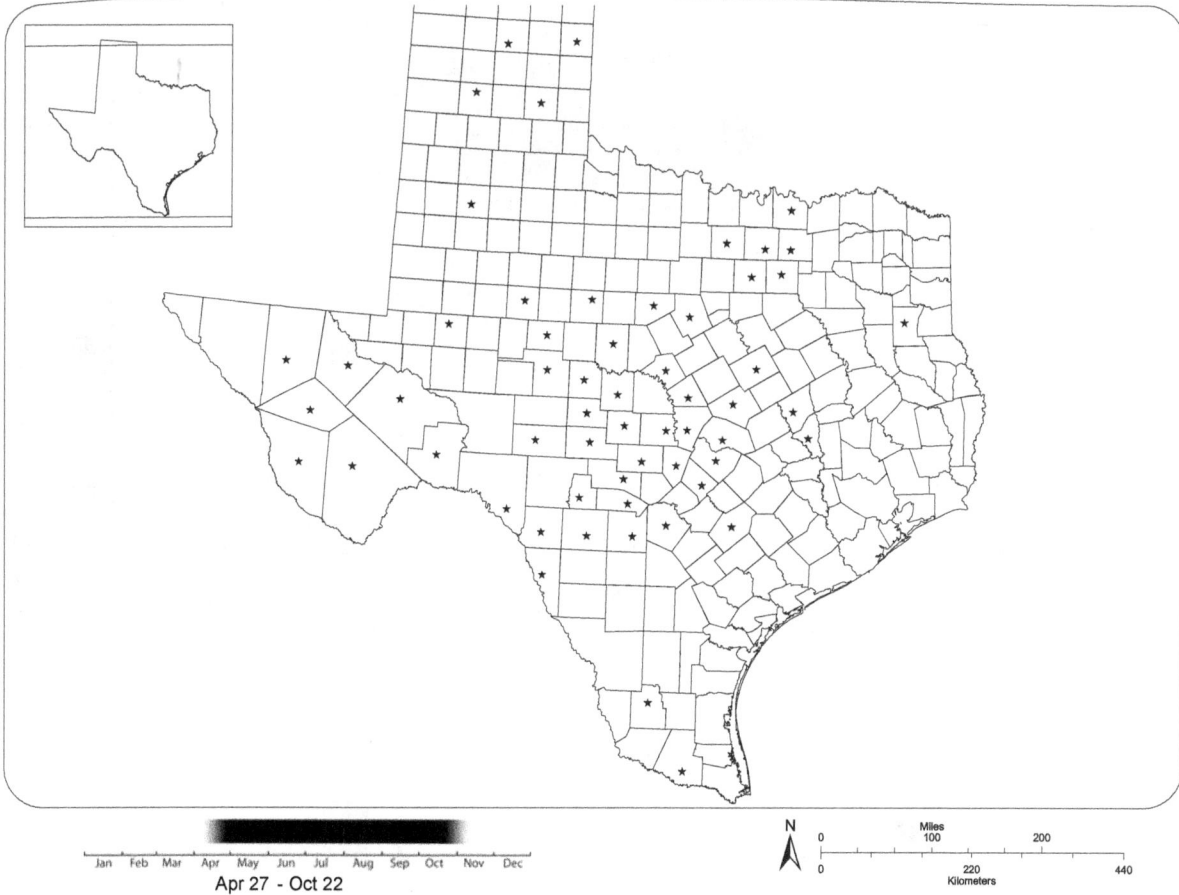

Jan Feb Mar Apr May Jun Jul Aug Sep Oct Nov Dec

Apr 27 - Oct 22

HABITAT: Ponds, lakes and sluggish streams.

Bandera	Kerr	Tom Green
Bell	Kimble	Travis
Bexar	Kinney	Uvalde
Blanco	Lampasas	Val Verde
Brazos	Llano	Williamson
Brewster	Lubbock	Wise
Burnet	Mason	
Coke	Maverick	
Coleman	McCulloch	
Collin	McLennan	
Concho	Medina	
Culberson	Menard	
Dallas	Midland	
Denton	Mills	
Donley	Mitchell	
Eastland	Pecos	
Erath	Presidio	
Gillespie	Randall	
Gonzales	Real	
Grayson	Reeves	
Hays	Robertson	
Hemphill	Rusk	
Hidalgo	Sutton	
Hutchinson	Tarrant	
Jeff Davis	Taylor	
Jim Hogg	Terrell	

Libellula composita (Hagen) / Bleached Skimmer

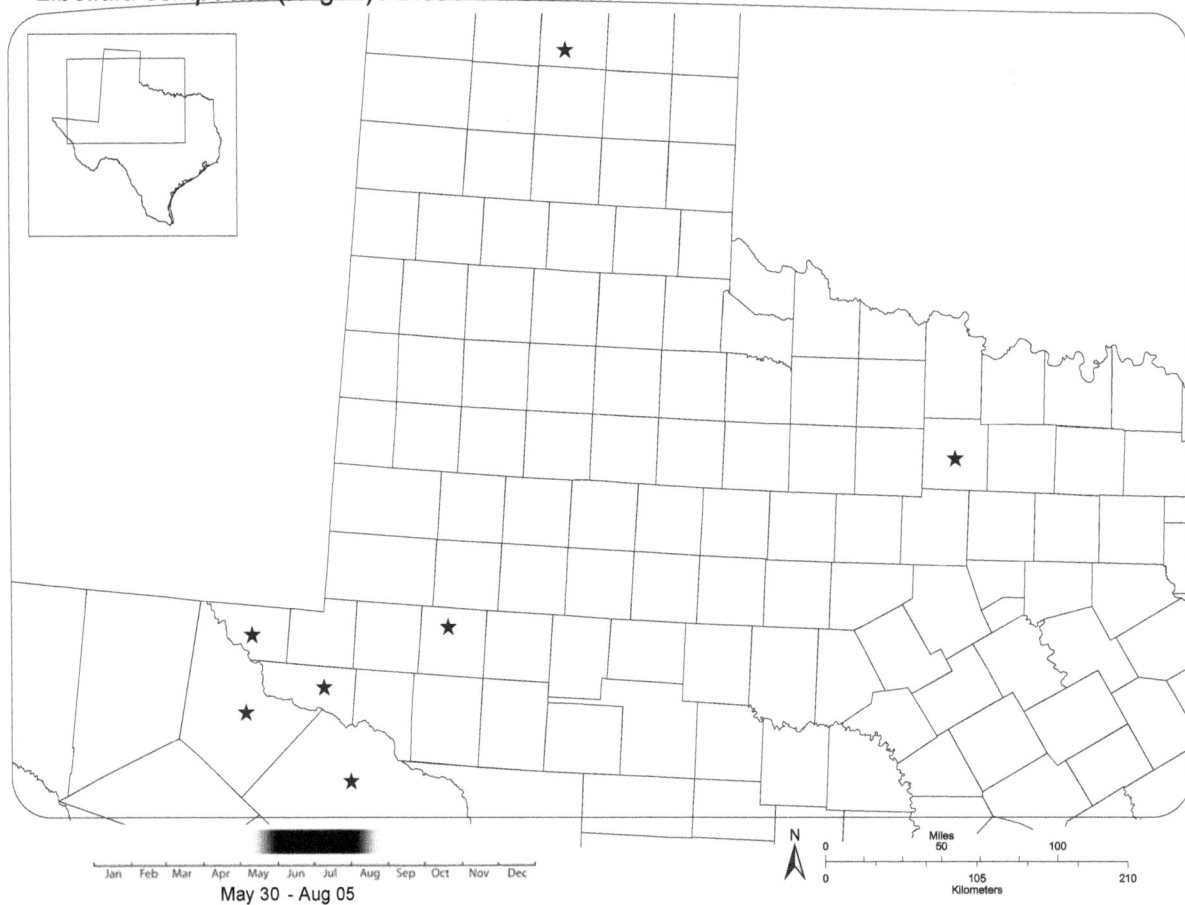

May 30 - Aug 05

HABITAT: Desert alkaline ponds and lakes, often associated with springs.

Hutchinson
Jack
Loving
Midland
Pecos
Reeves
Ward

Libellula croceipennis Selys / Neon Skimmer

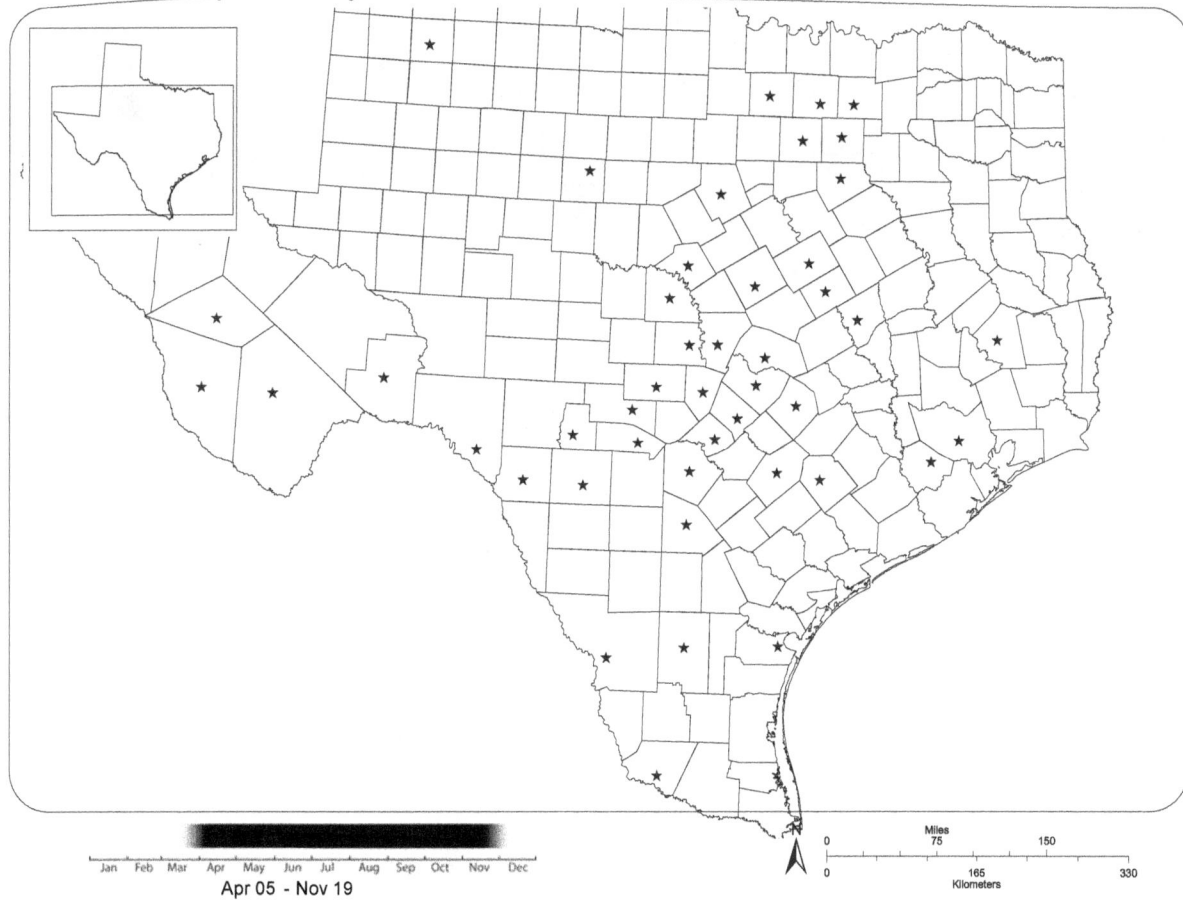

Apr 05 - Nov 19

HABITAT: Ponds, lakes and sluggish streams.

Atascosa	Lubbock
Bandera	McLennan
Bastrop	Mills
Bexar	Nueces
Blanco	Polk
Brewster	Presidio
Burnet	Real
Collin	Robertson
Comal	San Saba
Coryell	Starr
Dallas	Tarrant
Denton	Taylor
Duval	Terrell
Ellis	Travis
Erath	Uvalde
Falls	Val Verde
Fort Bend	Webb
Gillespie	Williamson
Gonzales	Wise
Harris	
Hays	
Jeff Davis	
Kerr	
Kinney	
Lavaca	
Llano	

Libellula cyanea Fabricius / Spangled Skimmer

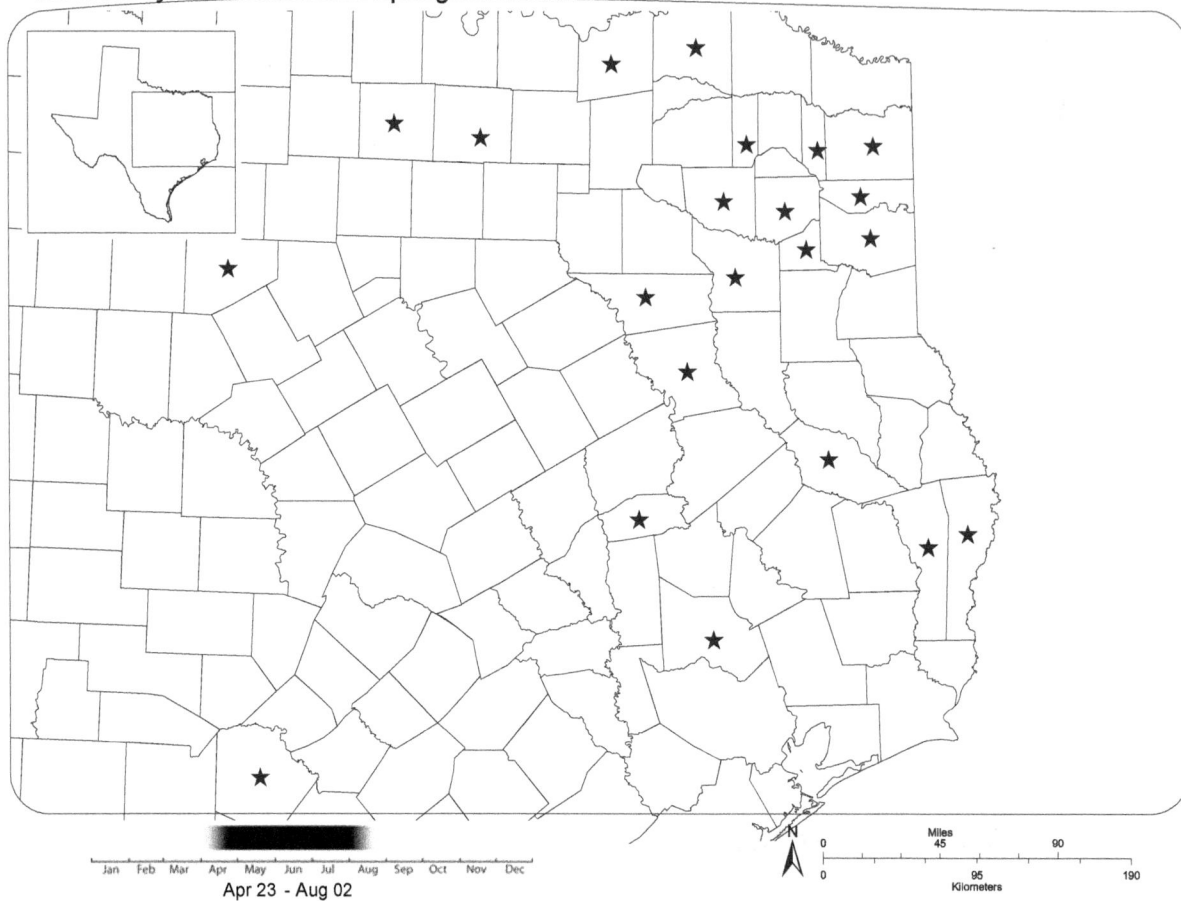

Apr 23 - Aug 02

HABITAT: Marshy ponds, pools and lakes.

Anderson
Angelina
Bexar
Cass
Denton
Eastland
Fannin
Franklin
Gregg
Harrison
Henderson
Jasper
Lamar
Madison
Marion
Montgomery
Morris
Newton
Smith
Upshur
Wise
Wood

Libellula flavida Rambur / Yellow-sided Skimmer

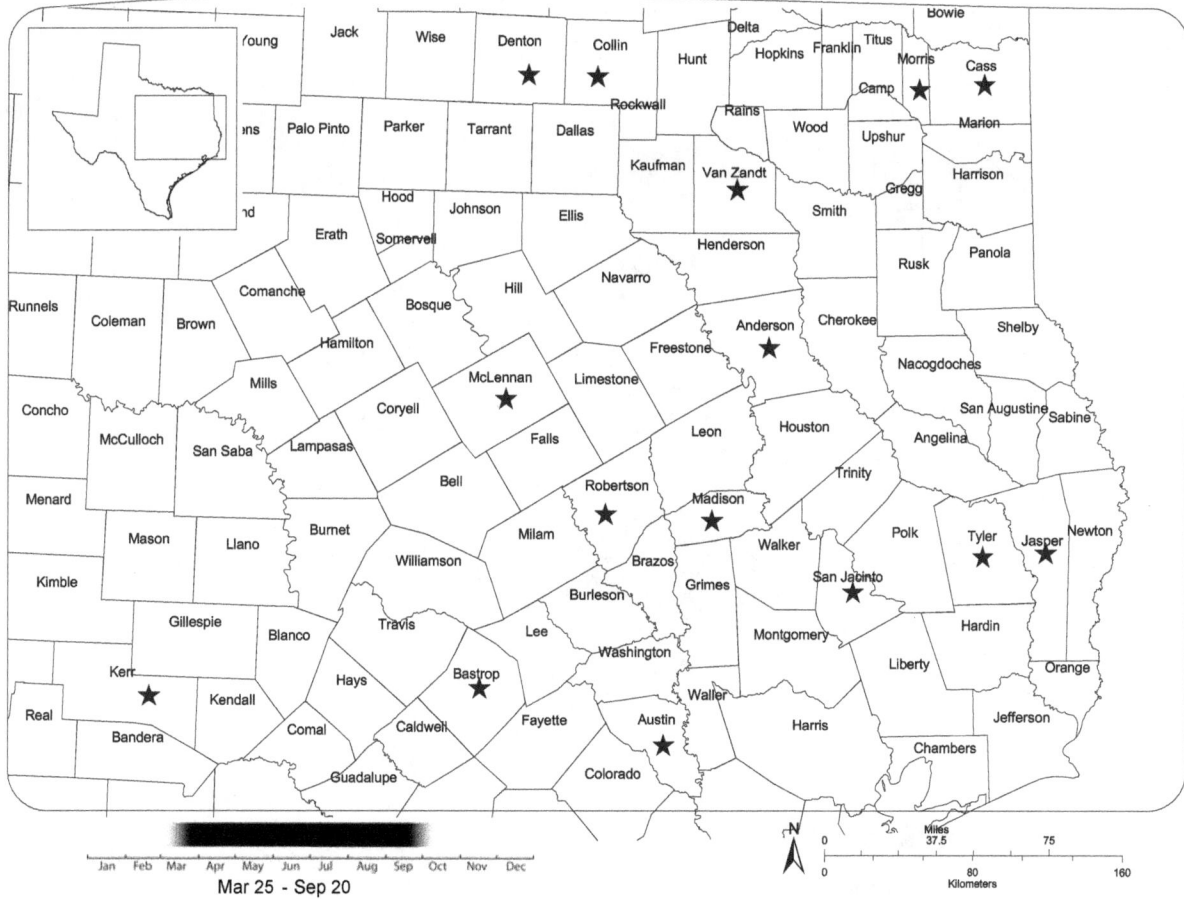

Mar 25 - Sep 20

HABITAT: Marshy ponds, lakes, borrow pits and slow flowing streams.

Anderson
Austin
Bastrop
Cass
Collin
Denton
Jasper
Kerr
Madison
McLennan
Morris
Robertson
San Jacinto
Tyler
Van Zandt

Libellula incesta Hagen / Slaty Skimmer

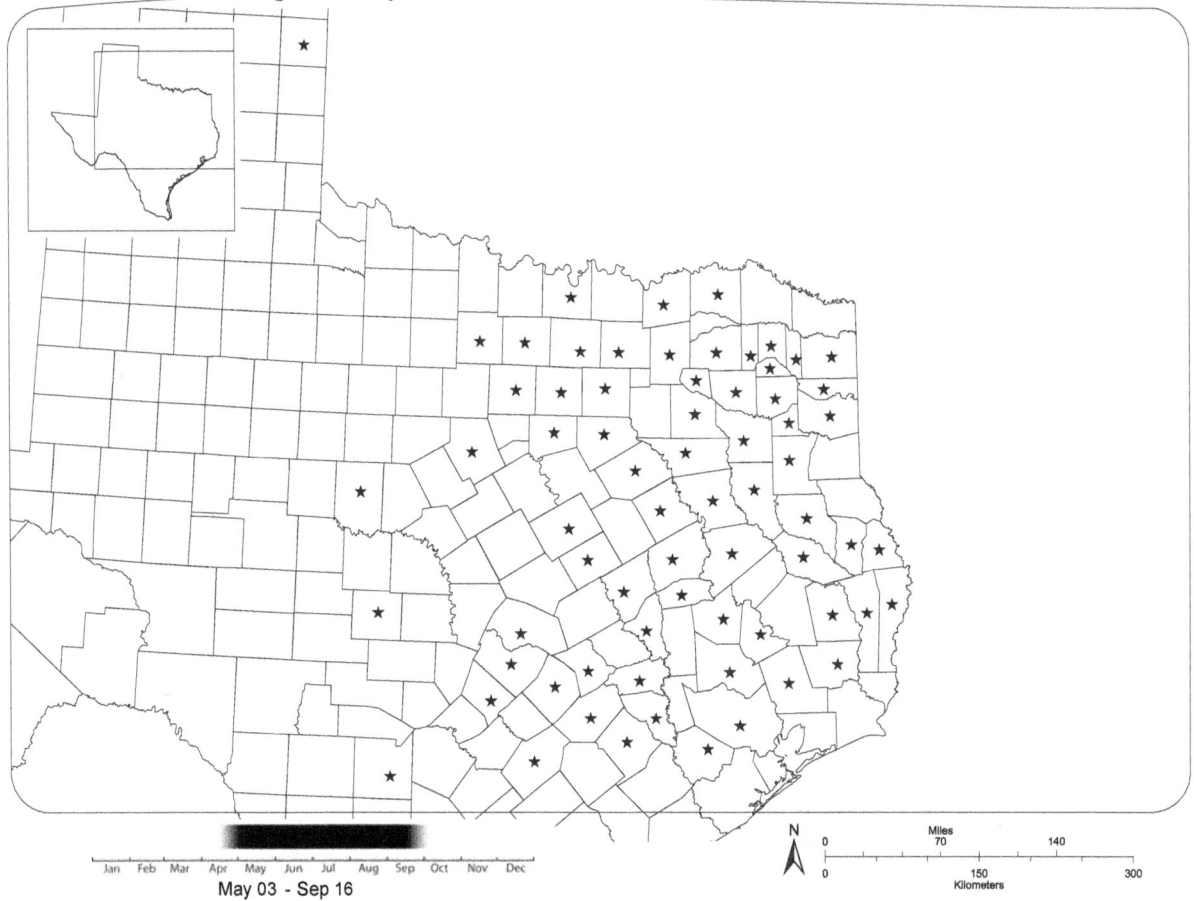

May 03 - Sep 16

HABITAT: Marshy ponds, lakes and slow flowing forest streams with muck bottoms.

Anderson	Harrison	Robertson
Angelina	Hays	Rusk
Austin	Hemphill	Sabine
Bastrop	Henderson	San Augustine
Brazos	Hopkins	San Jacinto
Camp	Houston	Smith
Cass	Hunt	Tarrant
Cherokee	Jack	Titus
Coleman	Jasper	Travis
Collin	Johnson	Tyler
Colorado	Lamar	Upshur
Cooke	Lee	Van Zandt
Dallas	Leon	Walker
Denton	Liberty	Washington
Ellis	Madison	Williamson
Erath	Marion	Wise
Falls	Mason	Wood
Fannin	McLennan	
Fayette	Medina	
Fort Bend	Montgomery	
Franklin	Morris	
Freestone	Nacogdoches	
Gonzales	Navarro	
Gregg	Newton	
Hardin	Parker	
Harris	Rains	

Libellula luctuosa Burmeister / Widow Skimmer

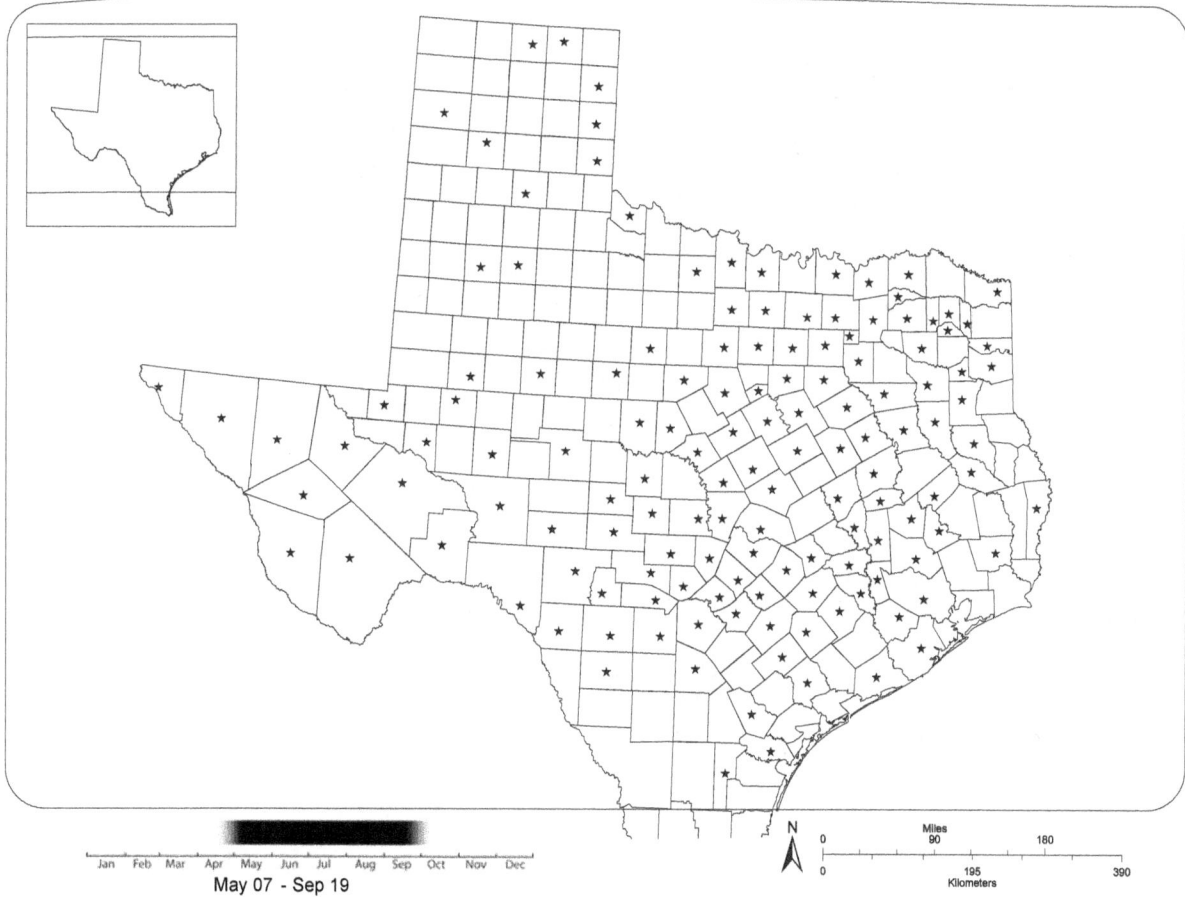

May 07 - Sep 19

HABITAT: Still bodies of water, including marshy ponds, lakes and borrow pits

Anderson	Colorado	Guadalupe	Lee	Parker	Walker
Angelina	Comal	Hamilton	Leon	Pecos	Waller
Archer	Coryell	Hansford	Limestone	Presidio	Washington
Atascosa	Crane	Hardeman	Llano	Randall	Wheeler
Austin	Crockett	Hardin	Lubbock	Reagan	Williamson
Bandera	Crosby	Harris	Madison	Real	Winkler
Bastrop	Culberson	Harrison	Marion	Reeves	Wise
Bee	Dallas	Hays	Martin	Robertson	Wood
Bell	Delta	Hemphill	Mason	Rockwall	Zavala
Bexar	Denton	Henderson	Matagorda	Rusk	
Blanco	DeWitt	Hill	McCulloch	San Jacinto	
Bosque	Eastland	Hopkins	McLennan	San Patricio	
Bowie	Edwards	Hudspeth	Medina	Shackelford	
Brazoria	El Paso	Hunt	Menard	Smith	
Brazos	Ellis	Jack	Midland	Somervell	
Brewster	Erath	Jeff Davis	Mills	Sutton	
Briscoe	Fannin	Jim Wells	Mitchell	Tarrant	
Brown	Fayette	Johnson	Montague	Taylor	
Burnet	Fort Bend	Kaufman	Montgomery	Terrell	
Caldwell	Franklin	Kendall	Morris	Titus	
Camp	Freestone	Kerr	Nacogdoches	Tom Green	
Cherokee	Gillespie	Kimble	Navarro	Travis	
Clay	Gonzales	Kinney	Newton	Trinity	
Coleman	Grayson	Lamar	Ochiltree	Uvalde	
Collin	Gregg	Lampasas	Oldham	Val Verde	
Collingsworth	Grimes	Lavaca	Palo Pinto	Victoria	

Libellula needhami Westfall / Needham's Skimmer

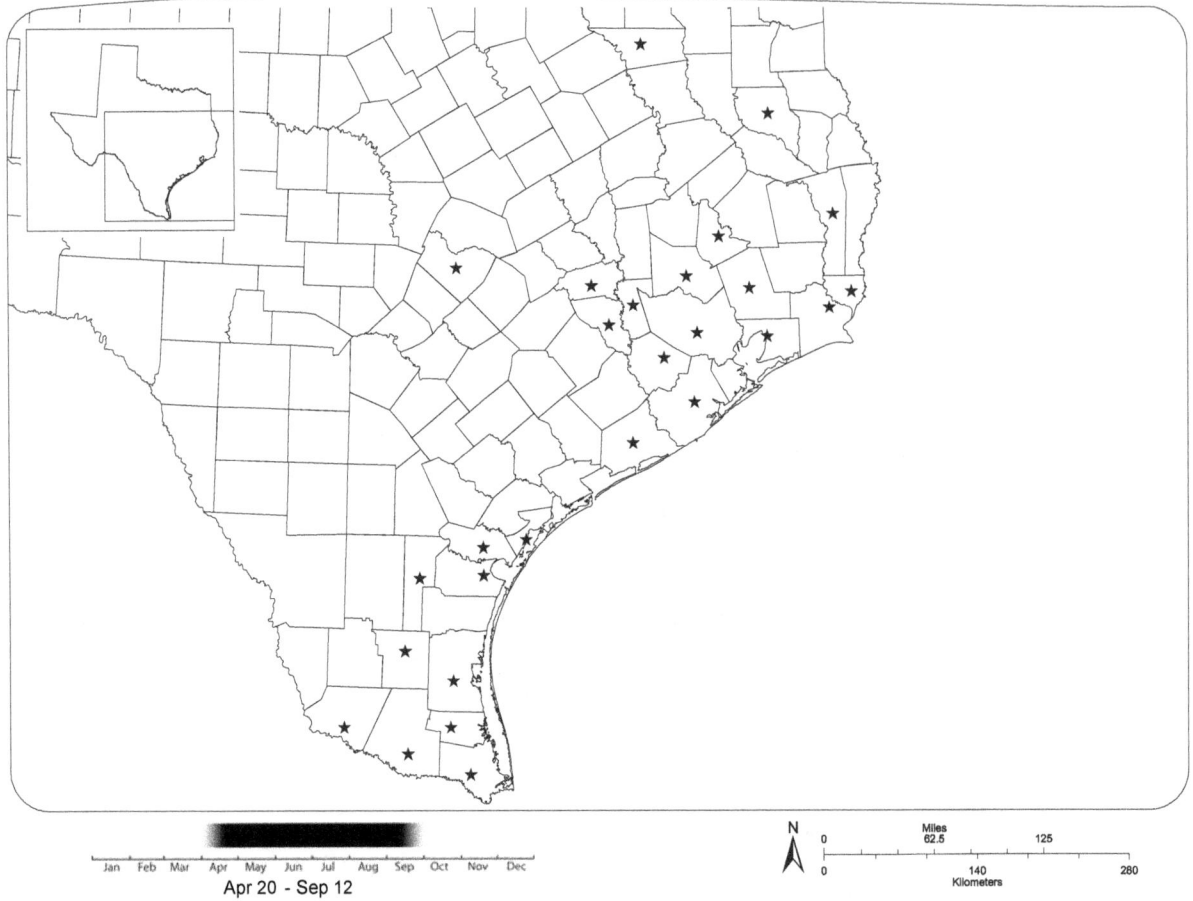

Apr 20 - Sep 12

HABITAT: Marshy ponds and lakes including brackish waters.

Aransas
Austin
Brazoria
Brooks
Cameron
Chambers
Fort Bend
Harris
Henderson
Hidalgo
Jasper
Jefferson
Jim Wells
Kenedy
Liberty
Matagorda
Montgomery
Nacogdoches
Nueces
Orange
San Jacinto
San Patricio
Starr
Travis
Waller
Washington

Willacy

Libellula pulchella Drury / Twelve-spotted Skimmer

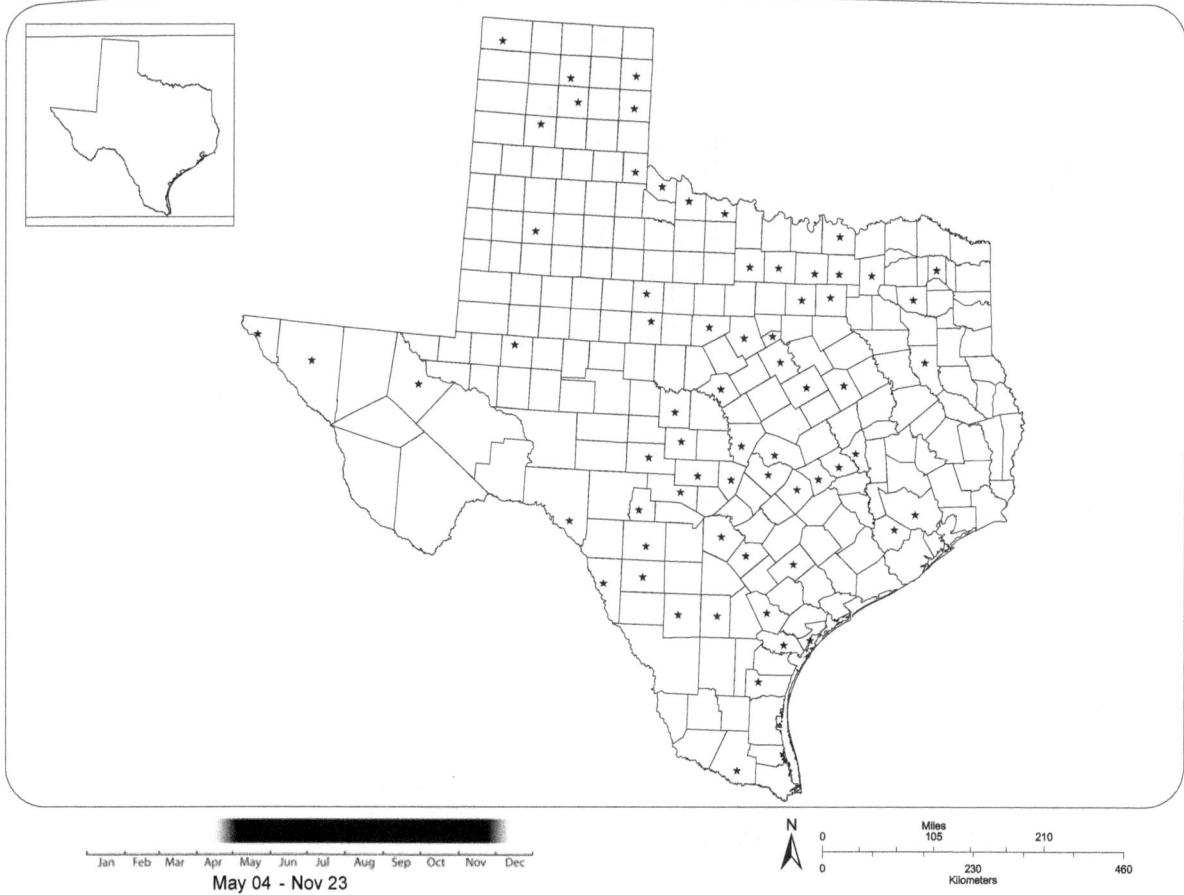

May 04 - Nov 23

HABITAT: Shallow ponds, lakes, marshes and slow streams.

Aransas	Hidalgo	Taylor
Bastrop	Hudspeth	Titus
Bee	Hunt	Travis
Bexar	Hutchinson	Uvalde
Blanco	Jack	Val Verde
Bosque	Jones	Wheeler
Brazos	Kerr	Wichita
Burleson	Kimble	Wilbarger
Burnet	Kleberg	Williamson
Carson	La Salle	Wilson
Cherokee	Lee	Wise
Childress	Limestone	Wood
Collin	Lubbock	Zavala
Dallam	Mason	
Dallas	Maverick	
Denton	McCulloch	
DeWitt	McLennan	
Eastland	McMullen	
El Paso	Midland	
Erath	Mills	
Fort Bend	Randall	
Gillespie	Real	
Grayson	Reeves	
Hardeman	San Patricio	
Harris	Somervell	
Hemphill	Tarrant	

Libellula saturata Uhler / Flame Skimmer

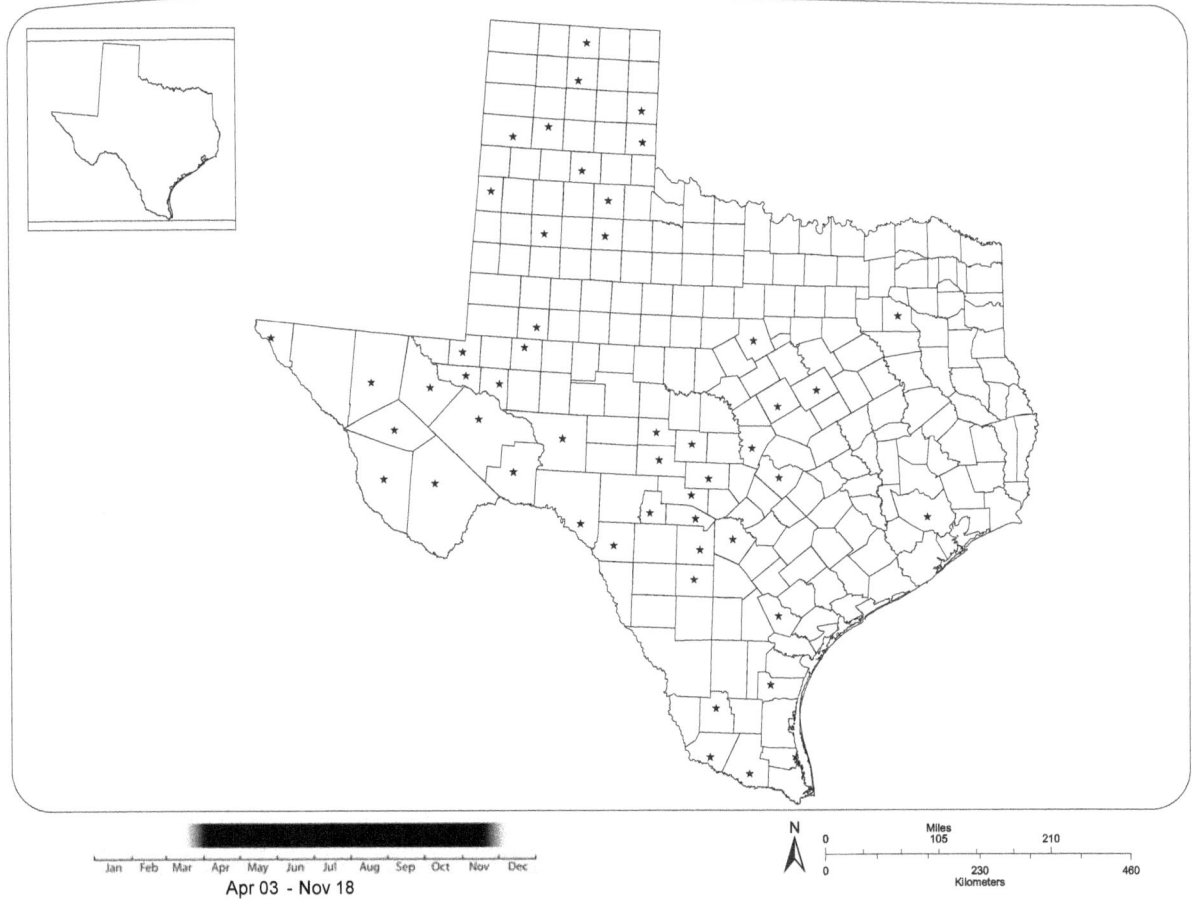

Apr 03 - Nov 18

HABITAT: Ponds, lakes and slow streams, including artificial ponds.

Bailey	Kinney
Bandera	Kleberg
Bee	Lubbock
Bexar	Martin
Brewster	Mason
Briscoe	McLennan
Burnet	Medina
Collingsworth	Menard
Coryell	Midland
Crane	Motley
Crockett	Pecos
Culberson	Presidio
Deaf Smith	Randall
Dickens	Real
El Paso	Reeves
Erath	Starr
Frio	Terrell
Gillespie	Travis
Hansford	Val Verde
Harris	Van Zandt
Hidalgo	Ward
Hutchinson	Wheeler
Jeff Davis	Winkler
Jim Hogg	
Kerr	
Kimble	

Libellula semifasciata Burmeister / Painted Skimmer

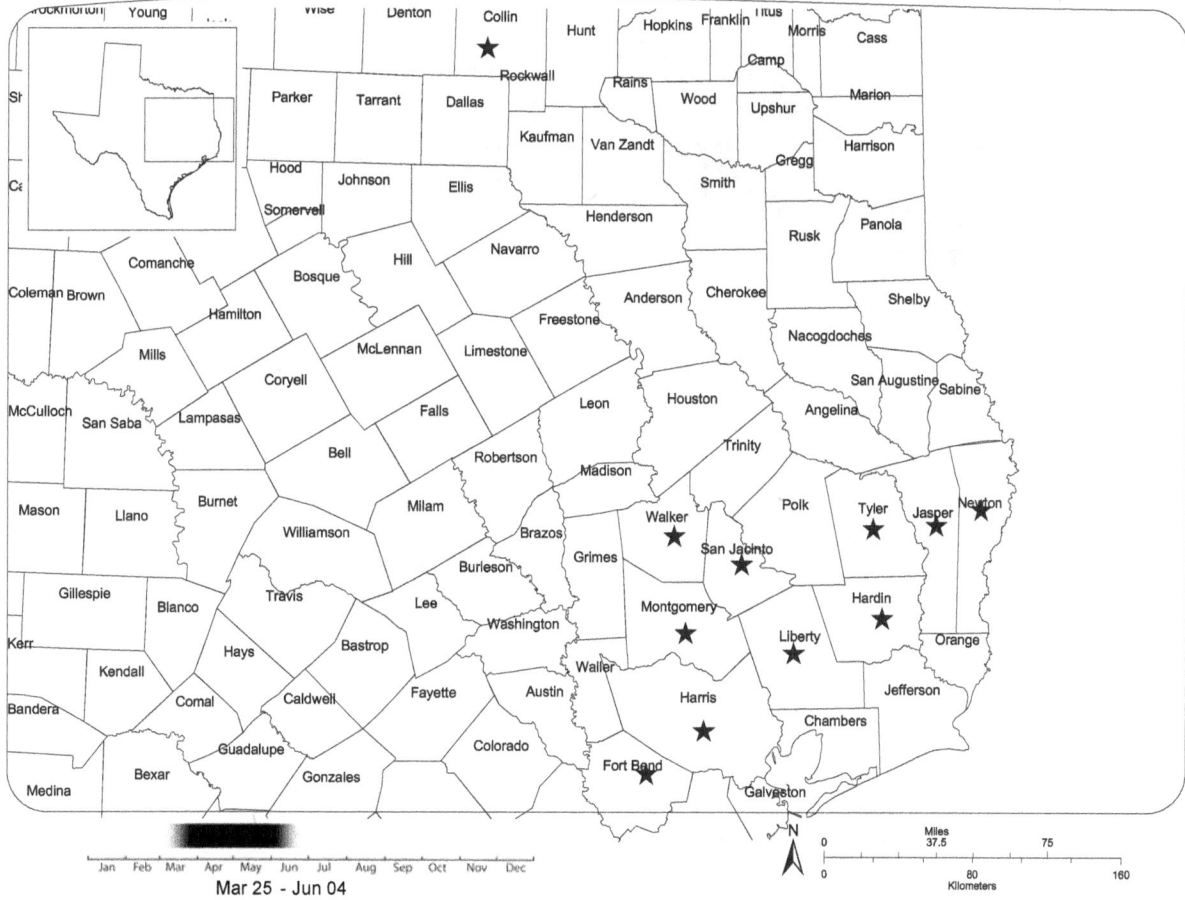

Mar 25 - Jun 04

HABITAT: Marshy forest seepages, ponds and slow streams.

Collin
Fort Bend
Hardin
Harris
Jasper
Liberty
Montgomery
Newton
Palo Pinto
San Jacinto
Tyler
Walker

Libellula vibrans Fabricius / Great Blue Skimmer

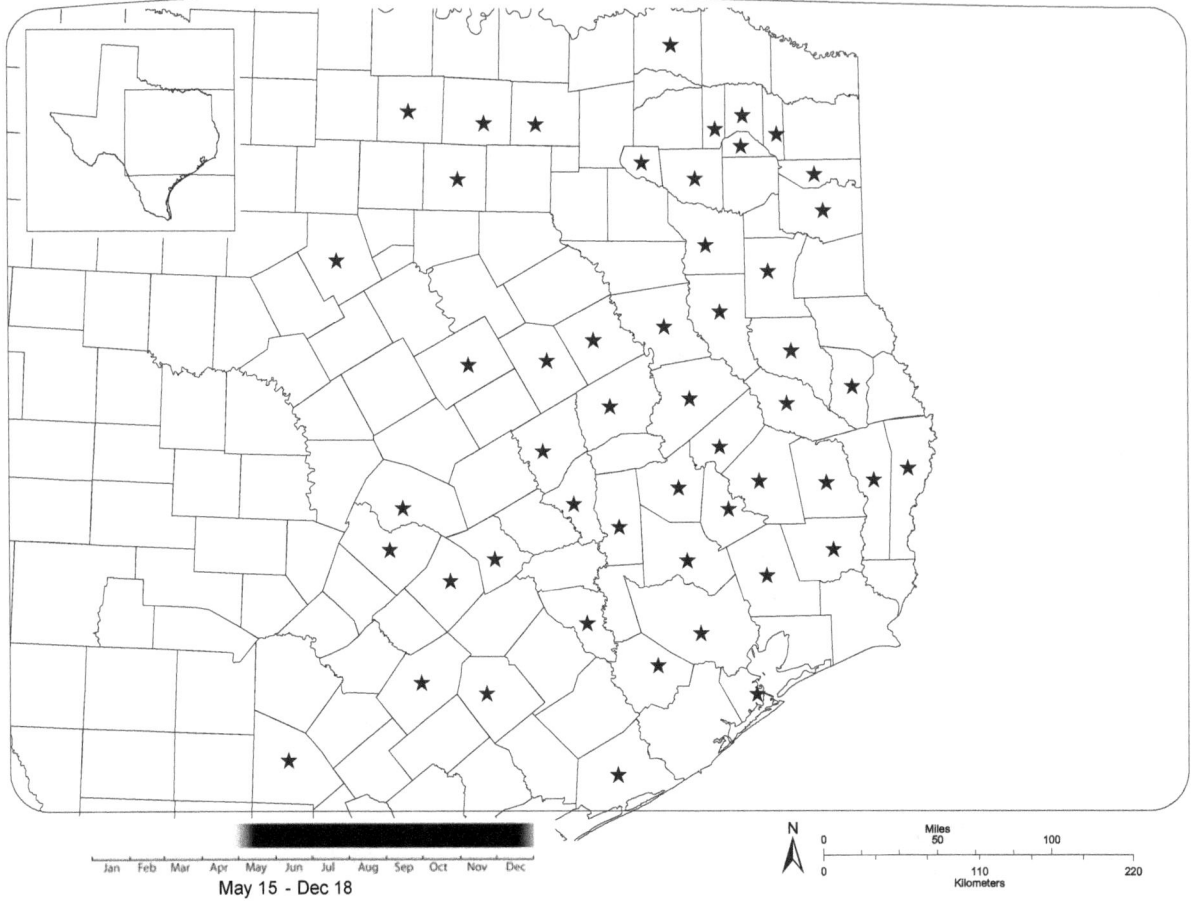

May 15 - Dec 18

HABITAT: Swampy ponds, lakes and slow forest streams.

Anderson	Liberty
Angelina	Limestone
Atascosa	Marion
Austin	Matagorda
Bastrop	McLennan
Brazos	Montgomery
Camp	Morris
Cherokee	Nacogdoches
Collin	Newton
Denton	Polk
Erath	Rains
Fort Bend	Robertson
Franklin	Rusk
Freestone	San Augustine
Galveston	San Jacinto
Gonzales	Smith
Grimes	Tarrant
Hardin	Titus
Harris	Travis
Harrison	Trinity
Houston	Tyler
Jasper	Walker
Lamar	Williamson
Lavaca	Wise
Lee	Wood
Leon	

Macrodiplax balteata (Hagen) / Marl Pennant

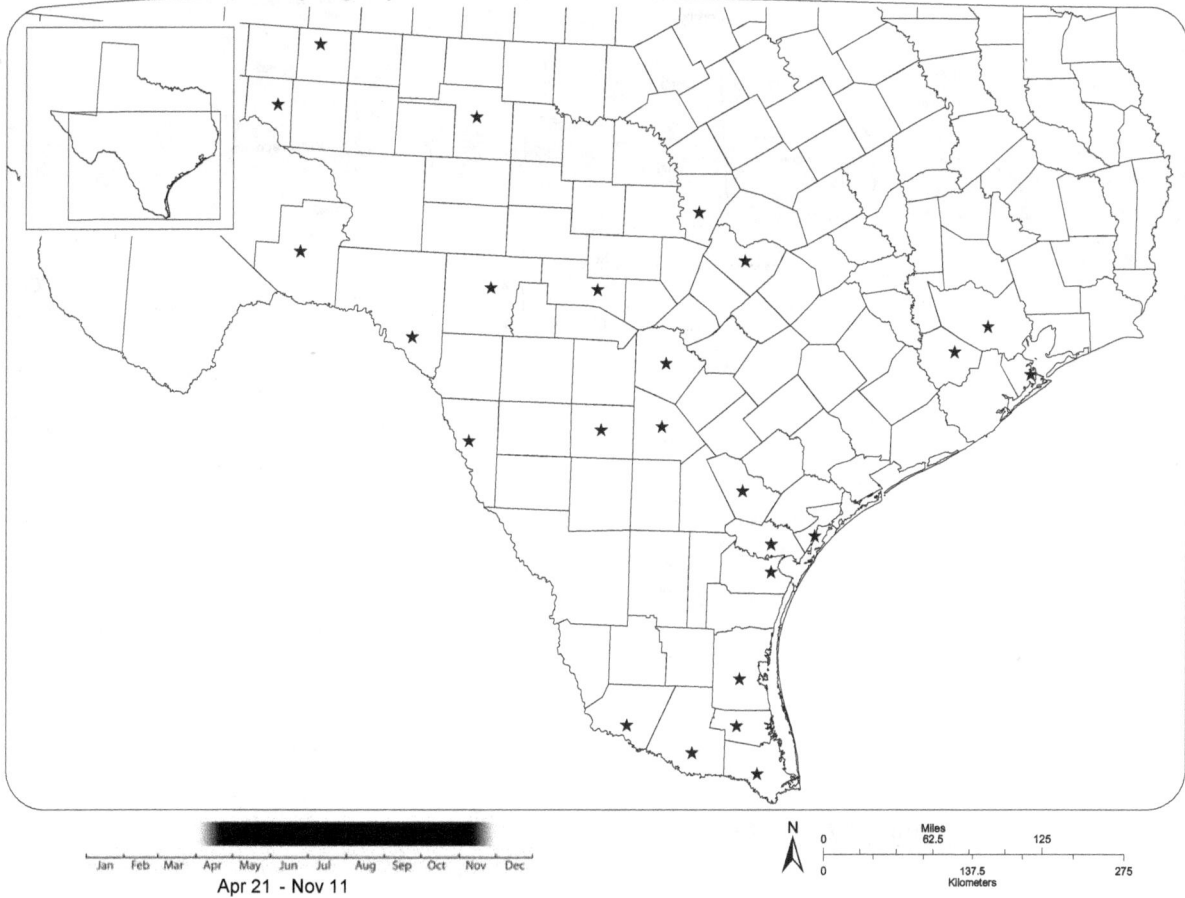

Apr 21 - Nov 11

HABITAT: Large brackish ponds and lakes.

Aransas   Willacy
Atascosa
Bee
Bexar
Burnet
Cameron
Crane
Edwards
Fort Bend
Frio
Galveston
Harris
Hidalgo
Kenedy
Kerr
Maverick
Midland
Nueces
Reeves
San Patricio
Starr
Terrell
Tom Green
Travis
Val Verde
Ward

Macrothemis imitans leucozona Ris / Ivory-striped Sylph

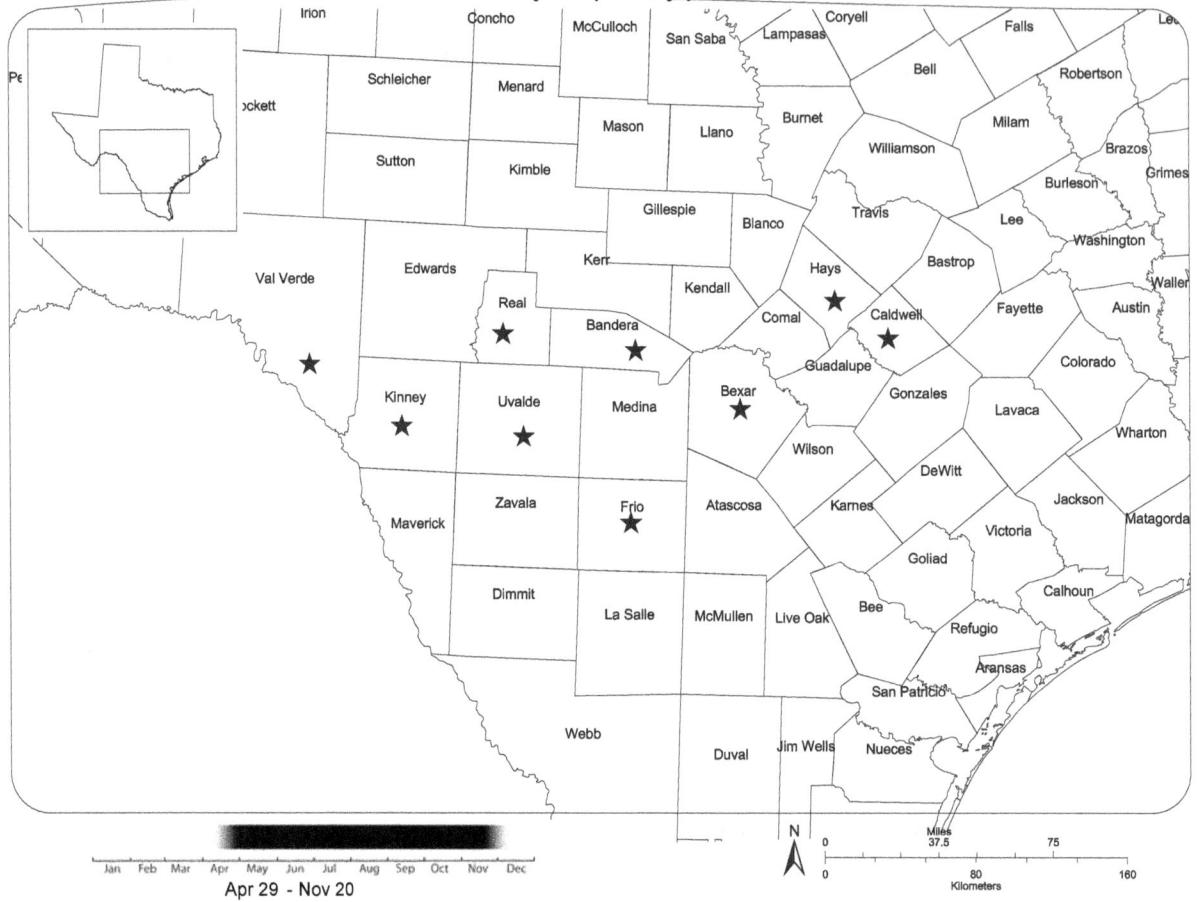

Apr 29 - Nov 20

HABITAT: Rocky streams and rivers

Bandera
Bexar
Caldwell
Frio
Hays
Kinney
Real
Uvalde
Val Verde

Macrothemis inacuta Calvert / Straw-colored Sylph

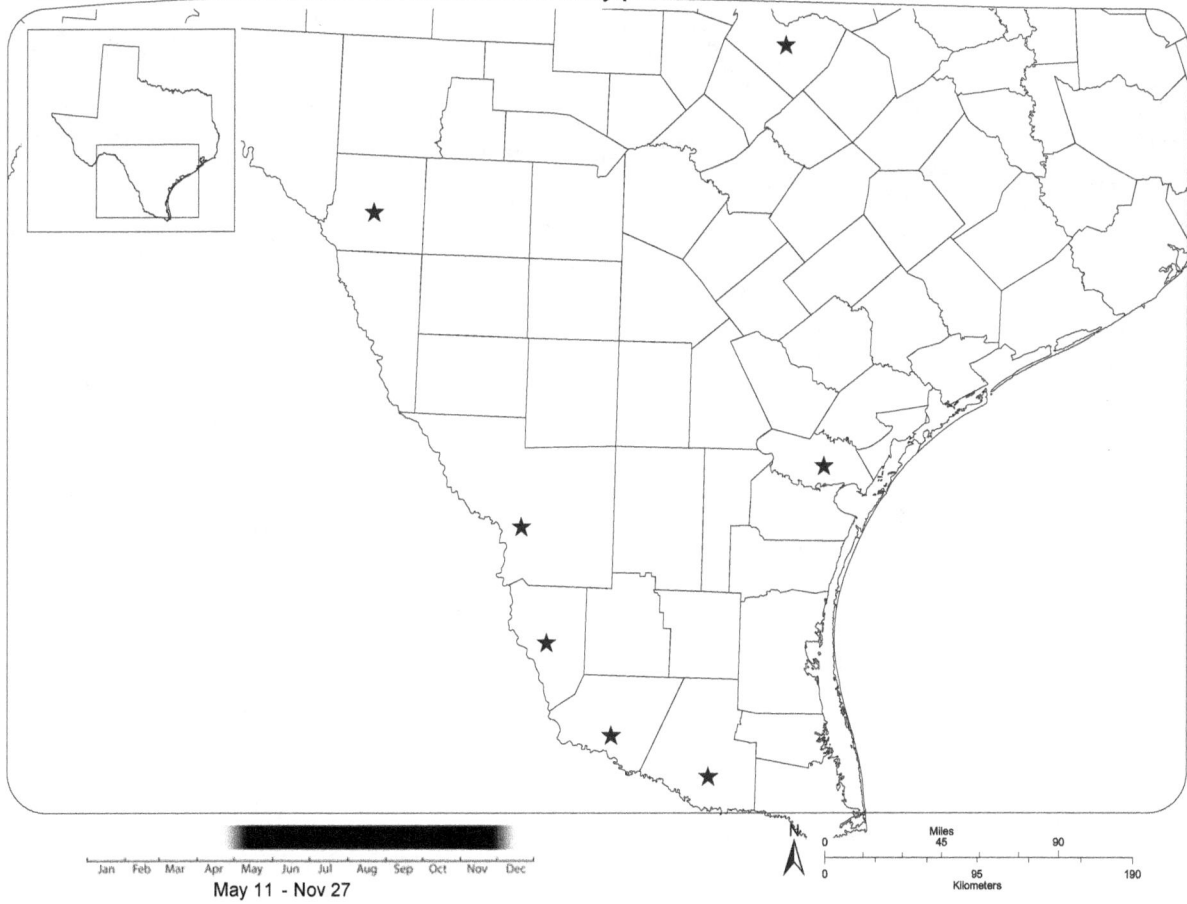

May 11 - Nov 27

HABITAT: Clear rocky streams and rivers.

Hidalgo
Kinney
San Patricio
Starr
Travis
Webb
Zapata

Macrothemis inequiunguis Calvert / Jade-striped Sylph

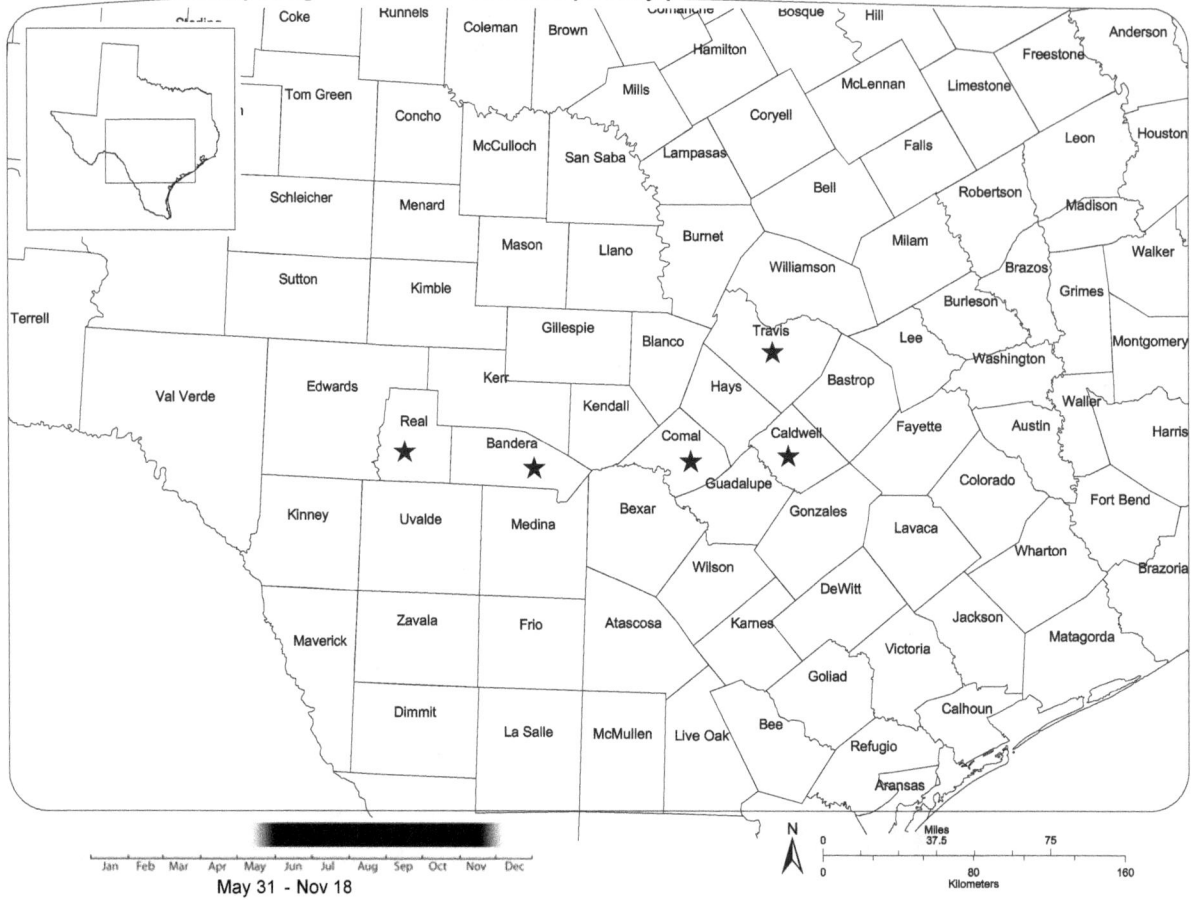

May 31 - Nov 18

HABITAT: Rocky streams and rivers.

Bandera
Caldwell
Comal
Real
Travis

Miathyria marcella (Selys in Sagra) / Hyacinth Glider

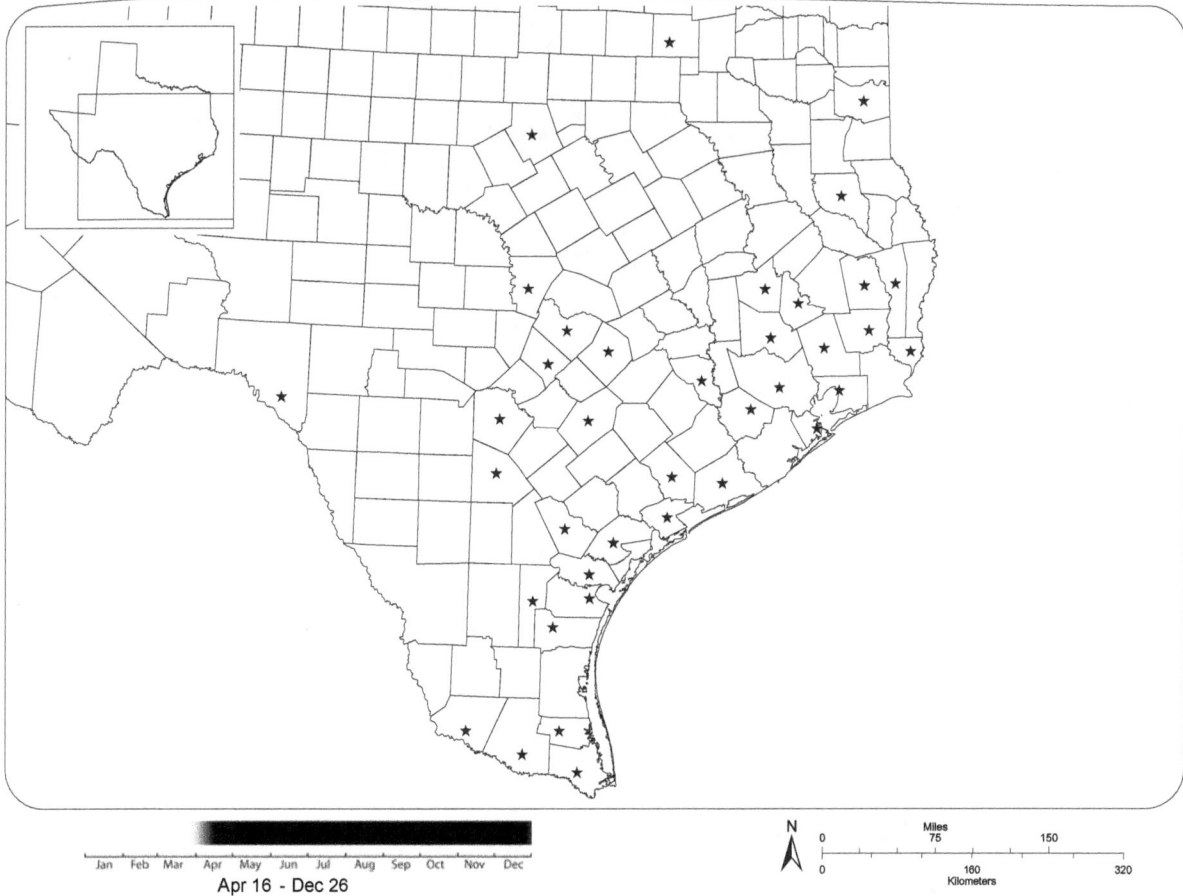

Apr 16 - Dec 26

HABITAT: Marshy ponds and lakes, including brackish waters, with water hyacinth.

Atascosa	Nacogdoches
Austin	Nueces
Bastrop	Orange
Bee	Refugio
Bexar	San Jacinto
Burnet	San Patricio
Calhoun	Starr
Cameron	Travis
Chambers	Tyler
Collin	Val Verde
Erath	Walker
Fort Bend	Willacy
Galveston	
Gonzales	
Hardin	
Harris	
Harrison	
Hays	
Hidalgo	
Jackson	
Jasper	
Jim Wells	
Kleberg	
Liberty	
Matagorda	
Montgomery	

Micrathyria aequalis (Hagen) / Spot-tailed Dasher

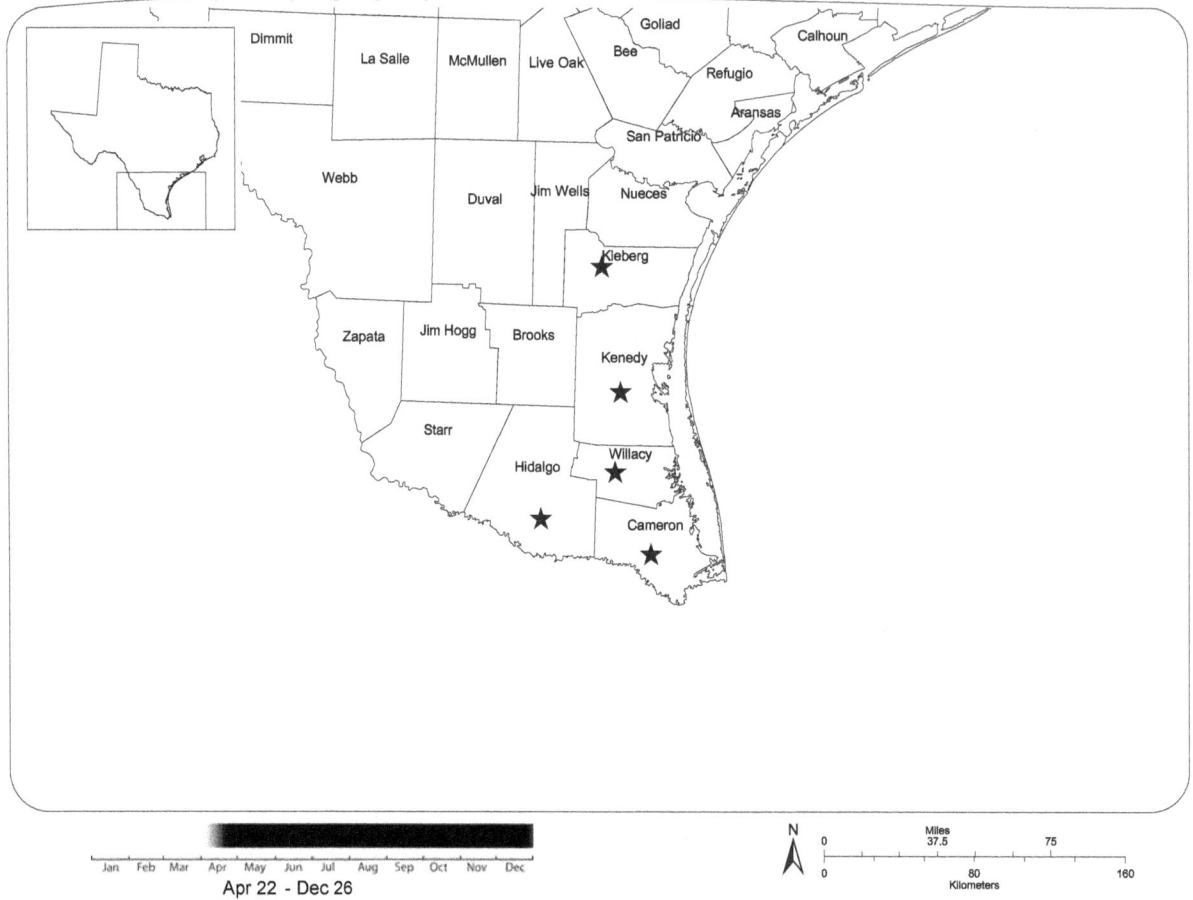

Apr 22 - Dec 26

HABITAT: Permanent and temporary ponds, sloughs and lakes.

Cameron
Hidalgo
Kenedy
Kleberg
Willacy

Micrathyria didyma (Selys) / Three-striped Dasher

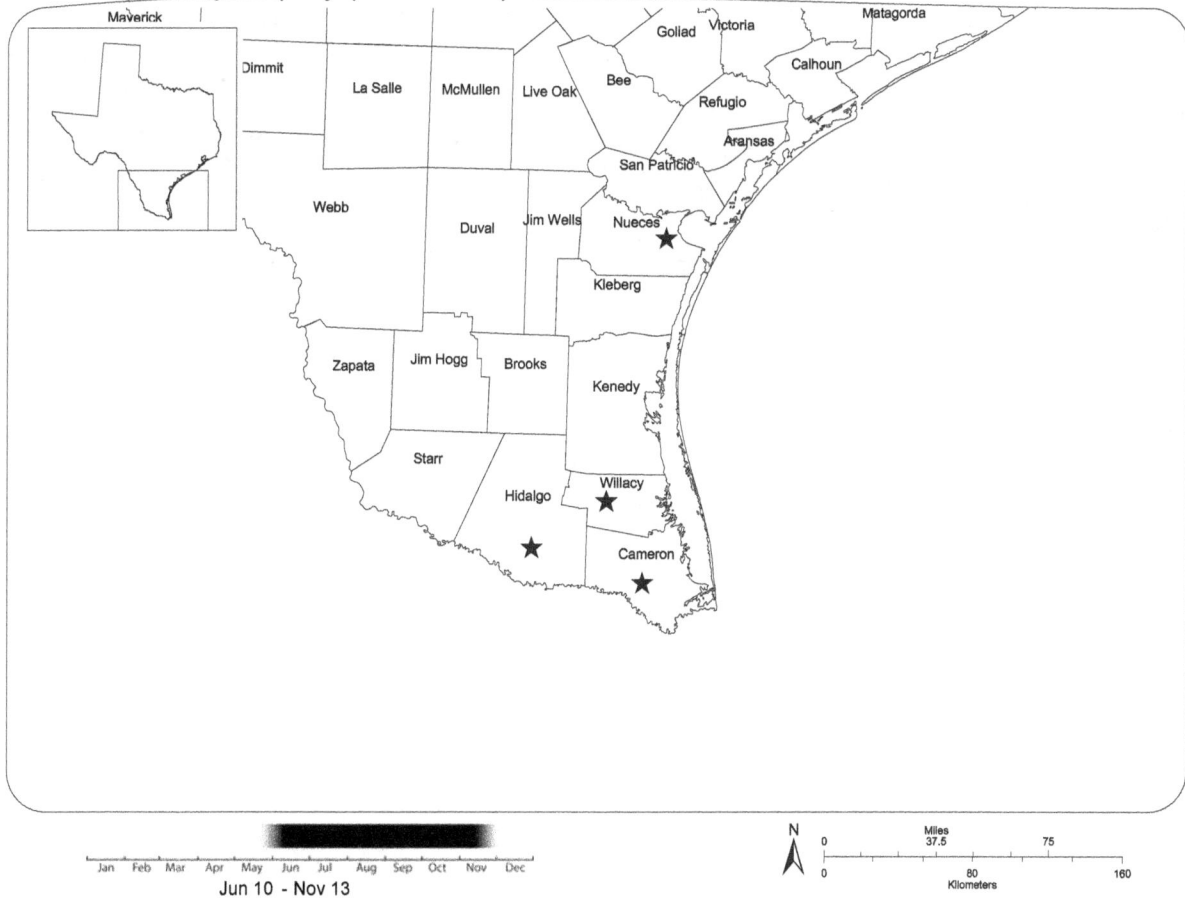

Jun 10 - Nov 13

HABITAT: Weedy pools, ponds, brooks and ditches in the shade.

Cameron
Hidalgo
Nueces
Willacy

Micrathyria hagenii Kirby / Thornbush Dasher

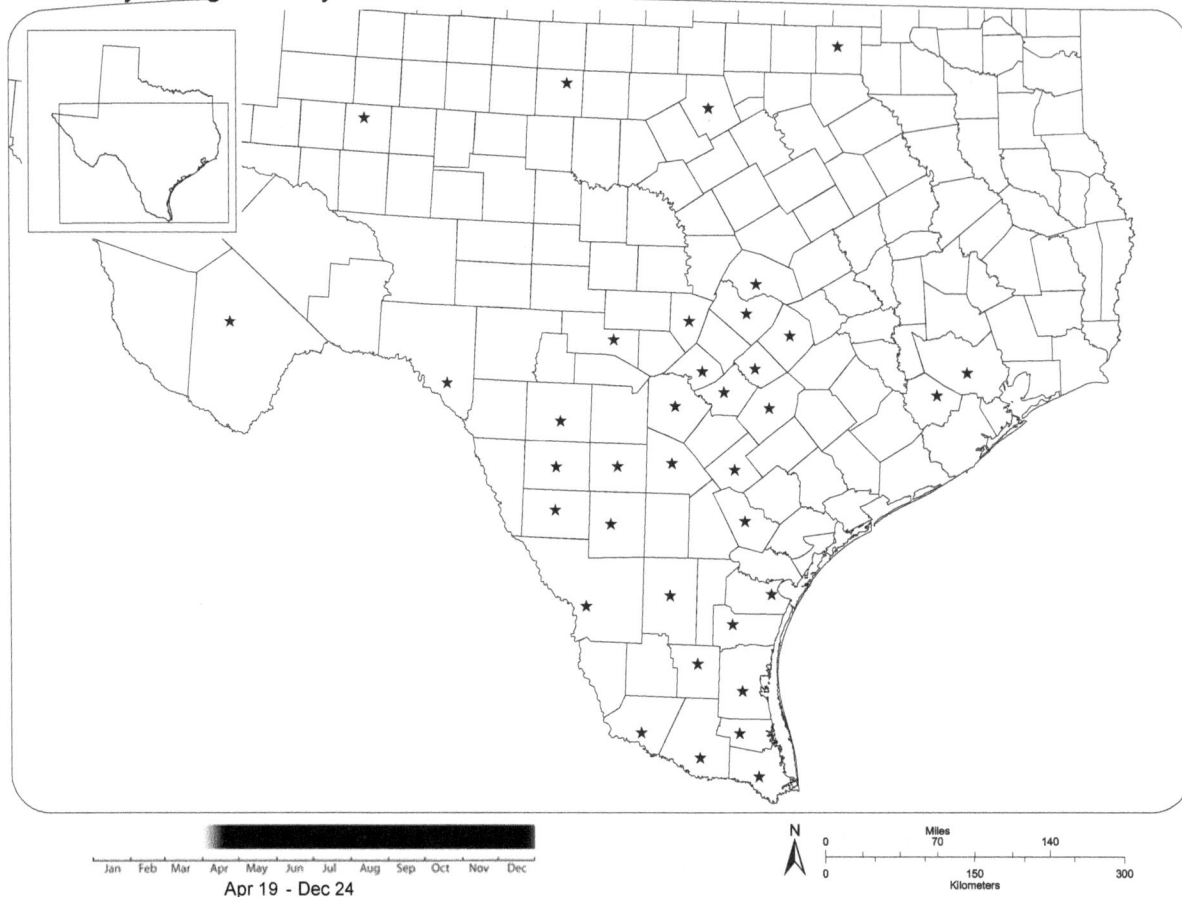

Apr 19 - Dec 24

HABITAT: Heavily vegetated ponds and lakes.

Atascosa
Bastrop
Bee
Bexar
Blanco
Brewster
Brooks
Caldwell
Cameron
Comal
Dallas
Dimmit
Duval
Erath
Fort Bend
Frio
Gonzales
Guadalupe
Harris
Hidalgo
Karnes
Kenedy
Kerr
Kleberg
La Salle
Midland

Nueces
Starr
Taylor
Travis
Uvalde
Val Verde
Webb
Willacy
Williamson
Zavala

Orthemis discolor (Burmeister) / Carmine Skimmer

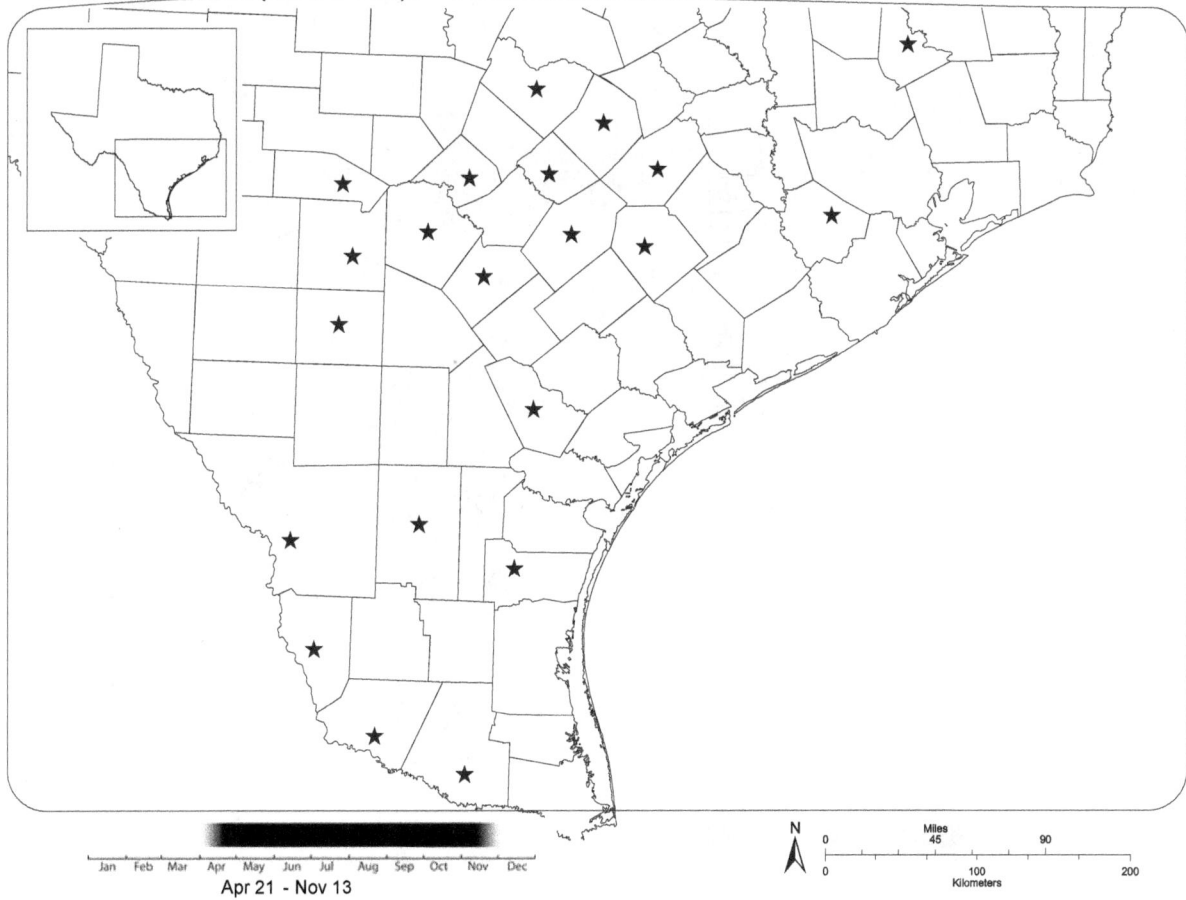

Apr 21 - Nov 13

HABITAT: Temporary and permanent ponds, lakes, ditches and slow streams.

Bandera
Bastrop
Bee
Bexar
Caldwell
Comal
Duval
Fayette
Fort Bend
Frio
Gonzales
Hidalgo
Kleberg
Lavaca
Medina
San Jacinto
Starr
Travis
Webb
Wilson
Zapata

Orthemis ferruginea (Fabricius) / Roseate Skimmer

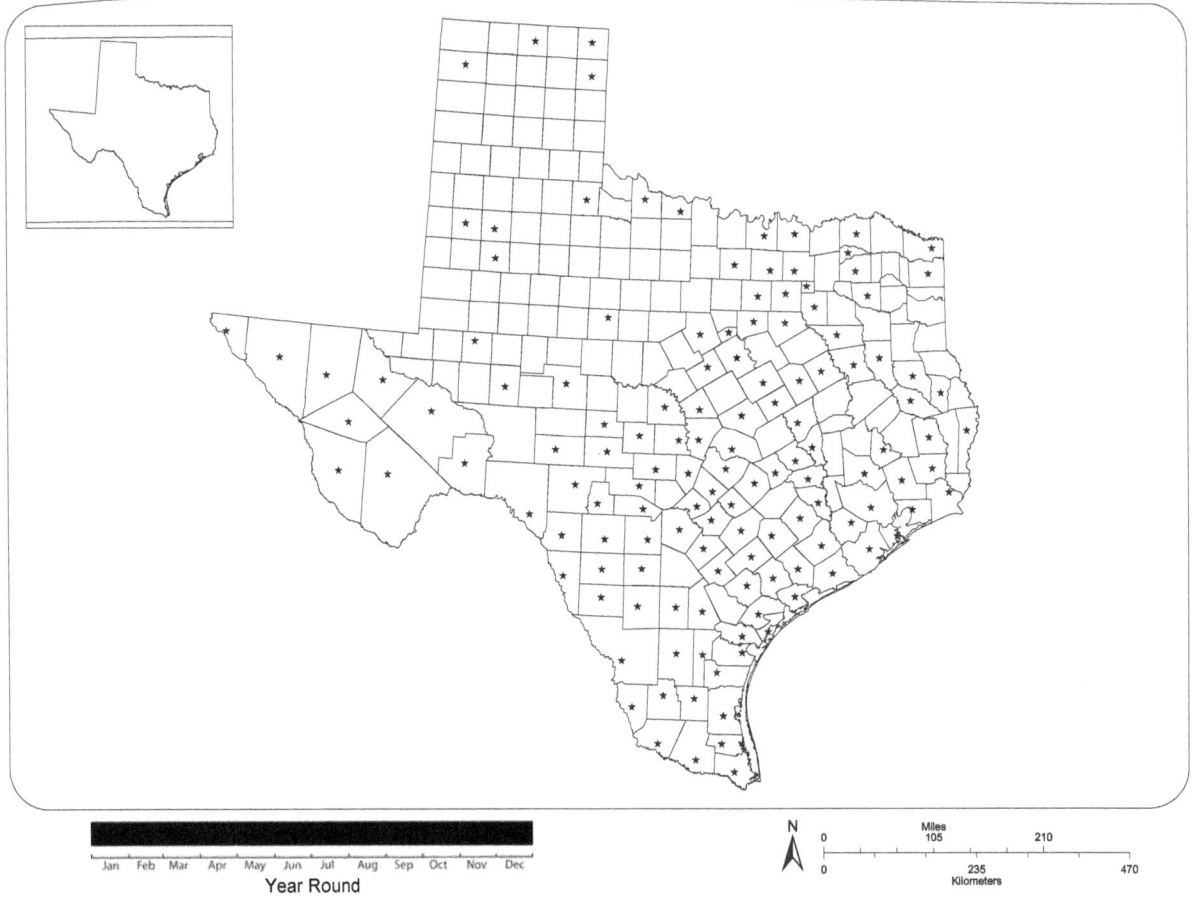

Jan Feb Mar Apr May Jun Jul Aug Sep Oct Nov Dec
Year Round

HABITAT: Temporary and permanent ponds, lakes, ditches and slow streams.

Anderson	Cooke	Harris	Lee	Robertson	Wise
Angelina	Cottle	Hartley	Liberty	Rockwall	Wood
Aransas	Culberson	Hays	Limestone	San Augustine	Zapata
Austin	Dallas	Hemphill	Lipscomb	San Jacinto	Zavala
Bandera	Delta	Henderson	Live Oak	San Patricio	
Bastrop	Denton	Hidalgo	Llano	San Saba	
Bell	DeWitt	Hockley	Lubbock	Somervell	
Bexar	Dimmit	Hopkins	Lynn	Starr	
Blanco	Duval	Hudspeth	Mason	Sutton	
Bosque	Edwards	Jackson	Matagorda	Tarrant	
Bowie	El Paso	Jeff Davis	Maverick	Taylor	
Brazoria	Ellis	Jefferson	McLennan	Terrell	
Brazos	Erath	Jim Hogg	McMullen	Tom Green	
Brewster	Falls	Jim Wells	Medina	Travis	
Brooks	Fort Bend	Johnson	Menard	Tyler	
Burleson	Freestone	Karnes	Midland	Uvalde	
Burnet	Frio	Kaufman	Montgomery	Val Verde	
Caldwell	Galveston	Kenedy	Nacogdoches	Victoria	
Calhoun	Gillespie	Kerr	Newton	Washington	
Cameron	Goliad	Kimble	Nueces	Webb	
Cass	Gonzales	Kinney	Pecos	Wharton	
Chambers	Grayson	Kleberg	Presidio	Wichita	
Cherokee	Guadalupe	La Salle	Reagan	Wilbarger	
Collin	Hamilton	Lamar	Real	Willacy	
Colorado	Hansford	Lampasas	Reeves	Williamson	
Comal	Hardin	Lavaca	Refugio	Wilson	

Pachydiplax longipennis (Burmeister) / Blue Dasher

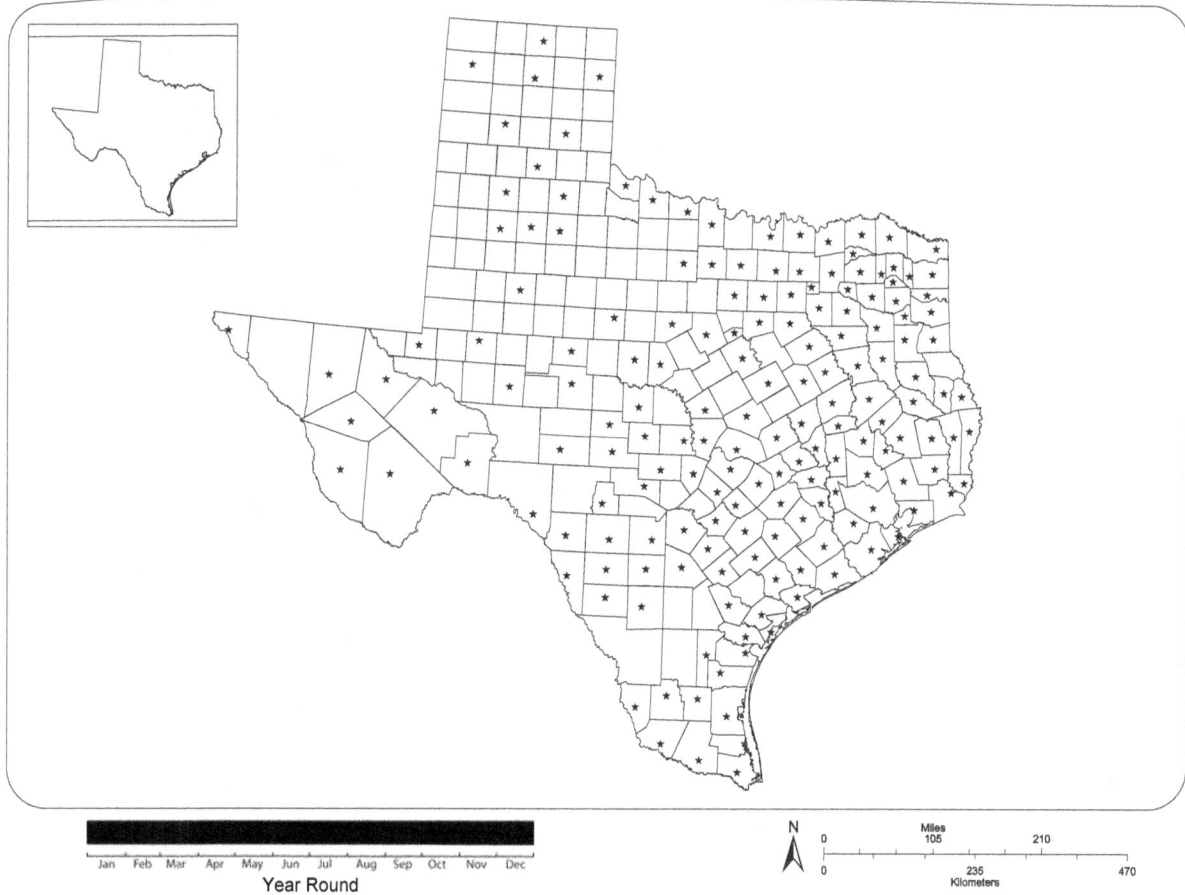

Jan Feb Mar Apr May Jun Jul Aug Sep Oct Nov Dec

Year Round

N

Miles
0   105   210

0   235   470
Kilometers

HABITAT: Ponds, lakes, marshes, ditches, slow streams and other quiet bodies of water.

Anderson	Chambers	Frio	Jeff Davis	Maverick	Refugio	Waller
Angelina	Cherokee	Galveston	Jefferson	McCulloch	Robertson	Washington
Aransas	Clay	Gillespie	Jim Hogg	McLennan	Rockwall	Wharton
Atascosa	Coke	Gonzales	Jim Wells	Medina	Rusk	Wichita
Austin	Coleman	Grayson	Johnson	Menard	Sabine	Wilbarger
Bastrop	Collin	Gregg	Karnes	Midland	San Augustine	Williamson
Bee	Colorado	Grimes	Kaufman	Milam	San Jacinto	Wilson
Bell	Cooke	Guadalupe	Kenedy	Montgomery	San Patricio	Winkler
Bexar	Crosby	Hale	Kerr	Morris	Smith	Wise
Blanco	Culberson	Hansford	Kimble	Motley	Somervell	Wood
Borden	Dallas	Hardeman	Kinney	Nacogdoches	Starr	Young
Bosque	Delta	Hardin	Kleberg	Navarro	Sutton	Zapata
Bowie	Denton	Harris	La Salle	Newton	Tarrant	Zavala
Brazoria	DeWitt	Harrison	Lamar	Nueces	Taylor	
Brazos	Dickens	Hartley	Lampasas	Orange	Terrell	
Brewster	Dimmit	Hays	Lavaca	Panola	Titus	
Briscoe	Donley	Hemphill	Lee	Parker	Tom Green	
Brooks	Eastland	Henderson	Leon	Pecos	Travis	
Brown	El Paso	Hidalgo	Liberty	Polk	Trinity	
Burleson	Ellis	Hopkins	Limestone	Presidio	Tyler	
Burnet	Erath	Houston	Llano	Rains	Upshur	
Caldwell	Fannin	Hunt	Lubbock	Randall	Uvalde	
Calhoun	Fayette	Hutchinson	Madison	Reagan	Val Verde	
Cameron	Fort Bend	Jack	Marion	Real	Van Zandt	
Camp	Franklin	Jackson	Mason	Red River	Victoria	
Cass	Freestone	Jasper	Matagorda	Reeves	Walker	

Paltothemis lineatipes Karsch / Red Rock Skimmer

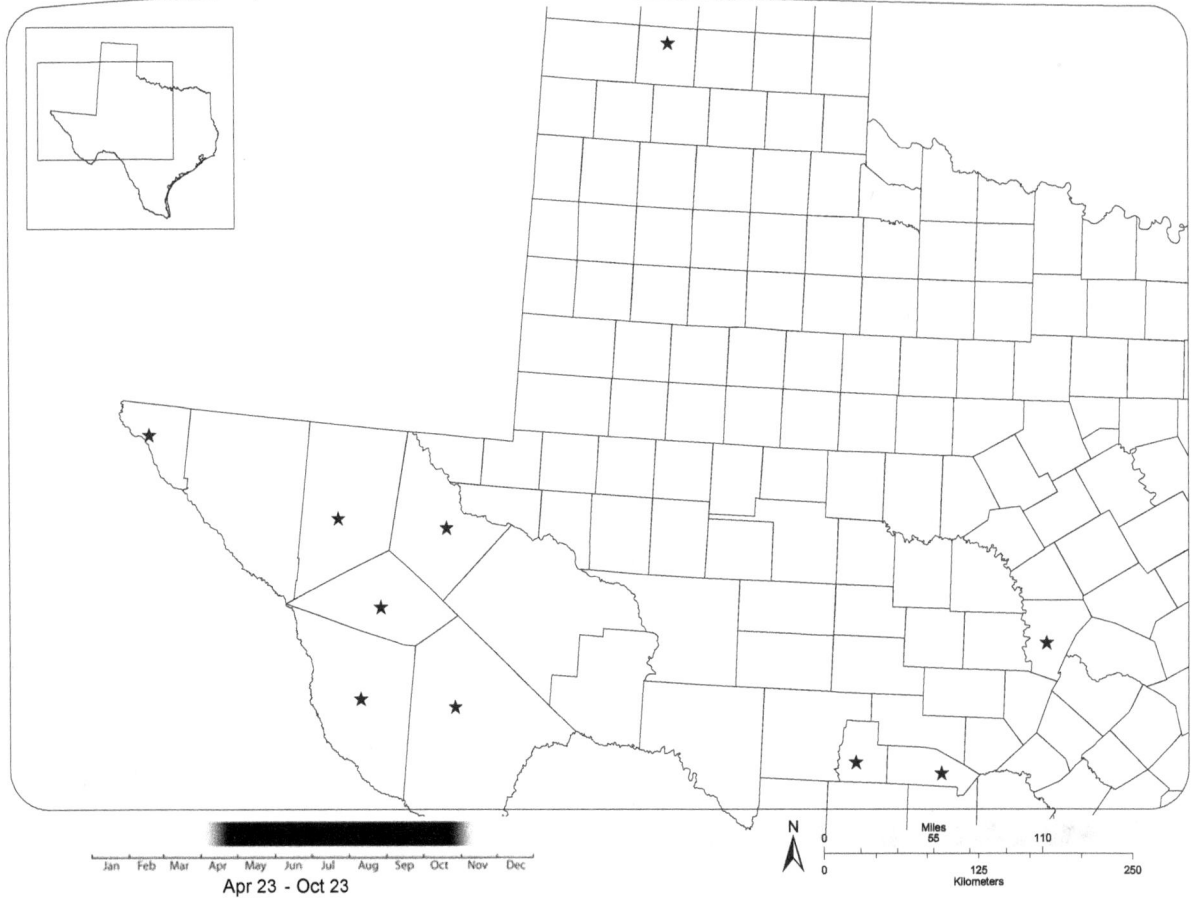

Jan Feb Mar Apr May Jun Jul Aug Sep Oct Nov Dec

Apr 23 - Oct 23

HABITAT: Small, sunlit, rocky, forest streams.

Bandera
Brewster
Burnet
Culberson
El Paso
Jeff Davis
Presidio
Randall
Real
Reeves

Pantala flavescens (Fabricius) / Wandering Glider

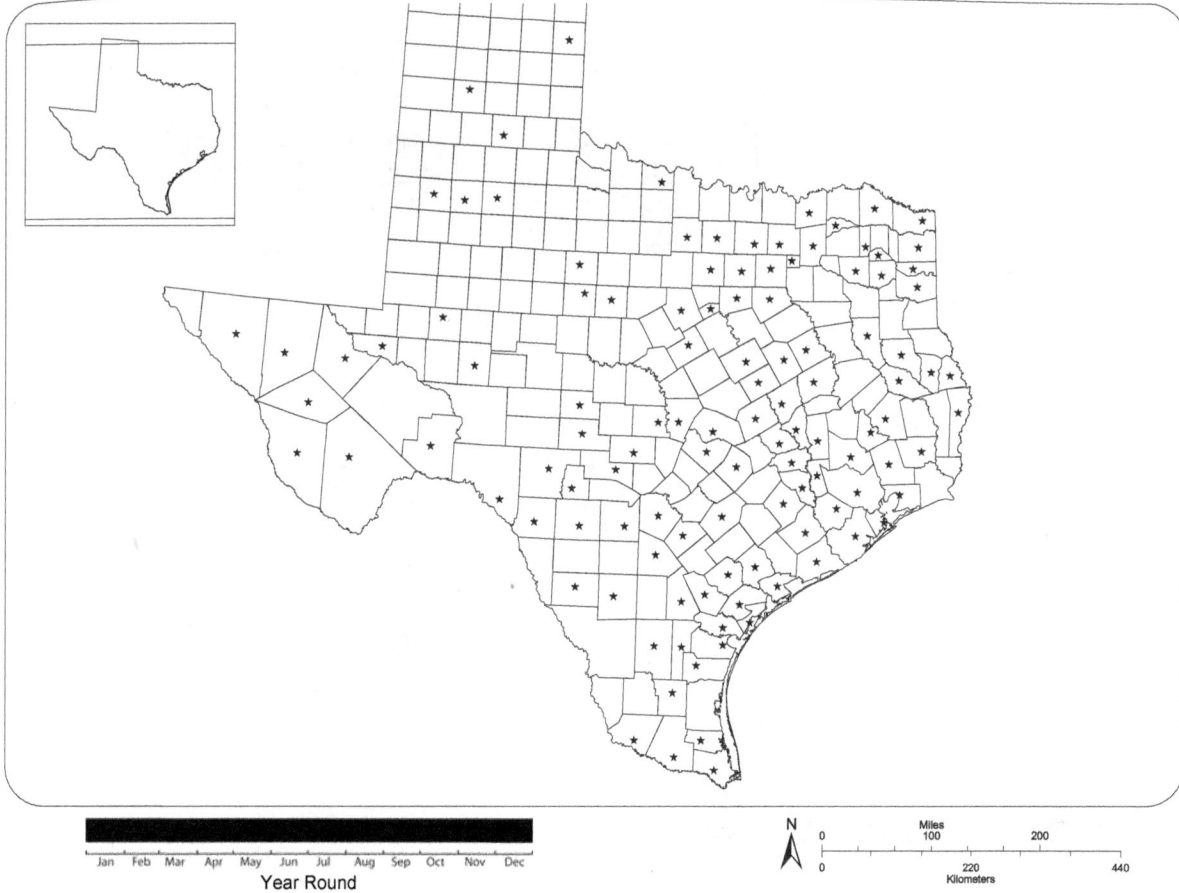

Year Round

HABITAT: Permanent and temporary ponds, pools and other water bodies, including brackish ones.

Angelina	Dallas	Hunt	Newton	Victoria
Aransas	Delta	Jack	Nueces	Waller
Atascosa	Denton	Jeff Davis	Parker	Ward
Austin	Dimmit	Jim Wells	Polk	Washington
Bastrop	Duval	Johnson	Presidio	Wharton
Bee	Edwards	Jones	Randall	Wichita
Bexar	Ellis	Kerr	Reagan	Willacy
Bowie	Erath	Kimble	Real	Williamson
Brazoria	Falls	Kinney	Red River	Wilson
Brazos	Fannin	Kleberg	Reeves	Wise
Brewster	Fort Bend	La Salle	Refugio	Wood
Briscoe	Franklin	Leon	Robertson	
Brooks	Freestone	Liberty	Rockwall	
Burleson	Galveston	Limestone	Sabine	
Burnet	Gillespie	Live Oak	San Augustine	
Calhoun	Goliad	Llano	San Jacinto	
Callahan	Gonzales	Lubbock	San Patricio	
Cameron	Grimes	Marion	Somervell	
Camp	Hamilton	Matagorda	Starr	
Cass	Hardin	McLennan	Tarrant	
Chambers	Harris	Medina	Taylor	
Cherokee	Harrison	Menard	Terrell	
Collin	Hemphill	Midland	Travis	
Colorado	Hidalgo	Milam	Upshur	
Crosby	Hockley	Montgomery	Uvalde	
Culberson	Hudspeth	Nacogdoches	Val Verde	

Pantala hymenaea (Say) / Spot-winged Glider

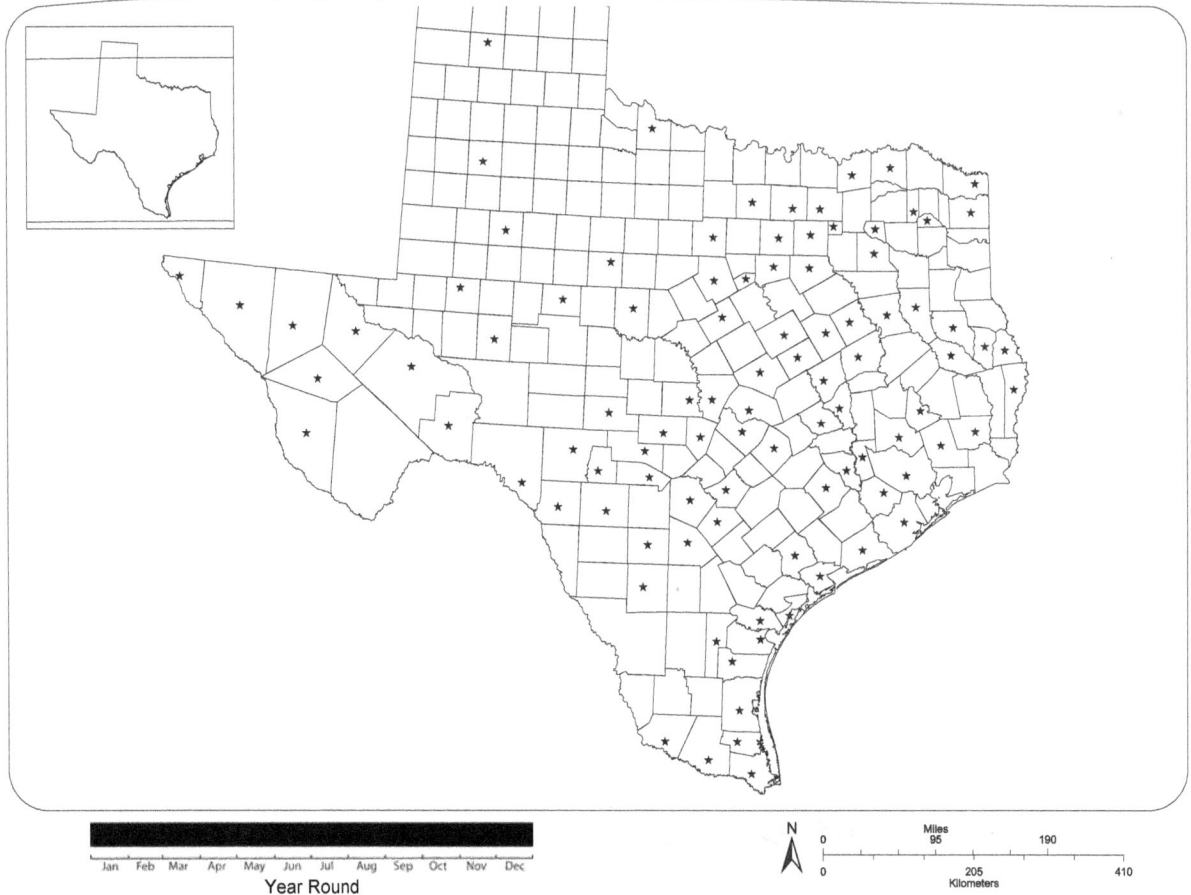

Jan Feb Mar Apr May Jun Jul Aug Sep Oct Nov Dec
Year Round

N

| Miles | | |
| 0 | 95 | 190 |

| 0 | 205 | 410 |
| | Kilometers | |

HABITAT: Open, temporary and artificial ponds and pools, including brackish waters.

Anderson	Dallas	Kleberg	San Augustine
Angelina	Denton	La Salle	San Jacinto
Aransas	Edwards	Lamar	San Patricio
Atascosa	El Paso	Leon	Somervell
Austin	Ellis	Liberty	Starr
Bandera	Erath	Limestone	Tarrant
Bastrop	Falls	Llano	Taylor
Bell	Fannin	Lubbock	Terrell
Bexar	Fort Bend	Matagorda	Travis
Blanco	Franklin	McLennan	Uvalde
Borden	Freestone	Midland	Val Verde
Bowie	Frio	Montgomery	Van Zandt
Brazoria	Gillespie	Nacogdoches	Victoria
Brazos	Guadalupe	Newton	Waller
Burleson	Hamilton	Nueces	Wilbarger
Burnet	Hardin	Palo Pinto	Willacy
Calhoun	Harris	Pecos	Williamson
Cameron	Hidalgo	Presidio	Wilson
Camp	Hudspeth	Rains	Wise
Cass	Jeff Davis	Randall	
Cherokee	Jim Wells	Reagan	
Coke	Johnson	Real	
Coleman	Kenedy	Reeves	
Collin	Kerr	Robertson	
Colorado	Kimble	Rockwall	
Culberson	Kinney	Sabine	

Perithemis domitia (Drury) / Slough Amberwing

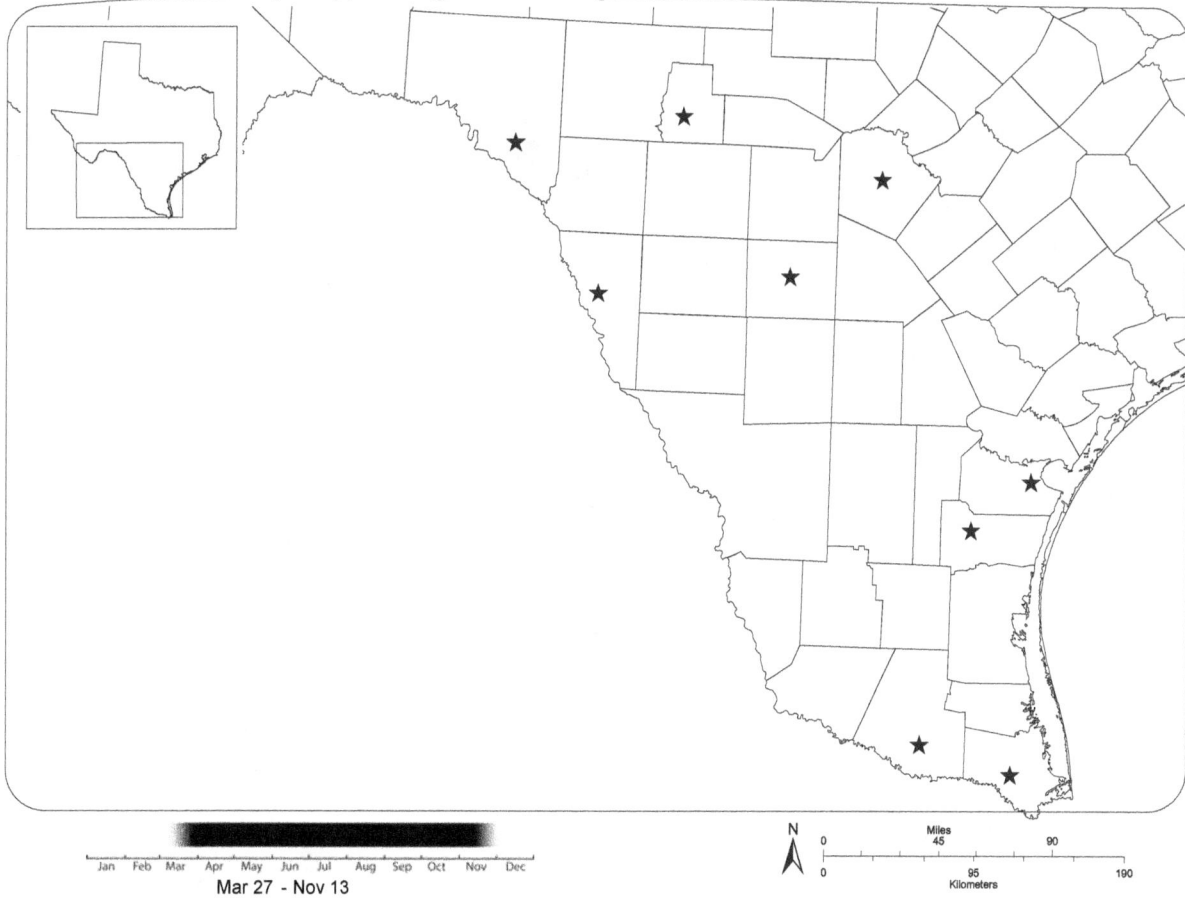

Mar 27 - Nov 13

HABITAT: Shaded sloughs, ponds, pools, roadside ditches and other still waters.

Bexar
Brewster
Cameron
Frio
Hidalgo
Kleberg
Maverick
Nueces
Real
Val Verde

Perithemis tenera (Say) / Eastern Amberwing

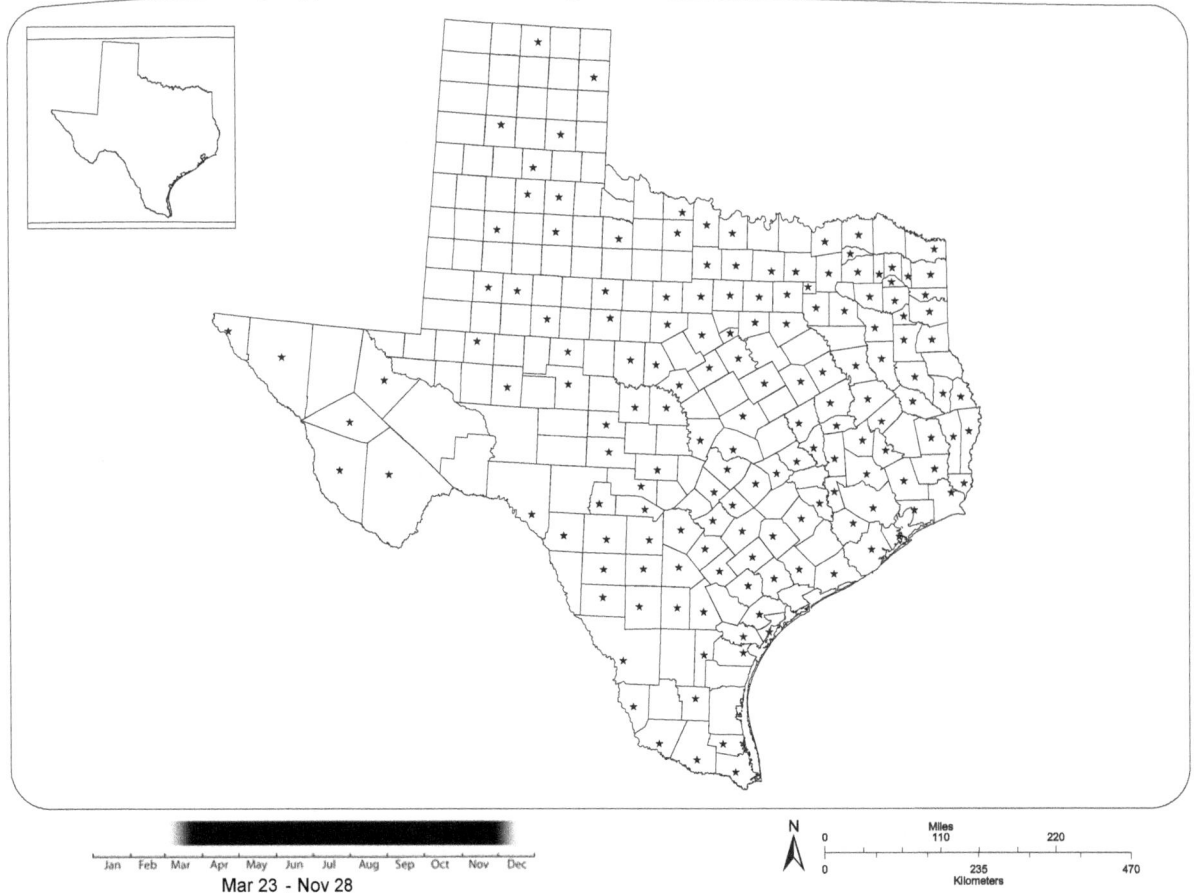

Mar 23 - Nov 28

HABITAT: Open sloughs, ponds, pools, roadside ditches and other still waters.

Anderson	Cherokee	Goliad	Kaufman	Montgomery	Starr
Angelina	Clay	Gonzales	Kerr	Morris	Stephens
Aransas	Coke	Gregg	Kimble	Motley	Tarrant
Archer	Coleman	Grimes	Kinney	Nacogdoches	Taylor
Atascosa	Collin	Guadalupe	Knox	Newton	Titus
Austin	Colorado	Hamilton	La Salle	Nueces	Tom Green
Bandera	Dallas	Hansford	Lamar	Orange	Travis
Bastrop	Dawson	Hardin	Lavaca	Palo Pinto	Trinity
Bell	Delta	Harris	Lee	Panola	Tyler
Bexar	Denton	Harrison	Leon	Parker	Upshur
Borden	DeWitt	Hays	Liberty	Presidio	Uvalde
Bosque	Dickens	Hemphill	Limestone	Randall	Val Verde
Bowie	Dimmit	Hidalgo	Live Oak	Reagan	Van Zandt
Brazoria	Donley	Hopkins	Lubbock	Real	Victoria
Brazos	Eastland	Houston	Madison	Reeves	Walker
Brewster	El Paso	Hudspeth	Marion	Refugio	Waller
Briscoe	Ellis	Hunt	Matagorda	Robertson	Webb
Brooks	Erath	Jack	McCulloch	Rockwall	Wichita
Brown	Fannin	Jackson	McLennan	Rusk	Willacy
Burleson	Floyd	Jasper	McMullen	Sabine	Williamson
Burnet	Fort Bend	Jeff Davis	Medina	San Augustine	Wilson
Caldwell	Franklin	Jefferson	Menard	San Jacinto	Wise
Cameron	Freestone	Jim Wells	Midland	San Patricio	Wood
Camp	Frio	Johnson	Mills	San Saba	Zapata
Cass	Galveston	Jones	Mitchell	Smith	Zavala
Chambers	Gillespie	Karnes	Montague	Somervell	

Plathemis lydia (Drury) / Common Whitetail

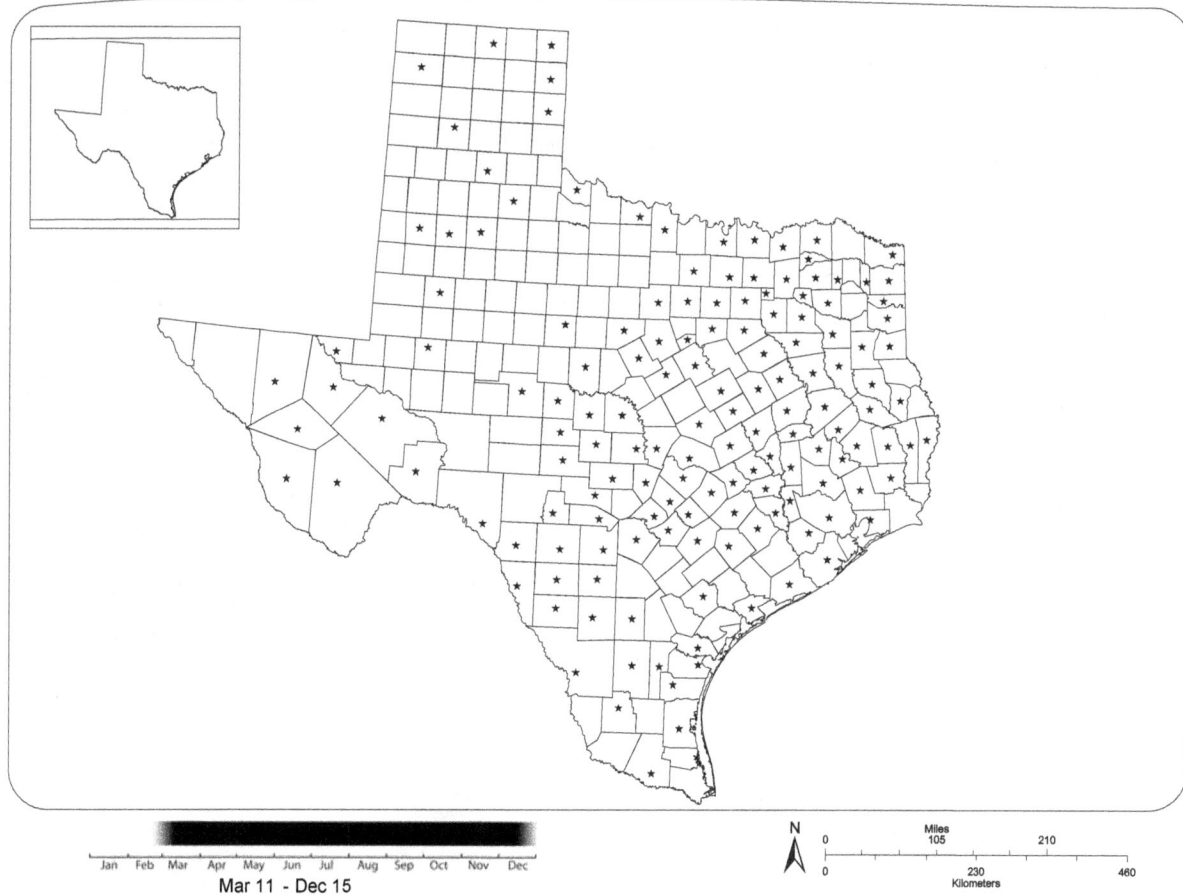

Mar 11 - Dec 15

HABITAT: Nearly any pool, pond, lake or quiet stream.

Anderson	Comanche	Guadalupe	Kleberg	Motley	Terrell
Angelina	Concho	Hamilton	La Salle	Nacogdoches	Tom Green
Austin	Cooke	Hansford	Lamar	Navarro	Travis
Bandera	Crosby	Hardeman	Lavaca	Newton	Trinity
Bastrop	Culberson	Hardin	Lee	Nueces	Tyler
Bell	Dallas	Harris	Leon	Palo Pinto	Uvalde
Bexar	Dawson	Harrison	Liberty	Panola	Val Verde
Blanco	Delta	Hartley	Limestone	Parker	Van Zandt
Bosque	Denton	Hays	Lipscomb	Pecos	Walker
Bowie	Dimmit	Hemphill	Llano	Polk	Waller
Brazoria	Duval	Henderson	Loving	Presidio	Washington
Brazos	Eastland	Hidalgo	Lubbock	Rains	Webb
Brewster	Ellis	Hockley	Madison	Randall	Wheeler
Briscoe	Erath	Hopkins	Marion	Real	Wichita
Burleson	Falls	Houston	Mason	Reeves	Williamson
Burnet	Fannin	Hunt	Matagorda	Robertson	Wise
Caldwell	Fayette	Jasper	Maverick	Rockwall	Wood
Calhoun	Fort Bend	Jeff Davis	McCulloch	Rusk	Zavala
Cass	Franklin	Jim Hogg	McLennan	San Augustine	
Chambers	Freestone	Jim Wells	McMullen	San Jacinto	
Cherokee	Frio	Johnson	Medina	San Patricio	
Clay	Gillespie	Kaufman	Menard	San Saba	
Coleman	Goliad	Kenedy	Midland	Smith	
Collin	Gonzales	Kerr	Milam	Somervell	
Colorado	Grayson	Kimble	Montgomery	Tarrant	
Comal	Grimes	Kinney	Morris	Taylor	

Plathemis subornata Hagen / Desert Whitetail

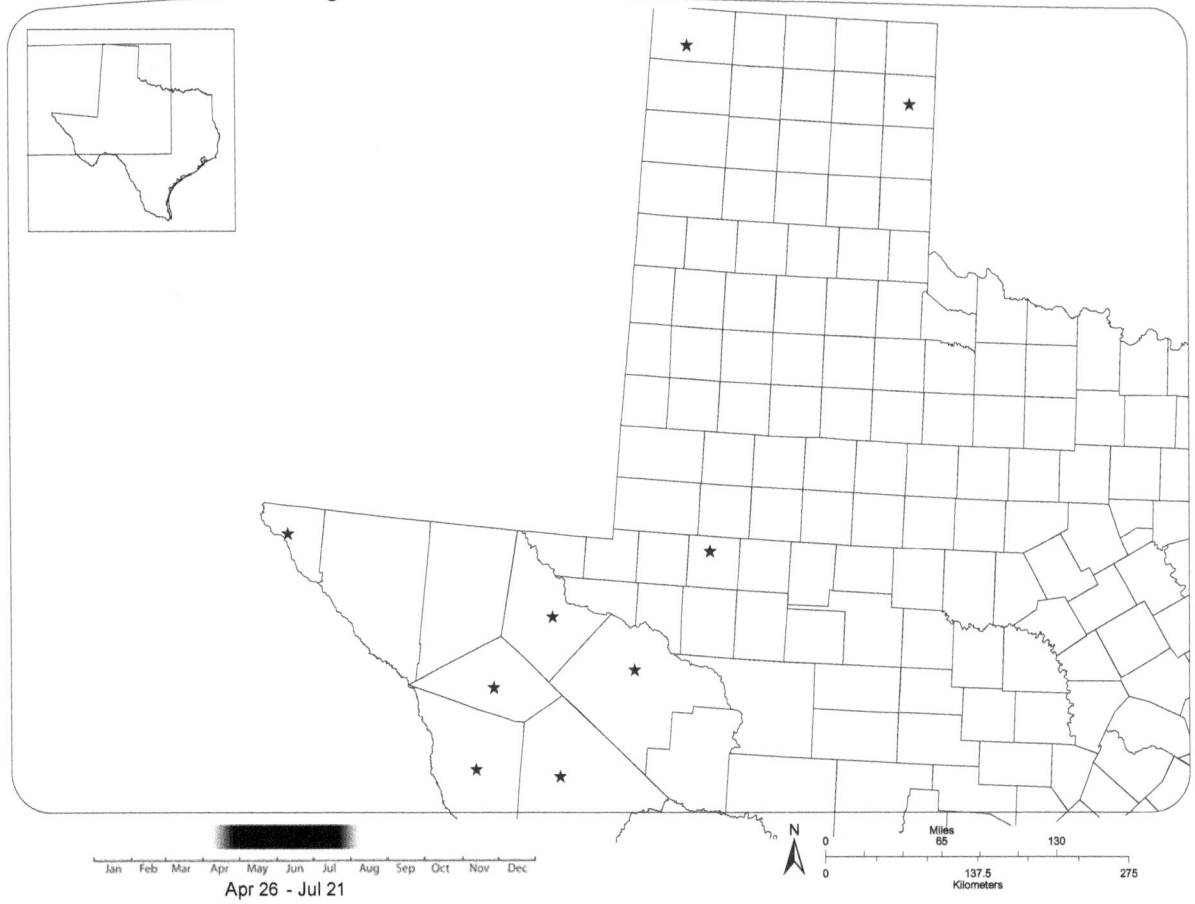

Apr 26 - Jul 21

HABITAT: Desert pools, ponds and slow streams with thick emergent vegetation and mud bottoms.

Brewster
Dallam
El Paso
Hemphill
Jeff Davis
Midland
Pecos
Presidio
Reeves

Pseudoleon superbus (Hagen) / Filigree Skimmer

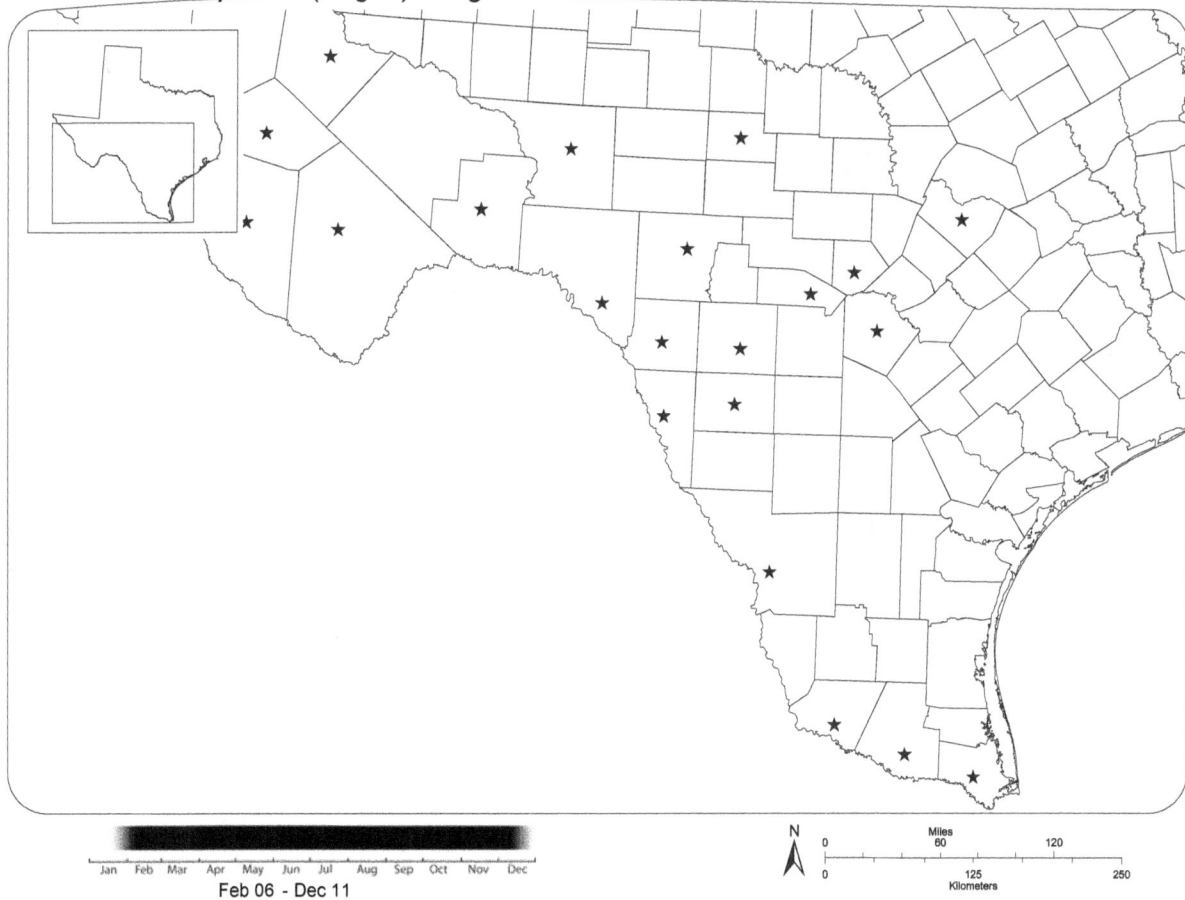

Feb 06 - Dec 11

HABITAT: Desert ponds and slow streams.

Bandera
Bexar
Brewster
Cameron
Crockett
Culberson
Edwards
Hidalgo
Jeff Davis
Kendall
Kinney
Maverick
Menard
Presidio
Reeves
Starr
Terrell
Travis
Uvalde
Val Verde
Webb
Zavala

Sympetrum ambiguum (Rambur) / Blue-faced Meadowhawk

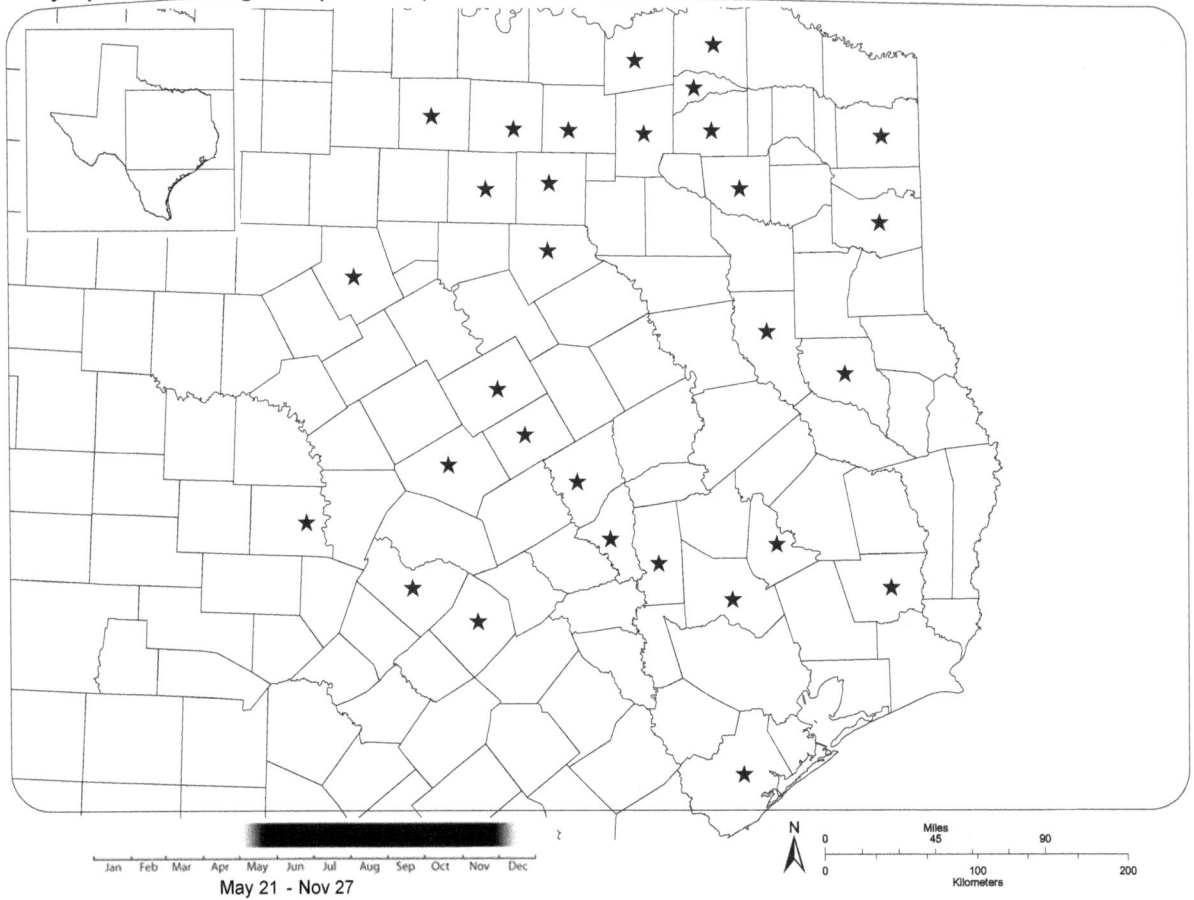

May 21 - Nov 27

HABITAT: Partially shaded temporary and permanent ponds, pools, marshes, swamps and sloughs.

Bastrop	Tarrant
Bell	Travis
Brazoria	Wise
Brazos	Wood
Cass	
Cherokee	
Collin	
Dallas	
Delta	
Denton	
Ellis	
Erath	
Falls	
Fannin	
Grimes	
Hardin	
Harrison	
Hopkins	
Hunt	
Lamar	
Llano	
McLennan	
Montgomery	
Nacogdoches	
Robertson	
San Jacinto	

Sympetrum corruptum (Hagen) / Variegated Meadowhawk

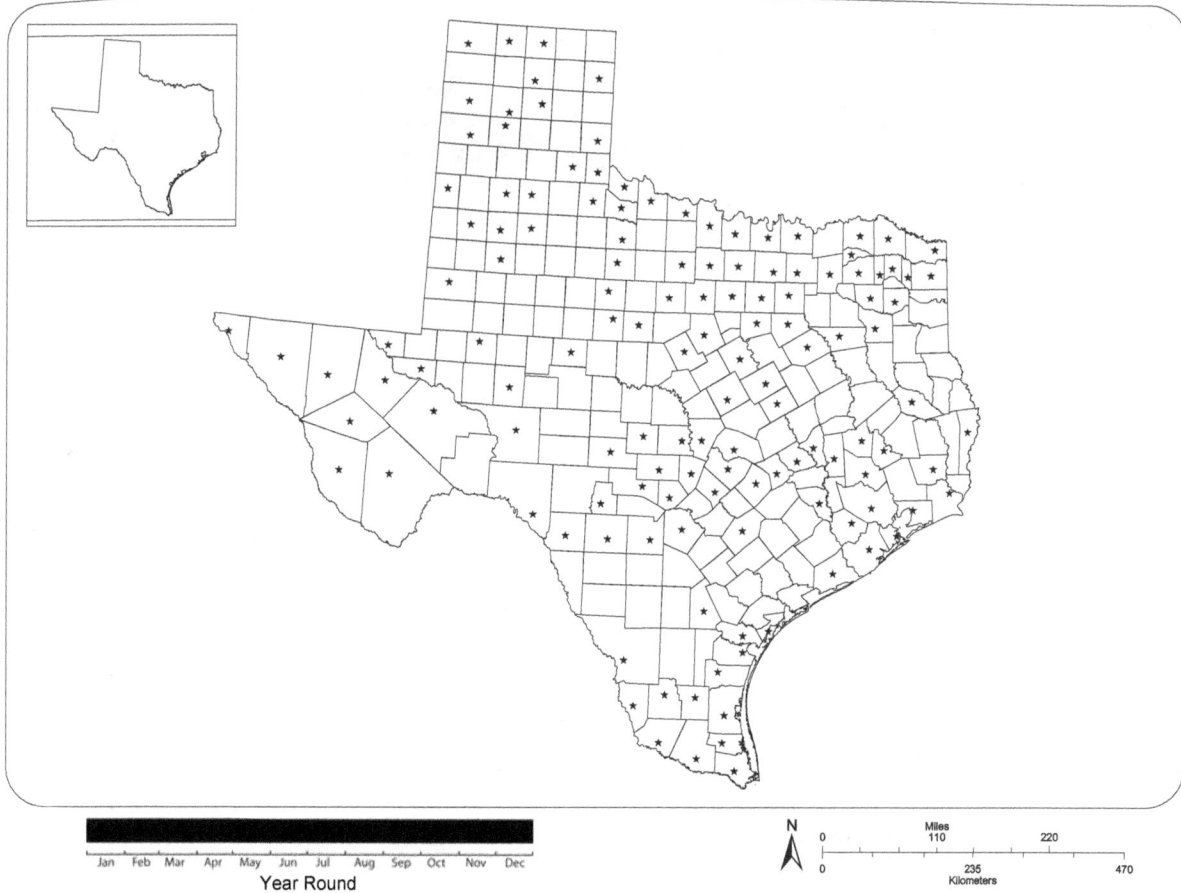

Jan Feb Mar Apr May Jun Jul Aug Sep Oct Nov Dec

Year Round

N

Miles
0　　110　　220
0　　235　　470
Kilometers

HABITAT: Ponds and slow streams, preferably with sandy or cobble bottoms, but occasionally including brackish waters.

Angelina	Cooke	Hall	Kleberg	Randall	Wood
Aransas	Coryell	Hansford	Knox	Reagan	Young
Austin	Cottle	Hardeman	Lamar	Real	Zapata
Bailey	Crockett	Hardin	Lee	Red River	
Bastrop	Crosby	Harris	Live Oak	Reeves	
Bexar	Culberson	Haskell	Llano	San Jacinto	
Blanco	Dallam	Hays	Loving	San Patricio	
Bosque	Dallas	Hemphill	Lubbock	Sherman	
Bowie	Deaf Smith	Henderson	Lynn	Smith	
Brazoria	Delta	Hidalgo	Mason	Starr	
Brazos	Denton	Hockley	Matagorda	Stephens	
Brewster	El Paso	Hopkins	McLennan	Tarrant	
Brooks	Ellis	Hudspeth	Medina	Taylor	
Burleson	Erath	Hunt	Midland	Titus	
Burnet	Falls	Hutchinson	Montague	Travis	
Callahan	Floyd	Jack	Montgomery	Upshur	
Cameron	Foard	Jeff Davis	Morris	Uvalde	
Carson	Fort Bend	Jefferson	Navarro	Val Verde	
Cass	Franklin	Jim Hogg	Newton	Walker	
Chambers	Gaines	Johnson	Nueces	Ward	
Childress	Galveston	Jones	Oldham	Webb	
Clay	Gillespie	Kendall	Palo Pinto	Wichita	
Coke	Gonzales	Kenedy	Parker	Wilbarger	
Collin	Grayson	Kerr	Pecos	Willacy	
Collingsworth	Grimes	Kimble	Potter	Williamson	
Comanche	Hale	Kinney	Presidio	Wise	

Sympetrum illotum (Selys) / Cardinal Meadowhawk

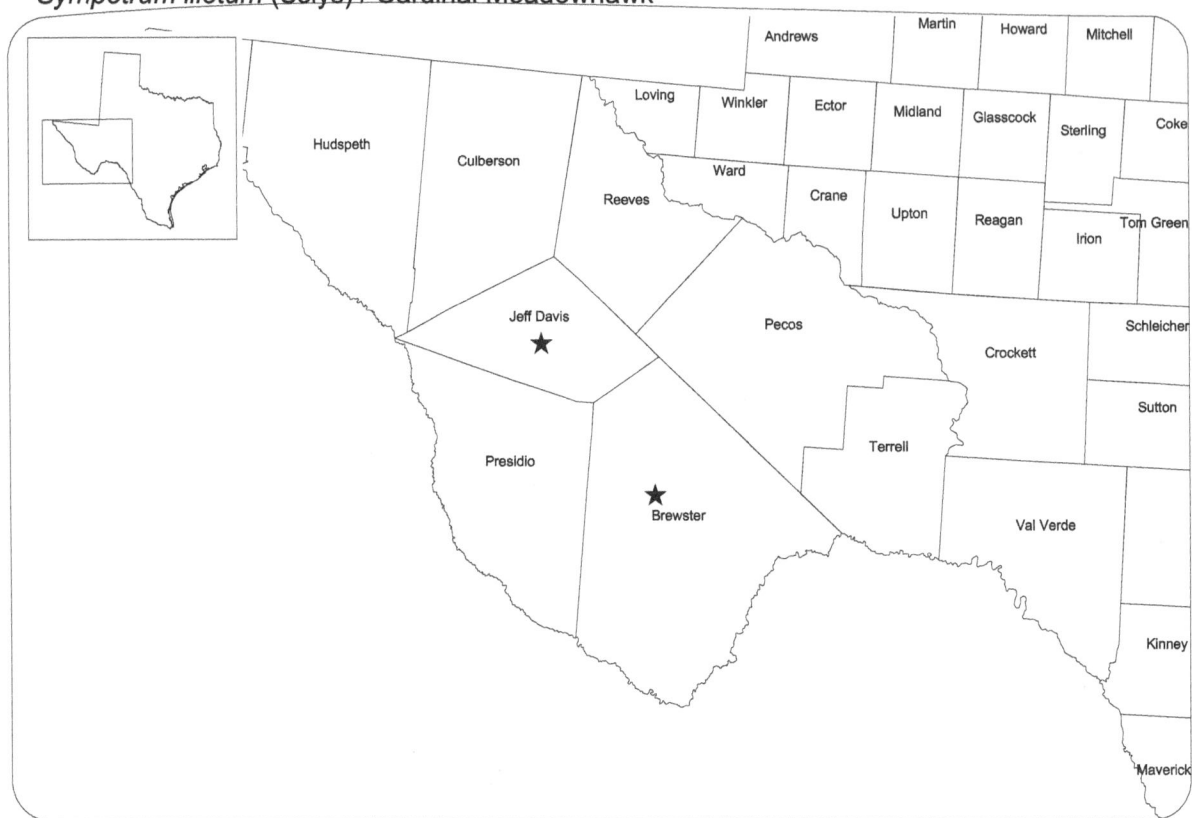

Apr 20 - Sep 13

HABITAT: Small ponds and slow streams.

Brewster
Jeff Davis

Sympetrum internum Montgomery / Cherry-faced Meadowhawk

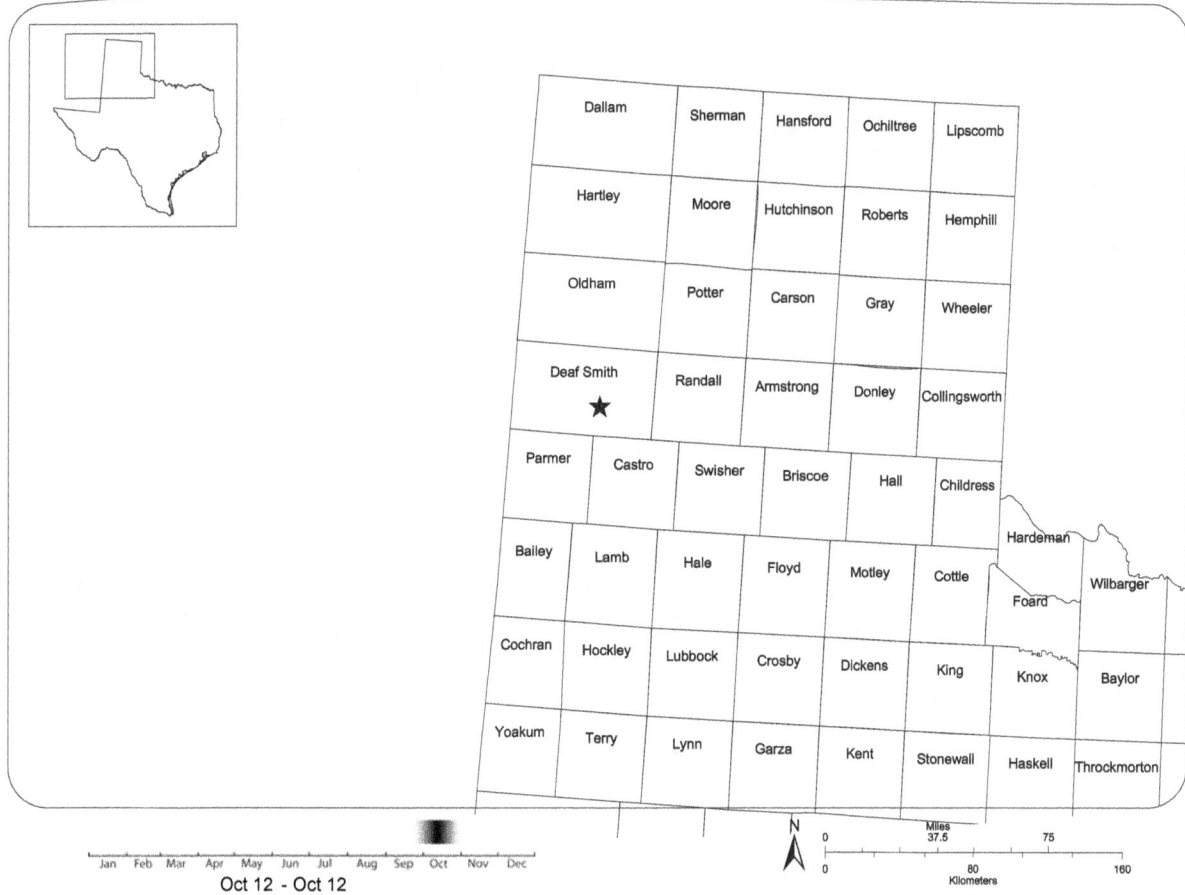

Dallam	Sherman	Hansford	Ochiltree	Lipscomb
Hartley	Moore	Hutchinson	Roberts	Hemphill
Oldham	Potter	Carson	Gray	Wheeler
Deaf Smith ★	Randall	Armstrong	Donley	Collingsworth
Parmer	Castro	Swisher	Briscoe	Hall	Childress
Bailey	Lamb	Hale	Floyd	Motley	Cottle	Hardeman	Wilbarger
						Foard	
Cochran	Hockley	Lubbock	Crosby	Dickens	King	Knox	Baylor
Yoakum	Terry	Lynn	Garza	Kent	Stonewall	Haskell	Throckmorton

Jan Feb Mar Apr May Jun Jul Aug Sep Oct Nov Dec

Oct 12 - Oct 12

Miles 37.5 75

Kilometers 80 160

HABITAT: Ponds, pools and slow shady streams.

Deaf Smith

Sympetrum semicinctum (Say) / Band-winged Meadowhawk

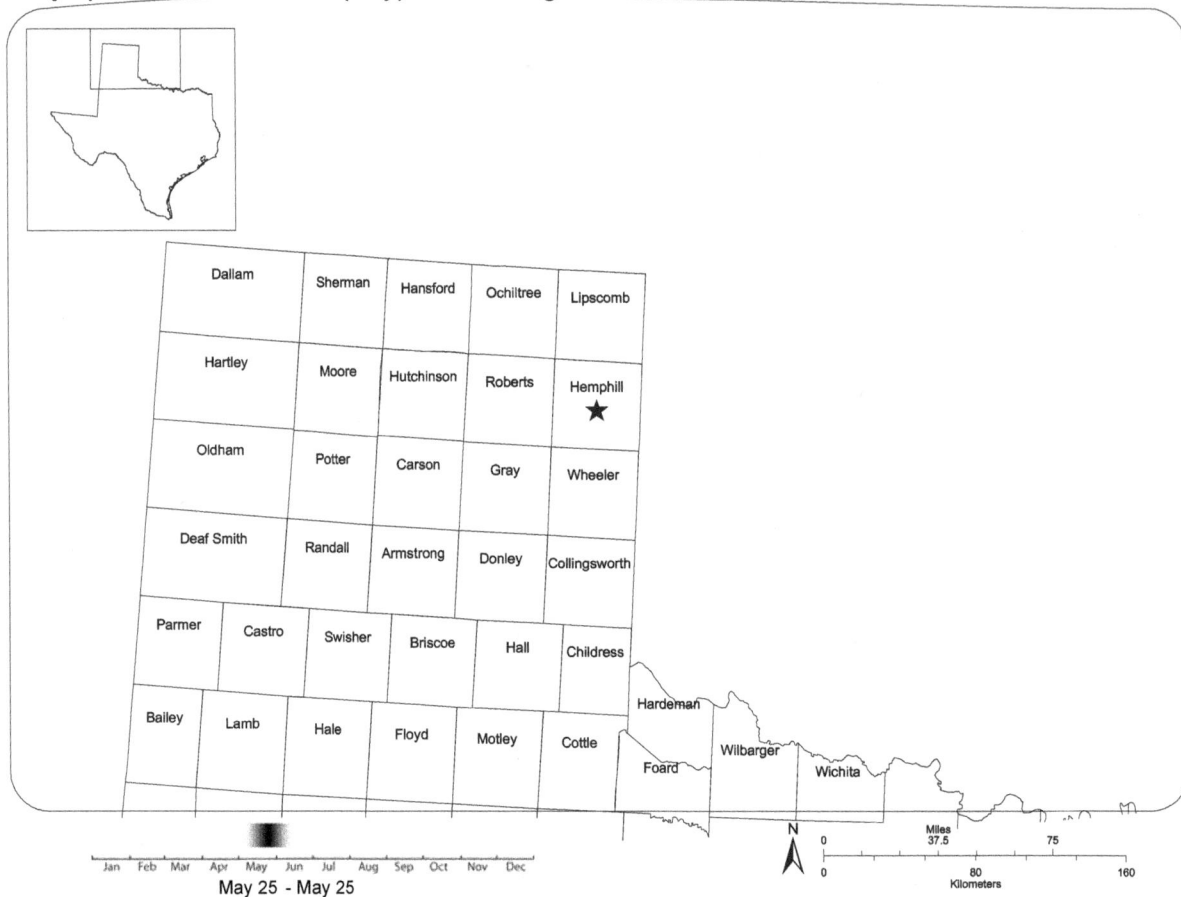

Dallam	Sherman	Hansford	Ochiltree	Lipscomb
Hartley	Moore	Hutchinson	Roberts	Hemphill ★
Oldham	Potter	Carson	Gray	Wheeler
Deaf Smith	Randall	Armstrong	Donley	Collingsworth
Parmer / Castro	Swisher	Briscoe	Hall	Childress
Bailey / Lamb	Hale	Floyd	Motley	Cottle

Hardeman

Foard

Wilbarger

Wichita

Jan Feb Mar Apr May Jun Jul Aug Sep Oct Nov Dec

May 25 - May 25

Miles
37.5   75

N

0   80   160
Kilometers

HABITAT: Marshy, sometimes spring fed, muddy bottomed ponds, sloughs and swamps.

Hemphill

Sympetrum vicinum (Hagen) / Autumn Meadowhawk

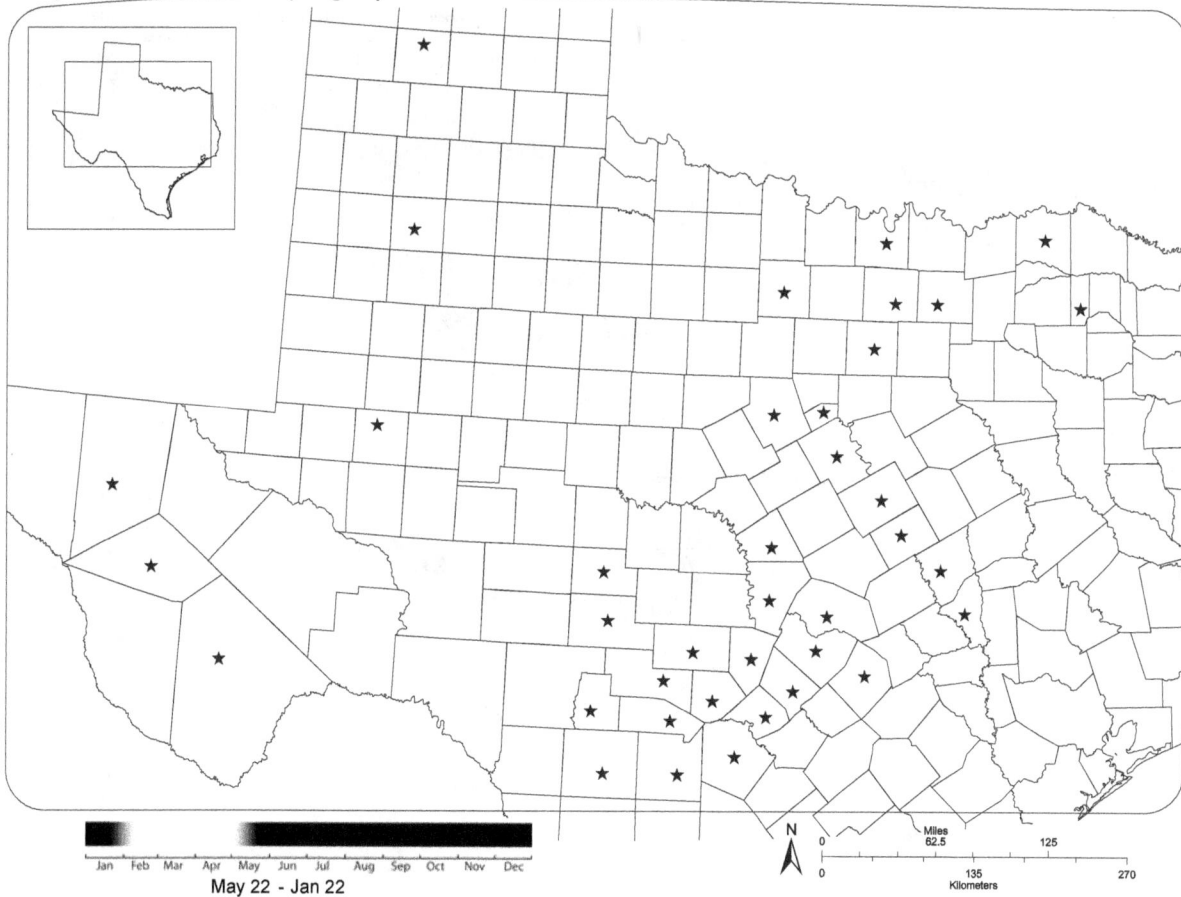

May 22 - Jan 22

HABITAT: Permanent ponds and slow flowing streams.

Bandera	McLennan
Bastrop	Medina
Bexar	Menard
Blanco	Midland
Bosque	Randall
Brazos	Real
Brewster	Robertson
Burnet	Somervell
Collin	Tarrant
Comal	Travis
Cooke	Uvalde
Culberson	Williamson
Denton	
Erath	
Falls	
Franklin	
Gillespie	
Hays	
Jack	
Jeff Davis	
Kendall	
Kerr	
Kimble	
Lamar	
Lampasas	
Lubbock	

Tauriphila azteca Calvert / Aztec Glider

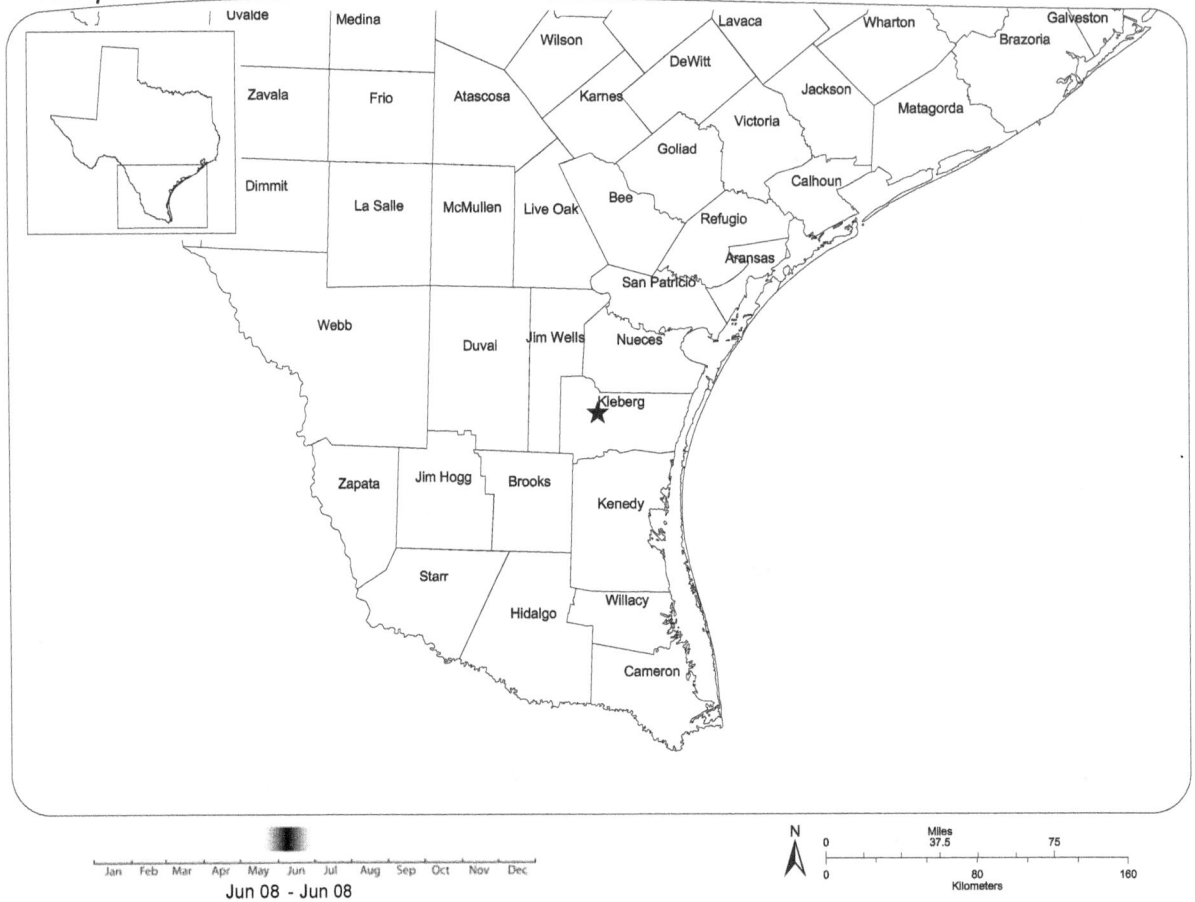

Jun 08 - Jun 08

HABITAT: Slow, calm waters with emergent or floating vegetation.

Kleberg

Tholymis citrina Hagen / Evening Skimmer

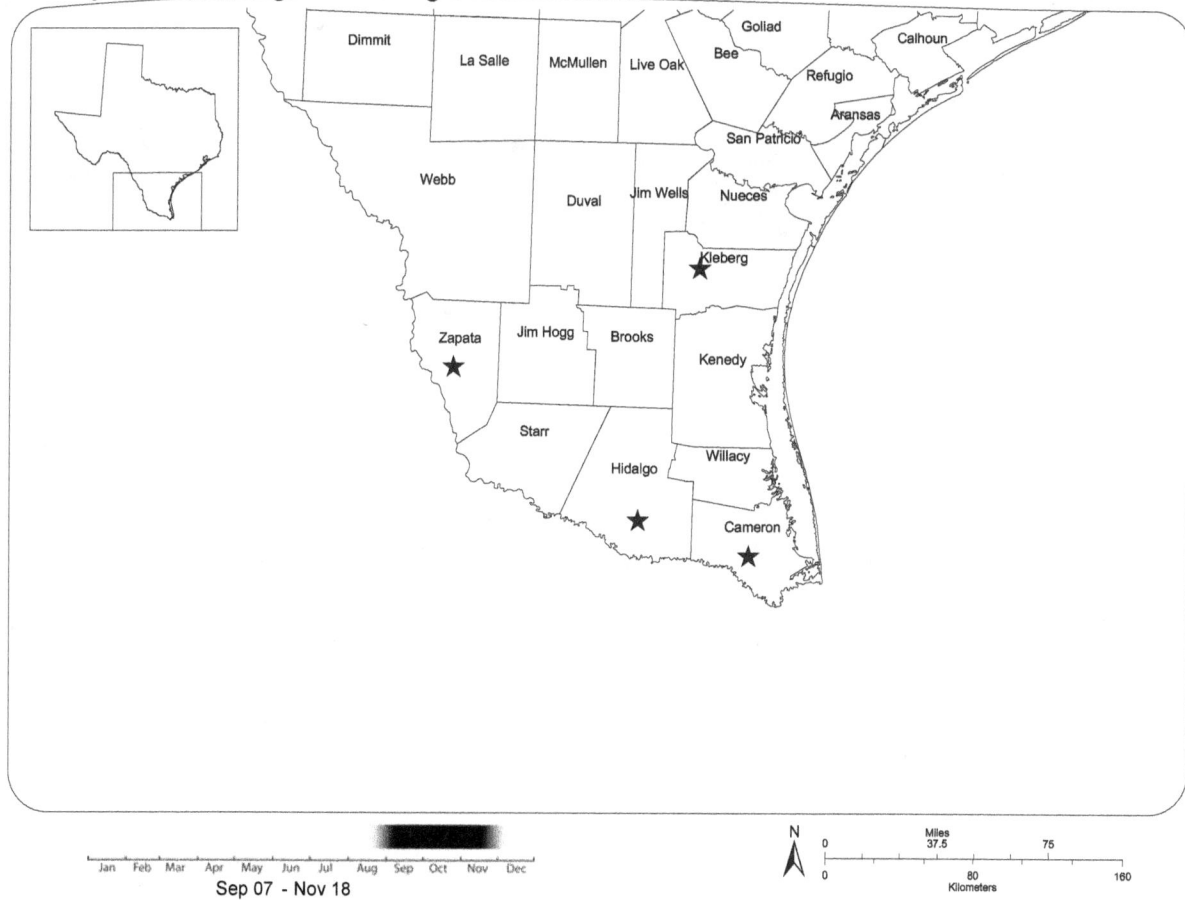

Sep 07 - Nov 18

HABITAT: Vegetated ponds and lakes

Cameron
Hidalgo
Kleberg
Zapata

Tramea abdominalis (Rambur) / Vermilion Saddlebags

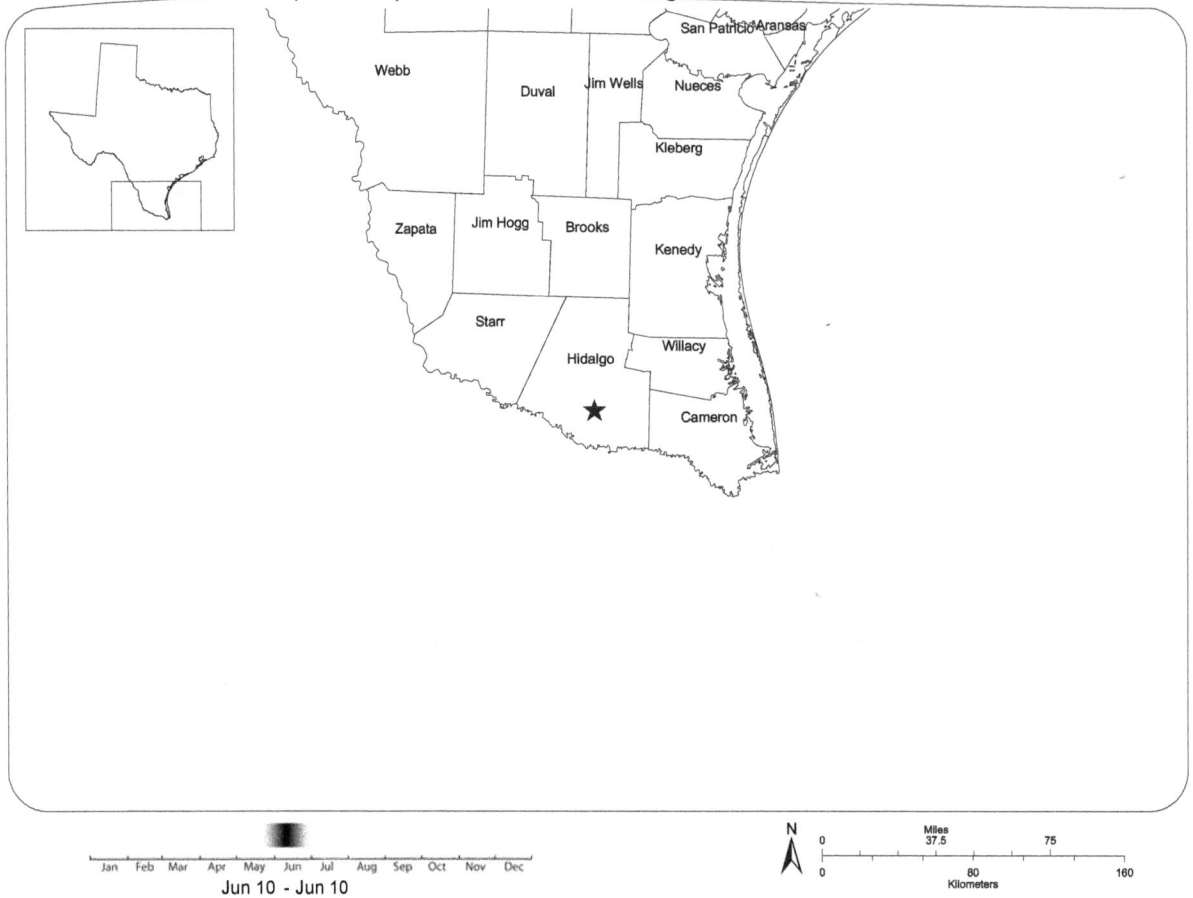

Jun 10 - Jun 10

HABITAT: Permanent and temporary ponds, lakes and slow streams.

Hidalgo

Tramea calverti Muttkowski / Striped Saddlebags

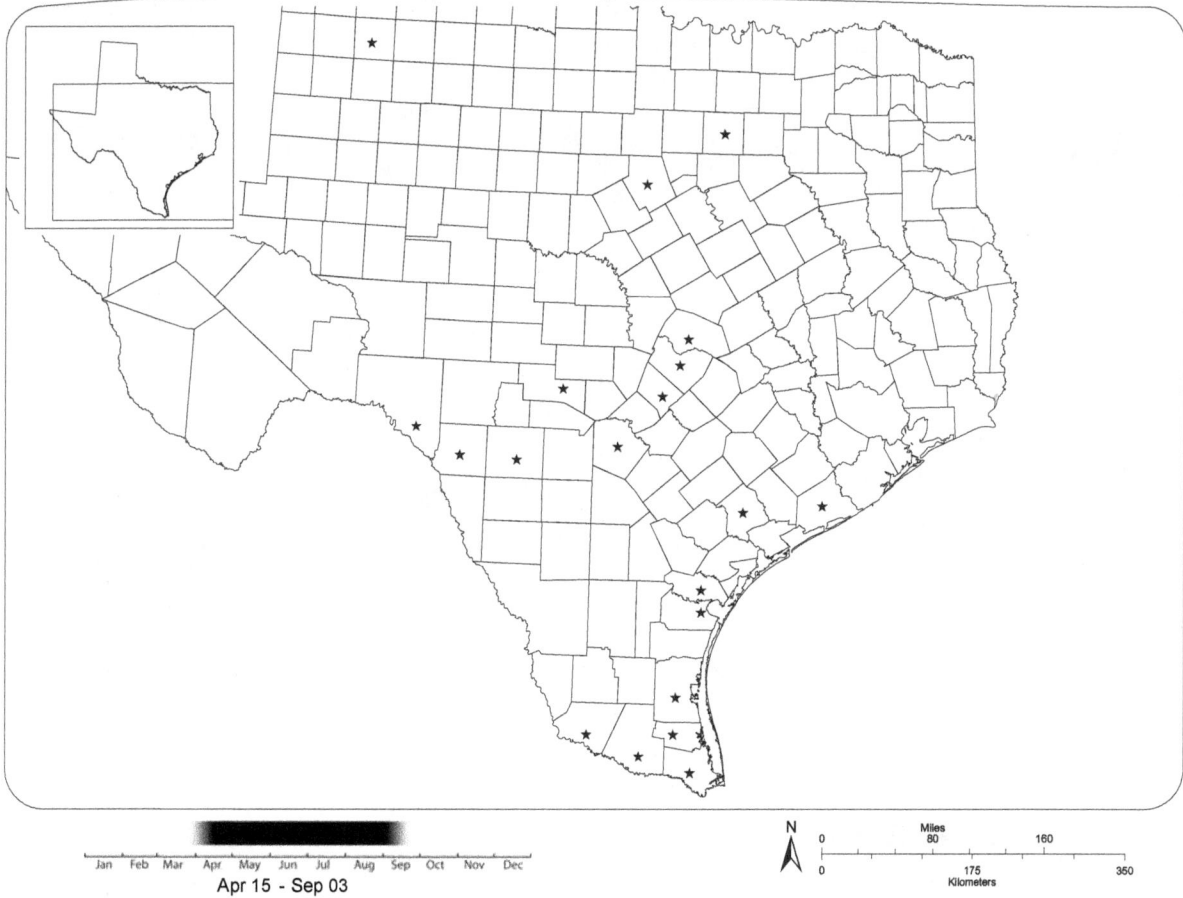

Apr 15 - Sep 03

HABITAT: Temporary and permanent ponds and slow streams.

Bexar
Cameron
Erath
Hays
Hidalgo
Kenedy
Kerr
Kinney
Lubbock
Matagorda
Nueces
San Patricio
Starr
Tarrant
Travis
Uvalde
Val Verde
Victoria
Willacy
Williamson

Tramea carolina (Linnaeus) / Carolina Saddlebags

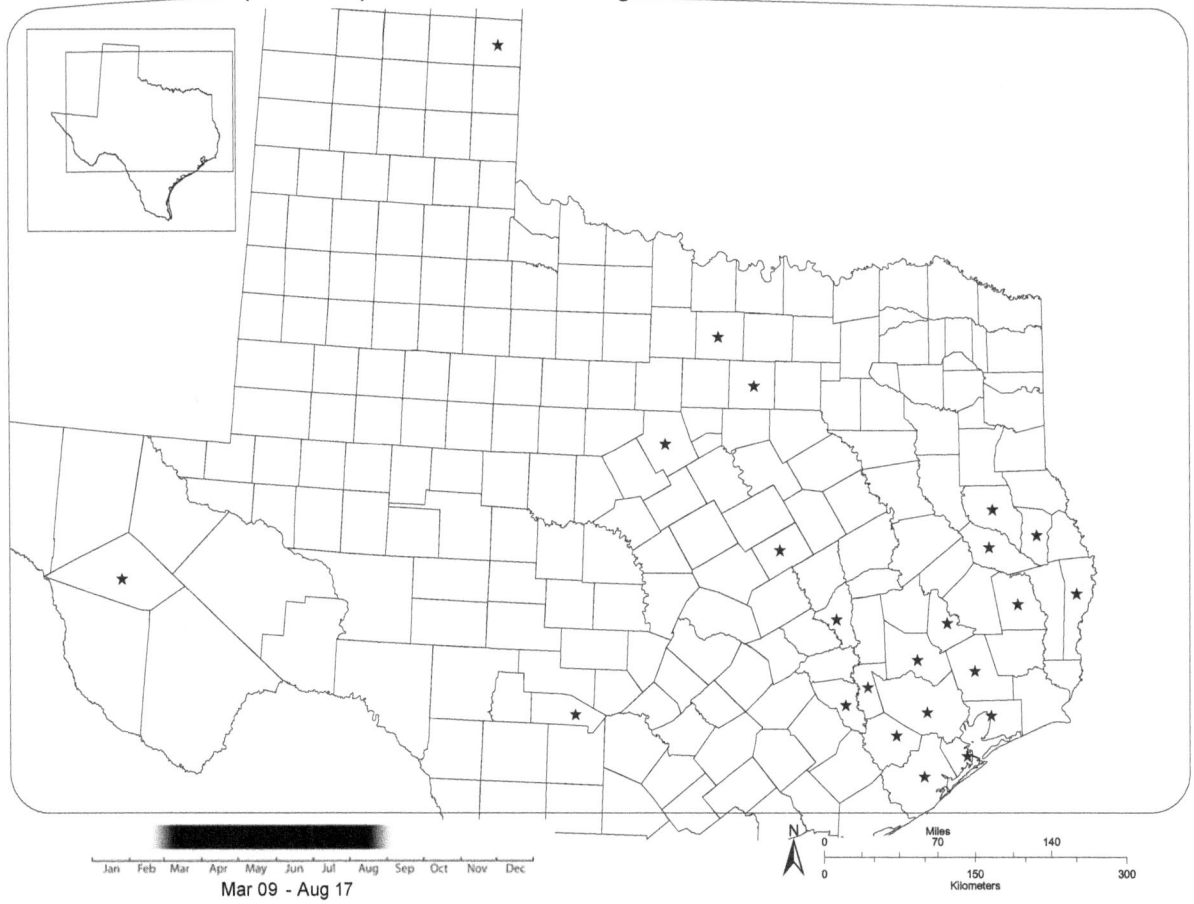

Mar 09 - Aug 17

HABITAT: Ponds, lakes and slow streams with thick emergent vegetation.

Angelina
Austin
Bandera
Brazoria
Brazos
Chambers
Erath
Falls
Fort Bend
Galveston
Harris
Hemphill
Jeff Davis
Liberty
Montgomery
Nacogdoches
Newton
San Augustine
San Jacinto
Tarrant
Tyler
Waller
Wise

Tramea insularis Hagen / Antillean Saddlebags

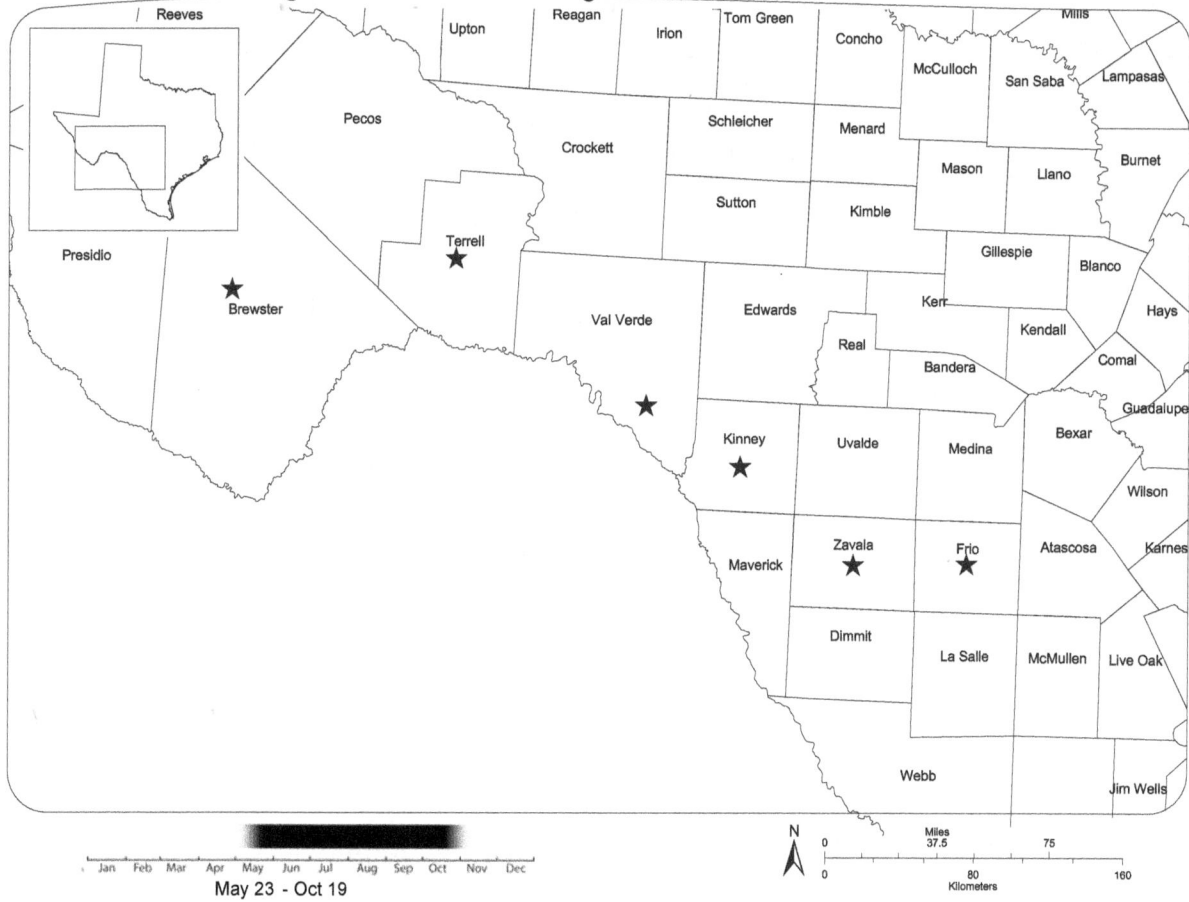

May 23 - Oct 19

HABITAT: Ponds, lakes and slow streams.

Brewster
Frio
Kinney
Terrell
Val Verde
Zavala

Tramea lacerata Hagen / Black Saddlebags

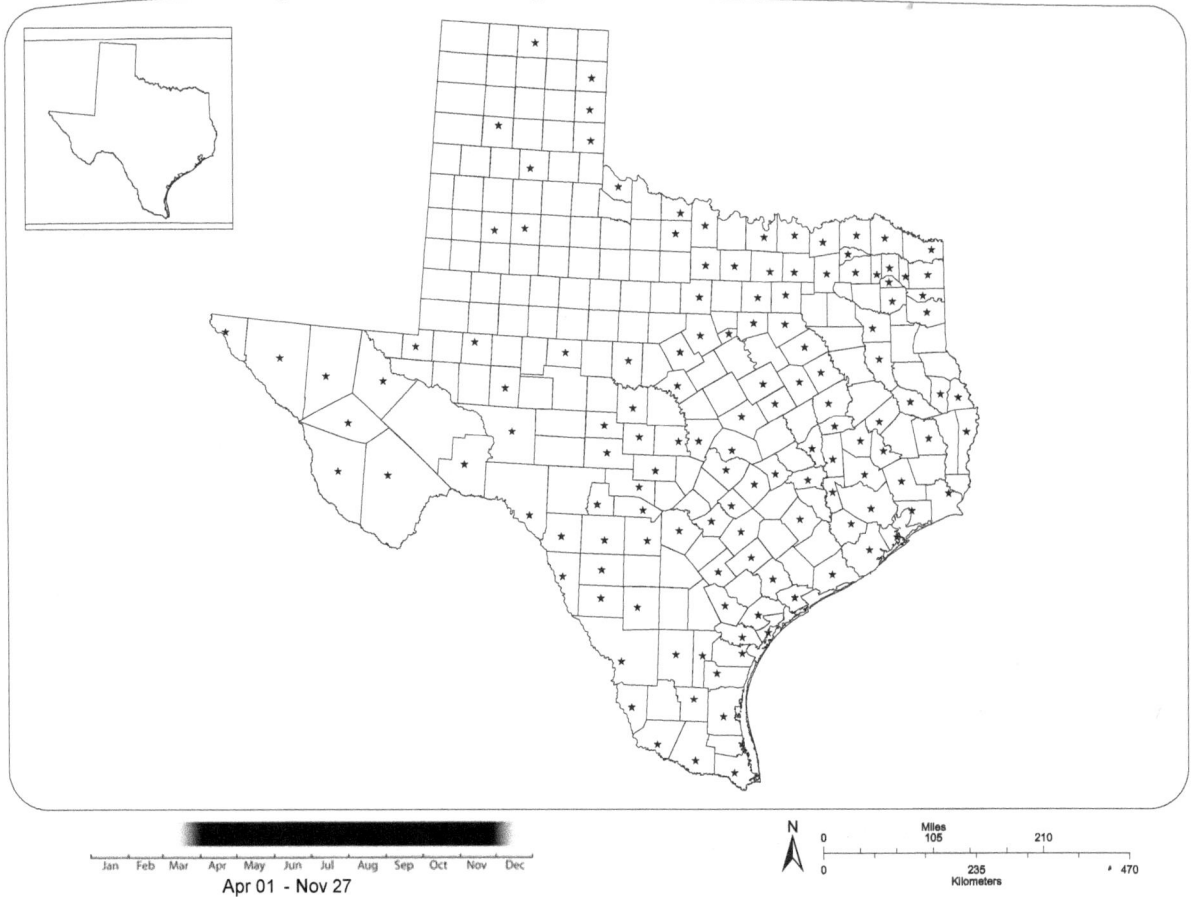

Jan Feb Mar Apr May Jun Jul Aug Sep Oct Nov Dec

Apr 01 - Nov 27

HABITAT: Marshy ponds, lakes, ditches and slow streams.

Angelina	Collingsworth	Guadalupe	Limestone	Refugio	Wise
Aransas	Colorado	Hansford	Llano	Sabine	Zapata
Archer	Comanche	Hardeman	Lubbock	San Augustine	Zavala
Bandera	Cooke	Harris	Madison	San Jacinto	
Bastrop	Crockett	Harrison	Marion	San Patricio	
Bee	Crosby	Hemphill	Mason	Smith	
Bell	Culberson	Hidalgo	Matagorda	Somervell	
Bexar	Dallas	Hopkins	Maverick	Starr	
Bowie	Delta	Hudspeth	McCulloch	Tarrant	
Brazoria	Denton	Hunt	McLennan	Terrell	
Brazos	DeWitt	Jack	Medina	Titus	
Brewster	Dimmit	Jeff Davis	Menard	Travis	
Briscoe	Duval	Jefferson	Midland	Trinity	
Brooks	El Paso	Jim Wells	Mills	Tyler	
Burnet	Ellis	Johnson	Montgomery	Upshur	
Caldwell	Erath	Karnes	Morris	Uvalde	
Calhoun	Falls	Kenedy	Navarro	Val Verde	
Cameron	Fannin	Kerr	Newton	Victoria	
Camp	Fort Bend	Kimble	Nueces	Walker	
Cass	Franklin	Kinney	Palo Pinto	Waller	
Chambers	Freestone	Kleberg	Presidio	Washington	
Cherokee	Galveston	La Salle	Randall	Webb	
Clay	Gillespie	Lamar	Reagan	Wheeler	
Coke	Gonzales	Lee	Real	Wichita	
Coleman	Grayson	Leon	Red River	Williamson	
Collin	Grimes	Liberty	Reeves	Winkler	

Tramea onusta Hagen / Red Saddlebags

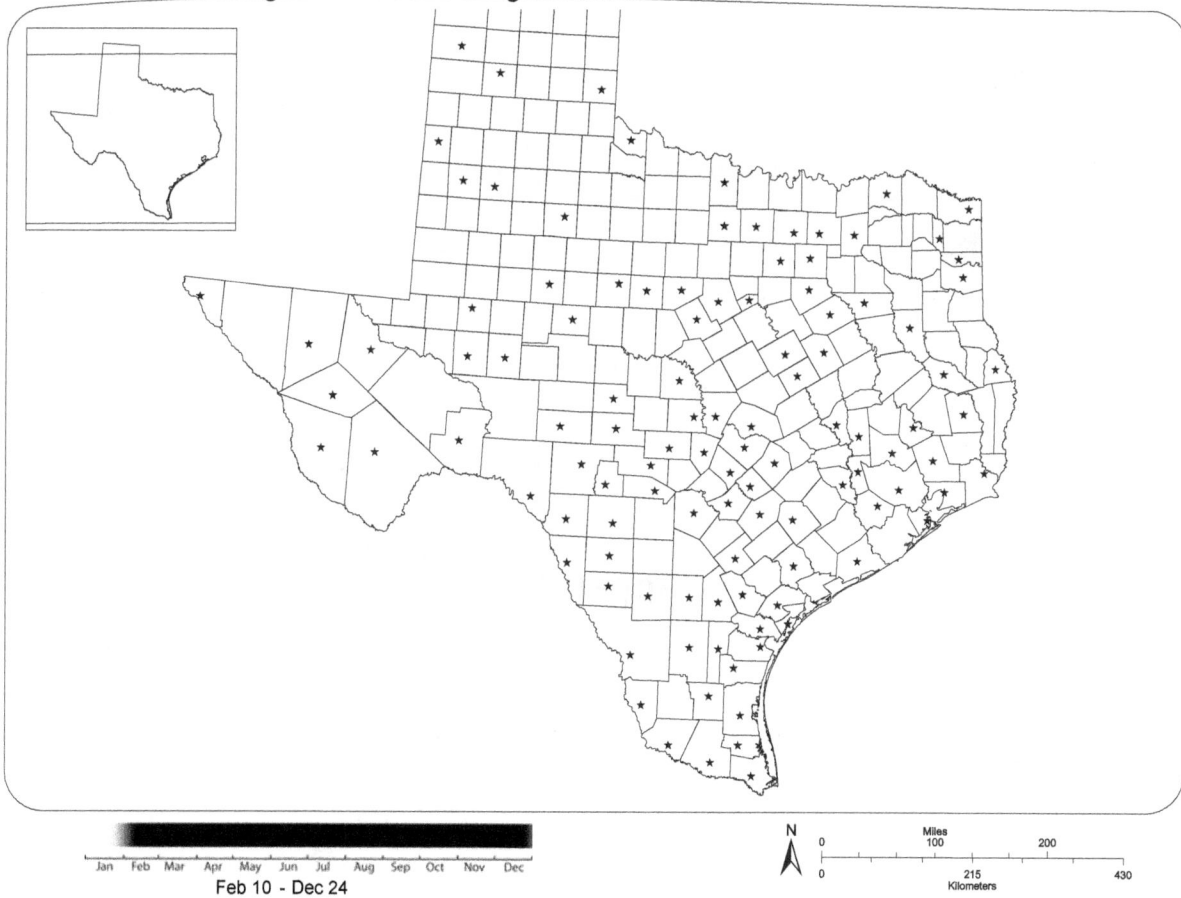

Feb 10 - Dec 24

HABITAT: Permanent and temporary ponds, lakes and slow streams.

Angelina	Denton	Jim Wells	Navarro	Webb
Aransas	Dimmit	Karnes	Nueces	Willacy
Austin	Duval	Kenedy	Oldham	Williamson
Bailey	Eastland	Kent	Presidio	Wise
Bandera	Edwards	Kerr	Randall	Zapata
Bastrop	El Paso	Kimble	Reagan	Zavala
Bee	Ellis	Kinney	Real	
Bexar	Erath	Kleberg	Reeves	
Blanco	Falls	La Salle	Refugio	
Bowie	Fort Bend	Lamar	Sabine	
Brazos	Galveston	Lavaca	San Jacinto	
Brewster	Gillespie	Liberty	San Patricio	
Brooks	Gonzales	Limestone	San Saba	
Burnet	Grimes	Live Oak	Somervell	
Caldwell	Guadalupe	Llano	Starr	
Callahan	Hardeman	Lubbock	Sutton	
Cameron	Harris	Marion	Tarrant	
Chambers	Harrison	Matagorda	Taylor	
Cherokee	Hays	Maverick	Terrell	
Clay	Henderson	McLennan	Travis	
Coke	Hidalgo	McMullen	Tyler	
Collin	Hockley	Menard	Upton	
Collingsworth	Hunt	Midland	Uvalde	
Comanche	Jack	Mitchell	Val Verde	
Culberson	Jeff Davis	Montgomery	Victoria	
Dallas	Jefferson	Morris	Waller	

Appendix

Collection Guidelines for the
Odonata Survey of Texas

John C. Abbott

The Odonata Survey of Texas (OST) is dedicated to providing a better understanding of the distribution, biology, and taxonomy of the dragonfly and damselfly fauna of Texas. To this end, many OST participants will be collecting in a wide variety of habitats across the state of Texas. Though high-quality photographs are often adequate to document a species, collected specimens are the preferred way to voucher the presence of a species at a particular locale. The collection of specimens may comprise adults, nymphs, exuviae, or all three life stages. Collections form the basis of a permanent record for the odonate fauna in a given place and time. The specimens and associated data can then be checked in perpetuity for authentication and verification. A collection of specimens is also invaluable for future taxonomic studies. My hope is that many members of the OST will make serious efforts to document the fauna of Texas through collections. With this in mind, I present the following to form the basis of the protocols for the OST.

Members collecting for the OST are encouraged to not collect any more specimens than necessary to adequately document a species in a given area. An exception to this would be the collection of exuviae. However, it is important to recognize that there are many similar-looking species in certain genera, such as *Argia, Enallagma*, and *Gomphus*, and more extensive sampling be required in these and other groups. No more than a single specimen is required for species that are unquestionably recogniz-able, such as Ebony Jewelwing (*Calopteryx maculata*). Collections of larvae and adults should always be limited to representative sampling so that depletion of populations in a area does not occur.

Photographic documentation is also acceptable for species that are readily recognizable. Photographs should be taken at the highest quality possible and take numerous photographs from different angles whenever possible. All photographs should be submitted through OdonataCentral (http://www.odonatacentral.com) to be included in the survey. Photographs of specimens in hand are perfectly acceptable as long as the species is recognizable.

Sight records may be accepted in cases where species are unmistakable and when made by trained, experienced, reliable observers. It is important to maintain a high-level of scientific integrity with this survey and appropriate documentation of records is paramount. Document your sightings thoroughly providing as much detail about locality, appearance and behavior of the individual(s) observed. These sight records, along with the accompanying documentation, should be submitted through OdonataCentral and will be evaluated for acceptance in the survey.

The OST requests that participants follow the Dragonfly Society of the Americas Guidelines for Collecting.

The Dragonfly Society of the Americas
Guidelines for Collecting

PREAMBLE

Our ethical responsibility to assess and preserve natural resources, for the maintenance of biological diversity in perpetuity, and for the increase of knowledge requires that Odonatologists examine the rationale and practices of collecting Odonata, for the purpose of governing their own activities. While we recognize that historically most threats to preservation of odonate species have been a consequence of habitat destruction, we believe that there is a need for responsible collecting practices. To this end, the following guidelines are outlined, based on these premises:

0.1. Odonata are a natural resource.

0.2. Any human interaction with a natural resource (e.g. Odonata and their environment) should be in a manner not harmful to the perpetuation of that resource.

0.3. The collection of Odonata:

0.3.1 is a means of introducing children and adults to awareness and study of their natural environment;

0.3.2 has an essential role in gathering of scientific information including the advancement of taxonomic knowl edge, both for its own sake and as a basis from which to develop rational means for protecting the environment, and maintaining the health of the bio sphere;

0.3.3 is an enjoyable educational or scientific activity which can generally be pursued in a manner not detrimental to the resource (e.g. Odonata and their environment) involved.

GUIDELINES

Purposes of Collecting:
(consistent with the above)

1.1 To create a reference collection for study and appreciation.

1.2 To document regional diversity, frequency and variability of species, and as voucher material for published records.

1.3 To document faunal representation in environments undergoing or threatened with alteration by human or natural forces.

1.4 To participate in development of regional checklists and institutional reference collections.

1.5 To complement a planned research endeavor.

1.6 To aid in dissemination of educational information.

1.7 To provide material for taxonomic studies.

1.8 To provide information for ecological studies.

Restraints As To Numbers:

2.1 Collection (of adults or of immature stages) should be limited to sampling, not depleting, the population concerned; numbers collected should be consistent with, and not excessive for, the purpose of the collecting.

2.2 When collecting where the extent and/or fragility of the population is unknown, caution and restraint should be exercised.

Collecting Methods:

3.1 Field collecting should be selective and should minimize harm to non-target organisms.

Live Material:

4.1 Rearing to elucidate life histories and to obtain series of immature stages and adults is encouraged, provided that collection of the rearing stock is in keeping with the guidelines.

4.2 Reared material in excess of need should be released, but only in the region where it originated, and in suitable habitat.

4.3 Because of such concerns as introduction of disease and adverse redistribution of genetic resources, release of excess reared material is not encouraged unless it is done in conjunction with a planned restoration program, and under supervision of knowledgeable biologists.

Environmental and Legal Considerations:

5.1 Protecting the supporting habitat must be recognized as essential to the protection of a species.

5.2 Collecting should be performed in a manner such as to minimize trampling or other damage to the habitat.

5.3 Property rights and sensibilities of others must be respected (including those of nature photographers and observers).

5.4 All collecting must be in compliance with regulations relating to public lands (such as state and national parks, monuments, recreational areas, etc.) and to individual species and habitats.

5.5 Importation and movement of exotic species must be in compliance with international, national, or regional laws prior to importing live or dead material.

Responsibility For Collected Material:

6.1 All material should be preserved with full data attached, including parentage of immatures when known.

6.2 All material should be protected from physical damage and deterioration, as by light, molds, and museum pests.

6.3 Collections should be made available for examination by qualified researchers.

6.4 Collections or specimens, and their associated written, electronic, photographic and other records, should be willed or offered to the care of an appropriate scientific institution, if the collector lacks space or loses interest, or anticipates death.

6.5 Type specimens, especially holotypes or allotypes, should be deposited in appropriate institutions.

Related Activities Of Collectors:

7.1 Collecting should include permanently recorded field notes regarding habitat, conditions, and other pertinent information.

7.2 Recording of observations of behavior and of biological interactions should be encouraged and receive as high a priority as collecting.

7.3 Photographic records, with full data, are also encouraged.

7.4 Education of the public about collecting and conservation, as reciprocally beneficial activities, should be undertaken whenever possible.

Traffic In Odonata Specimens:

8.1 Collections of specimens for exchange should be performed in accordance with these guidelines.

8.2 Rearing of specimens for exchange should be from stock obtained in a manner consistent with these guidelines, and so documented.

8.3 The sale of individual specimens or the mass collection of Odonata for commercial purposes (e.g. fish bait), and collection or use of specimens for creation of salable artifacts, are not included among the purposes of the Dragonfly Society of the Americas.

For more information about the Dragonfly Society of the Americas, and how to join DSA, write:

DSA, c/o Jerrell Daigle
2067 Little River Lane
Tallahassee, FL 32311

Specific Collecting & Preserving Instructions

John C. Abbott

COLLECTION

Always gather collection information in a field notebook and ensure that it is easily related to each specimen (see "Guidelines for Field Notes and Data Recording"). Record exact locality (including, if possible, lat/long or UTM coordinates), date, time, collector's name, and any pertinent habitat and behavioral information — data on the type of water body, dominant plants, reproductive behavior, etc. are extremely useful.

Larvae:
Collect live larvae by sweeping aquatic habitats with a dip net. Dragonfly larvae are best kept in containers with a bit of wet moss or other vegetation to keep the humidity high. Once you are back home, drop them in water that has been brought to a boil, but removed from the heat source; leave them for 30 seconds, remove and place on paper toweling to remove excess water. Store in 70% ethanol.

This heating coagulates the protein (preventing the disintegration of internal tissues) and preserves the color pattern much better than placement directly into alcohol does. Hot water can damage the caudal gills of damselflies, however. I put damselfly larvae directly into 70% ethanol in the field, then replace it with fresh ethanol after returning home.

Exuviae (cast larval skins) should also be collected from aquatic plants, rocks, logs and other supports. I prefer to keep exuviae in 70% ethanol. You can also keep them dry and pin them through the base of the wing cases, placing a small drop of white glue at the point where the pin exits the base of the thorax ventrally.

Adults:
Use a long-handled aerial net. A net opening of at least 18" makes the job easier (see BioQuip for purchasing a net). Some collectors feel that a dark net bag (black or green) is less conspicuous to dragonflies, and thus is more effective than a white one. White nets however, make it easier to find the specimen in the net and I generally prefer white because I have found it actually attracts certain groups, namely gomphids (club tails). A wide mesh (but small enough to hold the smallest specimens) is preferred because this reduces air resistance and allows a faster swing. A large mesh may not be satisfactory if you're also collecting insects other than dragonflies and damselflies.

Observing patrolling dragonflies before swinging at

them often pays off; positioning yourself in the most advantageous location, especially if it is somewhat concealed, is usually fruitful. When attempting to capture a dragonfly, move deliberately. Refrain from waving the net around; keep it as inconspicuous as possible. Swing at fast-flying agile species from behind as they fly by; many will easily dodge a net swung head-on.

When an adult specimen is captured, place it alive in a glassine envelope (available in several sizes — I prefer 3.125" x 5." These can purchased at local hobby stores or BioQuip. Clip a very small portion of one or both of the bottom corners of the envelope; this will allow the acetone to fill and drain from the envelope easier (see below). The wings should be together above the abdomen. In general, put only one specimen in each envelope, otherwise they can damage each other. However, place pairs caught in tandem or in copula in the same envelope if possible; face them away from each other. If they are too large to go together in a single envelope, make certain that you note they were mating or in tandem on both envelopes. You can write the locality or a collection number (cross-referenced to field notes) right on the envelope in soft pencil, india ink or other ink that is insoluble in acetone (see the acetone treatment describe below).

Teneral (recently emerged) adults are fragile and preserve poorly. Place them in paper bags for a day or two so that the cuticle hardens and the color pattern develops. If collected while emerging, the adult should be placed in a bag and the associated exuviae should be kept and cross-referenced to the adult specimen.

PRESERVATION

While in the field keep the envelopes containing live dragonflies as cool as possible and always out of the sun. Store them in a non-crushable box. Tupperware boxes of the proper size are excellent for this purpose. For convenience, specimens are usually kept alive in the envelopes until the collector returns to base. This also allows time for the specimens to empty their digestive tracts; color is usually preserved better in a specimen with an empty gut. The color pattern of some species, especially some blue ones (e.g. *Aeshna*'s, *Argia*'s & *Enallagma*'s) fades soon after capture. This fading can often be reversed by exposing the specimen to sunlight. However, if possible, such specimens should be treated in acetone immediately.

Acetone treatment:
Acetone dehydrates the specimen and dissolves fat, reducing the decomposition of color pigments. Acetone

is flammable, so use caution. Avoid breathing the fumes and absorbing the liquid through your skin. Acetone can be purchased at most hardware or paint stores. Keep the acetone in a wide-mouth plastic or glass jar or other container inert to the solvent. The wider the mouth, the better, but make sure that the lid is leak-proof. Pack the jar(s) in a box for stability if you take it in an automobile. The dragonfly must be killed and its body arranged in the proper configuration before treating it in acetone, otherwise it may dry with its abdomen curled and wings and legs askew. I prefer to do this by dropping the specimen in the acetone outside of the envelope. As soon as the insect is dead, I remove the specimen and position it in the envelope. The abdomen should be straightened (to make measuring it easier) and the legs should be brought forward (so that they do not obscure the base of the abdomen). The envelope containing the dragonfly then is immersed, on edge, in the acetone. Sometimes you may want to kill dragonflies, especially larger species, by injecting acetone into the thorax and base of abdomen with a hypodermic needle. This introduces the acetone into the muscles and organs faster than simple soaking, and it seems to improve color retention in the larger specimens.

Leave the dragonflies and damselflies in the acetone for about 8-24 hours depending on size. I usually process my specimens at night removing them from the acetone the next morning. You may also find it easier to take the envelopes out for drying when the next day's catch is ready to go into the jar(s). The acetone should be replaced after a few batches have been processed, i.e. when it becomes pale yellow (indicating considerable dissolved fat). Drain the acetone out of the envelopes back into the jar (this is where snipping a little bit off the bottom corners of the envelopes will facilitate drainage). Dry the specimens while still in the envelopes in a well-ventilated place as quickly as possible. This is the critical step. To get the best specimens, they need to be dried as quickly as possible. When possible, I use a dryer such

as a convection oven. If I'm in the field, I simply place the envelopes directly in the sun. The faster the drying occurs, the better the color preservation. A desiccant can be used in an enclosed container to help with the drying. I find it best to leave the specimens in the acetone for a longer period, rather than remove them when I have no way of drying them quickly.

When the envelopes are dry, store them in Tupperware or cardboard boxes that will withstand crushing. Store the envelopes vertically, like a card file, and DO NOT pack them tightly. At this stage specimens can be broken and flattened if jammed too closely together. Tupperware containers are good because their tight seal prevents most pests from attacking your specimens. In humid climates, the inclusion of silica gel desiccating packets helps keep the specimens dry.

If using acetone is impossible, simply dry the specimens as rapidly as possible after they have been killed. Placing the boxes containing specimen envelopes at close range over or under electric lights is helpful for drying. You can also put them on the dashboard of your car in the sun. The faster the drying occurs, the better the color preservation. In the tropics, acetone is invaluable, since air-drying specimens is difficult and specimen damage is common. Acetoned dragonflies may even resist damage by pests (ants, carpet beetles, psocids, mice) more than untreated ones.

Permanent Storage:
Once the specimens arrive at a museum or other collection, they are taken out of the glassine envelopes and stored permanently in clear envelopes made of cellophane, mylar or polypropylene. These can be purchased from the International Odonata Research Institute. The identification and collection data are typed on 3" x 5" cards, which are inserted in the envelopes behind the specimen. The envelopes are then stored like a card file in drawers and cabinets.

Dromogomphus spoliatus ♂
det. J.C. Abbott 2002

TEXAS: Terrell Co.
Oasis Ranch; Independence Creek
Sheffield, 15.1 mi S 11-14.VI.2002
N30.466 W101.800   595 m
J.C. Abbott #1022 & Field Entomology Class

COLLECTION OF BRACKENRIDGE FIELD LABORATORY

Guidelines For Field Notes & Data Recording

John C. Abbott

It is imperative that accurate and complete data be collected with a record whether a specimen or photograph is obtained. A specimen is worthless without its associated data. At a minimum, you should record the following for purposes of the OST.

1. **State** and **County**.
2. **Locality**. Include distance to nearest town or road intersection. If at all possible, take a GPS reading to provide **Latitude** and **Longitude** coordinates. Though these coordinates can be converted, the OST database uses decimal degree format (N 98.4562° W 103.3476°; datum = WGS84).
3. **Habitat**. Stream, river, pond, lake, etc...
4. **Date.** Either spell out the month or use Roman numerals to designate the month. This avoids confusion. The OST database uses the following format DD/MM/YYYY.
5. **Collector.** Your name or whoever collected the specimen.
6. **Collection Number.** It is often very useful to associate a number with a specific collecting event (specific place and time). This can be tracked back to your field notebook where additional information about the collection and location may be obtained. I use a system of my initials followed by continuous numbers(e.g. JCA#2564). Some collectors will include a year in their system (e.g. JCA#2005-43). It is not useful for purposes of the OST to assign an individual number to each specimen.
7. **Notes.** Provide additional biological, behavioral, and ecological information here. This may include, but is not limited to, water temperature, air temperature, specific stream or lake habitat, pH of water, dominant vegetation, and weather conditions.

All of this information can then be submitted along with a photo or specimen to the OST database on OdonataCentral.com

Additional Information on Keeping a Field Notebook: Naturalists have kept field notebooks for centuries. Charles Darwin, on his epic voyage as naturalist on HMS Beagle, kept detailed field notes that stimulated his thinking about evolution and natural selection and formed the basis for much of his subsequent work. Traditionally, field notebooks follow a system developed by Joseph Grinnell and are divided into 3 sections: the Journal, the Species Accounts, and the Catalog. The Species Accounts of the Grinnellian field notebook include information about any species that is of particular interest to you. And if you were working for a museum, collecting specimens in the field, you would keep a personal Catalog, including detailed information about each specimen you collected, in addition to when it was collected, and exactly where it was collected.

The information in a field notebook should be entered in the field, neatness is not the priority here; accuracy is. Notes from indoor laboratories and visits to museum collections should be kept in a separate notebook. There should be an entry in your notebook for every trip you make into the field. Number the pages of your field notebook consecutively in the uppermost, right-hand corner of each page.

Dates should be written in the "continental" format, with day of the month (a number from 1 to 31), followed by month (spelled out or abbreviated as a roman numeral), followed by the year (all 4 digits). The information about the locations you visit should be written so that someone not familiar with the area could return to the exact, same location, using maps and your location description. For location information, reference to the appropriate U.S. Geological Survey map sheet ("quad") is also very useful. Your objective is to record date and location information in the most detailed, least ambiguous way possible so that others can know exactly when, where, and under what conditions your observations were made.

The recommended procedure is to write your notes directly into your field notebook in the field. However, some like to take temporary notes in a second, smaller "field" notebook, and then transcribe these notes to your permanent notebook. Transcription should be done as soon as possible after you leave the field, and always the same day as your field trip. Preferably, use a bound notebook, not loose-leaf and use only permanent, black ink or pencil in your notebook. Modern writing instruments have replaced the original quill pen and permanent, black, India ink used for the field journals of early naturalists. "Pigma" pens and Fiber-tip pens made by "Sharpie," with "ultra fine" points seem to work well and are widely available. Another, suitable pen is the "uni-ball vision micro," made by Sanford, with waterproof/fade-proof ink. To test for waterproof qualities, scribble a few notes on water-proof paper and hold it under running water for a few minutes (or let it soak in water overnight). If the ink disappears from the paper, or becomes too faint to read, it is not suitable for our purposes. Ball-point pens are not acceptable, since their ink tends to fade with age and they sometimes fail to work when air temperature is below freezing.

In keeping a field journal, your objective is to create
a permanent, accurate, written record of your field
activities and observations.

Reference

Herman, Steven G. 1986. The Naturalist's Field Journal.
Buteo Books, Vermillion, SD. 200 pp.

THE PALEST INK IS BETTER
THAN THE BEST MEMORY
(Old Chinese Proverb)

Dragonflies and Damselflies of Texas and the South-Central United States:
Texas, Louisiana, Arkansas, Oklahoma, and New Mexico

John C. Abbott

This is the first guide to dragonflies and damselflies of the south-central United States. The book covers 263 species, epresenting more than half of the North American fauna. The area of coverage significantly overlaps with other regions of the country making this book a useful aid in identifying the dragonflies and damselflies in any part of the United States, Canada, or northeastern Mexico.

More photographs of damselflies in North America appear here than in any other previously published work. All 85 damselfly and 178 dragonfly species found in the region are distinguished by photographs, numerous line drawings, keys, and detailed descriptions to help with identifications.

Features include:

- Discussions of habitats, zoogeography, and seasonality
- Details on dragonfly and damselfly life history and conservation
- An introduction on studying and photographing dragonflies and damselflies
- An entire section devoted to the external anatomy of dragonflies and damselflies
- Species accounts organized by family into sections on size, regional and general distribution, flight season, identification, similar species, habitat and biology and ecology
- Range maps for each species, as well as an extensive bibliography and a list of resources for further study

John C. Abbott is Curator of Entomology for the Texas Natural Science Center and Senior Lecturer in the Section of Integrative Biology at The University of Texas, Austin.

Paper | 2005 | $35.00 / £22.95 | ISBN: 0-691-11364-5
Cloth | 2005 | $79.50 / £51.95 | ISBN: 0-691-11363-7
424 pp. | 7 3/8 x 9 1/4 | 384 color photographs. 32 line illus. 6 tables. 263 maps.

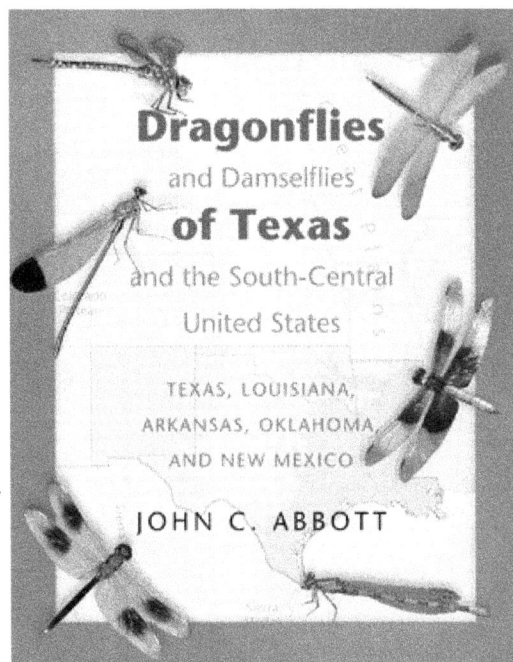

Celebrating 100 Years of Excellence
PRINCETON
University Press 1905-2005

Odonata Field Guides, References, Societies, & Suppliers

References & Field Guides

Abbott, J.C. 2005. Dragonflies and Damselflies of Texas and the South-central United States: Texas, Louisiana, Arkansas, Oklahoma and New Mexico. Princeton University Press. 344 pp.

Abbott, J.C. 2006. Dragonflies and Damselflies (Odonata) of Texas, Volume 1. Odonata Survey of Texas. 320 pp.

Abbott, J.C. 2007. Dragonflies and Damselflies (Odonata) of Texas, Volume 2. Odonata Survey of Texas. 311 pp.

Acorn, J. 2004. Damselflies of Alberta: Flying neon toothpicks in the grass. The University of Alberta Press, 156 pp.

Beaton, G. 2007. Dragonflies and damselflies of Georgia and the southeast. Univ. of Georgia Press.

Beckemeyer, R.J. and D.G. Huggins. 1997. Checklist of Kansas dragonflies. Kansas School Naturalist 43:1-15.

Beckemeyer, R.J. and D.G. Huggins. 1998. Checklist of Kansas damselflies. Kansas School Naturalist. 44:1-15.

Berger, C. 2004. Dragonflies. Wild Guide. StackPole Books. 124 pp.

Biggs, K. 2000. Common dragonflies of California: A beginner's pocket guide. Sebastopol, CA: Azalea Creek Publishing. 96 pp.

Biggs, K. 2004. Common Dragonflies of the Southwest: A beginner's pocket guide. Sebastopol, CA: Azalea Creek Publishing. 160 pp.

Brooks, S. 2003. Dragonflies. Smithsonian Books. 96pp.

Cannings, R.A. 2002. Introducing the dragonflies of British Columbia and the Yukon. Royal British Columbia Museum. 96 pp.

Corbet, P.S. 1963. A biology of dragonflies. Chicago: Quadrangle Books. 247 pp.

Corbet, P.S. 1980. Biology of Odonata. Annual Review of Entomology 25:189-217.

Corbet, P.S. 1999. Dragonflies, behavior and ecology of Odonata. Ithaca: Cornell University Press. 829 pp.

Curry, J.R. 2001. Dragonflies of Indiana. Indiana Academy of Science. 304pp.

Dunkle, S.W. 1989. Dragonflies of the Florida Peninsula, Bermuda, and the Bahamas. Gainesville, Florida: Scientific Publishers. 154 pp.

Dunkle, S.W. 1990. Damselflies of Florida, Bermuda and the Bahamas. Gainesville, Florida: Scientific Publishers. 148 pp.

Dunkle, S.W. 2000. Dragonflies through binoculars: A field guide to dragonflies of North America. New York: Oxford University Press. 266 pp.

Garrison, R.W., N. von Ellenrieder, and J. Louton. 2006. Dragonfly Genera of the New World: An illustrated key to the Anisoptera. Johns Hopkins University Press. 368 pp.

Glotzhober, R.C. and D. McShaffrey. 2002. The dragonflies and damselflies of Ohio. Columbus: Ohio Biological Survey. 364 pp.

Legler, K., D. Legler, and D. Westover. 1998. Color guide to common dragonflies of Wisconsin. 429 Franklin St., Sauk City, WI 53583.

Manolis, T. 2003. Dragonflies and damselflies of California. California Natural History Guides. 201pp.

Merritt, R.W. and K.W. Cummins. 2008. An Introduction to the Aquatic Insects of North America. 4th Ed. Kendall/Hunt Publishing Co. [Larval Identifications]

Mitchell, F.L. and J.L. Lasswell. 2005. A Dazzle of Dragonflies. Texas A&M University Press, 224 pp.

Needham, J.G., M.J. Westfall, Jr. and M.L. May. 2000. Dragonflies of North America, Revised Edition. Gainesville, Florida: Scientific Publishers. 939pp.

Nikula, B., J. Sones, D. Stokes and L. Stokes. 2002. Beginner's guide to dragonflies. Boston: Little, Brown and Company. 159 pp.

Paulson, D.R. 1999. Dragonflies of Washington. Seattle Audubon Society. 32pp.

Paulson, D.R. and S.W. Dunkle. 1999. A checklist of North American Odonata. University of Puget Sound Occasional Papers 56:1-86.

Rosche, L. 2002. Dragonflies and damselflies of northeast Ohio. Cleveland Museum of Natural History. 94 pp.

Silsby, J. 2001. Dragonflies of the world. Washington, D.C.: Smithsonian Institution Press. 216 pp.

Walton, R.K. and R.A. Forster. 1997. Common dragonflies of the northeast. (Video, NHS, 7 Concord Creene #8, Concord, MA 01742).

Westfall, M.J., Jr. and M.L. May. 1996. Damselflies of North America. Sci. Publishers: Gainesville, FL 649pp.

Westfall, M.J., Jr. and M.L. May. 2006. Damselflies of North America. Revised Edition. Sci. Publishers: Gainesville, FL 502 pp.

Westfall, M.J., Jr. and K.J. Tennessen. 1996. Odonata in R.W. Merritt and K.W. Cummins, eds. An introduction to the aquatic insects of North America, 3rd ed. Kendall/Hunt, Dubuque, Iowa, 862 pp.

Odonata Societies

Dragonfly Society of the Americas
Jerrell Daigle
2067 Little River Lane
Tallahassee, FL 32311
jdaigle@nettally.com
http://www.odonatacentral.org/

Societas Internationalis Odonatologica (S.I.O.) Foundation
Editor of Odonatologica, Bastiaan Kiauta
P.O. Box 256
7520 AG Bilthoven
The Netherlands

Worldwide Dragonfly Association
Jill Silsby
1 Haydn Avenue
Purley
Surrey, CR8 4AG United Kingdom
http://ecoevo.uvigo.es/WDA

ListServes

Texas - TexOdes
http://groups.yahoo.com/group/TexOdes/

North America/World Odonata-l
https://mail.ups.edu/mailman/listinfo/odonata-l

California - CalOdes
http://groups.yahoo.com/group/CalOdes/

Great Lakes - Great Lakes Dragonflies
http://groups.yahoo.com/group/gl_odonata

Northeast - Northeast Odonates
http://groups.yahoo.com/group/NEodes/

Southwest - SoWestOdes
http://groups.yahoo.com/group/SoWestOdes/

South East - SE Odes
http://groups.yahoo.com/group/se-odonata/

Texas Web Pages

OdonataCentral - Odonata of North America and the
South-central U.S.; Dragonfly Society of the Americas (John
Abbott)
http://www.odonatacentral.org/

Damselflies of Texas (F. Mitchell, J. Lasswell)
http://stephenville.tamu.edu/~fmitchel/damselfly

Digital Dragonflies (F.Mitchell, J.Lasswell)
http://www.dragonflies.org/

Dragonflies & Damselflies of Houston, TX (R. Orr, R. Honig,
R. Behrstock)
http://texasnaturalist.net/dragon/dragon.htm

Dragonflies and Damselflies of the Lower Rio Grande Valley
(T. Eubanks & R.Behrstock)

http://www.fermatainc.com/nat_odonates.html

Dragonfly and Damselfly Scans from Louisiana (Gail Strickland)
http://public.fotki.com/gstrick3/

Odonata of Tarrant County (O. Bocanegra)
http://www.uta.edu/biology/robinson/bocanegra/images.
htm

Odonate research at the University of Texas at Arlington
(Robinson)
http://www.uta.edu/biology/robinson/odonate_research.htm

Photographs of Texas Odonates by John Abbott
http://www.aeshna.com

Photographs of Texas Odonates by Greg Lasley
http://www.greglasley.net/dragonix.html

Photographs of Texas Odonates by Martin Reid
http://www.martinreid.com

Photographs of Texas Odonates by Bob Thomas
http://www.bthomasphotography.com/

Other Useful Web Pages

Dragonfly Biodiversity (D. Paulson)
http://www.ups.edu/x5666.xml

International Odonata Research Institute
http://www.iodonata.net

Suppliers

Bioquip Products
17803 LaSalle Ave.
Gardena, CA 90248
http://www.bioquip.com

Rose Entomology
P.O. Box 1474
Benson, AZ 85602
Local: (520) 586-7586
Fax: (520) 586-8741
Toll Free: (877) 249-1623
sales@roseentomology.com
http://www.roseentomology.com/

The International Odonata Research Institute
Bill Mauffray
PO Box 147100
Gainesville FL 32614-7100
http://www.iodonata.net
iodonata@bellsouth.net

Glossary of Terms Relating to Odonata

John C. Abbott

abdomen – posterior section of body; long, slender and comprised of 10 segments.

aestivation – a state in which an organism is metabolically inactive or physically dormant during the summer.

anal loop – group of cells in basal area of hindwing that may be distinctively shaped as a circle or boot.

andromorphic – color form of females that is similar to males of same species.

Anisoptera – suborder to which dragonflies belong.

antealar carina – ridge anterior to the wing bases on the meso- and metathorax.

antehumeral stripe – longitudinal stripe on the mesepisternum between middorsal carina and mesopleural suture.

antenodal crossveins – crossveins connecting costa, subcosta and radius that are proximal to the nodus.

auricle – ear-shaped projection on each side of second abdominal segment of male dragonflies except Libellulidae.

brace vein – slanted crossvein proximal to the posterobasal corner of the pterostigma.

caudal appendages – structures at the end of abdomen; comprised of cerci, epiproct and paraprocts.

cerci – appendages at tip of abdomen that can sometimes be useful in identification of males; often broken in females.

circumtropical - distributed throughout the tropics.

claspers – appendages at the end of a male Odonata used to grasp the female's head (Anisoptera) or prothorax (Zygoptera) during mating; made up of cerci and epiproct in Anisoptera and cerci and paraprocts in Zygoptera.

club – enlarged area of abdominal segments 7 – 9; particularly in Clubtails (Gomphidae).

clypeus – anterior sclerite above labrum and below frons on head.

copulation – act of mating; male holds female as in tandem, but tip of female's abdomen swings up to contact second segment of male, where accessory genitalia are located.

costa – longitudinal vein running along leading edge of wing.

crepuscular – active at dusk or sometimes daybreak.

damselfly – member of suborder Zygoptera; characterized by fore– and hindwings of same shape; small, widely separated eyes; generally smaller, more slender body.

degree days – number of degrees above a minimum temperature acceptable for growth, multiplied by time in days.

diapause – programmed period of metabolic inactivity; usually occurring as a mechanism of survival through environmentally unfavorable conditions.

dimorphic – having two forms; often referring to differences in sexes.

dorsal – top or back side.

dragonfly – member of suborder Anisoptera; characterized by broad wings of different shape (hindwing broader basally); large eyes, touching in most groups; generally larger and more robust body.

emerge(nce) – larva leaves water to undergo metamorphosis into adult; exuviae is left behind.

endemic – having a distribution restricted to a particular region.

endophytic oviposition – laying eggs into plant tissue.

epiproct – dorsal projection from 10th abdominal segment; in dragonflies = inferior appendages; in damselflies usually reduced to small button-like structure.

exophytic oviposition – laying eggs onto water or land.

exoskeleton – outer hard part of an insect, including legs and wings.

exuviae (sing. and pl.) – cast skin from any larval molt (including transformation into adult).

femur – first (basal) long leg segment.

flight season – period during which adults occur.

forage – actively searching for food.

frons – front of head, essentially face.

gaff – short region beyond posterior angle of triangle in hindwing of Anisoptera, where distal branch of A_1 is fused with posterior branch of Cu_2.

guarding – male defends female against attack by other males while she lays eggs.

gynomorphic – color form of female that is distinctly different from males of same species.

hamules – ventrally projecting, paired structures housed in capsule under second abdominal segment; hold female abdomen in place during copulation.

hastate – spear-shaped.

humeral stripe – dark stripe along mesopleural suture.

immature – adult past teneral stage but still not with mature coloration; often seen some distance from water.

in copula – collected while copulating.

instar – individual larval stage.

interspecific – between different species.

intraspecific – between individuals of same species.

labium – lower lip of Odonata.

labrum – upper lip of Odonata.

larva (pl. larvae) – immature stage of Odonata.

lentic – bodies of standing water; ponds, lakes, pools.

lotic – bodies of running water; rivers, streams, creeks.

mandible(s) – large tooth like structure used for chewing.

mature – reproductive age with full coloration.

mesepisternal tubercle – small tubercle on the mesepisternum just behind the mesostigmal plates in some damselflies (*Argia* and *Enallagma*).

mesepisternum – dorsum of thorax; area anterior and dorsal to the wings on the mesothorax.

mesostigmal plate – small, often triangular-shaped sclerite located on the dorsal anterior edge of the mesepisternum.

metamorphosis – process of changing from larva to adult; happens within larval exoskeleton.

molt – each time exuviae is shed; permits additional growth.

naiad – larval stage; technically referring only to Ephemeroptera (mayflies), Odonata, and Plecoptera (stoneflies).

nodus – indention or notch along front margin of wing, generally centrally located.

nominate – a subordinate taxon (subspecies, subgenus) which bears the same name (e.g. *Lestes disjunctus disjunctus*).

nymph – immature stage of hemimetabolous insects; commonly used when referring to Odonata.

obelisk – position used for thermoregulation by some Skimmers and Clubtails to lessen exposure of the body to the sun, thereby helping to keep cooler on hot summer days.

occiput – area of head between vertex and neck region.

ocellus (pl. ocelli) – simple eyes between the large compound eyes used for light detection.

Odonata – order containing dragonflies and damselflies.

odonate – term referring to both dragonflies and damselflies.

ommatidium (pl. ommatidia) – one division of a compound eye.

oviposit – to lay eggs.

oviposition – act of laying eggs.

ovipositor – complex structure at posterior end of some female Odonata (damselflies, Darners (Aeshnidae), and Petaltails (Petaluridae)) that functions in endophytic oviposition.

polymorphism – more than one distinct form within a species.

postnodal crossveins – crossveins connecting the costa,

vein R_1 and vein M_1 that are distal to the nodus.

prothorax – small first segment of thorax; just after head and before larger thoracic area with wings.

pruinose – waxy or powdery covering that exudes from cuticle and turns light blue, gray or white.

pruinosity (pruinescence) – waxy or powdery covering on some species of mature odonates.

pterostigma – thickened blood–filled cell at front of wingtip in most Odonata.

pterothorax – combined meso- and meta-thoracic segments; those thoracic segments bearing wings.

quadrangle – cell in Zygoptera wings bounded by M, Cu, Arculus and a crossvein between M and Cu.

riparian – relating to the bank of streams and rivers.

secondary genitalia – structures beneath the second abdominal segment of male odonates where females attach their abdomens for mating.

sigmoid – s-shaped, particularly referring to the caudal appendages of male damselflies.

spine – immovable projection on legs.

spur – movable projection on legs.

tandem – male and female linked either in flight or at rest.

tarsus – third leg segment, made up of several subsegments.

teneral – adult after it has just emerged, soft, pale and often with shimmer to wings.

territoriality – active defense of a small area.

thorax – second body section bearing wings and legs.

tibia – second leg segment, usually longer and thinner than femur.

torus (tori) – dorsoapical, median, protuberance on tenth abdominal segment in male Dancers (*Argia*).

torifer – basal area dorsally on abdominal segment 10 in male Dancers (*Argia*).

triangle – small cell or group of cells in basal area of each wing.

vein – hollow tubes in wings, providing strength and framework; blood is pumped through these at emergence.

ventral – underside of insect.

vertex – top of head between eyes.

vulvar lamina – plate under 9th abdominal segment of female odonates that serves to hold eggs in place during exophytic oviposition; often distinctive of species.

wheel position – term often used for the copulation position.

Zygoptera – suborder of Odonata that damselflies belong to.

INDEX TO MAPS

What is the Odonata Survey of Texas (OST)?

The OST, centered at The University of Texas at Austin (UT), includes a group of people with a shared interest in the study of the distribution, biology, behavior, and enjoyment of dragonflies and damselflies occurring in Texas. The purpose of the OST is to act as an official organization whose job it will be to encourage, solicit, and maintain the Texas database for dragonfly and damselfly distributional information. The OST is chaired by John C. Abbott of the University of Texas and Brackenridge Field Laboratory.

How to join the OST?

Membership is free and the OST is a volunteer organization. We need as many volunteers to assist in the documentation of the state fauna as possible. Primarily we are in need of field assistants; individuals who can assist in the discovery, collection, and otherwise documentation of species. We welcome collaborators from other institutions and agencies who are interested in sharing biological and environmental data, as well as specimens from sampling activities.

The OST currently has no funding and no paid positions. Monetary donations are accepted made out to The University of Texas at Austin.

Goals of the OST

The primary goal of the OST is to solicit, catalog, and make available the most up-to-date and definitive information on the dragonflies and damselflies of Texas. Additional goals and priorities of the OST include:

1. Promotion of the study and appreciation of Odonata at various levels of understanding.

2. Survey the state of Texas as thoroughly as possible and practical to document the current status of all Odonata species occurring in the state.

3. Provide an organized clearinghouse for odonatological information pertaining to Texas.

4. Recognize and critically examine habitats where species of limited distribution or threatened status are found.

5. Increase our knowledge of poorly known species.

6. Provide an internet-based source of information including, website, listserve, and printed materials.

7. Cooperate with agencies and groups where common goals are shared.

Contacts for Additional Information

Odonata Survey of Texas
c/o John C. Abbott, Ph.D.
Section of Integrative Biology
1 University Station #L7000
The University of Texas at Austin
Austin, Texas 78712 USA

Office Phone: (512) 471-5467
Lab Phone: (512) 232-1896
Fax: (512) 475-6286
jcabbott@mail.utexas.edu

Official Website
http://www.odonatacentral.org

Official Listserv
http://groups.yahoo.com/group/TexOde

www.ingramcontent.com/pod-product-compliance
Lightning Source LLC
Chambersburg PA
CBHW080512220326
41599CB00032B/6056

* 9 7 8 0 6 1 5 1 9 4 9 4 3 *